An Illustrated
Dictionary of Glass

HAROLD NEWMAN

An Illustrated
Dictionary of Glass

2,442 entries, including definitions of
wares, materials, processes, forms, and
decorative styles, and entries on
principal glass-makers, decorators, and
designers, from antiquity to the present

with an introductory survey of the
history of glass-making by
Robert J. Charleston

625 illustrations, 17 in color

THAMES AND HUDSON

For Gwynneth

© 1977 Thames and Hudson Ltd, London

First published in the United States in 1977 by
Thames and Hudson Inc., 500 Fifth Avenue,
New York, New York 10110
First paperback edition 1987

Library of Congress Catalog Card Number 86-50983

Printed and bound in Hong Kong

Preface

This dictionary is intended primarily to define terms relating to glass and glassware, such as the constituent elements, the methods of production and decoration, and the styles in various regions and periods, and also to describe some pieces that bear recognized names. But (unlike its companion book, *An Illustrated Dictionary of Ceramics*, which is limited to defining such terms), this book also lists the names of countries, regions, glassworks, and individuals that have developed or produced various types of glass or have created styles of glassware, including shapes, colours, and methods of decoration. Certain glassworks and individual makers, designers, and decorators of glassware are so closely identified with their products that a comprehensive dictionary of glass would be seriously deficient if it did not also include entries naming them and giving brief historical and biographical data.

The range of the entries here is wide, both in time and place. They relate to glassware made from the pre-Christian era down through the heyday of glass-making in the Middle East and the Roman Empire, and onward to the period, after the Dark Ages, when glass-making in Venice, Germany, and Bohemia was in the ascendancy, the later days of superior quality English, Irish, and American glass, and the early 20th-century Art Nouveau and Art Deco styles and recent new techniques and styles. Geographically, the coverage extends from the great early Middle Eastern centres of Syria, Persia, and Egypt, eastward to China and westward to Rome and almost all the countries of Europe and beyond to the United States.

Embracing such a wide scope, it has been manifestly impossible to mention all the glass-makers known to the art historians and to the writers of treatises and articles on glass. Hence a selection was imperative, and – due to the overall limit of available space – many meritorious factories, designers, and decorators of the past, and especially of recent years, have had to be omitted. Some readers will inevitably disagree with the selections made, cavilling at some omissions and dissenting from some inclusions. The judgment has had to be a personal one of the author, guided by a consensus of views of the many writers in the field and the recognition accorded by museums in various countries.

Another factor in deciding upon the range of the entries was the likely use of the book. Should it be an exhaustive dictionary with all the technological and esoteric terms that a specialist might seek, or a limited glossary useful mainly to the neophyte collector? The former course would have rendered a single volume both unwieldy and far too cluttered for most interested persons; the latter would have merely extended the many glossaries already in existence in other books. A middle road has been chosen, aiming for a reasonably comprehensive dictionary, with enough basic terms and names to meet the needs of those just entering the field or with some knowledge of glass, and also – for established collectors and for museums and dealers – a wide spectrum of specialized English-language and foreign terms that are seldom defined elsewhere. As the book is also intended to interest all those who admire and wish to become better acquainted with fine examples of glassware, it includes descriptions and photographs of many important (and sometimes little-known) pieces in the museums of the world.

The function of a dictionary is normally to state briefly the meaning of words as understood by the preponderant number of contemporary users, with occasional references to changes of meaning that have occurred over periods of time. Although some terms are occasionally qualified as obsolete or regional, it is not customary for a lexicographer to criticize usage or to seek to change the accepted meaning of an oft-used word. Here, however, I have gone beyond this usual practice and have sought to achieve more precision in the vocabulary of glass. All too often books, magazine articles, and auction catalogues about glass loosely use terms that are without historical background or that ambiguously relate to more than one type of object. This is especially deplorable when other words are available whose use would convey a clearer image and would enable collectors and museums to avoid being misled by a vague or inaccurate term. For example, *putto* and *amorino* are often used interchangeably when each word could be reserved for its own proper meaning. Such regrettable usage is frequently due to blind repetition of terms as they have been used by too many writers in the past; for example, the term *vetro di trina* has been used so loosely that I concur with leading Italian authorities who now urge its complete abandonment in favour of the two definite and traditional terms *retorti* and *reticello*. It is also advocated that the term *latticino*, often contradictorily used to include glassware with coloured threads, be confined to *filigrana* decoration made exclusively with white threads, making 'white *latticino*' tautological.

However, some terms have, by long usage, been accorded a broad meaning that, in the light of later research, goes beyond the original restrictive nomenclature. These are now so established that they must be tolerated, provided their wider significance is well understood. Examples are 'Damascus glassware', 'Stiegel glassware', and 'Nailsea glassware'; but even with these terms the addition of the suffix '-type' would add clarity in any references in which specific attribution is not intended.

On the other hand, even long usage can hardly justify perpetuating misnomers, such as 'sand-core glass' for ancient core-glass that is no longer believed by authorities to have been made over a core of sand, or the term 'Michael Edkins decoration' for much English milk glass no longer thought to have been enamelled by him, or 'Spessart glass' for beakers with chequered spiral trailing now authoritatively said to have been made in a different region. More recent unjustified uses are 'monteith' for a bonnet glass, 'cassolette' for a candlestick-vase, and 'Kit-Cat glass' (often mis-spelled 'Kit-Kat') for an English wine glass having a trumpet bowl on a drawn stem rather than a rounded funnel bowl on a baluster stem. It is hoped that this dictionary will help towards ending the use of such misnomers.

In some cases some recognized type of glassware has not yet been designated by any name which would be helpful to those wishing to refer to it. So in a few such instances an appropriate name has been suggested here, e.g., 'guttular jar' or 'millefiori mosaic'.

In pursuing this general ideal for clarity and accuracy in nomenclature, it has sometimes been necessary to depart from the usage of some recognized writers or even to disagree in some instances, but it is hoped that the definitions as used here will win acceptance and support from writers, museum experts, dealers, auction houses, and especially collectors, so that they will become the basis for a greater degree of precision and better mutual understanding in the world of glass.

The 608 black-and-white illustrations (each placed as near as practicable to the relevant entry) have been selected from two standpoints. Some show objects that are typical of a class with regard to: form (e.g., cage-cup or *Römer*); decoration (e.g., stippling or *Hochschnitt*); surface texture (e.g., Cypriote glass or ice glass); maker (e.g., Gallé or Marinot); or decorator (e.g., Beilby or Schaper). Others show a unique object that has a recognized name, ranging from the Portland Vase, the Barovier Cup, and the Lycurgus Cup, to pieces that are not generally so well known or as available for viewing by collectors but which are mentioned by name in books (e.g., the Vendange Vase, the Reformation *Humpen*, and the Goblet of the Eight Priests), and the unique modern presentation pieces made by Steuben and others. The colour plates have been chosen to show glassware whose colour is a technological achievement or has particular aesthetic appeal. Thus readers will be enabled to become visually acquainted with a wide spectrum of glassware.

Cross-references (in small capitals) are used extensively in the text so that definitions can be kept concise by confining them to essentials and by listing separately qualifying terms, related subjects, terms that are more basic than the word defined, or some specific example.

In preparing a book of this type and scope, it is normal, indeed inevitable, to draw on the writings and knowledge of predecessors and contemporaries, whether they be the authors of earlier dictionaries and glossaries, of treatises on glass factories, types of glassware, and important collections, or of relevant magazine articles. However, no bibliography is included here, as the literature on glass and glassware is too vast, and much of it is in books in foreign languages not available to or readily understandable by the average English-speaking reader, and besides many books on glass already do include extensive bibliographies. But in many cases a reference will be found, at the end of the entry, to a source book or article that discusses authoritatively the particular subject or that includes an exceptionally large number of illustrations. I gratefully acknowledge my indebtedness to all such prior writers, albeit in some cases I have felt compelled to modify or disagree with some authors' views.

Appreciation is expressed to the many museums which have made available photographs of their pieces with permission to reproduce them. To those museums which have supplied a large number of photographs, special thanks are due. These include:

Adria, Italy: Museo Civico
Coburg, Germany: Kunstsammlungen der Veste Coburg
Cologne, Germany: Kunstgewerbemuseum;
                  Römisch-Germanisches Museum
Copenhagen, Denmark: Rosenborg Castle (Royal Collection)
Corning, N.Y.: Corning Museum of Glass
Düsseldorf, Germany: Kunstmuseum
Frankfurt, Germany: Museum für Kunsthandwerk
Leningrad, U.S.S.R.: Hermitage Museum
London: British Museum;
        Percival David Foundation of Chinese Art;
        Science Museum;
        Victoria and Albert Museum
Murano, Italy: Museo Vetrario
New Orleans, La.: New Orleans Museum of Art
Prague, Czechoslovakia: Museum of Decorative Arts
Toledo, Ohio: Toledo Museum of Art
Washington, D.C.: Hillwood Museum.

A number of glass dealers have also generously supplied photographs of pieces that they own or have handled, and I wish to thank particularly W. G. T. Burne, Delomosne & Son, Spink & Son, Ltd, Derek Davis, Hugh Moss, Ltd, and Maureen Thompson, as well as Sotheby's and Christie's, all of London. I am also grateful for many photographs supplied by the Cinzano Collection, London, and by Steuben Glass Co., New York City. In all cases, acknowledgement of the photographs used has been made in the respective captions.

I particularly express my great indebtedness and deep appreciation to Mr Robert J. Charleston, for 28 years with the Victoria and Albert Museum, London, and its Keeper of

Ceramics and Glass from 1963 until 1976; in addition to writing the Introduction he has read much of the manuscript and made a great many extremely helpful and invaluable suggestions. I am also grateful to Mr T. H. Clarke, of Sotheby's, London, who has read and made comments as to the entries on paper-weights, to Mr Hugh Moss, London, for his assistance regarding Chinese glassware and particularly snuff bottles, and to Mr Phelps Warren, New York, for his suggestions about Irish glassware, as well as to Signor Luigi Zecchin, of Murano, and Mr Paul Perrot, of the Smithsonian Institution, Washington, D.C., for their letters regarding Italian *filigrana* glassware. In all cases, however, the responsibility for the content of the entries is solely mine.

London 1977   H.N.

Introduction

A BRIEF SURVEY OF THE HISTORY OF GLASS AND GLASS-MAKING
by ROBERT J. CHARLESTON

Glass as a substance may be looked at in many ways. In its beginnings it was regarded as a substitute for natural stones, an attitude of mind which is echoed in the Greek term meaning 'poured stone'. Most of the earliest glasses were opaque and coloured, but when transparent glasses were developed, and finally a decolourized glass-material perfected, it was natural to compare it with rock-crystal, and this aspiration to the status of rock-crystal is a theme which recurs repeatedly in the history of glass. It was natural enough that such a material should be shaped and decorated in the same ways as semi-precious stones, by turning, cutting and engraving, by means of swiftly rotating wheels and abrasives of varying degrees of fineness. Seen in this perspective, Ruskin's well-known condemnation of cut-glass may be dismissed as pure prejudice, caused by a natural revulsion from the excesses perpetrated by the glass-cutters of the mid-19th century. Glass when hot, however, is plastic, and can be manipulated in a number of ways, of some of which Ruskin probably would have approved, others perhaps not. It can be freely worked by the blow-pipe and a number of simple tools into a vast variety of shapes, provided they are basically circular in section. With the aid of moulds these possibilities can be infinitely extended, but there is the ever-present danger that the creative ability of the glass-maker may be surrendered to the fantasy of the mould-maker, sometimes with painful results. Perhaps the most successful use of moulds is that which employs them to impart basic simple designs (such as vertical ribbing, or mesh-patterns) which are subsequently, by the mere process of working the glass, invested with a truly glassy quality impossible in, say, a metal cast in similar moulds. To the basic ability of the glass-maker to impart a desired shape by swelling the glass bubble by blowing and constraining this outward pressure by the use of his tools, may be added the decorative techniques of trailing threads (sometimes in almost infinitely extended fine spirals) or applying blobs of hot glass, whether in the same 'metal' as that of the glass itself, or in contrasted colours or different degrees of transparency.

Nor do these infinitely variable methods of decoration exhaust the available possibilities. It was early discovered that glass, once cooled, could be painted with pigments (themselves of a glassy nature) which, when refired in a furnace, gave a permanent resistant coating of colours, their range being limited only by the number of suitable metallic oxides available to produce them (enamelling). This technique, perhaps originally suggested by similar processes used to decorate metalwork, could be supplemented by gilding, another resource of the metal-worker. Gilding on glass could be executed either by the use of gold-leaf or by treating precipitated gold in fine particles, mixed with a suitable painting-medium, as a kind of pigment.

In addition to these modes of decoration, which required the use of heat, the old techniques of wheel-grinding were, with the passage of time, developed to make possible the finest engraving, in which circular or oval cuts of varying depth were used to build up an illusion, called to life by transmitted light, of modelling and depth – for instance, in the rendering of the human figure. Another possibility was exploited by the use of layered glasses of contrasting character, cutting through the outer layer and exposing the glass below. This could be done with layered combinations of opaque glass to simulate the effects of cameo-carving on natural stones; or with crystal glass and coloured or opaque (or opaque coloured) 'overlays', sometimes in more than one layer. Sometimes all these techniques could be used together – often with disastrous results, as in some Bohemian glassware of the mid-19th century, where cut double overlays are combined with enamelling and gilding, to horrible effect.

To take sand and ashes and, by submitting them to the transmuting agency of fire, to produce an infinity of forms, colours and textures, is the magic of the glass-maker's art.

* * *

The earliest known glass, found in archaeologically datable Middle Eastern contexts of the 3rd millennium BC, consists of beads, and these never thereafter ceased to be made. The first glass vessels, however, occur in the 16th–15th centuries BC, mainly in the areas of Syria and Mesopotamia (between the rivers Tigris and Euphrates) associated with the kingdom of the Mitanni. These vessels (often beakers) were made by forming glass over a core (probably ceramic) held on a metal rod, and were decorated by means of applied threads of contrasting colours in wave-patterns: a second method was to shape the vessel over a core using the cross-sections of coloured glass rods, forming a pattern ('mosaic'). The former technique was transplanted to Egypt probably early in the 15th century BC under Tuthmosis III, and thereafter it was practised with great success in that country. It continued to flourish in Mesopotamia and probably spread from there to the countries of the Eastern Mediterranean in the course of the 15th century BC. The industry, however, probably declined with the general breakdown of Bronze Age civilization about this time, although the core-technique was never lost, reappearing again in the Eastern Mediterranean (but not Egypt) in the 9th century BC, with dark-blue vessels of Greek ceramic forms, decorated with bright yellow and other coloured threads in zigzag patterns. Mosaic vessels also recur in Mesopotamia. There, however, from the 8th century onwards, vessel-making from solid blocks of plain greenish glass, hollowed and shaped by means of abrasion, gained in importance. Dishes with petal-decoration made by this means probably continued right through to the Hellenistic period, to be taken up, with other metal-imitating bowl-forms, as one of the main types made in Alexandria, which from its foundation in 332 BC apparently became a prime centre of glass-manufacture. Here mosaic glass was revived, both for vessels and for decorative plaques of great refinement; and fine bowls were made with etched gold-leaf designs sandwiched between two layers of glass. All these techniques were carried over to the Roman period (34 BC onwards), but about this time the techniques of glass-making were revolutionized by the invention of glass-blowing. This probably occurred in Syria, whose workmen henceforth excelled in both free- and mould-blowing, and in the manipulation of hot glass in the form of decorative threads and blobs.

The Roman period is characterized by the spread of Oriental glass-makers, from Egypt probably to southern Italy, and from Syria to north-eastern Italy, Gaul and Germany, so that Roman glass has a truly international character. Glass was probably more widely used in the Roman Empire than at any time before the 19th century in Europe, with widespread employment of glass containers and windows. During the 1st century AD colourless glass outstripped the earlier fine coloured glasses in popularity, and proved particularly suitable for cutting and engraving in formal and figural designs. Objects made of 'gold sandwich' glass recurred in the 3rd–4th centuries AD, and alongside them 'vasa diatreta', on which the decoration was completely undercut, leaving the design attached by fine struts to the underlying body of a vessel.

With the division of the Roman Empire in 330 AD, glass-making became less international and more provincial, and regional types of mould-blown, cut, and thread-decorated glasses are found with a limited range of distribution. In due course these merge into the glass-types of the Teutonic North on the one hand, and the Syrian, Iranian and Egyptian glass-styles of the Islamic period on the other. In the northern countries glass-making tended to move away from the centres of population into the forests which supplied fuel for the furnaces, and latterly the ash produced in the furnace was substituted for the ashes of marine plants which had been the almost universal fluxing agent used in Roman glass-making. In the countries of southern and eastern Europe, and in the Islamic lands of the Near East, glass-making continued in centres of population, perpetuating the Roman tradition. The northern 'forest' glass-makers, conditioned by their raw materials, made mainly green glass and decorated it with furnace-wrought embellishments of simple rib-moulding, applied trails and blobs, mostly in the same material as the vessel itself. In the course of the later Middle Ages, this material was improved to produce a substantial glass of beautiful quality and a variety of green tones, used in a characteristic range of shapes of great originality and charm. In the Near East the technical possibilities were more varied and extensive. Whereas the Egyptians mostly adhered to furnace-decorated glasses in a continuation of the green 'metal' of the Roman period, in Iran (and probably also in Iraq) a tradition of cutting – from powerful relief-work in the form of bosses to delicate intaglio figural engraving – permitted the development in the 9th–10th centuries of a brilliant Persian-Mesopotamian school of relief-cutting on mainly colourless glass, the designs outlined by deep notched lines. This engraving was occasionally executed on a glass 'cased' with an overlay of emerald-green or blue. Parallel with this luxurious relief-engraving went a simpler and rougher style of intaglio engraving. In Egypt in the 8th century lustre-painting in metallic pigments on a usually greenish glass reached great heights of sophistication, and an original type of decoration was produced by tongs applied on both sides of the vessel-wall. In both areas glasses of individual character continued to be produced by mould-blowing, free-blowing and the manipulation of applied threads and blobs, sometimes with white threads inlaid in a dark-blue, brown or purple matrix. From Egypt may have spread the art of gilding glass which formed the basic element in the technique of gilt and enamelled glass which developed in Syria during the late 12th–13th centuries. Two main styles of glass-enamelling are discernible, one using thick enamels for figures and inscriptions, the other a fine linear style in red, both on a gold ground. The conquest of Syria by Timur towards 1400, however, put an end to this brilliant school of glass-decoration.

The glass-making of south-eastern Europe in the medieval period is as yet only imperfectly understood, but it is evident that a good-quality crystal glass had evolved by the 13th century at the latest. This was shaped into objects of often considerable finesse and delicacy, the fragility of which has usually ensured their destruction. It was most typically decorated by mould-blowing and the use of applied threads and 'prunts' of blue glass. The forms made were long-stemmed wine-glasses and 'prunted' beakers. Considerable use was made of opaque-red, both for whole vessels and for decoration. Glass of this character was certainly made in Italy, and also in Byzantine Greece. It is possible that a number of Italian centres were involved, but by the 15th century Venice had become the most important of these, and the eastern predominance in glass-making was in this period usurped by Venice. A glass industry was already established there in the 10th century, but in 1292 the furnaces were moved to the nearby island of Murano for safety's sake. Glass-painters are recorded in Murano as early as 1280; and colourless glass was almost certainly made there in the 13th century. From the middle of the 15th century these two branches of the art experienced an unparalleled expansion. The glasses with which Venice first entered the world market were enamelled and gilded, often on a blue, green or purple metal, but latterly also on clear colourless 'cristallo'. These glass objects were painted with themes of typical Renaissance inspiration – triumphs, allegories of love, etc. Inspired by classical Rome, the Venetian glass-makers also produced mosaic, millefiori, aventurine (with gold particles) and *calcedonio* glass (this last imitating Roman 'agate' glass, itself imitating the natural stone). The material with which they conquered the world, however, was *cristallo*, which captured the taste of the 15th century. Enamelled glasses gradually went out of favour, except for customers in Northern Europe, and mould-blown or exquisite plain forms in thin glass succeeded, sometimes decorated by bands or cables of opaque-white (*lattimo*) threads, occasionally by gilding, diamond-point engraving, cold-painting behind the glass, or a surface *craquelure* ('ice-glass'). As the century progressed, furnace-wrought decorations of applied threads or bosses, often of fantastic forms, became more popular, and in the 17th century this tendency was sometimes carried to extravagant lengths.

These stylistic elements – in the '*façon de Venise*' – were rapidly disseminated through Europe in the numerous glasshouses set up (despite the stern interdict of the Venetian Senate) by migrant glass-workers from Venice itself, and from L'Altare, near Genoa – beginning with Vienna in 1428. A Venetian-style glasshouse was at work in Antwerp before 1550, and this city became a secondary centre of diffusion for Northern Europe. In the 17th century Venice began to lose ground to these numerous rivals, and taste was beginning to change, looking towards more solid, colourless 'crystal' truly resembling the natural substance. This was achieved in two different countries by different means. In the Bohemian lands a potash-fluxed glass had long been the staple material: somewhere about 1680 calcareous additions began to be made to the 'batches', to produce a water-clear and solid crystal which proved ideal for decorating by wheel-engraving and -cutting; these techniques had been first developed in Prague before 1600, and were then transplanted to Nuremberg, where a school of engraving flourished throughout the second half of the 17th century. Wheel-engraving was greatly developed in Bohemia and Silesia in this period, a speciality being imposing goblets engraved in high relief (*Hochschnitt*), the ground of the design being cut back by means of water-driven wheels. From Silesia this art was transplanted to Potsdam, where one of the greatest glass-engravers – Gottfried Spiller – worked for the Elector of Brandenburg. In the course of the 18th century the art spread throughout Germany and Central Europe, centres of special importance being Nuremberg, Kassel, Gotha, Weimar, Dresden and Brunswick, with Warmbrunn and others in Silesia.

In England a solid 'crystal' glass was developed *c*. 1675–6 by George Ravenscroft, using lead-oxide as a flux in a purified potash-lime glass. Used first in Venetian-derived forms of some complexity, it was perfected in monumental glasses of great simplicity which evolved *c*. 1690–1700. Only after about 1740 did engraving, cutting, gilding and enamelling come to be used on English glass, which depended for its charms primarily on its brilliant metal and harmonious proportions. Lead-glass was often copied on the Continent, but without much success until the end of the 18th century, by which time it was realized that lead-crystal, with its high refractive-index, was the ideal material for cut decoration. Cut-glass dominated the international scene until about 1825, after which date the Bohemian glass-makers began to capture the imagination of Europe with coloured and opaque-white glass objects having enamelled and gilt decoration, or colourless glass 'cased' with red or blue layers through which decoration was cut or engraved. These 'Biedermeier' glasses were made in an infinite variety of combined techniques, and the glass-makers of England and France soon began to copy them. Wheel-engraving, however, still continued to be practised in Bohemia with the utmost skill, by artists such as Dominik Bimann, August Böhm, Karel Pfohl and the Pelikan and Simm families. Many Bohemian engravers at this period took their skills abroad, notably to England, where a flourishing glass-industry centred on the Midlands town of Stourbridge.

The second half of the 19th century was a period of enormous technical inventiveness, with England and Bohemia leading the field in the middle of the century, but France coming markedly to the fore in artistic glass-making towards the end of it, and the great industrial power of America gradually making itself felt, albeit mainly in styles imported from Europe. Apart from the traditional techniques of enamelling and gilding, wheel-engraving and -cutting, this period saw the introduction of transfer-printing and acid-etching; surface-treatments such as a revived 'ice-glass', and acid-etched 'satin' finishes; silver- and gold-leaf decoration, 'aventurine', and an electrolytic silver deposit process; bubbles of air trapped within the glass; glass shading from one colour to another owing to heat treatment ('Burmese', 'Peach Blow', etc.); wrought decoration of every sort, with twisted and ribbed elements, enclosed opaque and coloured twists and threads; and applied drops, and threading (now done by a mechanical process). All these methods were used in an infinite variety of combinations, reflecting the eclectic taste of the period. Of perhaps the greatest significance, however, was the perfection of press-moulding in America about 1825, for this brought decorated glass within the means of the poorer classes of society.

In France in the 1870s this multiplicity of available techniques was harnessed to the making of individual works of art in glass by two men, Eugène Rousseau and Emile Gallé. Rousseau was greatly inspired by Japanese art and his early glass pieces show this in their forms, their decorative compositions and themes; later, however, he devised original shapes and decorative techniques of enclosed red, green and black markings, of crackle, and of deep wheel-cutting. Rousseau abandoned glass-making in 1885, and he was overtaken in importance by Gallé, who had started manufacturing glass in 1867. Gallé, after a period of experimentation with every available decorative technique in a variety of historic styles, developed about 1885, under the influence of Japanese art, a new lyrical style in which vases with mainly floral polychrome designs were made in complicated casing-techniques, with much use of wheel-cutting and acid-etching. Such vases epitomize the 'Art Nouveau' in this material and were highly influential until the First World War. Of equal complexity and sophistication, however, were the glasses of the American, Louis Comfort Tiffany, who used embedded drawn threads and other forms, combined with an iridescent surface-treatment, to produce glass objects equally characteristic of this period.

After the disruption of war, a neutral country – Sweden – took the lead in glass-fashion. Two artists, Simon Gate and Edvard Hald, employed by the Orrefors factory, devised a variant of Gallé's cased glass, covering the cut and coloured layers with a colourless coating, thus embedding them and giving them a 'glassier' feeling ('Graal' glass). They also designed for wheel-engraving on fine crystal. The Swedish ideal generally was to produce (apart from these luxury wares) well-designed goods for everyday use, and in this their lead was followed by other countries in Northern Europe. A more personal style of glass-making, however, was developed in France by Maurice Marinot, a 'Fauve' painter who not only designed glass but made it with his own hands, using techniques of trapped bubbles, powdered oxides, and deep acid-etching in glass objects which reflected the forms and styles of the 1930s. This spontaneous approach to glass-making was reflected in the work of other French contemporaries (Henri Navarre, André Thuret, Jean Sala), even the commercial firm of Daum abandoning the traditions of Gallé for glass made under Marinot's influence; while René Lalique applied to mechanical production a highly sophisticated taste, designing deeply moulded glasses which could be enlivened by acid-etching and enamelling.

The northern European striving after functionalism was echoed in Murano after 1920, notably in the work of Paolo Venini, Ercole Barovier, and Flavio Poli. After the decline of Venetian glass-making in the 18th century, the skills and traditions were kept alive through the 19th, mainly by the enterprise of Antonio Salviati, and were at the service of the new style when it came. The simple shapes of Venini were often decorated by the thread- and mosaic-techniques, the opaque-white (*lattimo*) and coloured glasses traditional to Venice, while both

Venini and Barovier invented new surface-treatments to give their various glass products mottled, granulated or dewy effects.

The work of these masters continued after the Second World War, but this period has been chiefly characterized by the emergence of the studio glass-maker, first in America (Harvey K. Littleton, Dominick Labino, and others), then in Europe. Working single-handed at small one-man furnaces, these artists have produced glassware in an infinity of shapes (often non-functional and sculptural) decorated with bubbles, embedded coloured metallic powders, and other furnace-made ornaments, latterly supplemented by wheel-cutting.

*   *   *

As will already have become apparent, glass-making and its history have a wide variety of technical terms – to say nothing of an extensive collectors' jargon – devoted to their description and elucidation. The present book has been devised to throw light on this sometimes arcane world of words, and to provide in alphabetical sequence a wealth of general information on glass. Mr Newman has broken much new ground, as well as covering relatively well-charted territory; and the reader, whatever the extent of his familiarity with the subject, will welcome him as a guide.

Dictionary of Glass

Alphabetization of entries in the dictionary is by the operative letter or word: thus, 'G S Goblet' precedes 'gadget' and 'colour etching' precedes 'colourband glassware'. Cross-references to related subjects are indicated by the use of small capitals within entries. In the case of subjects embracing a wide range of different styles or types (e.g., paper-weights, bowls, stems, etc.), all such variations are listed under the major entry.

Illustrations in black and white will as a rule be found on the same page as the relevant entry or on the facing page, though a few appear on adjacent pages due to considerations of space.

The colour plates illustrate the following entries:

A

Aalto, Alvar (1898–1976). The well-known Finnish architect who, in 1937–8, designed glassware for IITTALA GLASSWORKS. His vases are of clear green glass, featuring curling vertical sides in an asymmetrical form.

aborigene, vetro (Italian). A type of ornamental glass developed by Ercole Barovier for BAROVIER & TOSO, *c.*1954, characterized by an overall haphazard pattern of small and vari-shaped spots of opaque coloured glass embedded in the colourless glass forming the body of an object; it was used for FREE-FORMED bowls, vases, etc.

abraded decoration. A type of decoration done by shallow grinding with the wheel, as a speedier and less exacting process than facet-engraving, and with designs in curved lines, circles, and ovals. It was practised in Egypt and Syria in the 4th and 5th centuries BC, and also in Bohemia in the 18th century.

Absolon glassware. Glass pieces said to have been enamelled by William Absolon, an independent enameller and gilder, of Great Yarmouth, England, from the end of the 18th century to the early 19th century. He decorated mainly CREAM JUGS, MUGS, TUMBLERS, and RUMMERS, of colourless, coloured, and OPAQUE WHITE GLASS, with gilt ENAMELLING, and engraving. Some pieces were inscribed, e.g. 'A Trifle from Yarmouth', and sometimes Absolon's name or initials added under the foot.

acanthus. A southern European plant, the spiked leaf of which has been used as a decorative motif since very early times; on glassware, as on ceramic ware, it is used as painted or relief decoration. In Greek and Byzantine forms it was usually stylized; in Roman and Renaissance decoration it was often used naturalistically. In profile, it is termed an 'acanthus scroll'. *See* ELGIN JUG; ELGIN VASE; CANOSA BOWL.

acetabulum (Latin). A small Roman cup used as a measure and for serving vinegar. They are semi-globular, sometimes with vertical or sloping collar, and rest on a FOOT-RING. Usually made of pottery, some were made of ROMAN GLASS in the 1st century AD.

Achaemenian bowl. A type of glass bowl, colourless or pale green, attributed to the period of the Achaemenid dynasty in Persia (559–330 BC). They are hemispherical and shallow, with an everted rim and a plain interior; exterior decoration is in low relief on the lower half of the bowl, either fluting or a rosette of pointed leaves, radiating from the base, and with a relief encircling ridge below the rim. An example with relief leaves was sold at Sotheby's, London, on 13 July 1976 for £62,000.

acid cutback glassware. A type of ART GLASS having the appearance of CAMEO GLASS; it consists usually of two layers of CASED GLASS of contrasting colours (sometimes of one layer only) with a design produced by acid etching (rather than carving) on the outer, lighter-coloured layer. It is so called in the United States, where a process for making such ware was developed *c.*1932 by FREDERICK CARDER at the STEUBEN GLASS WORKS; this involved transferring a design to the body of a piece by a print made with a wax-based ink, and then painting the surface area not covered by the design with a protective wax before immersing the piece in acid to etch the design.

acid polishing. The process of giving to CUT GLASS a glossy polished surface by dipping it into HYDROFLUORIC ACID mixed with sulphuric acid. *See* BLANKÄTZUNG.

Aalto, Alvar. Vase designed by Aalto, 1937, Iittala Glassworks, Finland.

acetabulum. Blue glass, Roman, *c.*100 AD. Ht 5·5 cm. Römisch-Germanisches Museum, Cologne.

Achaemenian bowl. Pale green glass with stylized leaf decoration in low relief, *c.*5th century BC; diam. 17·2 cm. Private collection. Photo courtesy, Mahboubian Gallery, London.

Absolon glassware. Wine glass with enamelled inscriptions ('A trip to Camp' and 'A trifle from Yarmouth'), *c*.1800. Ht 9·1 cm. Norfolk Museums Service (Norwich Castle Museum).

acorn goblet. Greyish glass; bowl and cover coil-wound; lion-masks on stem. Netherlands, 16th/17th century. Ht 23·5 cm. British Museum, London.

acorn finial. A type of FINIAL shaped like an acorn with the pointed end upward. Such finials were popular in the United States.

acorn goblet. A rare type of GOBLET with a wide hemispherical bowl on which rests a tall COVER in the form of a high narrow beehive, so that the total silhouette resembles that of an acorn and cup. The bowl rests on a short lion-mask STEM, with a flat folded FOOT. The bowl and cover are ornamented with closely spaced horizontal ridges made by winding coils of glass threads. The shape is derived from metal prototypes.

acrolithic statuette. A small composite figure made of wood and stone but the head of which was often made of sculptured glass in Egypt during the XVIIIth and XIXth Dynasties of the New Kingdom.

Adam style. The English version of the NEO-CLASSICAL STYLE, introduced soon after 1760 by the architect and designer, Robert Adam (1728–92). He believed that architecture should take into consideration the applied arts, and designed many objects in various fields, including some pieces of glassware. Ornamentation in this style became simpler, similar in many ways to the French version, with FESTOONS, MEDALLIONS, and URNS commonly being employed for decorative purposes. CHANDELIERS, CANDELABRA, LUSTRES, and SCONCES became popular, affording opportunity to use glass DROPS with diamond and prismatic CUTTING, often in conjunction with Wedgwood's jasper ware.

Adams, L. A Dutch decorator of glassware by STIPPLING, in the manner of DAVID WOLFF, during the first part of the 19th century. Five pieces signed by him are known, one dated 1806.

Agata Glass. A type of ART GLASS developed by JOSEPH LOCKE at the NEW ENGLAND GLASS CO. and patented by him *c*.1887. It was a variety of WILD ROSE GLASS, made by coating the object, wholly or partially, with a coloured metallic stain and then spattering it with a volatile liquid that would evaporate and leave a glossy mottled surface that was fixed by a light firing. It was produced for less than a year, before the factory moved to Toledo, Ohio.

agate glass. (1) A type of decorative opaque glass made by mingling METAL of several colours (most often green and purple) before shaping the object, so as to imitate natural semi-precious stones, such as agate, chalcedony, onyx, malachite, lapis lazuli, and jasper. This variegated or marbled glass was called *calcedonio* when developed in Venice in the late 15th century, but later became incorrectly known as *Schmelzglas* ('enamel glass') when made in Germany. It had been made as early as ROMAN GLASS in Alexandria. Some early examples made in Germany have been attributed to Count E. W. von Tschirnhausen (d. 1708) who did experimental work in porcelain at Meissen, and later German pieces are known. There are English specimens from the 17th century (sometimes called 'jasper'). In China in the 18th century such a marbled glass in imitation of semi-precious stones was produced, although some examples were made by layering glass of different colours and then carving it, as in CAMEO GLASS. Agate glass is analogous to solid agate pottery. The French term is *jaspe*. See AVENTURINE GLASS; PÂTE-DE-VERRE; ONYX GLASS; MOSS AGATE GLASS. (2) A type of ART GLASS developed by LOUIS C. TIFFANY that resembles agate stone. It was made by mixing in the POT different coloured opaque glass and stirring until a striated appearance resulted but not long enough to allow the colours to become unified; it was then polished. It is sometimes incorrectly called 'laminated glass'. Sometimes when the METAL became blackened it was cut and engraved in geometric or leafy patterns to reveal the interior agate effect. Some such glass became reddish-brown, and the pieces were lustred on the interior and exterior; this was termed 'metallic glass'.

Ages of Man, The. A decorative motif depicting man in the different decades of his life. Each decade (sometimes from age 10 to 100) is usually represented by a progressively aging man (rarely, a woman or a couple) in either of two styles: (1) ascending and descending a stepped bridge, or (2) shown in eight or ten panels set in two encircling bands; each decade is accompanied by an appropriate inscription. BEAKERS and TUMBLERS (occasionally the motif appears on a PASSGLAS) were so decorated by ENAMELLING or ENGRAVING in Germany and Bohemia from the 16th to the 19th centuries. The motif is from a series of woodcuts from the 16th century. The German term is *Lebensalterglas*.

aggry (aggri) bead. A BEAD made by cutting obliquely a glass CANE of many colours resulting in a zigzag pattern. They are of ancient manufacture and have been found buried in Africa.

aiguière. The French term for a EWER.

air-locks (or **air-traps**). A style of decoration in the form of embedded air pockets, larger than the usual air bubbles or tears, arranged in a decorative pattern, often in criss-crossed diagonal rows. Sometimes the air pockets occur in circular form (known as the HOBNAIL pattern), sometimes in diamond or lozenge form. The process involved blowing a GATHER of molten glass into a mould lined with shaped projections in the desired pattern, and then covering the piece with a second gather of glass that trapped the air within the indentations. The technique was patented in 1857 by Benjamin Richardson, of W. H., B. & J. RICHARDSON. Sometimes the piece was further decorated with an extra layer of opaque white or coloured glass, occasionally treated with acid to produce a satin surface with an etched decoration.

agate glass. Ewer; green, brown, and violet, Germany, early 19th century. Ht 32·5 cm. Kunstgewerbemuseum, Cologne.

Akerman, John (fl. 1719–55). A London glass-seller (at the Rose and Crown, Cornhill, and after 1746 in Fenchurch St) who made CUT GLASS and who was a principal officer of the GLASS SELLERS' COMPANY, *c.*1740–48. He employed a Bohemian glass-cutter named HAEDY. Akerman was the first person in England to advertise cut glass for sale, *c.*1719, stating that it was made by the first person to bring to England the German art of glass-cutting and engraving. His firm sold glass in London and in the provinces. He was succeeded by his son, Isaac Akerman, whose successors were THOMAS BETTS and then Jonathan Collet.

Alabaster glass. A type of translucent ornamental glass developed by FREDERICK CARDER at STEUBEN GLASS WORKS in the 1920s. It resembles the mineral (usually white) after which it was named, with IRIDESCENCE produced by spraying with stannous chloride before reheating AT-THE-FIRE. It was used to make ornamental JARS, VASES, and DISKS, as a CASING or as a background for motifs made by applied TRAILING, sometimes dragged in the manner of ancient COMBED DECORATION, or for motifs made by acid ETCHING. Sometimes it was used as a middle layer of glass between two casings of coloured glass (*see* AURENE GLASS; PLUM JADE GLASS), the outer layer sometimes having etched designs. It was also used in conjunction with Aurene applied decoration. *See* DOUBLE ETCHING.

alabastron (Greek). A small BOTTLE or FLASK for holding ointments, perfume or oil. It is elongated in form, early glass ones being cylindrical but later ones tapering inward toward the neck and being slightly rounded toward the base. It usually has two small loop-handles or two lugs at the sides (some later FUSIFORM marbled specimens have no handle). The MOUTH is a flat disk with a small central orifice. The *alabastron*, sometimes made in the form of a figure, was made to be suspended from the wrist by a string. Examples were generally made of Greek pottery, varying from 3 cm. to 35 cm. in height, but some were also made of CORE GLASS at Alexandria, *c.*300 BC–AD 100. These are about 8–12 cm. high, but one known example made by COLD CUTTING (with

The Ages of Man. Beaker with ten enamelled scenes, Bohemian, 16th century. Staatliche Kunstsammlungen, Kassel.

Aldrevandini beaker. Clear glass with encircling enamelling, attributed to a Syro-Frankish workshop, *c.* 1260–90. Ht 13 cm. British Museum, London.

ale glass. Jacobite type; inscribed 'Radiat' for Oak Society; plain stem; folded terraced foot; English, *c.* 1740. Ht 15·8 cm. Courtesy, Maureen Thompson, London.

small lugs instead of handles) is 21 cm. high. They usually have COMBED DECORATION, of white threads on a dark-blue ground, in a spiral, zigzag or palm-leaf pattern.

albarello. A DRUG-JAR for holding solid or viscous substances, but seldom liquids. They are more or less of cylindrical shape, but slightly WAISTED. Some have a low foot, and sometimes a COVER. The waisted form is to facilitate removal from a shelf of similar jars side by side. Generally made of tin-glazed earthenware, a few have been made of plain or enamelled glass, e.g. in Venice or in the *façon de Venise.*

Albert. A type of DROP made in the Victorian period.

Aldrevandini Beaker. An enamelled BEAKER decorated with an encircling pattern with three shields bearing Swabian coats-of-arms, partly painted on the inside of the glass and partly on the outer surface, with interspersed painted leaves. It bears the inscription *Magister Aldrevandin me Feci*[t]. There are undecorated areas above and below the decoration, and the glass has a KICK BASE. It is one of a group of some twenty-five related enamelled pieces that once were attributed to Venice but more recently to a Syrian workshop at a Frankish Court, *c.* 1260–90. The British Museum (the owner) in 1968 stated that the origin and date of the group are yet to be determined. *See* ENAMELLING, ISLAMIC; SYRO-FRANKISH GLASSWARE; HOPE BEAKER.

ale glass. A type of English drinking glass intended for drinking ale or beer. It developed from the early 17th-century short-stemmed DWARF ALE GLASS to the later, *c.* 1685–90, taller glass (three to five ounces capacity), with a stemmed FOOT (the stem being plain, knopped, air twist or opaque twist) (*see* TWIST). The shape approximated that of a contemporary elongated wine glass with an ogee or rounded funnel BOWL. Some, *c.* 1735–45, were decorated with engraved, enamelled or gilded designs of hops and barley motifs; and a few have BEILBY enamelled decoration. In addition to the 'dwarf ale glass' there are the 'small' and 'full' types and even the 'giant ale glass' (over 30 cm. high). The modern ale glass is in the form of a tall, thin, stemless TUMBLER, slightly tapering toward the bottom, on a solid base; its capacity is about 12 ounces. *See* CIDER GLASS; WINE GLASS ALE.

ale jug. A type of JUG for serving ale, characterized by engraved decoration of hops and barley motifs.

Alemayne, John (fl. 1351–55). The lessor to John Schurterre (*see* SCHURTERRE FAMILY) of property near Chiddingfold, Surrey, who acted as agent in selling glass, including much uncoloured window glass probably made by the Schurterres.

alembic. An apparatus often made of glass, formerly used in distillation. It is in the shape of an inverted bowl with an internal gutter along the rim into which the condensed vapour descends and from which extends a long horizontally sloping tube to the receptacle for the distilled liquid. The alembic was placed over a bottle (CUCURBIT) containing the liquid being heated over a flame; sometimes the alembic and cucurbit were made as one integrated piece. The alembic and cucurbit were made in ancient times of Roman and Islamic glass, and later in medieval times. *See* RETORT.

Aleppo glassware. A type of Syro-Islamic glassware, *c.* 1250–65, identified with, but not necessarily produced at, Aleppo, in Syria, but possibly also at Raqqa and Damascus. It is often blue or violet, decorated with gilding and with enamelling in thick rich blue, red and other colours and in elaborate style, sometimes painted on both sides of the glass. The ware included tall flared BEAKERS with a KICK-BASE and pyriform FLASKS. *See* LUCK OF EDENHALL; WÜRZBURG FLASK; ENAMELLING, ISLAMIC; SYRIAN GLASS.

Alexandrian glassware. Glassware made at Alexandria (and the neighbouring Nile Delta area) in Egypt from the founding of the city in 332 BC throughout the Hellenistic period and for several centuries after the conquest by the Romans in 27 BC, until being overrun by the Huns in the 4th century. The Alexandrian craftsmen probably devised new techniques, such as the making of embedded MOSAIC GLASS and the decolorization of glass with MANGANESE, and developed the making of luxury coloured glass and CAMEO GLASS. The ware was exported to Italy and farther afield, and emigrant workers carried their skills abroad, so that it is difficult to ascribe much of the glass of the period to any particular glass centre.

Alexandrit. *See* under MOSER, L. KOLOMAN.

Alexandrite. A type of ART GLASS made in one layer but having three blending colours; it was first produced *c*.1900. It was made by successive reheating of parts of the object, the colour varying from amber in the main part to blue and fuchsia shading towards the rim. Such glass was made in England by THOMAS WEBB & SONS. A similar looking ware, made by STEVENS & WILLIAMS, LTD, has a body of transparent amber, cased with rose and blue coloured glass, the outer layers of coloured glass being sometimes cut away to reveal a pattern on the amber ground. *See* MOSER, L. KOLOMAN.

alkali. A soluble salt, one of the essential ingredients in making glass, being about 15–20% of the BATCH; it serves as a FLUX to reduce the fusion point of the SILICA, the other necessary ingredient. It is supplied mainly by the use of SODA or POTASH; the former was used in making Spanish, Roman, Egyptian, Syrian, Venetian, and some English glass, the latter in glass made in Germany and Bohemia. Today both potash and soda are used in making glass.

allegorical subjects. Decorative symbolic subjects of figures; on glassware such subjects are usually WHEEL-ENGRAVED or ENAMELLED. The fashion for allegory on ceramic ware and glassware was especially popular in the 17th and 18th centuries, when many figures were depicted symbolizing a wide variety of subjects (e.g. the FOUR SEASONS; FOUR ELEMENTS; FOUR CONTINENTS; FIVE SENSES; TWELVE MONTHS; VIRTUES, TWELVE CARDINAL), identifiable by the attributes accompanying them. *See* MYTHOLOGICAL SUBJECTS.

Alloa Glassworks. A glasshouse established in 1750 at Alloa, Scotland, making over a long period mainly glass bottles, but also FRIGGERS in the style associated with NAILSEA GLASSWARE. A glass-worker named Timothy Warren migrated from Nailsea to make glass at Newcastle-upon-Tyne and later went to Alloa. The factory is known to have made double-necked FLASKS decorated with QUILLING, and later table-glass and other wares, decorated with engraved motifs, usually by the process of STIPPLING.

almond-bossed beaker. A type of mould-blown BEAKER, characterized by encircling rows (usually six) of almond-shaped (sometimes called 'lotus-bud') bosses; the spacing between the bosses in each row alternates with that of the adjacent row(s). The beaker is of inverted conical shape, with a raised encircling rib just above the base. The bosses are sometimes smooth, sometimes with raised outlines. Beakers of this type were made in Syria in the 1st and 2nd centuries AD. The German term is *Mandelbecher*.

almorratxa or **almorrata** (Spanish). A type of ROSE-WATER SPRINKLER intended to be used on ceremonial occasions, exuberantly modelled and decorated. The BODY rests on a pedestal FOOT, and the vessel is characterized by four tapering SPOUTS extending vertically upwards from the shoulder of the body and around a central NECK. They were made in Catalonia in the 16th to 18th centuries, but similar objects

allegorical subjects. Beaker with figures representing *Potestas, Nobilitas*, and *Liberalitas*, Caspar Lehmann, dated 1605. Museum of Decorative Arts, Prague.

almond-bossed beaker. Possibly Syrian, 1st century AD. Ht 20·7 cm. Kunstmuseum, Düsseldorf.

almorratxa. Clear glass, Barcelona, 17th century. Ht 22 cm. Victoria and Albert Museum, London.

alzata. Salvers with glass fruit, Venice, 19th century. Museo Vetrario, Murano.

of DAMASCUS GLASSWARE were imported into Spain in the 14th century. Vessels of similar form were made in Venice for the Spanish market. Another type, similar in shape and with a like neck and three tubes, is called a BOUQUETIÈRE; examples were made at HALL-IN-TYROL (Austria) in the 16th century. *See* SPRINKLER.

altar. A portable altar, made on a wood structure and decorated with elements made AT-THE-LAMP and sometimes with panels of glass painted on the reverse; these were made in MURANO in the 17th and 18th centuries. Others were made of VERRE ÉGLOMISÉ in the Netherlands and Italy from the 14th century.

Altare, L'. The site, near Genoa, of an important glass-making centre that rivalled Venice. Originally started in the 9th century by glass-makers who had fled from Normandy or Flanders (opinions differ), it became independently important by the second half of the 15th century. Unlike the Venetians, who sought to enforce secrecy, the Altarists had guilds that encouraged the spreading of the craft of glassmaking and sent its members abroad to found new centres, e.g. at Liège. These new centres of production greatly impinged on the exportation of glass from Venice. It has been said that no identified glass from L'Altare exists, and that glass in the fashion of Venice and L'Altare is difficult, if not impossible, to distinguish, as is that made outside Italy by emigrés from both places.

Altenburg Passglas. *See* SPIELKARTENHUMPEN.

alternate panels cutting. A decorative pattern on CUT GLASS; it consists of pairs of mitred diagonal grooves cut at right angle by another series of such pairs, leaving an uncut space between the diagonal pairs of grooves so that a flat area results, and then cutting horizontal and vertical single grooves so as to make the flat areas or 'panels' octagonal (or square with chamfered corners) in shape. Such octagonal areas, which are in a chequerboard pattern, are never further cut. This pattern is sometimes called 'cross-cut squares'; it is sometimes erroneously called a 'hobnail pattern'. It is often found on square Irish DECANTERS of the type placed in a TANTALUS.

alzata (Italian). Literally, elevation. A serving piece with a series of two to four circular serving trays placed one above another, the higher ones diminishing in diameter. The trays appear to be extending from a central post, but the piece is composed of several SALVERS or TAZZE with footed stem or spreading foot, stacked vertically and sometimes attached. On the trays are usually several small glass dishes, cups, etc., or sometimes decorative glass replicas of various fruits. They were made at Murano in the 19th century.

amalgam. An alloy of mercury with another metal. When made with gold, the amalgam is used for gilding porcelain or glass. When made with tin it is used as the backing for MIRRORS.

Amberina. A type of uncased ART GLASS featuring varying colours shading (usually from the bottom of a piece to the top) from light amber (straw) to rich ruby. When an object (mould-blown or free-blown) is first made, the overall colour is amber, but when part is reheated that part (due to particles of gold in the METAL) develops the ruby colour (*see* GOLD RUBY GLASS), and if reheated longer it develops a purple (fuchsia) colour. When, occasionally, the ruby part occurs in the bottom half of the piece, it is termed 'Reverse Amberina'. The glass was developed by JOSEPH LOCKE at the NEW ENGLAND GLASS CO. whose owner EDWARD DRUMMOND LIBBEY coined the name. It became popular when the entire factory supply was bought from Drummond by Tiffany & Co., the New York jewellery store. It was patented by Locke in 1883, and after 1888 was made by LIBBEY GLASS CO. When made by the New England Glass Co., it was marked with its initials; when made by Libbey, with the etched name 'Libbey'. Its popularity waned, but Amberina was revived by Libbey Glass Co. in 1917 to celebrate its 100th anniversary, and all pieces (about 40 types) then made, during a period of about two years, bear the etched trade-mark of Libbey. Similar glass was made by many other factories, including the MT. WASHINGTON GLASS CO., but when that company was sued by Libbey for infringement, the firm changed the name used by it to 'Rose Amberina'. The New England Glass Co., in 1886, licensed HOBBS, BROCKUNIER & CO. and, in 1883, SOWERBY'S ELLISON GLASS WORKS, LTD, of Gateshead-on-Tyne, to make 'Pressed Amberina'. *See* PLATED AMBERINA.

ambrosia dish. A type of DISH, asymmetrically shaped in the form of an oval BOWL with an irregularly lobed RIM and a small curling HANDLE at the small end, and having a stemmed FOOT. Such dishes were ornately decorated in ROCOCO style with FACETING and with polished and unpolished INTAGLIO engraving. Examples were made by CHRISTIAN GOTTFRIED SCHNEIDER at Warmbrunn, in Silesia, *c.* 1730–60. The German term is *Ambrosiaschälchen. See* SHELL DISH.

Amelung Glassworks. *See* NEW BREMEN GLASSWORKS.

Amen glass. A type of English WINE GLASS with a drawn STEM (usually a plain stem, but sometimes an air-twist stem), the BOWL being decorated with DIAMOND-POINT ENGRAVING of verses of the Jacobite Hymn (a version of which preceded the British National Anthem) and a crown or motifs associated with the Jacobite uprising of 1715. In addition, there is the word 'Amen'. These glasses were made in England from *c.* 1745. Fewer than twenty are recorded. *See* JACOBITE GLASSWARE.

American glassware. Glassware made in the United States from Colonial days to the present time, at innumerable glasshouses located mainly during the 19th century in Massachusetts, New Jersey, Pennsylvania, New York, and Ohio. Among the most important were the BOSTON & SANDWICH GLASS CO., the NEW ENGLAND GLASS CO., and the factories of HENRY WILLIAM STIEGEL and John Frederick Amelung (NEW BREMEN GLASS CO.), and later the LIBBEY GLASS CO. and the CORNING GLASS CO. (STEUBEN GLASS). The production included all types of glassware, but most characteristic of 19th-century ware was PRESSED GLASS (LACY GLASS), PATTERN-MOULDED GLASS, BLOWN-THREE-MOLD GLASS, and certain types of ART GLASS. For a full discussion of early American glassware, the factories, wares, and patterns, *see* George S. and Helen McKearin, *American Glass* (1945), and Ruth Webb Lee, *Sandwich Glass* (1939).

American System, The. A rare type of HISTORICAL FLASK that bears on one side a picture of a side-wheel river-boat flying the United States flag, encircled by an inscription 'The American System' and on the reverse a sheaf of rye. The phrase is associated with Henry Clay and the Congressional fight over the protective tariff in 1824. The first example is believed to have been made at BAKEWELL'S glasshouse.

Amiens Chalice. A vase of thick blue glass, urn-shaped with a spreading BASE, a wide MOUTH, and two massive vertical scroll HANDLES in grooved rope pattern. Its former attribution to the 5th century is now in doubt as a result of its showing CRISSELLING. The present tentative attribution by the British Museum, the owner, is perhaps to an unknown French glasshouse of the late 17th century. The piece is so-called because it is said to have been found near Amiens. *See* SAVOY VASE.

amorino (Italian). A winged infant Cupid employed as decoration of glassware, especially Italian, French, and German wares. Roman in origin, *amorini* (and *putti*) are to be seen in the wall paintings in the House of the Vettii at Pompeii (1st century AD). They were popular in the art of the Renaissance as a decorative motif. Notable examples in allegorical guises may be found on porcelain, but they are also on glassware, especially in engraved or stippled decoration. *See* PUTTO.

Ambrosia bowl. Faceted and with engraving in *intaglio* by Christian Gottfried Schneider, Warmbrunn, *c.* 1750–60. Ht 14 cm. Kunstmuseum, Düsseldorf.

Amen glass. Drawn trumpet glass with diamond-point engraving of the complete Jacobite hymn, the word 'Amen', and the date 6 March 1725, commemorating the birth of Prince Henry. Ht 19·5 cm. Cinzano Glass Collection, London.

Amiens Chalice. Blue glass, perhaps French, late 17th century. Ht 16 cm. British Museum, London.

amphoriskos. Blue core glass with yellow, blue, and white combed decoration, and disk base-knob; Roman glass, *c.* 500 BC. Ht 8·4 cm. Römisch-Germanisches Museum, Cologne.

amphora. A large ovoid JAR characterized by two vertical loop-handles, the ends of which are attached at the shoulder and below the rim of the mouth. There are two basic shapes: (1) the 'neck amphora', of which the neck, set off from the shoulder, meets the body at an angle, and (2) the more familiar 'continuous-curve amphora', of which the neck and the body merge in a continuous curve. The purpose of the amphora (apart from the Panathenaic type made as prizes) was for storing oil, wine or other liquids, this type being usually undecorated and sometimes tapering to a point at the bottom to be inserted into a metal tripod or into soft earth. Amphorae are generally of Greek pottery, but examples of ROMAN GLASS, *c.* 200–100 BC, are known, about 60 cm. high; however, most glass examples in the form of an amphora are small and are generally termed an AMPHORISKOS. *See* PORTLAND VASE; VENDANGE VASE.

amphoriskos (Greek). A small glass vessel, for oil or unguents, in the form of a Greek AMPHORA (of either the 'neck amphora' or 'continuous-curve amphora' type), made in Egypt or Cyprus, *c.* 1350 BC and, after a lapse of time, in the eastern Mediterranean region, *c.* 600–100 BC. Such vessels are of CORE GLASS, usually from 5 to 15 cm. in height, and have a bulbous or ovoid body, a pointed bottom, and a domed foot or a disk base-knob. The BODY is opaque, usually coloured turquoise or dark lapis-lazuli blue, with orange-yellow, light-blue or white COMBED DECORATION.

ampulla (Latin). A Roman vessel similar in purpose to the AMPHORA, but smaller in size; the name may be a diminutive of the latter term. The BODY is ovoid, usually with a tall slender neck, and usually one or two vertical loop side-handles, the ends being attached to the shoulder and just below the rim of the mouth. It rests on a circular flat base. They were generally made of pottery; some were made of ROMAN GLASS in the 1st to 4th centuries AD, but two, without handles, regarded as Venetian glass, *c.* 1463, were recently found at Cremona. The term has sometimes been applied to the type of EWER with a high curving spout or the type with two spouts and an overhead ring handle.

amulet. An ornament worn as a charm against evil, often with an inscription or inscribed symbol. Early examples were made of Egyptian glass, *c.* 1375 BC.

Anglo-Irish glassware. Glassware made in Ireland, *c.* 1780–1825, by English glass-makers who migrated to work in factories financed by Irish capital. The products (luxury glassware of LEAD GLASS) are considered by some writers to be a logical development of English glassware, but a contrary view is that a peculiarly Irish style was developed. Originally many vessel forms were derived from classical precedents (e.g. the boat-shaped FRUIT BOWL and the HELMET JUG), but toward the mid-19th century classic purity was replaced by heavy forms with deep overall CUTTING. *See* IRISH GLASSWARE.

Anglo-Venetian glassware. Glassware made in England in the 16th and 17th centuries by Venetian glass-workers who emigrated to England, including JACOPO VERZELINI. The work, in the FAÇON DE VENISE, included moulded KNOPS, DIAMOND-POINT ENGRAVING, and WINGED GLASSES.

Angster (German). The same as a KUTTROLF. The term is derived from the Latin *angustus*, meaning 'narrow', from the characteristic narrow tubes.

animal figures. Figures made in glass representing a wide variety of animals, made from earliest times (e.g. crocodile, Egyptian, 200 BC–AD 100) to the present, including a lion, mule, pig, ram, lizard, etc. Some OIL LAMPS were made in the form of an animal, e.g., a running horse made at Murano. *See* TIERHUMPEN; STEM, *figure.*

Annagrün (German). A type of glass of a yellowish-green fluorescent colour made by adding URANIUM to the BATCH. It was developed by a German, JOSEF RIEDEL, and made *c*. 1830–48. It was named after his wife, as was also the type called '*Annagelb*' which was of a greenish-yellow colour. Pieces were made in Germany during the BIEDERMEIER period, and have CUT and GILDED decoration.

annealing. The process of subjecting glass to heat and subsequent gradual and uniform cooling for the purpose of toughening the glass and making it less likely to crack or break. Glassware when first made retains heat in some parts while other parts cool, so that internal stresses are created. Such glass is prone to cracking or breaking either spontaneously or when subjected to changes of external temperature. To remove such stresses, glass articles must be reheated uniformly and then allowed to cool slowly and uniformly. The annealing is done in an annealing oven (LEHR) that today is automatically controlled and through which the glass passes on a conveyor belt. *See* STRAIN-CRACKING.

annulated. A type of decoration on English DRINKING GLASSES, consisting of a series of equal or graduated rings or MERESES vertically placed, used in decorating a FOOT, a STEM or a KNOP.

antimony. A metallic agent used in colouring glass, producing yellow (*see* YELLOW GLASS). It was also used in ancient times, *c*. 2000–500 BC, to produce opacity in glass of certain colours and, especially in Roman times, from the second to the fourth century AD, as a decolourant.

Apostelglas (German). Literally, Apostle glass. A tall HUMPEN decorated with the enamelled portrayal of the Twelve Apostles, usually arranged in two encircling rows beneath ribbons or arcades bearing their names. The motif, based on old woodcuts, was probably derived from the Apostle tankards (*Apostelkrügen*) made of stoneware at Kreussen, in Franconia, in the 16th/17th century. The glass examples were made in the late 16th and early 17th centuries. Comparable pieces depicted the Four Evangelists. *See* ENAMELLING, GERMAN.

applied relief. Decoration in RELIEF attached to the surface of an object after it has been formed rather than being an integral part of the object. Although found less often on glassware than porcelain or pottery, examples are applied trailed threads of glass, as on wares since ancient times, and PRUNTS, as on English, Bohemian, and German glassware.

Aqua Marine glass. A type of ART GLASS developed by FREDERICK CARDER at STEUBEN GLASS WORKS that is similar to and resembles VERRE DE SOIE but has a greenish tint.

Aquileia. A seaport on the Adriatic, east of Venice, that is said to have been a glass-making centre when it was overrun in 450 AD by Attila and his Huns. The refugees fled to Venice and are reputed to have introduced glass-making there.

arabesque. (1) In Islamic art, a flat decoration of intricate interlaced lines and bands and abstract ornaments adapted largely from classical sources. As a popular Moorish decoration, its Spanish form excluded close representations of animal forms, these being the work of Christian artists. The strapwork of arabesques of this kind later influenced northern BAROQUE ornament, and in 19th-century England such designs were often termed 'Moresques'. (2) In Renaissance and later European art, an ornament with flowing curved lines and fanciful intertwining of leaves, scrolls, and animal forms. Ornament of this kind was popular during the NEO-CLASSICAL period, after *c*. 1760 and until *c*. 1790. The term is sometimes applied to GROTESQUES. The Italian term is *arabesco* (pl., *arabeschi*).

ampulla. Polychrome combed decoration, Roman glass, 1st–4th centuries. Museo Civico, Adria.

animal figure. Porcupine, Vera Liškova, Prague, *c*. 1973, L. 40 cm. Kunstsammlungen der Veste Coburg.

Apostelhumpen. The twelve Apostles as decoration, Bohemian or Silesian, 16th/17th century. Ht 27·6 cm. Kunstsammlungen der Veste Coburg.

Argy-Rousseau glassware. Vase of *pâte-de verre*, Paris, 1920–25, signed 'G. Argy-Rousseau'. Ht 14 cm. Kunstmuseum, Düsseldorf.

Ariel glass. Vase by Edvin Ohrström. Courtesy, Orrefors Glasbruk.

Art Deco style. Mould-blown vase, amber coloured, by Lalique, 1925. Ht 20·2 cm. Courtesy, Sotheby's Belgravia, London.

arcaded rim. A style of RIM that is decorated with encircling glass loops or arcades, sometimes having a PRUNT at the joins. It is found on SWEETMEAT GLASSES, *c*.1760.

arched decoration. A decorative pattern on CUT GLASS in the form of a series of adjacent arches, each sometimes enclosing a mitre-cut pattern. It was used in England in the 1840s.

arched rectangle pattern. A decorative CUT GLASS pattern in the form of rounded arches placed above short vertical rectangles of the same width. It was used on WATERFORD GLASSWARE in the late 18th and early 19th centuries, both sections being covered with diamond cutting.

architectural glass. Flat glass used for utilitarian purposes, such as was originally used for WINDOWS, MIRRORS, and PANES for carriages, and in recent times more extensively, e.g. strengthened and used for shelves, table-tops, etc. In modern times some architectural glass is made in solid or hollow blocks and is used for making building walls; it is often made in irregular shapes and decorated by etching, SAND-BLASTING or other modern techniques.

Argental, d'. A name sometimes used by the Compagnie des Verreries et Cristalleries de Saint-Louis (*see* ST-LOUIS GLASS FACTORY) located in Alsace-Lorraine (during 1871–1918 in Germany) which hence signed some of its ware 'St-Louis Münzthal' or 'd'Argental' (the latter being an adaptation of 'Münzthal' into French).

Argy-Rousseau, Gabriel (1885–). A French innovator who produced a type of PÂTE-DE-VERRE called PÂTE-DE-CRISTAL, which is almost transparent and very resonant. He used it frequently in making lamps and vases, decorated with relief flowers. He also made figurines by the CIRE PERDUE method.

Ariel glass. A type of ornamental glass made at ORREFORS GLASSBRUK. It was created *c*.1936 by EDVIN ÖHRSTRÖM and developed by VICKE LINDSTRAND, and later also used by INGEBORG LUNDIN. It is a type of CASED GLASS with embedded air bubbles (hence the punning name) that form abstract and figurative patterns. It is sometimes combined with the technique for GRAAL GLASS.

Aristeas. A glass-maker, in ancient Greece, probably a follower of ENNION and thought to have had a glass factory in Syria in the 1st century AD. Only one known piece bears his signature.

armlet. A bracelet worn on the upper arm. They were made of ROMAN GLASS, in clear glass or glass of various colours, some very dark. *See* BANGLE.

armorial glassware. Glassware (including plates, goblets, wine-glasses, flasks, etc.) decorated with heraldic coats-of-arms and devices. It was made in Germany, France, Italy, England, and elsewhere, with decoration in ENAMELLING or ENGRAVING, during the 17th to 19th centuries. It is sometimes called 'heraldic ware'. The German term is *Wappengeschirr*. *See* BEILBY GLASSWARE.

arrowhead cane. A type of CANE used in making a PAPER-WEIGHT. It has a central rod surrounded by petal-shaped rods in each of which there is a 3-pronged motif resembling an arrow pointing toward the centre, creating a flower-like effect. It is used in some GROUNDS and also scattered in MILLEFIORI patterns. The motif and its background may be of various colours. They were used at BACCARAT and ST LOUIS. Sometimes termed 'crow's-foot'.

arsenic. A substance that is constituent, in very minute quantity, of almost all modern colourless glass, serving to improve the colour, transparency, and brilliance. It is introduced as oxide of arsenic, a white powder made by heating, in a current of air, minerals containing arsenic. In the melting pot some remains in the molten glass, although some is lost in vapour. It is also used in making OPAQUE WHITE GLASS and, sometimes, OPALINE.

Art Deco style. A decorative style that developed in the 1920s and 1930s out of the ART NOUVEAU style and later art movements. It emphasized geometric patterns, and examples are found in many branches of the arts, including glassware. The name is derived from the *Exposition Internationale des Arts Décoratifs et Industriels Modernes*, held in Paris in 1925. In glassware, leading exponents of the style were MAURICE MARINOT, his disciples HENRI NAVARRE and ANDRÉ THURET, and RENÉ LALIQUE.

Art Glass. (1) Several types of glass made by some United States glassworks from *c.*1870 which were produced in newly developed (and sometimes patented) surface textures and colour shades, some with several colours shading into each other, and often of layers of CASED glass. It was also produced in several European countries and was most popular in the 1880–90s. *See* Ray and Lee Grover, *Art Glass Nouveau* (1975) and *European Art Glass* (1970). (2) A general term applied to glassware made for ornamental rather than utilitarian purposes, with primary regard to the quality of the METAL and the artistic nature of the form and decoration. It was used usually in connection with modern glassware from *c.*1850.

Art Nouveau. The style of decoration current in the 1890s and early 1900s, the name being derived from a gallery for interior decoration opened in Paris in 1896, called the '*Maison de l'Art Nouveau*'. It was introduced in England *c.*1890, mainly as a product of the movement started by William Morris and the Pre-Raphaelites, which spread to the Continent and America. It came to an end with the outbreak of World War I. The same style in Germany was called *Jugendstil*, after a magazine called *Die Jugend* (Youth), and in Italy called *Floreale* or *Liberty* (after the London store that featured it). Applicable to all the decorative arts, it was adapted to glassware by EMILE GALLÉ and LOUIS TIFFANY, making pieces of FREE-FORMED GLASS and using as decorative motifs, in RELIEF and ENAMELLING, floral patterns with elaborately twining tendrils.   [*Plate I*]

Artigues, Aimé-Gabriel d' (1778–1848). The director from 1791 to 1795 of the ST-LOUIS GLASS FACTORY in Lorraine, where he learned about making crystal glass. In 1802 he purchased the VONÊCHE GLASSWORKS in the Ardennes region near Namur, Belgium, where he operated successfully until 1815 when the Treaty of Vienna ended French sovereignty in the region and new tariffs burdened imports of glass into France. Consequently, in 1816 he bought the Verreries de Sainte-Anne at Baccarat and transferred his operations there from Vonêche, renaming the factory Verrerie de Vonêche à Baccarat. He obtained in 1817 from Louis XVIII a tariff exemption in consideration for agreeing to build a crystal-glass factory and a decorating plant. The factories were sold by him in 1823 to his associates, including Pierre-Antoine Godard. *See* BACCARAT GLASS FACTORY.

aryballos (Greek). A small flask or bottle used for holding oil for the bath. It has a squat globular BODY, a short NECK, a flat disk MOUTH with a small orifice, and a loop or scroll-like handle (occasionally two) extending from the shoulder to just below the disk of the mouth. Usually a cord was tied to the handle, to be looped over the user's wrist. Examples usually of bronze or of Greek pottery, range in height from 3 cm. ($1\frac{1}{4}$ in.) to 20 cm. (8 in.) or more, but there are also other examples made in the Eastern Mediterranean region, *c.*600 BC–AD 100, of glass. These are CORE GLASS vessels, usually having two handles and COMBED DECORATION on a dark-blue body. *See* LENTOID FLASK; DOLPHIN BOTTLE.

armorial glassware. Wine glass with rounded funnel bowl and knopped balustroid stem, English, *c.*1745. Ht 17 cm. Courtesy, Maureen Thompson, London.

Art Nouveau. Vase by Burgun, Schverer & Cie, Meisenthal, Lorraine, *c.*1895–1900; marks 'B S & Cie', and 'Verrerie d'Art de Lorraine'. Ht 19·8 cm. Kunstmuseum, Düsseldorf.

aryballos. Blue Roman glass. Rhineland, 1st/2nd century, Ht 5·4 cm. Römisch-Germanisches Museum, Cologne.

ash-tray. A small receptacle for disposing of cigarette or cigar ashes. They have been made of ART GLASS, e.g. by DAUM FRÈRES and other modern glassworks. The German term is *Aschenbecher*, the French *cendrier*.

ashes. The residue of various materials after being burned, used in glass-making as the source of the required ALKALI. Early SODA GLASS made in Venice contained ashes of marine plants obtained from Spain (BARILLA) or Egypt (ROQUETTA). Ashes of KELP were used especially in northern France, Scandinavia, and England. Bohemian and German POTASH glass made use of ashes of beechwood (*see* WALDGLAS), while ashes of bracken (FERN-ASH) were employed in France (*see* VERRE DE FOUGÈRE). BONE-ASH was used mainly as a FLUX, and in Germany for opacifying glass.

askos (Greek). A small vessel, ranging in length from 7 cm. to 25 cm., used in ancient Greece and Rome for pouring oil into lamps. It is in the form of a low, fully enclosed bowl with an overhead handle, at one end of which is a cylindrical vertical or diagonal SPOUT. It was so-called from the fact that some examples were said to have been made in the shape of a wine-skin or leather bottle (*askos*). A few are in the form of an animal, bird or lobster claw. It was generally made of bronze or Greek pottery, but some have been made of ROMAN GLASS from *c.* 400 BC–AD 100.

aspersorium. A basin for holy water. They were made of glass in Venice and Spain, in the 16th century, shaped like a low BOWL with an overhead bail handle. They were used for both religious and secular purposes. The Spanish term is *calderata*. *See* BÉNITIER; BUCKET; SITULA.

astragal. A die (pl., dice), originally made from the astragalus, the ankle bone, by the Greeks, and used in ancient games played with various GAME PIECES. Some were made of glass.

at-the-fire. The process of reheating at the GLORY HOLE a glass object already formed. A blown object is so reheated to permit further blowing to enlarge it or alter its shape or to permit manipulating it with tools. Some objects are so reheated to affect the basic colour (as GOLD RUBY GLASS) or the surface colouring (as pieces sprayed with stannous chloride; *see* AURENE GLASS). Certain glass objects are FIRE-POLISHED in this manner. *See* AT-THE-LAMP.

askos. Roman glass of various colours, 1st century AD. L. 23 cm. Museo Civico, Adria.

at-the-lamp (or **at-the-flame**). The technique of shaping glassware from rods and tubes, made of a readily fusible glass, by heating and softening them at the flame, originally of an oil lamp and later of a Bunsen burner; the glass was then pincered, drawn out or otherwise manipulated. The process was probably discovered in the Roman era. It has been used for objects not made at a factory furnace, e.g., in Venice in the 17th century to make glass beads and small figures. It was used extensively in France (*see* NEVERS FIGURES) and the Netherlands, and in the 18th century in Bohemia and Thuringia (*see* LAUSCHA). Much COUNTRY-MARKET GLASS was made by this method at country fairs in many places of Europe. Certain moulded parts of vessels, e.g., a handle, were so heated in order to be fused to the object. In recent years the technique has been revived by French, Venetian, Viennese, and German makers of STUDIO GLASS, and much glassware of modern styles (especially of ART NOUVEAU style) has been made at-the-lamp. The term 'lamp-work' is sometimes applied to objects so made. The Italian term is *a lume. See* AT-THE-FIRE.

Atkins, Lloyd (1922–   ). An American glass-designer who joined the design department at STEUBEN GLASS WORKS in 1948. He has designed a number of important pieces of ornamental glassware with the engraving executed by others. One piece, 'Quintessence', in abstract form, has attached to it fifty gold rods designed by Patricia Davidson;

one example (from the edition of five) was presented to the Shah of Iran on the 2,500th anniversary of the Persian Empire. *See* SEA CHASE.

Auldjo Jug. A JUG (OENOCHOË) made of blue glass with white CAMEO decoration in the manner of the PORTLAND VASE, with vine leaves, grapes, and ivy in a border of doves and foliage. It is named after a former owner, Richardson Auldjo. It was found in fragments in the ruins of Pompeii, but has been re-assembled and restored. It has a flat base and a moulded BASE-RING. *See* VENDANGE VASE.

Aurene glass. A type of ART GLASS developed by FREDERICK CARDER at STEUBEN GLASS WORKS within a year of his joining the company. It is of two types, known as 'Gold Aurene' and 'Blue Aurene'. The earliest examples of Gold Aurene have a purplish cast, but from *c*. 1905 the fully developed glass has a lustrous gold appearance. The metallic surface was produced by spraying the glass AT-THE-FIRE with stannous (tin) chloride or lead chloride solutions under controlled atmospheric conditions. Blue Aurene (made from *c*. 1905 to 1933) has added cobalt oxide (hence sometimes called 'Cobalt Blue Aurene') and shades from dark blue to a pale silvery hue, with other hues occasionally found. In both types there are sometimes fine wavy lines in the surface that were made intentionally, radiating on the sprayed metallic areas that did not expand with the glass when the object was reheated; objects fully formed before being sprayed have no such lines. Some such glass (made *c*. 1904–10) was further decorated with TRAILING in the form of leaves and vines. Related to such Aurene glass are so-called 'Red Aurene', 'Green Aurene', and 'Brown Aurene', also made *c*. 1904–10, but these are not true Aurene glass, as the only Aurene sprayed surface is on some applied decoration or on some lining or feet; the body of these pieces is of ALABASTER or CALCITE GLASS, and is generally CASED, in whole or partially, with ruby or green glass (hence the names) or sometimes is of solid RUBY GLASS. Some 'Blue Aurene' glass that developed a silvery hue has been erroneously called 'Platinum Aurene'; IVRENE GLASS and VERRE DE SOIE have sometimes been erroneously called 'White Aurene'; and 'ROUGE FLAMBÉ' has sometimes been erroneously called 'Red Aurene'. The name 'Aurene', registered in 1904, is derived from the first three letters of the Latin *aurum* (gold) and the last three letters of *schene* (a Middle English word for 'sheen'). The pieces are signed by Carder on the underside of the base or foot.

Austrian glassware. Glassware (apart from the very early HALLSTATT CUPS) made in the former Austrian Empire from the 16th century when craftsmen from Venice worked at glasshouses at Vienna, HALL-IN-TYROL and INNSBRUCK, and when native workers had glasshouses at Graz (Styria) and Henriettental (Carinthia). From the 16th to the 19th centuries, there developed the great glass-making centres in Bohemia (*see* BOHEMIAN GLASSWARE) and Silesia (*see* SILESIAN GLASSWARE), and glass enamelling was done at Vienna (*see* ANTON KOTHGASSER and GOTTLOB SAMUEL MOHN). In 1823 J. & L. LOBMEYR was founded in Vienna and continues today as a leading maker of ornamental glassware and chandeliers. Also still operating is STÖLZLE GLASINDUSTRIE, founded in 1843. Other glasshouses are Riedel, at Kufstein, and Swarovski, at Wattens, and Salzburger Cristallglas. Vienna was a centre of the ART NOUVEAU movement and later the development of modern functional and industrial glass. *See* BAKALOWITS, E., & SÖHNE.

aventurine glass. A type of translucent glass flecked throughout with sparkling oxidized metallic particles, simulating the appearance of brownish aventurine quartz that is flecked with mica or other minerals. The earliest form, of brownish colour flecked with copper (called 'gold aventurine'), was first recorded at the beginning of the 17th century, attributed to a MURANO glass-making family named MIOTTO; glass rods or CAKES of such glass were exported by the Venetians and used by foreign factories after crushing them into various shapes and sizes. The Venetian product resulted from metallic copper particles being formed

Auldjo Jug. Blue glass with white cameo decoration, Roman glass, *c*. 100 BC–AD 100. Ht 22·8 cm. British Museum, London.

Aurene glass. Vase of 'Red Aurene', designed by Frederick Carder for Steuben Glass Works, *c*. 1905–10. Ht 25 cm. Rockwell Collection, Corning Museum of Glass, Corning, N.Y:

chemically by including in the BATCH copper oxide and forged scales. A later process, developed *c.* 1865, produced a greenish colour due to particles of CHROMIUM (called 'green aventurine'). French chemists in the 1860s developed other processes, and the technique was understood in China in the 18th century. A greenish-coloured aventurine has been made in the United States by the FOSTORIA GLASS CO. It has been said that the name is derived from the Italian word *avventurine*, meaning chance, due to the alleged accidental discovery of the process; however, the Italian word *avventurina* means a type of brown quartz, which the glass resembles and is hence also so called. In English the glass is also called 'goldstone'. *See* AGATE GLASS.

Awashima Glass Co. A glasshouse in Tokyo founded in 1956 by Masakichi Awashima (b. 1914), a glass-designer. It makes moulded blown glassware, some having a roughened surface texture intended to resemble raindrops, hence called 'Shizuku' (waterdrop). *See* JAPANESE GLASSWARE.

Ayckbowm glassworks. A glasshouse in London owned until 1780 by Herman and Dedereck Ayckbowm. It exhibited engraved and gilded glassware for sale at Bath in 1772/73 and also at Limerick. J. D. Ayckbowm, perhaps a son of one of the partners, represented the firm in Limerick in 1774 and migrated to Ireland in 1783. He established there his decorating firm, making cut glass from blanks made by others, but from 1799 he had a glass manufactory; he operated until 1802, when he transferred it to J. L. Rogers & Co., of Dublin, which is listed until 1808. J. D. Ayckbowm & Co. kept its glass warehouse in Dublin until *c.* 1820.

aventurine glass. Bowl, Venetian, 17th century. Ht 5·6 cm. Kunstsammlungen der Veste Coburg.

B

Baccarat Glass Factory. A glassworks, now the most important maker of crystal glass in France, that was founded in 1764 at Baccarat, near Lunéville, in Lorraine, by Monseigneur de Montmorency-Laval, Bishop of Metz, as a royal project under the patronage of Louis XV. Initially directed by Antoine Renaut, it was named the Verreries Renaut & Cie., after 1768 the Verreries de Baccarat, and after 1773 the Verreries de Sainte-Anne (after the patron saint of the chapel of the glassworks). In 1802 it was transferred to the three sons of Renaut, and in 1806 to M. Lippmann-Lippmann. In 1816 it was purchased by AIMÉ-GABRIEL D'ARTIGUES, who transferred to it the operations of his VONÊCHE GLASSWORKS and directed it, under licence from Louis XVIII, as the Verreries de Vonêche à Baccarat. Under the licence he built a new factory for making crystal glass and a decorating plant. The factories were sold by him in 1823 to three associates, including Pierre-Antoine Godard, and under successive owners it has operated to the present time, having changed its name in 1823 to Compagnie des Cristalleries de Baccarat, as it is now known. It originally made glassware in the English style. Since 1816 it has made crystal glass, since 1823 AGATE GLASS and OPALINE, since 1846 MILLEFIORI glassware and PAPER-WEIGHTS, and since 1848 SULPHIDES. Under the direction of Jean-Baptiste Toussaint, from 1851 to 1858, it introduced coloured glassware, and under Paul Michaut, from 1867 to 1883, it started WHEEL-ENGRAVING. Today it makes fine stemmed tableware, some from designs by Georges Chevallier, stressing harmony of form rather than elaborate decoration, and of remarkable thinness and resonance. It also makes objects of freely formed CUT GLASS. Some early pieces are identified by the initial 'B' but this refers to two craftsmen named Battestine. Moulded glass bottles decorated with portraits, etc., have been made since 1824, and are now made at a branch factory at Trelon (Nord). *See* HARCOURT; STAG HEAD.

Bacchus, George, & Son. An English glass-making firm at Birmingham; the firm specialized in wares made by FLASHING. Similar work was done by other firms in Newcastle-upon-Tyne and Stourbridge. Bacchus was among the early decorators of glass by TRANSFER-PRINTING, *c.*1809, the process having been introduced in the 18th century.

Bachmetov Glass Factory. A privately owned glasshouse founded in 1763 by A. Bachmetov (d. 1779) at Nikol'sko, near St Petersburg, that originally specialized in coloured glass. In the 1790s his heirs were operating three glasshouses which by 1802 made large quantities of crystal glass dishes, sheet glass, and inexpensive ware. One heir, Nikolai Alexeivich Bachmetov, tried in 1802 to lease the ST PETERSBURG IMPERIAL GLASS FACTORY. With the aid of a large Government loan in 1808, he greatly expanded the business so as to compete with imported glassware, making crystal glass, opaque white glass (especially with inserted medallions), and coloured and opalescent glass. After 1861 production decreased so that the widow of A. N. Bachmetov (d. *c.* 1861), who had taken control, had to reduce quality. When she died in 1884, ownership passed to Prince Obolensky who revived the making of high-quality glassware, and specialized in making SCENT BOTTLES. *See* RUSSIAN GLASSWARE.

back painting. Decoration by painting on the back of glass. It is to be found on (1) MIRRORS, (2) PANELS, and (3) some CHINESE SNUFF BOTTLES. *See* PICTURE ENGRAVING.

Baccarat glassware. Decanter and wine-glass, 'Harcourt' pattern, made since 1842. Cristalleries de Baccarat.

ball-stopper jug. Henry Ford Museum, Dearborn, Mich.

baluster, light. Newcastle wine glass, coat-of-arms probably engraved in Holland, 1734. Ht 18·3 cm. Courtesy, Delomosne & Son, London.

Bacon Goblet. A giant GOBLET (30·5 cm. high) of colourless English LEAD GLASS, having a rounded funnel bowl on a stem with two KNOPS (each enclosing an air bubble) and resting on a folded FOOT. It was made *c.* 1690–1700, and is so-called because it was given to the British Museum in 1949 by the Circle of Glass Collectors (London) in memory of its founder, John Maunsell Bacon (1886–1948).

bag beaker. A drinking vessel of elongated bag shape, with a waisted neck below a wide mouth, and usually decorated with vertical trailing extending from below the neck to the pointed bottom. One known example is decorated in the manner of a CLAW-BEAKER. All known examples were presumably made in England (as they are not known on the Continent except in Scandinavia) during the 7th century; and as they have been found at Faversham, in Kent, they were probably made there. *See* POUCH BOTTLE.

Bakalowits, E., & Söhne. A glasshouse in Vienna that featured in the early 1900s iridescent glassware, making pieces designed by R. Bakalowits of Graz, Austria, and by JOSEF HOFFMANN (b. 1870) and L. KOLOMAN MOSER.

Bakewell's. A glasshouse at Pittsburgh, Pennsylvania, established in 1808 by Benjamin and Edward Ensell, who were joined in 1809 by Benjamin Page. It acquired *c.* 1813 another glasshouse, and took the name Pittsburgh Flint Glass Manufactory. After several changes of ownership, all including members of the Bakewell family, and successful operation, it closed in 1882. It initially made bottles and flasks, and was the first in the region to make FLINT GLASS tableware; from 1810 it made cut and engraved glass, and later coloured and pressed glass. Cutting was introduced by William Peter Eichbaum, glass cutter to Louis XVI. It was the first American glassworks to supply the White House, making a service for President Monroe in 1817. It is known for its HISTORICAL FLASKS, including that called 'THE AMERICAN SYSTEM', and also for tumblers with SULPHIDES in the base portraying leading citizens.

ball-stopper. A glass ball that rests atop a glass jug, its diameter being slightly larger than the diameter of the mouth. Such jugs and stoppers were made in the mid-19th century in the United States, but earlier ones (called *flacon à boule*) had been made of opaque white glass in France, *c.* 1709.

ball-stopper bottle. A type of bottle for effervescent soft drinks that had a rubber washer in the neck and a loose glass ball (like a MARBLE) that was forced against the washer by pressure when a bottle was full and that sealed the contents. To release the pressure and permit pouring, a wooden dowel was used. It has been said that these bottles were invented in 1872 by Hiram Codd for a soft drink known as Codd's Wallop. Many were made by firms in Yorkshire.

balsamarium. The same as UNGUENTARIUM.

baluster. A term used to describe a type of DRINKING GLASS with a baluster stem (*see* STEM) made in England; the baluster motif was adopted from Renaissance architecture, and was used on glasses from Venice made in the early 17th century, these being in inverted baluster form. Glasses of this type were first made in England soon after GEORGE RAVENSCROFT's introduction of LEAD GLASS, *c.* 1676, and were widely used. Early examples, made *c.* 1685–1710, feature the inverted baluster, and later ones with true baluster stem date from the period *c.* 1710–1735. Glasses with baluster stems are very varied, with different types, numbers, and arrangements of KNOPS, as well as different forms of BOWL. *See also* BALUSTER, HEAVY; BALUSTER, LIGHT; BALUSTROID.

baluster, heavy. A drinking glass with STEM made in England *c.* 1685–1710, called 'first-period baluster'. Heavy balusters appear both as WINE GLASSES and GOBLETS. They are of particularly fine METAL and were perhaps inspired in part by the baluster columns of silve

I *Art Nouveau glassware*. Upper row: (left) overlay glass vase decorated with wheel-engraved orchids, ht 20 cm. (7¾ in.), by Daum Frères; (centre) night light, with wheel-engraved decoration and *marqueterie de verre*, on a silver base, ht 19 cm. (7½ in.), by Emile Gallé; (right) iridescent vase, ht 18 cm. (7 in.), by Lötz. Lower row: (left) iridescent vase, ht 23·5 cm. (9¼ in.), by Louis Comfort Tiffany; (centre) floriform vase, with pulled decoration on an iridescent opal ground, ht 31 cm. (12 in.), by L. C. Tiffany; (right) cameo glass jug, etched with iridescent design and fire polished, ht 25 cm. (10 in.), by Emile Gallé. Courtesy, Editions Graphiques Ltd, London.

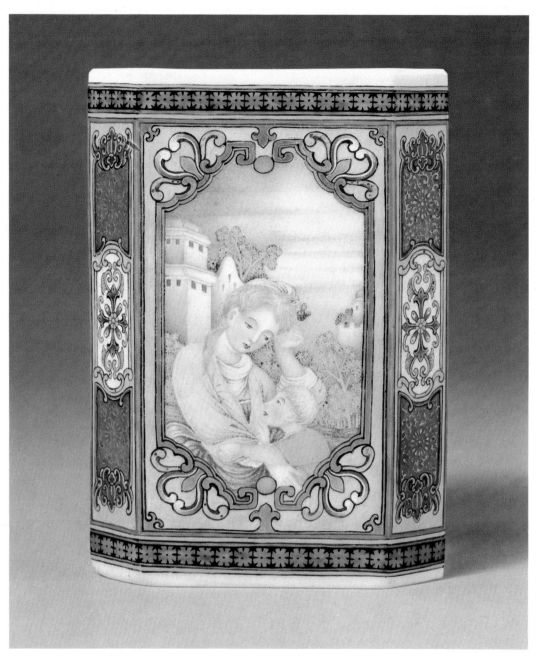

II *Brush-pot*. Opaque white glass with enamelled decoration and Ch'ien-Lung four-character reign-mark (1736–95), made at the Peking Imperial Palace workshops; ht 8·5 cm. (3½ in.). Courtesy, Hugh Moss Ltd, London.

candlesticks and the turned work of cabinet-makers. *See* BALUSTER, LIGHT; BALUSTROID.

baluster, light. A drinking glass with STEM made in England, *c.* 1710–35, that is on a smaller scale than a heavy baluster and with a stem composed of a series of small KNOPS of different form, although with rarely any of the earlier forms. COLLARS, or flattened knops, were incorporated into the stemmed designs. These light balusters are sometimes known as 'Newcastle glasses' as many were manufactured there. *See* BALUSTER, HEAVY; BALUSTROID.

balustroid. A type of BALUSTER glass developed in England, *c.* 1725–60, the stem being lighter and more elongated than the earlier heavy balusters. They have simple KNOPS, or are in inverted baluster form without knops. They are especially found in NEWCASTLE GLASS. *See* STEM, *balustroid*.

Bandwurmglas (German). Literally, tapeworm glass. A type of tall BEAKER, cylindrical or CLUB-SHAPED (a type of STANGENGLAS), which has on its exterior a notched glass trail or thread in spiral form and has a BASE with a high KICK. They were made in the Rhineland in the 15th and 16th centuries. Sometimes they are called a PASSGLAS, but this is erroneous as they have no horizontal equidistant dividing lines that are the characteristic of the *Passglas*.

Bang, Jacob E. (d. 1965). A Danish glass-designer who was chief designer at the Holmegaard Glasvaerk from 1925 until 1942 and, after some years in the ceramic industry, became employed in 1957 by Kastrup Glassworks (*see* KASTRUP & HOLMEGAARDS GLASSWORKS).

bangle. A ring-shaped bracelet. They are of graduated sizes, and sometimes are worn several along an arm. They have been made in many places and at many times, from the Middle La Tène period (300 BC–AD 100) of the Iron Age onward. They are found in India and are said to have been made there in the 19th century from glass material sent from Austria. They are also made in Nigeria in modern times; there the manufacturing process is to rotate a soft glass ring around an iron rod until the join disappears. *See* ARMLET. .

bank. A receptacle for small savings. Examples in glass made in the United States were very elaborate, made as presentation pieces. They are in the form of a tall (28 to 46 cm. high) thin bowl on a stemmed base, decorated with vertical tooled ribbing and made so that the ribbing extends upward an equal height as struts to support a complicated FINIAL (and some rare ones having three groups of struts, one above the other). Some have a coin enclosed in the hollow finial or foot. They were made entirely by hand manipulation, with much pincered decoration. They were made by the BOSTON & SANDWICH GLASS CO. and the NEW ENGLAND GLASS CO., *c.* 1815–50.

barbarico vetro (Italian). A type of ornamental glass developed by Ercole Barovier for BAROVIER & TOSO, *c.* 1951, with a transparent glass casing covering an overall pattern of numerous small spots of metallic blue colour.

Barbe, Jules. A Frenchman who decorated glassware with enamelling and gilding. After his work was shown at the 1878 Paris Exhibition, he was brought by THOMAS WEBB & SONS to its DENNIS GLASSWORKS where in the late 1880s he did enamelling on Queen's Burmese Glass (*see* BURMESE GLASS).

Barberini Vase. *See* PORTLAND VASE.

Barbini, Alfredo (1912–   ). An Italian glass-maker, descended from 17th-century Venetian glass-makers, who has worked since boyhood in glass at MURANO, including a period *c.* 1933 with Zecchin & Martinuzzi, then *c.* 1938 as a partner of V.A.M.S.A., and from 1947 to

Bandwurmglas. Beaker with continuous spiral notched trailing, Lauscha, Thuringia, *c.* 1700. Kunstsammlungen der Veste Coburg.

barbaricos, vetro. Bowl and vase designed by Ercole Barovier. Barovier & Toso, Murano.

Barbini glassware. Vase, orange with smoky exterior, designed and made by Alfredo Barbini, Murano. Ht 26 cm.

barometer. Water-barometer, Liège(?), 17th century. Ht 20 cm. Kunstgewerbemuseum, Cologne.

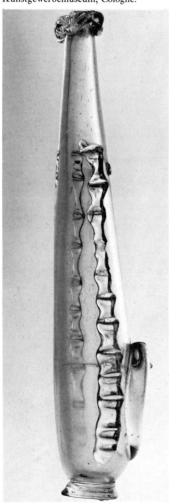

1950 with CENEDESE. Since 1950 he has had his own furnace, developing a new technique called *a masello*, which involves sculpting a solid block of molten glass without moulding or blowing. *See* MANO SUL SENO, LA; SCULPTURED GLASSWARE.

barilla. A marine plant, *Salsola soda* (*see* GLASSWORT), that grows in the salt-marshes near Alicante, Spain, and elsewhere in the Mediterranean region; it was used as a source of SODA, the alternative to POTASH as the ALKALI required in making glass. *See* ROQUETTA.

barium. An ingredient sometimes added to glass METAL for making MOULD-PRESSED GLASS that imparted a silvery ice-like appearance.

barium glass. Glass which includes BARIUM as an ingredient. Barium was used in China in the Han Dynasty (206 BC–AD 220), earlier than elsewhere, and its presence helps to identify certain early specimens as being examples of CHINESE GLASSWARE.

barometer. An instrument for measuring atmospheric pressure and forecasting weather conditions. The type known as a 'water barometer' was made in Germany and at Liège and Murano in the 17th century; it is an elongated tear-shaped glass bottle with a long graduated tube projecting upward from near the bottom.

baroque. A style in art which emerged shortly before 1600, of which the principal exponent was Giovanni Lorenzo Bernini. It remained current in Europe until the emergence of the ROCOCO style, *c.* 1730. It was a development of the Renaissance style, and is characterized by lively, curved, and exuberant forms, by vigorous movement, and by ornament based on classical sources which is symmetrical, as distinct from the asymmetry of the following rococo style. As applied to glassware, it is to be noticed in Venetian glasses where stems of GOBLETS were given extravagant forms, in the German use of LAUB- UND BANDELWERK, and in some English glasses by GEORGE RAVENSCROFT, with a great variety of designs for the bowls and stems.

Barovier, Anzolo (Angelo; d. 1460). A MURANO glass-maker concerned with technical aspects of glass-making and who is associated with the introduction of many new processes, including perhaps reviving the use at Venice of enamelling in the Islamic technique and probably contributing to the development of CRISTALLO glass. His descendants have been connected with glass-making, including Ercole Barovier (1889–1974) who furthered in Venice the development of modern glass varieties and forms, and the latter's son Angelo (1927–   ). *See* BAROVIER & TOSO; BAROVIER CUP.

Barovier & Toso. A glassworks at MURANO founded in 1936, and owned today by descendants of two ancient families of Murano glassmakers. Descendants of ANZOLO (Angelo) BAROVIER owned furnaces in Murano from the 16th century; in 1878 Benvenuto Barovier formed Artisti Barovier. In the 15th century the Toso family owned several furnaces at Murano; in 1936 the brothers Artemio and Decio Toso, together with Benvenuto's son, Ercole Barovier (1889–1974), formed Barovier & Toso, now directed by Ercole's son, Angelo (b. 1927), with Decio, Mario, and Piero Toso. The firm specializes in large chandeliers and custom-made industrial lighting fixtures, also making high-quality ornamental glassware and creating new glass techniques. Wares include VETRO ABORIGENE; VETRO BARBARICO; VETRO DAMASCO; VETRO DIAFANO; VETRO GEMMATO; VETRO PRIMAVERA; VETRO RUGIADOSO; VETRO SIDONE.

Barovier Cup. A WEDDING CUP made at Murano, *c.* 1470–80, of dark blue glass; the bowl rests on a hollow spreading BASE. Enamelled medallions on opposite sides of the bowl, portraying each of the parties, divide an encircling frieze showing on one side a procession of horsemen and ladies, and on the other side women bathing at the Fountain of Love. The enamelling is mainly of dark green, with some blue and red,

Barovier Cup. Wedding cup (standing bowl) of blue glass with enamelling, attributed to Anzolo Barovier, Murano, *c.* 1470–80. Ht 18 cm. Museo Vetrario, Murano.

and touches of gold. For many years the cup has been attributed by tradition, on the basis of his reputation at the time and similarity to other pieces, to ANZOLO BAROVIER; but recently Luigi Zecchin, Murano scholar, has disagreed, stating that there is no proof that Barovier ever did enamelling and that the cup is in Renaissance style, prevalent *c.* 1470–80, ten years after Barovier's death. *See* PROCESSION CUP; STANDING CUP; STANDING BOWL.

barrel-flask. A table FLASK, usually for brandy, in the form of a horizontal barrel, tapering toward each end and with trailed hoops, a short NECK at the top having a small MOUTH, and four small supporting feet. They were made of ROMAN GLASS in the late 2nd to early 3rd century, and of clear or coloured glass in the 17th and 18th centuries in Germany, Venice, France and Spain (Catalonia). The German term is *Fassflasche*; the Spanish is *barralet*.

basal rim. (1) The same as a FOOT-RING. (2) The edge of the concave base of some PAPER-WEIGHTS, on which the piece rests.

Basdorf glass. A type of glass having the appearance of OPAQUE WHITE GLASS, made at two factories in Basdorf, near Potsdam, Germany, from 1750 until *c.* 1783. It was said at the time to have been made from crushed green glass mixed with gypsum and lime, and then remelted to develop its opacity; it had a rough surface and required grinding. Since it was not made with oxide of tin or calcined stag's horn, it was not true MILCHGLAS, as it has frequently been erroneously called. The factories had been given a privilege to make glass but denied the right to make porcelain in competition with the porcelain factory at Berlin, so they sought to achieve a glass with the appearance of porcelain that could be decorated with enamelled flowers. It was first made experimentally by a workman named Schopp at the Tornow factory at Basdorf; soon a glass-engraver there named Friedrich Schackert obtained permission from Tornow to make glass, and in 1750 was authorized to open a new glass factory at Stennewitz, near Basdorf. He there produced Schopp's glass, which has been called 'glass porcelain', making sugar bowls, cream jugs, mugs, and tea services; the factory continued this ware under him and his successors until *c.* 1783 when it received permission to make porcelain. Recent analysis has not revealed the exact constituents of the glass.

base. The bottom part of a vessel. Bases have been made in many styles, and the style often helps in dating the piece. *See* FOOT; KICK BASE; SPLAYED BASE; STAR-CUT BASE.

base-ring. A ring of glass added to the base of a glass vessel after its body has been made. It is usually flattened so that the edge of the ring extends beyond the bottom of the vessel. *See* PAD-BASE; COIL-BASE; FOOT-RING.

basilisk. A fabulous serpent, lizard or dragon whose hissing sound drove away all other serpents and whose breath or look was fatal. A fanciful representation was made in glass in FAÇON DE VENISE in the 17th century.

basket. (1) A basket made of glass, especially those with a MOULDED base and the sides built up with threads of glass PINCERED together into an open-work mesh. Some have a glass overhead (bail) handle. Such baskets were made in MURANO, the Netherlands, and Spain in the 18th

Barovier & Toso. Group of objects in *vetro aborigene*, with greenish opaque spots, designed by Ercole Barovier, *c.* 1954, Murano.

basket. Crown-shaped basket in colourless and blue glass, Germany or Netherlands, 17th century. Ht 11·7 cm. Kunstgewerbemuseum, Cologne.

century. Some have an accompanying stand EN SUITE. They appear to be more ornamental than useful, but may have held sweetmeats or fruit. *See* TRAFORATO. (2) The funnel-shaped decoration surrounding the motif of some PAPER-WEIGHTS, such as those made of STAVES or of swirling white FILIGRANA (LATTICINO) glass in which is arranged the motif of fruit paper-weights.

batch. The aggregate of the various ingredients weighed out and mixed, ready to be placed in the melting pot and fused in the process of making glass, including a certain amount of CULLET. Sometimes called the 'mixture'. *See* METAL.

battle scenes. Decoration depicting battles, camp scenes, and cavalry skirmishes used on glassware and ceramic objects. It was sometimes done in SCHWARZLOT by JOHANN SCHAPER, *c.* 1665. The German term is *Schlachtenszenen.*

battledore. A glass-making tool in the form of a square wooden paddle with a projecting handle. It was used to smooth the bottom of vases and other objects. *See* TOOLS, GLASS-MAKERS'.

battute, vetro (Italian). Literally, beaten glass. A type of ornamental glass developed by VENINI, the characteristic feature of which is that the entire surface is ground by being placed lightly in contact with a revolving wheel, thus producing innumerable small, irregular, and adjacent markings running in the same direction, giving the object the appearance of having been beaten.

Bavarian glassware. Ware made in Bavaria, a province of southern Germany, adjoining Bohemia, with Munich as its capital and including the glass-decorating centres of Nuremberg and Ratisbon (Regensburg). *See* NUREMBERG GLASSWARE.

bead. A small object, usually globular (but sometimes oblate, cylindrical, polyhedral or irregularly shaped) and generally pierced for stringing, made from earliest times for personal adornment or to embellish other wares, usually in the form of strands. The earliest material having the character of glass was used in Egypt, before *c.* 4000 BC, as a glaze to cover beads of stone or clay in imitation of coloured precious stones. Later this glaze was used in Egypt to make beads as the earliest objects wholly of glass, *c.* 2500 BC. Beads are known from Mycenaea from the 16th to 13th centuries BC. The early decorative patterns on beads, from *c.* 1500 BC, are stripes and spots; later developments are EYE BEADS and beads with zigzags and chevrons. Egyptian beads were exported to many countries. These early beads were normally of opaque glass, frequently blue, with decoration in yellow, etc. The Roman beads followed the Egyptian tradition. In Venice in the 11th century and subsequently glass beads were made for trade. Later they became popular in Bohemia and elsewhere, being made in many forms and styles, including TRANSLUCENT, IRIDESCENT, FACETED, and ENAMELLED in various colours. Today the principal sources of glass beads are Venice, Japan, and the Gablonz region of Czechoslovakia; some are of good quality and are used for costume jewellery, but vast quantities serve merely as tourist souvenirs. Coloured beads may be either inferior varieties coloured on the surface only, or made entirely of coloured glass. *See* AGGRI BEAD; LEECH BEAD; MATAGAMA BEAD; BUGLE; CRUMB BEAD; CONTERIE; MYCENAEAN GLASS-PASTE BEADS; NECKLACE.

beaded edge. An EDGE decorated with BEADING.

beaded ware. Objects of which the entire surface is covered with tiny polychrome threaded beads making a design having the appearance of *petit point* embroidery. It was used on GOBLETS made in Bohemia, *c.* 1850 and on some SCENT-BOTTLES, jewel boxes, hand-bags, lip-stick cases, vanity-boxes, etc. made in France. The French term is *broderie de perles dites sablées.*

beaded ware. Casket, French, 18th century. W. 25 cm. Musée des Arts Décoratifs, Paris.

beading. A decoration in the form of a continuous row of small RELIEF beads.

beak. A type of LIP on a JUG or other pouring vessel that extends to a point, usually in the form of the lower mandible of a sparrow.

beaker. A drinking glass, often with a flared MOUTH but sometimes cylindrical, and basically of circular section down to a flat BASE. Some are supported by a stemmed foot or three bun feet (*see* FOOT). It has no HANDLE, although the term originally sometimes referred to a vessel with a handle and a SPOUT (BEAK). Some have a COVER. *See* TUMBLER; MUG; TANKARD; ARMORIAL BEAKER; HUMPEN; FAIRFAX CUP; LUCK OF EDENHALL; NUPPENBECHER; CONE BEAKER; SHELLFISH BEAKER; NETWORK BEAKER; BELL BEAKER; CONCAVE BEAKER; FIGURED BEAKER.

Beilby, William (1740–1819) and **Mary** (1749–97). Noted enamellers of glassware who were the children of William Beilby (1705–65), a goldsmith and silversmith who moved with his family of seven children from Durham to Newcastle-upon-Tyne in 1760 where another son Ralph (1743–1817) was a heraldic and general engraver. William and Mary worked probably for the glasshouse of Dagnia-Williams, doing enamelling in ROCOCO style from *c*. 1762 until 1774 when Mary had a paralytic stroke. In 1778 their mother died and they moved to Fife in Scotland. Until his father died in 1765, the son added 'junr' on his signed pieces. *See* BEILBY GLASSWARE; DAGNIA FAMILY.

Beilby glassware. English glassware of the mid-18th century noted for its enamelling by WILLIAM and MARY BEILBY. The early work, from 1762, presumably by William, shows the influence of their brother Ralph, a heraldic engraver, and bears heraldic motifs; these include eight recorded examples (three signed) of GOBLETS with bucket BOWLS and opaque twist stems, with the Royal Arms of George III, some with the Prince of Wales's feathers, made presumably to celebrate the birth of the future George IV, the first such dated piece being 1763. Later pieces, from *c*. 1774, depict in ROCOCO style floral subjects, rustic scenes, gardens, landscapes, and classical ruins and buildings, with often a butterfly, bird, or obelisk. Their enamelling is in white (bluish or pinkish) monochrome or in enamel colours. They decorated wine glasses, and goblets (usually with opaque twist stems) and DECANTERS. Some glasses have gilt on the rim. Some pieces are signed by the surname only, so it is not possible to know which Beilby enamelled them; but some early examples, before 1765, are signed 'W Beilby junr'.

Beinglas (German). Literally, bone glass. The German term for MILK-AND-WATER GLASS.

Belfast glassware. Glassware attributed to Belfast is generally accepted as having been made from 1776 at a glasshouse at the Long Bridge in Belfast by BENJAMIN EDWARDS, SENIOR. Especially known are tall thin DECANTERS with two neck-rings, many of which are marked 'B Edwards Belfast'; most are undecorated but a few are wheel-engraved. *See* IRISH GLASSWARE.

Belgian glassware. Glassware made in the region of present-day Belgium from the 5th century (*see* FRANKISH GLASS) to the present. In 1541 glass-workers from MURANO were recorded at Antwerp, and in 1569 workers from L'ALTARE were recorded at Liège, and glasshouses continued at both cities. It is difficult to distinguish the early native glass from the Lower Rhenish glass, or the later ware made in the *façon de Venise* from glass made in Murano, or that in the English style from contemporary English glass. (*See* BONHOMME GLASS FACTORY.) As to later glass, *see* VONÊCHE GLASSWORKS and the VAL ST LAMBERT factory. *See also* SOUTH NETHERLANDS GLASSWARE.

Beaker. Colourless glass with engraved Masonic insignia and inscription, Germany, *c*. 1800. Museum für Kunsthandwerk, Frankfurt.

Beilby glassware. Beaker (or bowl of goblet). Newcastle glass, with enamelled decoration, *c*. 1762. Victoria and Albert Museum, London.

bell beaker. Frankish glass, *c*.400–700. Ht 10·8 cm. Victoria and Albert Museum, London.

Berkemeyer. Type of *Römer* with funnel mouth and prunts, German, 16th/17th century. Ht 15·7 cm. Kunstgewerbemuseum, Cologne.

bell. A hollow object usually of inverted cup shape, with everted rim and equipped with a clapper, so as to emit a musical note when struck. They were made of glass in many varieties and styles. *See* HAND-BELL; TABLE-BELL; BELL GOBLET.

bell beaker. A BEAKER in the form of an inverted bell with concave sides and a wide mouth. They have a rounded bottom, occasionally with a button FINIAL. They were made in the Frankish period (*c*.400–*c*.700), and resemble certain cups of ROMAN GLASS without foot or handle.

bell goblet. A GOBLET that, instead of having a stemmed FOOT, rests on a bell attached to the bottom of the bowl. They were made in the Netherlands and Germany. Some FAKES exist, made of 17th-century glass but decorated by a Dutch engraver in the 1860s.

Bellingham, John (fl. 1665–85). A mirror-maker from Haarlem and Amsterdam who came to England and managed, from 1671 to 1674, the glasshouse at Vauxhall of the DUKE OF BUCKINGHAM, who had obtained an exclusive patent in 1663 for making MIRRORS. Bellingham later became the holder of the English patents for NORMANDY GLASS (1679) and CROWN GLASS (1685).

bellows flask. A type of flask shaped in the form of a bellows, with an ovoid body and long tapering vertical neck and having a base in form simulating handles. They were made in the United States in the 19th century, and also in England (usually of so-called NAILSEA GLASS, often with white festoons, but usually without a base).

belt hook. A small elongated object used in China as a belt fastener, made with an openwork slot for the belt. Usually made of jade, examples in imitation of jade were also made of glass in the Han dynasty (200 BC–AD 220).

Ben(c)kert(t), Hermann (1652–81). An enameller at Frankfurt-am-Main, who came as a follower of JOHANN SCHAPER to Nuremberg, where he also did enamelling in SCHWARZLOT.

bénitier (French). The French term for a HOLY-WATER STOUP. Examples in glass were made in France from the 16th century.

Bérain motifs. A style of decoration originating with Jean Bérain *père* and followed by his son, Jean Bérain *fils*. Taken from his design books, it features an elaborate pattern of mythological and grotesque figures and half-figures, vases of flowers, drapes, scrollwork, baldacchini, balustrades, and urns, all seemingly floating in mid-air. Motifs based on the work of Bérain occur on the marquetry furniture of Boulle, on faience from Moustiers, and on French porcelain. The motifs have been occasionally engraved as decoration on glassware, e.g., in Bohemia, *c*.1730. They reflect the transitional stage between BAROQUE and ROCOCO.

Bergh, Elis (1881–1954). A Swedish glass-designer for KOSTA GLASSWORKS from 1927 to *c*.1960, noted for his designs in CUT GLASS.

Berkemeyer (German). A type of BEAKER of WALDGLAS that is a variation of a RÖMER, the BOWL being funnel-shaped rather than globular. The lower part is cylindrical and decorated with PRUNTS, and sometimes has a FOOT made like that of a *Römer*, with encircling threading. They were made in the Rhineland and the Netherlands in the 16th and 17th centuries.

Berluze. The name given by DAUM FRÈRES to a vase shape introduced in 1900; it has an ovoid body and a tall neck which extends upward almost twice the height of the body. It was decorated with varied floral patterns.

Bertolini. A family of glass-makers of MURANO in the early decades of the 18th century who taught the MIOTTO family at the AL GESÙ GLASSWORKS.

Betts, Thomas (d. 1767). A London glass-grinder and polisher of MIRRORS, first mentioned in 1738 as having a shop in Bloomsbury and later at Charing Cross. He employed a Bohemian glass-cutter named Andrew Pawl who left him in 1744 after Betts had learned the technique. He made CRUETS, SPIRIT BOTTLES, etc.

bevelled glass. Flat glass with a chamfered corner, made by cutting away the edge where two flat surfaces meet at a right angle. The edges of MIRRORS are usually bevelled as protection against chipping. *See* FACETED GLASS; DIAMONDING.

biberon. A feeding bottle having a globular or ovoid BODY resting on its bottom or on a stemmed FOOT, having one HANDLE (usually a scroll loop) and a thin NECK, and characterized by a SPOUT opposite the handle, generally in the form of a long curved tube extended from the body. Some have a COVER or STOPPER. They were generally made of pottery (some with several spouts and an overhead handle), but some were made of Venetian glass or Spanish glass in the 17th century.

Biblical subjects. Subjects based on people named in or events occurring in the Old Testament or the New Testament of the Bible. Such subjects are found engraved on rare ROMAN GLASS, *c.* AD 300–500, e.g. a dish found at Homblières, near Abbeville, France, with an encircling engraved decoration depicting *The Fall of Man*, the *Prophet Daniel in the Lions' Den*, and *Susanna and the Elders*. Such subjects are also found enamelled on Venetian glassware of the 15th/16th centuries, e.g. the BOLOGNA GOBLET, and on PANELS with scenes painted on the reverse. Many such subjects are also depicted on enamelled glassware made in Germany and Bohemia in the 16th and 17th centuries. *See* RELIGIOUS SUBJECTS.

Bicheroux process. A process for making SHEET GLASS by modern machinery which produces the glass between rollers, requiring little GRINDING and POLISHING.

Biedermeier style. A German style of decoration which derived its name from satirical verses and drawings by Ludwig Eichrodt in a German journal featuring two smug bourgeois characters called Biedermann and Bummelmeier. The term 'Biedermeier' came to be applied to the style, current *c.* 1820–40, which followed the Continental EMPIRE STYLE, and is somewhat similar to the William IV and early Victorian styles in England. Although principally a furniture style, the term is also applied to porcelain and glassware which harmonizes with it. This period (known as the '*Biedermeierzeit*') of middle-class prosperity following the Napoleonic Wars brought a revival in the art of glass-making and glass-engraving in Bohemia. Clear cut glass was superseded by vivid coloured glass (*see* HYALITH; LITHYALIN; ANNAGRÜN) and engraved FLASHED glass; the forms became massive, intricately cut, and coarse generally, although the better pieces show talent, especially in pictorial ENGRAVING. The ablest German engravers worked to the order of private clients visiting at the leading spas; these include DOMINIK BIEMANN, AUGUST BÖHM, and FRANZ ANTON PELIKAN. Of this same style are some of the RANFTBECHER made at Vienna by ANTON KOTHGASSER.

Biedermeierzeit. The German term for the period of the BIEDERMEIER STYLE, *c.* 1820–40.

Biemann, Dominik (1800–57). A glass-engraver known mainly for his engraved portraits but who also engraved many pieces with landscapes and other subjects. He trained for many years at Karkachov in Silesia

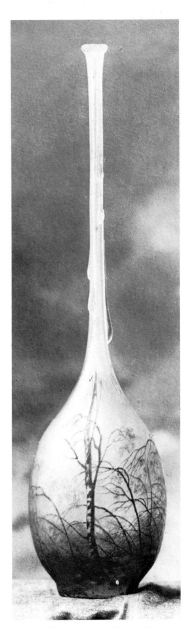

Berluze. Vase by Daum Frères, Nancy, 1900. Ht 25 cm.

until 1825, and later worked mainly at Prague. He did portraits for private clients, during the season at the spa of Franzensbad in western Bohemia, on glass MEDALLIONS and on GOBLETS and BEAKERS. He signed his works, sometimes with initials, and sometimes with his name (variously spelled: 'Biman'; 'Bieman'; or 'Bimann'). For many illustrations of his work, see *Journal of Glass Studies* (Corning), VIII (1965), p. 83.

Bigaglia, Pietro. An Italian glass-maker who worked at MURANO and in 1845 showed at Vienna the earliest 'millefiori' PAPER-WEIGHT, which was the forerunner of the weights made at the great French factories. It is a 5 cm. cube decorated in white FILIGRANA, with some CANES bearing the initials 'P B' and the date 1845. Other examples by him in circular form are known, also initialled and dated 1845.

bird feeder. A hollow receptacle having, projecting horizontally from the bottom, a small spout with an opening on its top for the birds to feed on the seeds that drop down from the container. They were made of glass in England and elsewhere in the 18th and 19th centuries.

Bird-fountain, The. An elaborate glass mantelpiece ornament or CENTRE-PIECE made AT-THE-LAMP in England in the mid-19th century, with two large exotic birds resting on two pedestals and two more higher on the rim of a fountain; the birds and also several small ones, are of coloured glass with GLASS FIBRE tails, and at the base of the fountain there is a mass of glass fibre. Some small trees are of coloured glass. *See* FOUNTAIN.

birds, exotic. A decorative motif introduced on porcelain at Meissen (called *Fantasie Vögel*) and later used at Sèvres and Worcester. It appears on some glass SCENT BOTTLES, VASES etc. with gilded decoration attributed to JAMES GILES.

Bishopp, Hawley (fl. 1676–85). The assistant and successor to GEORGE RAVENSCROFT as director of the Henley glasshouse of the GLASS SELLERS' COMPANY in 1676. He continued the experiments to make LEAD GLASS until its perfection. In 1682 he took over the Savoy glasshouse of the company upon Ravenscroft's death *c*.1681.

bitters bottle. A type of small bottle (10–12·5 cm. high) used for holding bitters (a tonic popular in the United States), and having a metal dropper. The bottles were made in the 1860–70s in large quantity and in hundreds of varieties, often globular or pyriform. Many glassworks made them, including the Whitney Glass Co., Glassboro, New Jersey, which made coloured PATTERN-MOULDED bottles with a pineapple relief pattern.

black bottle. A collector's term for a glass bottle made from the mid-17th century, first in England and later in many countries on the Continent, that was not really black in colour but dark green or brown. Its advantages were that its colour protected the contents and the metal afforded a durable glass suitable for objects intended for transportation. The English bottles were used for spa waters, beer, and cider, and for wines (after corks were used to make the bottles air-tight) in the 17th century, especially after the acceptance of the corkscrew which was introduced *c*.1686; they were also used for storing wine with the bottles laid horizontally so that the cork remained moist. *See* SEALED BOTTLE; DECANTER BOTTLE.

black glass. (1) Glass coloured black by the deliberate inclusion in the BATCH of oxide of MANGANESE blended with oxide of iron, i.e. not merely darkened blackish glass such as results from an excess of any colouring oxide. An early type of opaque black glass was produced in England in the early 17th century by Haughton Green, near Manchester; it resulted from including oxides of iron, manganese, and

Biemann, Dominik. Cut-glass beaker with engraved portrait of Duke Ernst of Coburg-Gotha, Bohemian, 1830–31; signed 'Biman'. Ht 13·4 cm. Kunstsammlungen der Veste Coburg.

bird-fountain. Centre-piece of fountain with opaque white and blue birds, and glass fibre, English, mid-19th century. Ht 43 cm. Victoria and Albert Museum, London.

sulphur and the use of a REDUCING ATMOSPHERE. A black glass, enamelled in red and black in imitation of Greek pottery, was made by J. F. Christy, of Lambeth, London, in the mid-19th century. *See* HYALITH GLASS; ZECHLIN; OPALINE. (2) A type of glass of a dark-green or purple colour that is so dark as to appear almost black; so-called in the United States.

Blackfriars Glasshouse. A glassworks dating from 1596 when SIR JEROME BOWES leased the former Blackfriars monastery in London, and a glassworks was started there in 1598 by his assignees, WILLIAM ROBSON and William Turner. It was discontinued in 1614. *See* POTTERS, BARBARA, GOBLET.

blacks. Surplus glass removed during production from the BLOWING IRON and PONTIL, and hence having minute particles of iron attached. It is immersed in a solution of hydrofluoric acid that removes the iron, and then re-used with CULLET.

blank. An object made of glass intended to be, but not yet, decorated. Some glassworks made such blanks in the form of DRINKING GLASSES, etc., for engraving at a glass-engraving plant or for painting or gilding by a HAUSMALER, e.g. JAMES GILES. *See* BLANK CASED GLASS.

blank cased glass. An object in CASED GLASS made in two layers by the glass-maker but before the top layer has been engraved by the decorator. Such pieces were made by PHILIP PARGETER, e.g. an experimental piece for the later replica of the PORTLAND VASE. *See* BLANK.

Blankätzung (German). Literally, bright etching. A process for polishing glass with a chemical solution, consisting of HYDROFLUORIC ACID, sulphuric acid, and water.

Blankschnitt (German). Literally, polished cut. A style of decorating glassware by which the MAT surface of the ground part of INTAGLIO engraving is polished by using soft disks or brushes, e.g. of leather or cork. The result is that parts of the engraving are polished thereby enhancing the sculptural effect. This method was used on NUREMBERG GLASSWARE, and its invention is attributed to GEORG SCHWANHARDT. It enabled the thin POTASH glass to be decorated with a brilliant effect before the advent of the Bohemian heavier glass with HOCHSCHNITT decoration.

Blaschka, Leopold (1822–95). A German glass-worker and naturalist who, with his son Rudolph (1857–1939), created at Dresden the HARVARD GLASS FLOWERS. Leopold was born at Aicha, in northern Bohemia, and was the founder of the art of representing natural history objects in coloured glass. He made some reproductions in glass of marine fauna that were exhibited in the Natural History Museum at Dresden, and specimens of which are now in many museums. The father and son travelled extensively to obtain specimens and sketches of plants for reproduction. They did their work personally, with not even an apprentice.

blaze. A decorative pattern on CUT GLASS, in the form of COMBED FLUTING – either vertical or diagonal – with parallel incisions ascending and descending in groups, sometimes producing a zigzag effect.

bleeding glass. A small vessel used in the 18th and early 19th centuries for bleeding patients. Some were of cylindrical shape, with a rounded rim and stemmed FOOT, about 9 cm. high. The method of use was to warm the glass and then press it close to the skin, and as it cooled the air contracted and drew the blood to the surface (called 'dry cupping'). For 'wet cupping', the skin was pierced and the same operation was repeated, drawing blood into the bowl. The pieces were entirely

Blaschka glass flowers. Bouquet made for Elizabeth C. and Mary Ware by Leopold Blaschka, Dresden, 1888. Ware Collection, Botanical Museum, Harvard University, Cambridge, Mass.

different in shape and usage from the ceramic shallow bleeding bowls with side-handles. Also termed a 'cupping bowl'.

Blenko Glass Co. A glassworks at Milton, West Virginia, founded in 1922 by William J. Blenko (1855–1933), a British glass-maker who, after entering the United States in 1893, had previously started three glassworks that failed. He was succeeded by his son W. H. Blenko (d. 1969), and the latter's son, W. H. Blenko, Jr., is now the President. The company produced initially stained glass for church windows, but since 1929 it has made decorative and household glassware and also coloured-glass building slabs.

Blijk, Jacobus van den (1736–1814). A Dutch engraver who decorated many glasses by STIPPLING in the 1770s, copying pictures.

blobbed glassware. Glass objects with decoration in the form of irregularly shaped and variously sized white blobs scattered haphazardly over the entire surface of, e.g. a JUG or BOWL. Such ware was produced by applying to a coloured body numerous blobs of opaque white molten glass which were then rolled into the surface (*see* MARVER) and enlarged by blowing. The ware is to be distinguished from FLECKED GLASSWARE, ENCRUSTED GLASSWARE and JEWELLED GLASSWARE.

block. A block of wood hollowed out to form a spherical recess. It was used, after being dipped in water to prevent charring, to form a blob of molten glass into a sphere as an initial operation in making a globular PARAISON. Also called a 'blocking wood'. *See* GLASS-MAKERS' TOOLS.

bloom. A dull, somewhat opaque appearance on the surface of old glass, for which there are three distinct causes: (1) 'inherent deficiency bloom', due to an excess of alkali in the BATCH; (2) 'sulphur bloom', due to reheating of the glass at the GLORY-HOLE near sulphur-containing fumes; and (3) 'fixative bloom', due to wearing off, usually in distinct bands, of gilding or enamelling.

blowing. The process of shaping a mass of MOLTEN GLASS (PARAISON), after it has been rolled on the MARVER and reheated, by blowing air through a metal tube (BLOW-PIPE) into the mass attached at its end. The partially blown piece may be reheated and blown repeatedly to modify the size and form. The process was developed probably in Syria in the 1st century BC. *See* FREE-BLOWN GLASS; MOULD-BLOWN GLASS.

blown-moulded. *See* MOULD-BLOWN GLASS.

blown-three-mold glass. A type of MOULD-BLOWN glassware that is blown in full-size tripartite moulds that determine both the size and pattern of each piece. It was developed in the United States *c.* 1820 as inexpensive ware, to imitate and compete with ANGLO-IRISH CUT GLASS. It is characterized by: (1) a concave-convex effect by which the raised parts on the outside are opposite corresponding depressions on the inside (unlike PATTERN-MOULDED GLASS); (2) the faint character of the mould-marks; (3) the softly outlined patterns that are not sharp as with pattern-moulded glass; and (4) the great variety of the patterns peculiar to this type, ranging from styles with vertical ribbing to baroque styles. The type has been sometimes referred to by various other terms, including 'Blown Molded' and 'Three Mold', but 'blown-three-mold' has been adopted as most nearly describing the type and method of production. The varied ware includes decanters, pitchers, tumblers, mugs, toilet bottles, sugar bowls, salts, etc., made at many United States factories. It was usually made with three-piece moulds, but occasionally the term is extended to include pieces made from two- or four-piece moulds, and sometimes the final shape was not given solely by the mould.

Bohemian glassware. Goblet, cut and engraved at Neuwelt, Bohemia, *c.* 1830. Museum of Decorative Arts, Prague.

blow-pipe (or **blow-iron**). The hollow iron tube used to take a GATHERING of MOLTEN GLASS from the POT and then to shape it by blowing the mass (PARAISON) into a bubble. The tube (usually up to 1·75 m. long, and about 2·5 cm. in diameter) has a thickened end for gathering the METAL and sometimes wooden coverings at points where the workmen must handle it. The metal end is heated before use to cause the gathering to adhere. *See* PONTIL; TOOLS, GLASS-MAKERS'.

blue glass. Glass made with oxide of COBALT (ZAFFER) or COPPER. *See* BRISTOL BLUE GLASS.

boat. A piece of ornamental glassware made in the form of a boat or galley. Some were made of ROMAN GLASS in the 1st century AD. For more elaborate pieces in the form of a ship, *see* NEF and NEF EWER.

bobêche (French). The drip-pan of a CANDLE-NOZZLE.

bocca (Italian, mouth). A large opening in the arches on the sides of a FURNACE through which the POTS were put into the furnace and the BATCH put into the pots. A temporary screen of clay covered the opening while the batch was being melted, except for a narrow opening (SPY-HOLE) through which the melting could be checked, impurities removed, and samples taken for testing. *See* BYE-HOLE; GLORY-HOLE.

Bocksbeutel (German). A type of BOTTLE with a flattened spherical body, a tall thin tapering neck, and usually high KICK BASE. It is the traditional bottle for *Steinwein* from the Franconian wine region of Germany, used from the 18th century until today.

Boda Glassworks. A Swedish glasshouse founded in 1864 by two glass-workers, Scheutz and Widlund, who had previously worked at the KOSTA GLASSWORKS. In 1970 it joined the Kosta Glassworks in becoming a part of the Afors Group. Its production includes much unconventional and imaginative glassware, designed by a group of progressive designers.

Bode, Johann Christian (1675–1751). A German glass-engraver who worked at Potsdam between 1718 and 1751.

body. The main portion of a vessel as distinguished from the BASE, STEM, HANDLE, SPOUT, COVER, or applied decoration.

Bohemian glassware. Glassware made in Bohemia, an area adjoining BAVARIA and SILESIA, which in 1526 came under the Austrian Hapsburgs. Since 1918 it has been a part of Czechoslovakia, with Prague its capital. The earliest glassworks were recorded in the 14th century and at least twenty by the 15th century. By 1530 factories were started at Falknov and Krompach, and from the late 16th century wheel-engraving was widely practised; but after Rudolph II moved his Court from Prague to Vienna in 1612 and Royal patronage ended, production lapsed, especially after the Thirty Years' War. Later engravers and enamellers founded guilds from 1661 to 1684, and in 1685 the industry was revived by the invention of LIME GLASS, perhaps by Michael Müller (d. 1709) and by its development by GEORG VON BUQUOY. During the BIEDERMEIER period, in the mid-19th century, when coal supplanted wood as fuel due to the reduced supply of timber, many factories sprang up, especially at Haida (now Novy Bor), *c*. 1870, where Michlovka built a glasshouse in 1874, and many more were opened between 1900 and the 1920s. The main factories were at Haida and Karlsbad (now Karlovy-Vary), where LUDWIG MOSER made glass portraits in the style of DOMINIK BIEMANN. The making of engraved glass was always a Bohemian speciality and from the end of the 17th century Bohemian travelling salesmen and traders carried their product to most parts of the world. The glassware was often decorated to suit the buyers, and the salesmen were also engravers who filled this demand.

Böhm, August (1812–90). An outstanding Bohemian glass-engraver of the BIEDERMEIER period, a native of Meistersdorf, near Steinschönau.

boat. Cobalt-blue glass, Roman, mid-1st century AD. L. 17·7 cm. British Museum, London.

bollicine, vetro. Vase by Venini, Murano.

Bologna Goblet. Enamelled goblet depicting figures commemorating the marriage of Ginevra Sforza and Giovanni (II) Bentivoglio, 1465, and the Flight into Egypt and Adoration of the Magi. Ht 19·1 cm. Museo Civico Medievale, Bologna.

Booz bottle. Mould-blown flask with name 'E C Booz'. Henry Ford Museum, Dearborn, Mich.

He did mainly figure subjects. He travelled extensively in England and America.

Boldini, Andrea. An Italian glass-maker who worked for ANTONIO SALVIATI at MURANO, and later for LOUIS COMFORT TIFFANY at his factories in New York.

bollicine, vetro. A type of Italian ornamental glass developed by VENINI, the characteristic feature of which is innumerable tiny air bubbles just below the surface making an overall decorative effect. It is made by use of a metal base with sharp protuberances upon which the heated glass is placed and then drawn away suddenly so as to create the air bubbles.

Bologna Goblet. A STANDING CUP of blue glass, decorated with encircling enamelled scenes depicting the Flight into Egypt and (on the reverse) the Adoration of the Magi. The decoration has been associated with the style of the Murano master Antonio da Murano, called Vivarini. By tradition, the piece was probably made in the last quarter of the 15th century to commemorate the marriage of Ginevra Sforza and Giovanni Bentivoglio (1465). *See* BIBLICAL SUBJECTS.

bomb. A glass object, variously shaped (sometimes as a SPRINKLER), made of ISLAMIC GLASS and said to have been used in medieval times as a hand grenade, a precursor of the modern 'Molotov Cocktail'. *See* GRENADE.

bone-ash. Ashes of calcined bones. It was used as a FLUX in glass-making, and also as an opacifier in making OPAQUE WHITE GLASS, MILK-AND-WATER GLASS, and OPALINE.

Bonhomme Glass Factory. A glassworks at Liège, Belgium, taken over *c.* 1638 by Henri and Leonard Bonhomme, businessmen from the Netherlands, who made glassware in the English and Venetian styles from *c.* 1680, also utilitarian glassware in native fashion. The Bonhomme family opened branches in several other cities of the Netherlands and France, and prospered into the 18th century, when the factory succumbed to competition from other Liège glassworks.

bonne femme. The French name for a type of straight-sided BOTTLE made at Liège (Belgium) in the 18th century.

bonnet-glass. A type of small glass receptacle, occurring in a great variety of shapes, said to have been used for sweetmeats. The bowl is usually double-ogee-shaped or cup-shaped, plain or with moulded decoration; later examples have an ovoid bowl, more suitable for cut decoration. They have a knop stem, resting on a flat or hollow bottom that is often dome-shaped or terraced, and is circular, square or rarely scalloped. Such objects have been incorrectly termed MONTEITHS by E. Barrington Haynes (1941) – and consequently so-called by some later English writers and dealers – who did not recognize any necessity for a true monteith to serve as a glass-cooler, and so applied the term to the bonnet-glass with no justification other than that the circular foot is occasionally scalloped.

Bontemps, Georges (1799–1884). A glass-maker and Director at CHOISY-LE-ROI GLASSWORKS (1823–48), whose primary interest was the making of SHEET GLASS. He was the author of the *Guide du Verrier* (1868), an important technical treatise. He revived the Venetian art of making glass, which he exhibited at the Exhibition of French Industry of 1839 and the Exhibition of 1844. He also made CASED GLASS in Bohemian style from *c.* 1842 and MILLEFIORI glass from *c.* 1844. He was a friend and associate of ROBERT LUCAS CHANCE, and joined him at the SPON LANE GLASSWORKS in 1848, where he supervised the colour-glass department and did useful research.

Bonus Eventus plaque. A square moulded PLAQUE of cobalt-blue glass depicting in LOW RELIEF a partly draped youth holding a PATERA and ears of corn, and bearing the inscription, *'Bono Eventi'*. It is the upper slab of a larger piece. Bonus Eventus (literally, Good Success) personified a popular Roman deity who presided especially over agriculture. The plaque is attributed to ROMAN GLASS, 1st century AD.

boot glass. A drinking glass made in the form of a boot, sometimes spurred; some were calf-high with a flared mouth, others knee-high and leg-shaped. They have been made in Germany, the Netherlands, and Italy in the 17th century, and in England probably for the late 18th century. The smaller English examples (usually only about 10 cm. high) have been said to have been made in mockery of the unpopular Earl of Bute (1714–92) but such objects were made before his day. The Italian term is *stivale*. *See* JOKE GLASS; SHOE.

Booz bottle. A type of FLASK made in the United States during the 1860s in the form of a square two-story house with shingle roof, plain walls, and a chimney (the spout). They were made by the WHITNEY GLASS WORKS for Edmund G. Booz (d. 1870), Philadelphia, who made the whiskey so bottled. (The popular belief that the word 'booze' was derived from his name is erroneous; it is said to be from 'bouse', a 17th-century word meaning 'to carouse'.) The numeral '1840' on some examples is said to refer to the date of the contents. Reproductions of such bottles were made *c.* 1931 and again later in coloured glass. *See* CABIN BOTTLE; HISTORICAL FLASK.

borax. A crystalline salt that, as boric acid, was used originally as a FLUX in making LEAD GLASS. It was used in the 18th century in making PLATE GLASS. It is included in the BATCH to strengthen the glass against blows, chipping, and changes of temperature. It adds durability, hence is used today in heat-resistant laboratory and kitchenware glass and certain optical glass, replacing part of the SILICA. *See* PYREX.

Boston & Sandwich Glass Co. A glasshouse founded in 1825 by DEMING JARVES at Sandwich, near Cape Cod, Massachusetts, as the Sandwich Manufacturing Co., which in 1826 was incorporated as the Boston & Sandwich Glass Co. It specialized (under Hiram Dillaway) in MOULD-PRESSED GLASS, making plates, cup-plates, etc., in LACY GLASS, but also making paper-weights and ware in the Venetian and English styles, and after 1830 some OPALINE. Jarves left the company in 1858. The factory suffered from cheaper-glass competition of Mid-Western factories, and after a strike in 1888 it closed. Its mould-glass became known as 'Sandwich Glass', and its ware from the period 1825/40 is referred to as 'Early Sandwich Glass'. For a history of the company and illustrations of a great number of its forms and patterns, *see* Ruth Webb Lee, *Sandwich Glass* (1939).

Boston Silver Glass Co. A glasshouse founded by A. Young at Cambridge, Massachusetts, in 1857 and closed *c.* 1871. It specialized in making SILVERED GLASSWARE.

botijo (Spanish). A vessel for holding and pouring water. It has a globular BODY on a spreading BASE, with an overhead loop HANDLE and one short wide SPOUT. They were made of glass in Spain in the 17th century and are similar to a CRUCHE.

bottle. A vessel for holding a liquid and not primarily a drinking vessel. They have been made since Syrian and Roman days in a great variety of shapes, sizes, and styles, and for a multitude of uses. The BODY is often cylindrical, but may be globular, pear-shaped, or of square, hexagonal or octagonal section, with a great many variations. They may have flat or rounded sides, and a flat bottom or a KICK BASE. They usually have a STOPPER (of glass, cork or other material), but some have a screw cap. WINE BOTTLES vary in shape, size, and colour in different countries.

Bonus Eventus plaque. Blue glass with yellowish-white flecks, Roman, 1st century AD. Ht 18 cm. British Museum.

botijo. Clear glass with trailing and cresting, Spanish, 17th century. Ht 21 cm. Victoria and Albert Museum, London.

Some bottles are for holding potable liquids (e.g. FLASK, PILGRIM BOTTLE, CARAFE, DECANTER, DECANTER BOTTLE); some are small, for personal use (e.g. SCENT BOTTLE, SNUFF BOTTLE); some have special uses (e.g. CRUET BOTTLE, INK BOTTLE, SAUCE BOTTLE, TOILET BOTTLE, CARBOY). Some are MOULD-BLOWN manually (e.g. VIOLIN FLASK), but a machine for mechanically blowing bottles in a metal mould was perfected in the United States toward the end of the 19th century.

——, *casting*. A type of SCENT BOTTLE that was used in England for scattering scent in a room. *See* SPRINKLER.

——, *multiple*. A type of bottle or flask, with interior divisions and separate SPOUTS, made to contain two, three, or four different liquids. An example is known from Crete, late 2nd or 3rd century AD, in the form of a globular JUG with a HANDLE and three sections. Later CRUETS were made with two sections, the spouts often turned in opposite directions. In recent times, LIQUEUR BOTTLES have been made in France for four liqueurs; they are globular in form and divided into four equal sections.

bottle glass. A common naturally coloured (dark greenish or brownish) glass. The colour is characteristic of unrefined glass including traces of iron found in the natural SILICA that was used as an ingredient. Sometimes additional quantities of iron, in the form of iron oxide, were added to darken the colour. Such glass was used for making early glass BOTTLES in England in the mid-17th century. It is sometimes called 'green glass'. *See* NAILSEA GLASS.

bottle label. The name of the intended contents of a BOTTLE or DECANTER, painted or engraved on the surface. At least sixteen different names of alcoholic contents so painted or engraved are known, e.g. beer, cyder, ale, claret, Burgundy, hock, Madeira, mountain, port, red port, white port, white wine, white, sherry, spruce, and even 'champaign' in addition to the names of spirits, e.g. brandy, gin, Hollands, rum, shrub, spirits, whisky. Occasionally such a label is found on a GOBLET for cyder. Such decoration on bottles ceased *c*.1760 when BOTTLE TICKETS became popular.

bottle ticket. A small PLAQUE suspended by a chain around the NECK of a wine-bottle, DECANTER, or SPIRIT DECANTER, and lettered with the name of the contents, or left blank for a name to be inserted. Known in silver from *c*.1740 and in enamel on copper from *c*.1753, examples in porcelain and glass were made from the early 1760s, superseding the inscribing of the names on the bottles. (*See* BOTTLE LABEL). They were also known as 'wine labels'.

bouquet. A group of two or more flowers (natural or stylized), with leaves, that are arranged in a bouquet or garland, usually as the motif of one type of PAPER-WEIGHT, as distinguished from an individual embedded glass flower. *See* STANDING BOUQUET; FLAT BOUQUET; HARVARD GLASS FLOWERS.

bouquetière (French). A type of vase for cut flowers, characterized by multiple vertical tubes for inserting individual stems, and a central orifice for filling the vessel with water. The form, used at HALL-IN-TYROL in the 16th century, is akin to that of the ALMORRATXA. *See* FLOWER STAND.

bourdalou (French). A small oval urinary receptacle for female use, made generally of porcelain or pottery, *c*.1710–1850, by the leading Continental and English factories, but also made occasionally of glass in Murano and elsewhere. Sometimes confused with a SAUCE-BOAT, it is basically different in form in that the front end has an incurved rim rather than a pouring lip. There is usually a simple FOOT-RING (or none at all) rather than a stemmed FOOT. The single handle, at the end of the long axis, is usually a simple loop. An apocryphal explanation of the

bottle labels. Decanters with initialled lozenge stoppers, Bristol blue glass with gilding, *c*.1790. Ht 31·6 cm. Courtesy, Maureen Thompson, London.

bouquetière. Dark-blue glass with gilt decoration and diamond-engraving. Attributed to Hall-in-Tyrol, second half of 16th century. Ht 20·2 cm. British Museum, London.

origin of the name attributes it to Père Louis Bourdaloue (1632–1704), a Jesuit preacher at the Court of Louis XIV, whose long discourses detained the ladies of the Court so as to necessitate this practical receptacle. However, the first known examples are of Dutch delftware, made *c.*1710. The terms used in England are 'coach-pot', 'oval chamber-pot', and 'slipper'. *See* URINAL; CHAMBER-POT. *See* Harold Newman, 'Bourdalous', *The Connoisseur*, December 1970 and May 1971.

Bowes, Sir Jerome (d. 1616). An English soldier and diplomat who sought a fortune by securing in 1592 a monopoly, effective in 1595, to make glass and to import Venetian glassware for 12 years. He took over the remaining three years of the licence of JACOPO VERZELINI when the latter retired in 1592. When his new licence became effective in 1595 he built a new furnace at the BLACKFRIARS GLASSHOUSE which he leased to WILLIAM ROBSON and William Turner. He received a new patent in 1605, but in 1607 a reversion of his monopoly, effective on his death, was granted to Percival Hart and Edward Forcett. *See* MONOPOLISTS, THE.

bowl. (1) A concave vessel, wider than it is high, but deeper than a SAUCER. It usually, but not invariably, has a FOOT-RING. Small bowls for table purposes often have two side HANDLES, and some have a COVER. They vary in size, depending on the purpose, and are of many forms and styles. The French terms are *bol* and, when larger, *jatte*; the Italian is *scodella*; the German is *Schale*. *See* PUNCH-BOWL; SUGAR-BOWL; IRISH GLASSWARE, *fruit (salad) bowl*; BOWL, VENETIAN; STANDING BOWL. (2) The hollow part of a DRINKING GLASS that holds the contents. Such bowls were made in various shapes, sizes, and styles, and with respect to English WINE-GLASSES serve to identify the period of production. They are decorated with DIAMOND-POINT ENGRAVING; WHEEL-ENGRAVING; CUTTING; ENAMELLING; GILDING; STIPPLING.

TYPES OF BOWL OF DRINKING GLASSES

——, *bell.* A type of bowl that is shaped like an inverted bell. Variations: Some are slightly WAISTED, some have a solid base.

——, *bucket.* A type of bowl that is basically of cylindrical shape (slightly sloping sides), with a flat bottom where it joins the stem. Variations: 'square bucket' with vertical sides; 'curved bucket' with slightly curved sides; 'waisted bucket' with waisted sides; 'lipped bucket' with lipped rim. *See* bucket-topped bowl (below).

——, *bucket-topped.* A type of bowl, the upper part of which is shaped like a bucket bowl (see above), with a deep round funnel below the bowl. *See* pan-topped bowl (below).

——, *conical.* A type of bowl that is shaped like an inverted cone, with its MOUTH slightly narrower than a funnel bowl (see below).

——, *cup.* A type of bowl that is somewhat spherical, with an inverted RIM. *See* cup-topped bowl (below).

——, *cup-topped.* A type of bowl, the upper part of which is shaped like a cup bowl, with a deep round funnel below it. *See* pan-topped bowl (below).

——, *double ogee.* A type of bowl whose sides have a DOUBLE-OGEE shape.

——, *faceted.* A type of bowl of any of various shapes that has FACETED decoration.

——, *flexed funnel.* A type of funnel bowl (see below), having straight sides or a rounded bottom; it becomes broader near the rim and then extends vertically upward.

——, *funnel.* A type of bowl that is shaped like a funnel, with its MOUTH slightly wider than a conical bowl (see above). Variations: with a rounded bottom, or with a pointed bottom.

——, *hexagonal.* A type of bowl of hexagonal horizontal section, slightly concave as it joins the stem. Some bowls are of octagonal section.

——, *lipped.* A type of bowl, the RIM of which is slightly everted. Variations: Some are ogee bowls (below), some are bucket bowls (above).

bourdalou. French. L. 21 cm. Courtesy, H. Alain Brieux, Paris.

bowl, bell. Detail of wine glass, English, *c.*1750. Ht 16·5 cm. Courtesy, Sotheby's, London.

bowl, bucket. Detail of goblet with colour twist stem, *c.*1765. Courtesy, Delomosne & Son, London.

bowl, faceted. Detail of goblet with faceted bowl and knop, Germany, late 17th century. Ht 24·2 cm. Museum für Kunsthandwerk, Frankfurt.

bowl, pan-topped. Detail of wine glass with engraved flower decoration. Courtesy, Derek C. Davis, London.

bowl, thistle. Detail of goblet with opaque twist stem, English, *c.* 1760. Courtesy, Sotheby's, London.

——, *ogee.* A type of bowl, the shape of which is cylindrical but with a concave curve where it joins the STEM.

——, *ovoid.* A type of bowl, the shape of which is vertically OVOID.

——, *pan-topped.* A type of bowl, the upper part of which is shaped like a shallow pan, with a deep round funnel below the bottom of the pan. *See* DECANTER, *lipped pan.*

——, *saucer-topped.* A type of bowl, the upper part of which is shaped like a shallow saucer, with a deep round funnel below it. *See* pan-topped bowl (above).

——, *thistle.* (1) A type of bowl that is shaped like a thistle with the MOUTH being formed as the thistle-head. Variations: the bottom part is hollow or solid. (2) A large bowl similarly shaped, with an everted RIM and with the lower half bulging and usually with CUT diamonds in diagonal rows or cut mitres. Between the upper and lower parts there is usually a band with cut FLUTING or swirls.

——, *trumpet.* A type of bowl that is tall, conical, and slender, with its sides slightly concave.

——, *tulip.* A type of bowl that is globular or cylindrical, with an everted rim, and then tapers inward to the stem, having the appearance of an open tulip.

——, *turn-over.* A type of bowl, usually found on a fruit (salad) bowl of IRISH GLASSWARE, the RIM of which is curved outward and downward.

——, *U-shaped.* A type of bowl that is cylindrical with a rounded bottom, resembling a letter 'U'. They were sometimes WRYTHEN or decorated with THUMB-PRINTS, and rested on a knopped STEM. They were used only for an ALE GLASS.

——, *waisted.* A type of bowl whose sides are WAISTED. Variations: waisted bucket; waisted bell; waisted ogee.

bowl-gatherer. A member of the team (CHAIR or SHOP) assisting the GAFFER in making such glass objects as a wine-glass.

bowl stand. A support upon which to rest a BOWL. Examples were made of ANGLO-IRISH GLASS in various shapes, some as a waisted pedestal, some flat.

Bowles, Benjamin (fl. 1744). A glass-maker at Southwark, London, whose speciality was making OPAQUE WHITE GLASS.

Bowles, John (1640–1709). An English glass-maker who made CROWN GLASS at the RATCLIFF FACTORY at Southwark, London. His successor was BENJAMIN BOWLES who, among other things, made OPAQUE WHITE GLASS in the mid-18th century.

box. A receptacle with a COVER, or with a LID pierced and attached with metal hinges or mounted in a metal frame for the same purpose. Glass boxes are usually small; if large, they are usually termed a CASKET. The French term is *boîte*; the German terms are *Dose* when flat and *Böchse* when high.

Brandenburg glassware. Glassware made in Brandenburg, Germany, especially that made under the patronage of Elector Friedrich Wilhelm (1620–88) from 1679. Until the 17th century, Brandenburg imported glass, but in 1602 a glasshouse was founded near Grimnitz under the patronage of Elector Joachim Friedrich, who brought in Martin Friedrich and workers from Bohemia to operate it. It was moved in 1607 to Marienwalde, near Küstrin, closing in 1825, having made enamelled and engraved glassware; it employed GOTTFRIED GAMPE, *c.* 1668, and his brothers Daniel and Hans Gregor, and from 1750 Samuel Gottlieb Gampe. In 1653 the Elector founded a second glasshouse at Grimnitz, where GEORG PREUSSLER became director in 1658; no engraving was done, but coloured goblets were enamelled until its closure in 1792. Both the Grimnitz works made glass in Bohemian style. In 1674 a glassworks was founded by the Elector Friedrich Wilhelm at Drewitz, near

III *Cameo glass.* Four-colour oviform vase, the olive-green ground being decorated in triple overlay with leaves and sprigs in opaque white glass; the leaves are incised to reveal contrasting veins (blue underlay) and the peach and leaf-edges are tinged with pink; ht 26·5 cm. (10$\frac{1}{2}$ in.), probably made by Thomas Webb & Sons, *c.* 1870–90. Courtesy, Christie's, London.

IV *Chinese snuff bottles.* Top row (all *c.* 1730–1830): (left and right) single overlay on snowflake ground; (centre) realgar glass. Middle row: (left) single overlay with carved design, *c.* 1780–1880; (centre) triple overlay on snowflake ground, with carved lotus design, *c.* 1760–1880; (right) interior decoration, signed 'Yeh Chung-san', *c.* 1924. Bottom row: (left) five-colour overlay on white ground, *c.* 1780–1880; (centre) mottled green glass, in imitation of jadeite, *c.* 1760–1860: (right) enamelled decoration in the *famille rose* palette, Ku-yüeh Hsüan mark, Yeh family kilns, Peking, *c.* 1925–35. All $\frac{2}{3}$ scale. Courtesy, Hugh Moss Ltd, London.

Potsdam, with GEORG GONDELACH, of Hesse, as glass-maker, engaging Nuremberg workers and operating until 1688. In 1679 a new glasshouse was founded by the Elector at Potsdam which from 1679 to 1693 was operated under the direction of JOHANN KUNCKEL; its ware suffered from CRISSELLING until *c*. 1689 when Kunckel counteracted it by the addition of chalk. The Elector built and presented to Kunckel in 1685 a new glasshouse near Potsdam, for experimenting in crystal glass, and MARTIN WINTER became engraver there in 1680. The Elector built for Martin Winter a water-power engraving plant in 1688. After the death in 1688 of the Elector, the Kunckel glasshouse was destroyed by fire, and Kunckel's successors later moved the factory in 1736 to Zechlin, where it was run as a state operation until 1890. Among important engravers at Berlin and Potsdam were GOTTFRIED SPILLER, ELIAS ROSBACH, HEINRICH JÄGER, and JOHANN CHRISTIAN BODE. The engraving at these glassworks featured work in HOCHSCHNITT and TIEFSCHNITT, borders of formal flower motifs, and after 1730 gilt decoration; some of the glasses, featured *putti* and military and mythological scenes. At BASDORF, near Potsdam, in the 1750s, OPAQUE WHITE GLASS with enamelled decoration was made. *See* POTSDAM GLASS FACTORY.

box. Milk glass with green and brown flashing, by Emile Gallé, Nancy, *c*.1910, D. 14 cm. Kunstgewerbemuseum, Cologne.

brandy bottle. A small bottle for brandy, usually square or rectangular in section, with a short threaded neck for a screw cover. They were decorated with peasant-style opaque enamelling of flowers and figures. They were made in Bohemia in the 18th century.

brandy glass. A large balloon-shaped drinking glass, with a slightly inverted RIM and a stemmed FOOT; it is so shaped so that the BOWL can be warmed by being held in the drinker's clasped hands so as to release the brandy fumes. Some are accompanied by a metal rack that holds the glass horizontally over a flame for better warming, but this leads to overheating and is not favoured by brandy connoisseurs. Sometimes called a 'brandy snifter' or a 'balloon glass'. It has been said that early English glasses for drinking brandy were cylindrical tumblers.

breast glass. A hollow glass receptacle, made in various shapes, to fit over the breast of a nursing mother and into which surplus milk would be drained by the vacuum created when the receptacle was heated. They have a small tube at one side for emptying, but some have instead a long curved spout. Examples in transparent glass were made in Venice and elsewhere from the 16th century. *See* NIPPLE SHIELD.

Briati, Giuseppe (d. 1772). A glass-maker from MURANO who specialized in FILIGRANA glassware and also ornate CHANDELIERS and TRIONFI. After starting in Murano, he is said to have worked three years in Bohemia, returning in 1736 to Murano where he was given a patent for ten years; but in 1739 he moved his glassworks to Venice.

bridge fluting. A decorative pattern on some English CUT GLASS drinking glasses in the form of FACETED short wide vertical FLUTING extending from the upper part of the STEM on to the lower part of the bowl and also sometimes at the junction of the stem and the foot.

Brierley glassware. *See* STEVENS & WILLIAMS, LTD.

Brilliant style. An American term for a style of CUT GLASS used mainly on large pieces of glassware, with very deep, complicated, and highly polished cutting. It was developed in the 2nd half of the 19th century, and was first exhibited in 1876, especially by the NEW ENGLAND GLASS CO. The style spread to Europe, notably to Sweden, where it was used by the KOSTA GLASSWORKS, and to Bohemia. The style lost popularity, *c*.1915.

Britannia glasses. Wine glasses engraved with portraits of Frederick the Great and Britannia, English, *c*.1763. Victoria and Albert Museum, London.

Bristol blue glass. Blue-coloured glassware, made in England at Bristol and at many other factories during the 18th century, but often all such ware is erroneously attributed to Bristol. Among the objects

Brocard glassware. Standing bowl in Syrian style, Paris, *c.* 1880. Signed 'Brocard'. Ht 11·1 cm. Kunstmuseum, Düsseldorf.

bubbles. Bowl with cut honeycomb decoration, Persian, 5th/6th century. Ht 8·3 cm. C. L. David Collection of Islamic Art, Copenhagen.

brush pot. Semi-opaque white glass with enamel decoration in Ku-yüeh Hsüan style; Ch'ien Lung mark, 1736–95. Ht 10·1 cm. Percival David Foundation of Chinese Art, London.

produced at Bristol are SCENT BOTTLES, DECANTERS, TOILET WATER BOTTLES, PATCH BOXES, FINGER BOWLS, and DISHES. The decanters are pyriform, with tall neck and lozenge STOPPER, resembling those made at Belfast by Benjamin Edwards who had gone there from Bristol in 1776. *See* BRISTOL GLASSWARE; JACOBS, ISAAC.

Bristol glassware. Various types and styles of glassware made at Bristol, in western England, which was the site in the 17th and 18th centuries of many glassworks. The first recorded one was established *c.* 1651 by the DAGNIA FAMILY. The numerous factories included the REDCLIFF BACKS GLASSHOUSE, NON-SUCH FLINT GLASS MANUFAC-TORY, and TEMPLE STREET GLASSHOUSE. Originally Bristol made window glass and bottles, but later its factories achieved fame for many varieties of high-quality glass. It was outstanding for its OPAQUE WHITE GLASS. So-called BRISTOL BLUE GLASS and some green glass was made here (e.g. by ISAAC and LAZARUS JACOBS) but was also made in many factories elsewhere, although often all such blue glass is attributed to Bristol. Bristol factories also were well-regarded for CUTTING and ENGRAVING, and for decorating glassware with gilt and ENAMELLING; among the gilders here was MICHAEL EDKINS. Among the ware made in Bristol are TEA CADDIES, DECANTERS, BOTTLES (e.g., SCENT WATER BOTTLES, TOILET BOTTLES, CRUET BOTTLES), VASES, CANDLESTICKS.

Britannia glass. Glassware to celebrate the end of the Seven Years' War, 1756–63, engraved with ships, portraits of Frederick the Great of Prussia, and figures of Britannia, and often with patriotic legends. The pieces are sometimes FACETED, an early appearance of this technique used on drinking glasses. The popularity of portraits of Frederick the Great on contemporary glassware and porcelain was due to the close connection between England and Prussia because English kings were then also Electors of Hanover, which territory adjoined Prussia.

broad glass. Flat PANE GLASS made by the process of BLOWING a large glass bubble and swinging it on the BLOW-PIPE to form a long bottle, then cutting off both hemispherical ends; the resultant cylinder ('muff') – up to 5 ft (1·50 m.) in length – was then cut lengthwise with shears and reheated, after which it was flattened by a wooden plane or by being allowed to sink to a flat state. This glass showed straight ripples and took less polish than CROWN GLASS. The process was called the 'Lorraine method' because it had been used in Lorraine as early as *c.* 1100. The Venetians, *c.* 1500, used it to make MIRRORS in lieu of the wasteful 'crown glass'. The process was improved, *c.* 1830, by ROBERT LUCAS CHANCE, at the SPON LANE GLASSWORKS, at Birmingham, where larger cylinders were blown, then split by a diamond glass-cutter and flattened. The process is now mechanized so that a continuous sheet of glass is drawn and rolled; *see* FLOAT GLASS. The 'broad glass' was also called 'cylinder glass' or 'muff glass'. *See* LUBBERS PROCESS.

Broad Street Glasshouse. A glasshouse built *c.* 1617, by SIR ROBERT MANSELL, in the City of London, probably in the abandoned monastic church of the Austin Friars, for the making of crystal glass. It was started with Italian workers under the management of WILLIAM ROBSON. Soon afterwards James Howell became manager and then Captain Francis Bacon, with Lady Mansell actively participating. After her disagree-ments with the Italian workmen, English workers were engaged. The factory prospered, making drinking glasses in large quantities and also flat glass for mirrors and spectacles. There is no record of its end.

Brocard, Philippe-Joseph (fl. 1867–90; d. 1896). A French artist-craftsman whose early STUDIO GLASSWARE was in the style of Syrian glass of the 13th and 14th centuries, inspired by the Islamic MOSQUE LAMPS which he copied. His later work, *c.* 1885, was of original style, being MOULDED glassware with relief naturalistic floral motifs, arranged in a formalistic pattern and coloured with ENAMELLING of natural colours. Some of his pieces are signed.

Bröckchen. Literally, little crumbs. The German term for the decoration on ENCRUSTED GLASSWARE made in the 1st–4th centuries.

brocs à glace (French). *See* ICE-GLASS.

broken swirl. A decorative style on glassware with combined vertical and swirled ribbing. It is made by pressing the GATHER into a ribbed MOULD, then reheating and twisting it, and then inserting it again into a mould to make the vertical ribbing.

Broncit glassware. Glassware decorated in the style developed by JOSEF HOFFMANN *c.* 1910. It features geometrical patterns in a blackish mat pigment against a background of transparent or mat clear glass. The ware was executed by J. & L. LOBMEYR, of Vienna.

brooch. *See* FIBULA.

Brooklyn Flint Glass Works. *See* GILLILAND, JOHN L.

bruise. A flaw in a PAPER-WEIGHT that has developed during production and shows as a small irregular defect just below the surface of the CROWN, as distinguished from a weight that shows surface 'damage' after production through ill-handling. A bruise is sometimes removed by grinding and polishing the weight, but the size is consequently slightly reduced.

Brummer, Arttu (1891–1951). A Finnish glass-designer who was Director of the Central School of Arts and Crafts at Helsinki and also designed for RIIHIMÄKI GLASSWORKS.

Brunswick star. A decorative motif, cut in the form of a 12-pointed star, each point being joined by a line to the fifth point in either direction from it; when all the points are joined in this way, a nearly circular outline is produced in the centre.

brush-pot. The Chinese or Japanese receptacle for holding and washing the brushes used in painting and calligraphy. They are of various shapes, sometimes barrel-shaped. Usually made of porcelain, there are some examples in Chinese glass, from the reign of Ch'ien Lung (1736–95), occasionally decorated in KU-YÜEH HSÜAN style imitating painting on European porcelain.   [*Plate II*]

Brussa. A family of Murano (including Oswald and his son Angelo) who, in the 2nd half of the 18th century, did enamelled decoration on BEAKERS, CARAFES, BOTTLES, and table ware of clear glass, depicting flowers, birds, etc., or Biblical scenes in the style of folk art.

bubble. (1) An air- or gas-filled cavity in a glass body, resulting from: faulty fusing of the ingredients (*see* SEEDS); or intentional techniques to create a decorative effect, such as (a) manipulation (*see* TEAR; TWIST, *air*), or (b) injection into the BATCH of a chemical (e.g., potassium nitrate) which reacts to the heat of the FIRING to form bubbles (*see* CLUTHRA GLASS). (2) The PARAISON of MOLTEN GLASS attached to a BLOW-PIPE or a PONTIL is sometimes referred to as a 'bubble'.

bubble ball. A globular glass ornament made with very many air bubbles throughout in a patterned design. They were made in the 1930–40s by TIFFANY and KIMBLE GLASS CO., and were sometimes mounted on a metal stand to be used as a LUMINOR.

bucket. A receptacle for holding or carrying water, usually of cylindrical form, or with sides sloping outward, and having a bail HANDLE. They were made of ROMAN GLASS, *c.* AD 100–400, late examples being elaborately carved and engraved, and of Venetian glass in the late 16th and early 17th centuries. They were also made in

Bröckchen. Blue-glass *kantharos* with glass 'crumbs'. Roman, 1st century AD. Ht 10·3 cm. (without handle). Römisch-Germanisches Museum, Cologne.

Broncit glassware. Wine glass, mat etched and with black enamel decoration, designed by Josef Hoffmann and L. H. Jungnickel; J. & L. Lobmeyr, Vienna, 1910. Ht 18·8 cm. Museum für Kunsthandwerk, Frankfurt.

bucket. Roman glass with snake-threaded decoration, *c.* 100–400 AD. Ht 8 cm. Römisch-Germanisches Museum, Cologne.

Buckley Bowl. Ribbed bowl, crisselled; Chinese glass, K'ang Hsi period (1662–1722). D. 27·5 cm. Victoria and Albert Museum, London.

bucranium, Finger-bowl, opaque white glass with neo-classical decoration including bucranium motif; studio of James Giles, *c.* 1790, D. 12·5 cm. Courtesy, Delomosne & Son, London.

Catalonia at that time; some made there have an elongated vertical bulb inside rising from the bottom, the purpose of which is not apparent. Some were made with curved or ogee-shaped sides. The German term is *Eimer. See* SITULA; SITULA PAGANA; ASPERSORIUM.

Buckingham, George Villiers, 2nd Duke of (1628–87). The successor to the monopoly of SIR ROBERT MANSELL, he obtained patents from Charles II when he was restored to the throne in 1660 and for fourteen years dominated the English glass industry by virtue of his wealth and his influence at Court (notwithstanding his being imprisoned in the Tower of London in 1667 and 1677). He acquired *c.* 1660 a glasshouse at Greenwich, employing Venetians working with JOHN DE LA CAM to make glass in the Venetian style, and *c.* 1663 another glasshouse at Foxhall (Vauxhall) to experiment in making SHEET GLASS and MIRRORS. He was not active personally in the glass industry, and the operations of his glassworks were largely taken over by the GLASS SELLERS' COMPANY. *See* ROYAL OAK GLASS.

Buckley Bowl. A semi-globular Chinese glass BOWL with slightly everted rim, rounded bottom, and spiral ribbing. It is yellowish in colour and much decayed from CRISSELLING so that it has a greenish-brown appearance. It was once believed to have been made during the T'ang dynasty (618–906) but present opinion considers it to be an experimental piece of the Imperial Glass Factory founded in 1680 in the reign of K'ang Hsi (1662–1722). It is so-called from being in the Buckley Collection at the Victoria and Albert Museum, which also owns another similar bowl.

bucranium (Latin). A decorative motif depicting an ox skull adorned with wreaths. Taken from ancient Roman sources, it was employed on Italian maiolica of the 15th and 16th centuries, and is found on some glassware with gilt decoration attributed to the studio of JAMES GILES.

Buddha figure. Seated figures of Buddha made of glass were originally attributed to Siam, 10th/11th centuries or 17th/18th centuries, or to Burma, 17th/18th centuries; recent opinion, however, attributes such figures to China – made for the export trade to Siam – *c.* 1824–51. Many such figures have been made in Germany, Czechoslovakia, and Japan in recent years, but most come now from the glasshouses at Dhanapuri, across the river from Bangkok. *See Journal of Glass Studies* (Corning), V (1963), p. 75.

Buggin Bowls. A pair of BOWLS commemorating the marriage, in 1676, of Butler Buggin (1646–90), of North Cray, Kent, and Winifred Burnett, of Leys, Aberdeen. The bowls bear the arms of the couple. They are decorated with DIAMOND-POINT ENGRAVING, probably done in London at the Savoy glasshouse, *c.* 1676, and are of LEAD GLASS, now showing CRISSELLING. They have been attributed to GEORGE RAVENSCROFT, but they do not bear the RAVEN'S HEAD seal which was not authorized until 1677.

bugle. (1) A glass ornament, about 30 cm. long, in the form of a bugle, attributed to NAILSEA and BRISTOL, *c.* 1800. Similar pieces were made in the form of a long horn. Both have a loop near the end for the mouthpiece. It has been suggested that they might have been made for actual use but seem more ornamental than functional. See COACH-HORN; POST-HORN; FIREMAN'S HORN. (2) A type of BEAD in the form of an elongated hollow cylinder, generally black, used in embroidered decoration on ladies' dresses. They were imported into England from Venice. Some were made at a factory established in the late 1750s at Beckkey, near Rye, England, by Godfrey Delahay, a Frenchman, and Sebastian Orlandini, a Venetian, which in 1580 was bought and moved to Ratcliffe by John Smith. They were also made at a factory at Godalming, near Guildford, set up *c.* 1587 by an Italian named Luthery.

bulle de savon (French). Literally, soap-bubble. A type of OPALINE glass of pastel hues, made in France from *c*. 1822.

bullet-resistant glass. Glass, generally about 4 cm. thick, made of several layers of LAMINATED GLASS separated by layers of plastic or vinyl.

bull's-eye and diamond-point pattern. A pattern used on early PRESSED GLASS in the United States, consisting of a series of encircling raised 'bull's-eyes', attached to which are, in pointed triangular form, groups of raised diamonds. The pattern was used on various objects, e.g. tumblers, etc.

bull's-eye mirror. *See* MIRROR, *convex*.

bull's-eye pane. PANE GLASS with central circular ridges around a PONTIL MARK. It was the centre part of a large pane of CROWN GLASS.

bun foot. A type of FOOT in the form of a flattened ball; they are found on a SCHAPER GLASS, in threes.

Bungar, Isaac (d. 1643). The English-born son of Peter Bungar (Pierre de Bongard) who came with JEAN CARRÉ from Normandy to Alford in the Weald of England (*see* WEALDEN GLASS) and whose family included many English glass-makers of the 16th/17th centuries, especially makers of window glass. Isaac and his brother Ferrand made CROWN GLASS, and by *c*. 1605 secured a monopoly in the London window-glass market, aided by the developing shortage of wood as glass-making fuel. He later encountered difficulties resulting from the change from the use of wood to coal as fuel and from competition and litigation with SIR ROBERT MANSELL, whose monopoly he unsuccessfully fought.

Buquoy, Georg Franz August Longueval, Count von (1781–1851). The owner of a number of glasshouses in Glatzen in southern Bohemia founded in the late 18th century. He developed several types of opaque glass of strong colours, including one of sealing-wax-red (*c*. 1803) and later (*c*. 1817) HYALITH, a black glass; with these he made forms inspired by Wedgwood's *rosso antico* and black basaltes, with decoration painted in gold and silver. He also contributed toward the development of LIME GLASS.

burette (French). A CRUET, especially an altar cruet for sacramental wine, usually having an elongated curved SPOUT, and sometimes on the opposite side a loop handle.

Burghley, Lord, Tankard. A tankard in the form of a tall cylinder of clear glass mounted on a silver-gilt base and with a silver-gilt handle with thumb-piece and hinged LID. It has been attributed to JACOPO VERZELINI, *c*. 1572–75, and was possibly a test piece made for Lord Burghley (William Cecil, 1520–98), having his enamelled coat-of-arms (dating from 1572 or after) on the lid and the thumb-piece. Its form resembles contemporary silver and crystal tankards. *See* VERZELINI GLASSWARE.

Burgundy glass. A large WINE GLASS on a stemmed FOOT, intended for drinking red Burgundy. The BOWL is sometimes somewhat globular, sometimes tall and ovoid with a slightly inverted RIM.

Burmese glass. A type of ART GLASS that was opaque, shading from greenish-yellow colour upwards to light rose at the top of a piece, with either a dull or glossy surface. It was developed by MT. WASHINGTON GLASS CO. and patented for it in 1885 by FREDERICK S. SHIRLEY. The Company sent some examples to Queen Victoria who ordered a tea-set of such glass, enamelled with the 'QUEEN'S DESIGN'.

Buggin Bowl. Diamond-point engraved decoration, bowl attributed to George Ravenscroft, *c*. 1676. Ht 9·8 cm. Victoria and Albert Museum, London.

Lord Burghley's Tankard. Colourless glass (with silver mounts), probably by Verzelini, *c*. 1575. Ht 21·3 cm. British Museum, London.

bust. Constantine II, blue opaque bust set within glass imitation wreath, Roman, *c.*325 AD. Ht 8 cm. Römisch-Germanisches Museum, Cologne.

button decoration. Bottle with encircling wheel-cut rings. Persian 5th/7th century. Ht 17·5 cm. C. L. David Collection of Islamic Art, Copenhagen.

As a result, a licence was granted to THOMAS WEBB & SONS, who in 1886 patented the glass in England and produced it there (calling it 'Queen's Burmese Glass') in colours shading from lemon-yellow to pink, and decorated with painted flowers and birds. The glass was used for tableware and many ornamental objects, in a great variety of shapes and decorative patterns; some examples were further decorated with applied designs.

Busch, August Otto Ernst von dem (1704–79). Canon of Hildesheim, Germany, who decorated glassware, as well as porcelain, by DIAMOND-POINT ENGRAVING filled in with black pigment, usually depicting flowers and pastoral scenes.

bust. A full-relief representation (usually a portrait) of a person, including the shoulders and chest of the sitter. A bust in ROMAN GLASS, AD 325, exists of the Emperor Constantine II, and busts of Domitian and of Hadrian have been recorded. Moulded MAT busts of Queen Victoria and Prince Albert, of Shakespeare, and of other prominent persons, were made of frosted glass by E. and C. Osler, of Birmingham, in 1848–50; and a bust formerly thought to be John Wesley (now thought to be Milton) in such frosted glass is a copy of the pottery one made by the Wood family in the 18th century. The French term is *buste*; the German *Büste*; the Italian *busto*. *See* HEAD.

butter dish. A small dish used for serving butter; it usually has a COVER and often two protruding vertical lug handles. Examples in glass are of a variety of forms and styles. Some rest on an accompanying stand or saucer, and some have an interior pierced strainer on which the butter rests. Formerly sometimes called a 'butter basin'.

button decoration. A style of WHEEL-CUT decoration, on some bottles, ewers, vases, etc., in the form of raised rings within which there is a smaller raised or sometimes depressed circle. In some examples the rings are thin, with no central motif, and sometimes they are on faceted surfaces. The rings are usually arranged in three rows encircling the vessel. The style is found on vessels said to be Byzantine or Persian of the 6th/8th centuries or Syrian of the 9th century.

bye-hole. A small hole in the side of a FURNACE, used for withdrawing small amounts of MOLTEN GLASS or for reheating the BLOW-PIPE and the PONTIL, and for inserting small objects or parts of a vessel for reheating; it is sometimes called the 'nose-hole'. *See* GLORY HOLE; SPY-HOLE; BOCCA.

Byzantine glassware. Glassware made at Byzantium (Constantinople, now Istanbul) known mainly from written sources, except for a few pieces and fragments in museums. Pieces mounted for church use in Byzantine metal-work, now in the Treasury of St Mark's, Venice, and taken there after the capture and plunder of Byzantium by the Franks and Venetians after the Fourth Crusade in 1204, are of uncertain origin, but a few earlier pieces there are regarded as probably Byzantine, e.g., a violet SITULA, some HANGING LAMPS of the LAMPADA PENSILE form, and some pieces with BUTTON DECORATION, *c.*6th–8th centuries. In the 13th century there were glass-makers in Constantinople; after the city's fall in 1204 they went to Venice and influenced Venetian glassware.

C

cabin bottle. A type of FLASK made in the United States in the form of a rectangular or square log-cabin with a hipped roof, a chimney (the spout), and usually a door and two windows. They were for whiskey, and were inscribed with the word 'Tippecanoe', referring to the presidential election campaign slogan of William Henry Harrison in 1840. *See* BOOZ BOTTLE.

cable. A single TWIST produced where closely wound glass threads (usually opaque white but sometimes coloured or an air twist) extend vertically through the length of the STEM of some English drinking glasses to create a rope-like effect. It may be a 'single cable' or, when two cables intertwine, a 'pair of cables'. Sometimes a cable twist is encircled by one or two spiralling opaque white twists or air twists of different character (e.g. a corkscrew twist in a double series stem). Sometimes LATTICINO cables are found in a PAPER-WEIGHT, and occasionally parallel cables occur in some modern weights made at ST-LOUIS.

cable decoration. A decorative relief pattern on some American glassware in the form of vertical columns in rope-like pattern.

cabochon. Basically, a precious or semi-precious stone cut in convex form, highly polished and not faceted; and, by extension, blobs or drops of glass, of various colours, that simulated such stones in decorating glassware by being arranged in various patterns. The style is found on ROMAN GLASS of the 4th century AD, made probably throughout the Empire.

cachepot (French). An ornamental pot (the name being derived from *cacher*, to hide) to contain and conceal a utilitarian flower pot.

Cadmium Ruby glass. *See* RUBY GLASS.

cage cup. A cup, of VASA DIATRETUM type, in the form of an ovoid BEAKER with no foot, about 5 in. (12 cm.) high. The body is surrounded by a network of tangential glass circles attached to it by small glass struts. Sometimes there is an encircling Greek or Latin inscription in raised glass letters below the mouth. The body is of transparent glass, and the network or the lettering is sometimes of coloured glass, indicating the use of a FLASHING of coloured glass. The production of these cups was formerly attributed to the Roman DIATRETARII in the late 3rd or early 4th centuries AD; but, as several examples have been found at Cologne (and recently two in Greece, at Athens, 1956, and Corinth, 1962), a recent view is that they were made by Roman craftsmen in the Rhineland and possible also in the East. See TRIVULZIO BOWL; LYCURGUS CUP; SITULA PAGANA.

cage cup. Diatreta glass: pale greenish, circles green, bands amber, letters ruby; 4th century AD, found at Cologne. Ht 12·1 cm. Römisch-Germanisches Museum, Cologne.

Caithness Glass, Ltd. A Scottish glasshouse founded in 1960, with factories now at Wick (Caithness) and Oban (Argyll), specializing in paper-weights of original design and also making crystal glass tableware and ornamental pieces with engraved decoration. *See* PAPER-WEIGHTS, *Caithness*; SCOTTISH GLASSWARE.

calabash. A term used in the United States for a type of FLASK made there which has an ovoid body that tapers upward to a tall neck. Although not regarded as a true flask, it is classified with flasks as its technique of production and its PRESS-MOULDED patterns are similar to those of flasks made by the same factories there. Some examples of HISTORICAL FLASKS, e.g., the JENNY LIND FLASK, were made as a calabash.

calcar. A type of reverberatory FURNACE, used for calcination of part of the BATCH into FRIT.

calcedonio (Italian). Glassware made in Venice in imitation of earlier Roman AGATE GLASS.

Calcite glass. A type of ART GLASS developed by FREDERICK CARDER *c.*1915 at STEUBEN GLASS WORKS. It is translucent and creamy-white, and was named after the mineral, calcite, that it resembles. The translucency was produced by including in the BATCH calcium phosphate (bone ash). Each piece was made from three gatherings of molten glass, applied in layers and fused. The outer layer was often sprayed with stannous chloride AT-THE-FIRE to produce a slight IRIDESCENCE. The glass was used mainly for illuminating fixtures, due to the soft transmitted light, but also for ornamental ware; it was sometimes decorated with acid etching, or with engraved designs having the pattern accented by rubbing in an unfired brown oxide. The glass was also used as the inner or outer casing for AURENE GLASS (gold or blue) and as the matrix for red, green, and brown Aurene Glass; such pieces were sometimes loosely termed 'Aurene' in Steuben catalogues and hence that term has often been misapplied to such Calcite glass.

calderata (Spanish). The Spanish term for an ASPERSORIUM.

calligraphy. A decorative style of elegant script; inscriptions on glassware were added by the process of DIAMOND-POINT ENGRAVING, especially in Holland. The technique was perfected by ANNA ROEMERS VISSCHER in the second quarter of the 17th century, mainly in decorating large RÖMERS. Another outstanding exponent of the style was WILLEM JACOBSZ VAN HEEMSKERK.

calligraphy. Beaker with three bun feet; diamond-point engraving by Willem Jacobsz van Heemskerk, Holland, 1679. Ht 17·3 cm. Rijksmuseum, Amsterdam.

Callot figures. Grotesque figures of dwarfs inspired by engravings by or after Jacques Callot (1592–1635) of Nancy. In ceramic ware they occur as painted decoration and as figures. They have been used in glassware to decorate, e.g., a drinking glass in line engraving by FRANS GREENWOOD, *c.*1720, a BEAKER decorated in SCHWARZLOT by JOHANN SCHAPER, *c.*1665, and a TUMBLER, decorated in red enamelling, probably Nuremberg, *c.*1720. Some of the engravings were made *c.*1621 by Callot based on sketches made by him in Florence in 1612 when seeing a troupe of performing hunchbacks; the series is named '*Varie Figure Gobbi*'. Others were adapted from a series of 'The Months', one version of which was published in Amsterdam, 1716.

Cam, Jean de la (fl. 1660–68). A Frenchman who came to England and induced the DUKE OF BUCKINGHAM to invest in and obtain a patent for a glasshouse at Greenwich, in 1660, which Cam operated, with Venetian craftsmen, from 1660 until 1668.

Callot figures. Tumbler with red enamelling, probably Nuremberg, *c.*1720. Ht 11·8 cm. British Museum, London.

camaïeu, en. A French phrase describing painting in several tones of the same colour. Such decoration has been used on glassware made of OPAQUE WHITE GLASS. *See* GRISAILLE; MONOCHROME.

cameo. Originally, a stone (having layers of different colours) or shell carved in LOW RELIEF to show the design and background in contrasting colours. The same technique of carving to show the design in two or more colours was used in glassware to make CAMEO GLASS.

cameo glass. An object formed of FLASHED (CASED) GLASS of two colours, such as opaque white that forms an OVERLAY on coloured glass, where the outer layer was carved on a wheel so as to leave a design in white RELIEF on the coloured background, with the effect depending upon the extent and depth to which the white was removed. More than two colours could be used. The process was the same as carving a CAMEO, and was known to the ancient Egyptians and Romans, and also to the Chinese in the 18th century. It was revived in Bohemia, France, and England in the 19th century, the modern technique being basically similar, although sometimes HYDROFLUORIC ACID was used to assist the process (*see* ACID CUT-BACK GLASS). Its leading French exponent was EMILE GALLÉ, and in England THOMAS WEBB & SONS and other firms in Stourbridge worked from designs by JOHN NORTHWOOD and by GEORGE and THOMAS WOODALL, who also made cameo glass. *See* CARVING; PORTLAND VASE; MILTON VASE. [*Plate III*]

cameo incrustation. The term used in England for a SULPHIDE. Also called 'encrusted cameo' or 'crystallo-ceramie'.

can. An 18th/19th-century English term for a type of drinking glass in the form of a cylindrical TANKARD or MUG.

Canada, The Great Ring of. A massive symbolic glass piece representing Canada and its twelve provinces and territories. Held in a setting of rhodium-plated steel, twelve emerald-cut glass plaques, each with an armorial bearing and flowers of the respective regions, encircle the piece and converge to a glass ring with four smaller plaques showing the arms of Canada and the Maple Leaf of the flag. Above are twelve faceted spheres surmounted by a faceted sphere representing the nation. Around the steel base is the inscribed dedication to the 'People of Canada on the Centenary of Canada's Nationhood, from the People of the United States of America'. It was presented in 1967 to the Prime Minister, Lester Pearson, by President Lyndon B. Johnson. This piece was designed by DONALD POLLARD for STEUBEN GLASS, INC., the engraving designed by Alexander Seidel, and the engraving done by Roland Erlacher and Ladislav Havlik.

candelabrum. An ornamental candle-holder with at least two (sometimes four or more) branches with candle-nozzles. They were made, often in pairs, to be placed on a side-table, usually before a mirror to reflect the brilliance of the glass. Early English examples, from the early 18th century, consisted merely of the base, short shaft, and branches with the nozzles, without further ornament. Later ones were taller and decorated with FESTOONS of DROPS as well as with icicles, spires, etc., and sometimes also with SNAKES and a CANOPY at the top of the shaft, from which were suspended more drops, etc. In the REGENCY period the branches were simplified, but a fringe of LUSTRES hanging from the DRIP-PANS almost hid the arms and shaft. Early simple drip-pans became ornate, some scalloped or star-shaped. In the NEO-CLASSICAL period some were set on a base of Wedgwood's jasper ware, or a base of solid glass with ORMOLU MOUNTS. The early English term, *c.*1766, was GIRANDOLE or LUSTRE, and 'candelabra' (now often used for a single piece) first appeared *c.*1792. The Italian term is *candelabro*. *See* SCONCE; CHANDELIER; GARNITURE.

candle cover (or shade). A tall glass hood to be placed over a lighted candle. Some were made in the form of a circular vase with a flaring mouth, and were attached to the candle-holder through a hole in the rounded bottom. Some, of bulbous shape and open top and bottom, were made to fit over a single candlestick. *See* HURRICANE SHADE. Others, designed to rest on an ormolu stand, were made in France in the early 19th century.

candle-nozzle. The socket into which a candle can be inserted. It usually occurs as part of a CANDLESTICK, CANDELABRUM, or CHANDELIER (sometimes removable), and also on the cover of a CANDLESTICK VASE. The socket is sometimes completed by the addition of a surrounding small tray, either integrated or detachable, known as a DRIP-PAN (*bobêche*), to catch the wax drippings. Some nozzles are decoratively shaped, e.g. a 'tulip-nozzle'.

candlestick. A holder for a single candle, usually made with a flat base (circular, square, or of ornamental shape) having a central column

Canada, Great Ring of. Symbolic ornament designed by Donald Pollard for Steuben Glass Works, 1967. Ht 102 cm. Courtesy, Steuben Glass, Corning, N.Y.

candelabrum. Crystal glass candelabrum made for the King of Nepal, 1893, by the St Louis Glass Factory.

candlestick. Chamber candlestick, blue-green glass, with prunts and crested handle, Spain, 1961. Ht 10 cm. Kunstgewerbemuseum, Cologne.

candlestick vase. Pair, with reversible covers, English, *c.* 1800. Ht 24·4 cm. Courtesy, Spink & Son, Ltd, London.

(SHAFT) for a socket or nozzle, and usually a DRIP-PAN. The nozzles were sometimes removable, especially on later examples ornately decorated with FESTOONS of DROPS. They were made in England when small-diameter candles superseded the wide candles that were used on spikes of pricket-candlesticks. They are of many styles, and were made also in France, Venice, Spain, Germany, etc. In England from the 17th century the style of the column (usually about 25 cm. high) and of the base changed with the different stylistic periods for stemmed glassware; early 19th-century examples have wide drip-pans from which are suspended drops and ICICLES. *See* CANDELABRUM; SCONCE; TABLE LUSTRE; CANDLE-NOZZLE; TAPERSTICK; COLUMN CANDLESTICK; FIGURE CANDLESTICK; VASE CANDLESTICK; CHAMBER CANDLESTICK; TEA CANDLESTICK; CRUCIFIXION CANDLESTICK.

candlestick unguentarium. A type of UNGUENTARIUM in the form of a tall thin cylindrical neck spreading toward the wide flat base, so as to resemble a simple candlestick. They were made in large numbers in ROMAN GLASS.

candlestick vase. A type of vase with a reversible cover on one side of which is a CANDLE-NOZZLE; they were sometimes made of porcelain or pottery, and examples were made of glass. The term 'cassolette' is sometimes used in England to refer to such vases, but incorrectly so when there are no holes in the cover to allow scent to escape.

cane. A stick of glass intended usually to be (1) cut into slices (set-ups) for making MILLEFIORI or MOSAIC glassware, or (2) made into STEMS of certain types of drinking glasses.

(1) Canes made for millefiori glassware are usually solid, although some are hollow tubes. Occasionally a single coloured solid ROD or hollow tube is used, but more often a cane is made from a group of rods fused together in a mould and then drawn out (attached to a PONTIL at each end) to the required thinness, and then cut crosswise into slices which are set into millefiori glassware. Rods of various colour combinations are selected to make, when fused together, a cane with a decorative pattern in cross-section (e.g., a flower, rosette, etc.); they are placed in a corrugated mould, then fused with opaque, or coloured or colourless transparent, molten glass. The canes are initially thick but are drawn out of the desired thinness. To make a complicated pattern, such canes are used as constituents of a new cane. The colours in the rods and canes do not change their relative positions by drawing the glass to whatever length and thinness; some canes are of such tiny diameter that the motif in cross-section is no larger than a pin-head. Other types of canes are made by moulding a rod so that the cross-section is a star, a shamrock or other motif, or a SILHOUETTE of a person or an animal, and these shaped rods are often embedded in clear glass. Some canes have a scalloped shape, e.g., a PIE-CRUST CANE. Slices of many such canes are intermingled and fused to make millefiori PAPER-WEIGHTS or other such objects. Such canes have been made since Roman times; the earliest Venetian canes are said to have been made by Maria Barovier at Murano, *c.* 1496.

(2) Canes made for stems of drinking glasses are usually formed by incorporating in a colourless glass rod an opaque white or coloured glass thread or an air bubble, then drawing (and usually twisting) it to create a vertical pattern within the resultant cane; sometimes the process is repeated 2 or 3 times by encasing the initial cane in molten glass into which is fused an opaque white or coloured thread or threads (or air bubbles), so as to make the double-series and triple-series stems. (*See* TWIST.) *See* ARROWHEAD CANE; SHAMROCK CANE; SILHOUETTE CANE; STAR-DUST CANE.

canoe-shape. A shape of certain glass oval receptacles that tapers to a raised point at both ends, somewhat like a canoe; it is found on some fruit bowls of IRISH GLASSWARE and some SALT CELLARS.

canopy. An ornament of glass placed at the top of a CANDELABRUM or CHANDELIER made in the 18th century, and to which festoons of DROPS were normally attached.

Canosa Bowls. A pair of glass bowls, so called from having been found (before 1870) in a tomb at Canosa, in Apulia, Italy. They have gilt decoration, and are made in two layers, with the decoration applied in gold leaf to the outer surface of the inner bowl. Both layers were probably moulded, then carefully ground and polished so as to fit closely together; the outer one extends to about 3 cm. below the thick rim of the inner, and is possibly attached to the inner by light fusion, perhaps at the lip only. The decoration is an overall pattern of tendrils and ACANTHUS leaves, below two encircling bands of VITRUVIAN SCROLLS. The bowls have been attributed to the last quarter of the 3rd century BC, and their origin at Alexandria or Antioch has been suggested. Other bowls made in the same technique have been found at Gordion in 1955 and elsewhere. Such bowls were made in the manner of the 18th-century Bohemian ZWISCHENGOLDGLAS.

Canosa Bowl. Colourless bowl, 2-layer moulded with enclosed pattern in gold leaf, late 3rd century BC. Ht 11·4 cm. British Museum, London.

canted corner. The same as a CHAMFERED CORNER.

cantharus. The Latin term for KANTHAROS.

cántir or **cántaro** (Spanish). A type of closed JUG with an overhead vertical ring HANDLE, characterized by two SPOUTS extending upward from the BODY, one short and wide, for filling and pouring, and the other slender, tall, and attenuated, for use in drinking by squirting a stream of liquid directly into the user's mouth. The globular, ovoid or pyriform body has a KICK BASE, and the handle sometimes has a FINIAL in the form of a fanciful bird or a flower. They were made in Catalonia in the 17th and 18th centuries. Some were made in Venice, probably for export to Spain; and some were made in southern France, late-17th or 18th centuries, less ornate, and sometimes in the form of a teapot (without cover) with one pouring spout and with an overhead handle or a loop side-handle. *See* PORRÓ.

cap cover. A type of COVER, usually circular, for a JAR. It is low and flat-topped, and fits over a cylindrical collar on the jar. *See* DOME COVER; SCREW CAP.

Cape Cod Glass Co. A glasshouse at Sandwich, Cape Cod, Mass., founded by DEMING JARVES in 1858, after his withdrawal from the BOSTON & SANDWICH GLASS CO. Jarves planned it for his son John, but when the latter died in 1863 he lost interest and the plant closed on his death in 1869. The ware was similar to that of his former company, including PEACHBLOW GLASS and other coloured glass.

cántir. Greenish glass with white *filigrana* decoration, Spanish, 17th century. Ht 30·6 cm. Victoria and Albert Museum, London.

captain glass. A large glass for sweetmeats, placed on the centre of a glass SALVER, *en suite* with the SWEETMEAT GLASSES and/or JELLY GLASSES placed around it.

carafe. A type of BOTTLE, usually with a somewhat globular body and thin neck and without a STOPPER, for serving wine or water. They were made of glass in England from, or earlier than, the 2nd half of the 18th century (then called a 'water craft'). They were originally used at the dinner-table until *c*. 1800, often placed, with two GOBLETS, between each pair of diners; these were usually somewhat smaller than a DECANTER, and were sometimes decorated with TRAILING, RIBBING, ENGRAVING or CUTTING. Later they were used as a water-jug in the bedroom, with a matching TUMBLER inverted to fit over the mouth and serve instead of a stopper; some were made of enamelled or gilded WHITE OPAQUE GLASS. In recent years carafes, usually plain or simply decorated, are again being used in homes and restaurants for serving water or inexpensive wine.

Caranza, Amédée de. Vase in Art
Nouveau style, Noyon, France, *c.*1900;
signed 'A. de Caranz'. Ht 29·4 cm.
Kunstmuseum, Düsseldorf.

Caranza, Amédée de (fl. 1900–18). A French glass-maker who was
associated from 1890 with H. A. Copillet & Cie at Noyon until the
factory was destroyed in 1918. He made glassware in ART NOUVEAU style
with an iridescent surface.

carboy (derived from Persian, *qarabah*). A large glass BOTTLE used
primarily for storing or transporting acids, corrosive liquids or distilled
water. They were also used in pharmacy shops for display, sometimes
with painted decoration. English examples were originally globular but
during the 19th century became pyriform, frequently of one gallon
capacity, and with a tall thin NECK having a STRING RIM. They were
generally labelled. After *c.*1830 they were made of larger size. Similar
large bottles were made in the 3rd century AD, one known example being
56 cm. high and 46 cm. wide, weighing 7·72 kilograms.

carchesium. The Latin form of KARCHESION.

Carder, Frederick (1863–1963). An English glass-designer and
technician from Stourbridge, who studied with EMILE GALLÉ, learning
modern glass techniques. He was a designer for STEVENS & WILLIAMS,
LTD, from 1880 to 1903, when he went to the United States to establish a
glass factory to make BLANKS for Thomas G. Hawkes & Co., of Corning,
New York, a firm that had previously decorated and engraved glass
blanks bought elsewhere. Carder, inspired by the ART NOUVEAU style,
preferred to experiment with coloured glass, and soon founded, with
Hawkes and others, the STEUBEN GLASS WORKS to make ornamental
glass. The company was sold in 1918 to CORNING GLASS CO., where
Carder was made Art Director. Until his retirement from Corning in
1933 he created many varieties of glass in new techniques and with
varied colours and surface effects. *See* ACID CUTBACK GLASS;
ALABASTER GLASS; AQUA MARINE GLASS; AURENE GLASS; CALCITE
GLASS; CINTRA GLASS; CLUTHRA GLASS; CYPRIAN GLASS; DIATRETA
GLASS; FLORENTIA GLASS; GROTESQUE GLASS; INTARSIA GLASS;
IRIDESCENT GLASS; JADE GLASS; MANDARIN YELLOW; MATSU-NO-
KE; MILLEFIORI; MOSS AGATE GLASS; OPALESCENT GLASS; PLUM JADE
GLASS; QUARTZ GLASS; ROUGE FLAMBÉ GLASS; RUBY GLASS;
SILVERINA GLASS; TYRIAN GLASS; VERRE DE SOIE. While he was Art
Director at Corning Glass Co. he made glassware by an adaptation of the
CIRE PERDUE process. Many factory-made Steuben pieces bear his
signature, 'F Carder', but on some made in the 1950s and early 1960s it
was added (sometimes called his 'late signature') to indicate merely his
design, with the exception of castings from the *cire perdue* moulds which
were usually signed by him and dated. After 1932 he also designed some
glass to be engraved in ART DECO STYLE. For a definitive illustrated
study of Carder and his work, *see* Paul V. Gardner, *The Glass of
Frederick Carder* (1971).

Carli, Johann Anton (d. 1682). An enameller of glassware from
Andernach, in the Rhineland, who decorated in SCHWARZLOT.

Carnival glassware. Cheap garish glassware made of PRESSED GLASS
with an orange-gold iridescence, remotely imitative of TIFFANY
GLASSWARE and so-called in the United States because the pieces were
offered as prizes at country fairs. It was artificially iridized by a cheap
spray (instead of metallic oxides included in the BATCH). Such ware was
made from 1895 to 1924, by: Harry Northwood (son of JOHN
NORTHWOOD), at Indiana, Pennsylvania, and marked 'N'; Fenton Art
Glass Co., founded in 1904 at Williamstown, West Virginia; and
Imperial Glass Co., founded in 1902 at Bellaire, Ohio. It was sometimes
called 'taffeta' or 'Nancy'.

Carolingian glassware. Glassware made after the period of FRANKISH
GLASS, and from the mid-8th to the mid-10th century. Very little was
produced and that not of good quality. The prohibition against glass for
CHALICES for the Mass further reduced the output, although some

glasses were made as RELIQUARIES and for church oil-lamps. In this period stained glass for church windows was probably first developed.

Carré, Jean (d. 1572). A native of Arras who, while living in Antwerp, acquired some experience in glass-making. In 1567 he migrated to England and obtained a Royal licence to manufacture window glass, for which purpose he established two furnaces near Alford, in the Weald (*see* WEALDEN GLASS), using workers he brought from Lorraine and Normandy. In 1570, with Peter Briet, he opened the CRUTCHED FRIARS GLASSHOUSE in London, using Flemish workers to make CRISTALLO glass in the Venetian style, but replaced them in 1570/71 with Italian craftsmen brought to England by him, including JACOPO VERZELINI who was placed in charge. Carré planned a new glasshouse in the Weald to make drinking glasses, but died before it was started.

carrot. A glass cone used in making an English opaque twist STEM for a drinking glass. To make it, a steel cylinder is lined with RODS of OPAQUE WHITE GLASS, and then molten glass is spread over them so that the rods are incorporated, then the resultant CANE is MARVERED into a carrot shape, and another layer of molten clear glass is spread over it. The piece is annealed and kept available for use as needed in making stems. When the narrow end is attached to the bowl of a glass and is drawn and twisted, the stem develops and the excess carrot is cut away.

case bottle. A four-sided bottle made to fit with others in compartments in a case or box. Some examples, of good-quality glass, occasionally have a silver mounting whose hall-mark may indicate the earliest probable date of the piece. *See* DECANTER, *whisky.*

cased glass. Glassware made of two or more layers of glass of different colours. The process involves first making the outer casing by blowing a GATHER, knocking off one end, and opening the piece to form a cup-like shell; the shell is then placed in a metal mould and a second gather, of different colour, is blown into it, and the combined piece is taken from the mould and reheated so that the two layers fuse together. If another casing is desired, the process is repeated. The outer layer is thick, as distinguished from the thin outer layer of FLASHED GLASS. Such glass was produced in Roman times (*see* PORTLAND VASE), but became popular in Bohemia from *c*. 1815 and was produced in England from the 1830s (mainly 1845–70, after the removal of the EXCISE TAXES), being made by GEORGE BACCHUS & SON. During the 18th century it was produced in China and decorated in the style of CAMEO GLASS (*see* CHINESE SNUFF BOTTLES). It was produced in the United States in the 1840s and became popular after 1853 and into the 1860s. Modern cased glass made in MURANO has two or three (even more) very thick layers of different-coloured glass. *See* VERRE DOUBLÉ; OVERLAY; LAYERED GLASS.

casket. A BOX of ornamental character, usually used for jewels or as a RELIQUARY. Some made of wood or lead are large and are covered with glass decoration, sometimes with many tiny FIGURES made AT-THE-LAMP. Examples made of English LEAD GLASS are usually completely decorated with CUTTING and generally have a hinged LID.

cassolette (French). A covered VASE, sometimes made of glass but more often of porcelain or pottery, to contain perfume to scent a room and hence having holes in the COVER or SHOULDER, or pierced decoration, or being mounted in ORMOLU or GILT BRONZE with openwork decoration. Matthew Boulton, who made many mounts, called the pieces 'essence vases' or 'essence pots'. If a vase having a reversible cover with a candle-nozzle on one side (sometimes called in England a 'cassolette') has no such holes, it should preferably be designated a CANDLESTICK-VASE. *See* POT-POURRI VASE.

cast glass. (1) A type of glass made either by (a) fusing powdered glass in a one- or two-piece MOULD (i.e. open vessels or solid objects but not

cased glass. Multi-layer (grey-green-white) vase in the style of Gallé, St Petersburg Imperial Glass Factory, 1901. Ht 27·5 cm. Hermitage Museum, Leningrad.

Castle Eden claw-beaker. Frankish (Teutonic) glass, 5th/6th century. Ht 19 cm. British Museum, London.

chain handle. Green-glass jug with handle, Roman, 3rd century AD. Ht 8·5 cm. British Museum, London.

hollow vessels), or (b) the CIRE PERDUE process. Such processes were probably used from the 8th to the 3rd century BC. *See* EGYPTIAN GLASS. (2) Glass in flat sheet form, such as was first produced by the Romans for MIRRORS by the technique of casting, and later in England by the improved technique, introduced by BERNARD PERROT and developed in 1688 in France by LOUIS-LUCAS DE NEHOU, whereby the molten glass was poured on an iron table covered with sand where it was rolled before being ground and finally polished with abrasives on both sides. Cast PLATE GLASS was not made in England until 1773. For many years cast glass was used only for mirrors, coach-windows, and windows in the homes of the aristocracy, e.g., Versailles, Blenheim, etc.

castellated. Similar to CRENELLATED.

Castello (or Castellan) Jean. A glass-maker at Orléans, France, descended from glass-makers from L'ALTARE, whose factory was under the patronage of Philippe, Duke of Orléans, brother of Louis XIV. He worked at Nevers from 1647, and in 1662 founded a glasshouse at Orléans, in association with his nephew, BÉRNARD PERROT.

Castle Eden claw-beaker. A CLAW-BEAKER so called from having been found in 1775 at Castle Eden, Co. Durham, England. It has two encircling rows of hollow PRUNTS and encircling bands of trailing below the rim and above the FOOT-RING. *See* RÜSSELBECHER.

castor. A small receptacle with holes for sprinkling salt, pepper or sugar. They are usually of cylindrical or baluster shape. Some have a NECK with interior threading for securing a threaded STOPPER or external threading for a SCREW CAP; the cover is often of silver. Others are of one piece with holes in the domed top; these are filled through a corked opening in the bottom. The holes, sometimes of ornamental shape, vary in size depending on the intended contents. They are sometimes called a 'cruet' or 'SIFTER', and sometimes, especially in the United States, the term is spelled 'caster'. *See* CRUET; SUGAR-CASTOR; PEPPER-BOTTLE; MUFFINEER.

Castro, Daniel Henriquez de (d. 1863). A Dutch chemist of Amsterdam who was a collector of glasses engraved in STIPPLING by DAVID WOLFF and who, as an amateur, decorated glasses in the manner of Wolff but who also did ETCHING in conjunction with stippling and line engraving.

caudle cup. A vessel for drinking caudle (a beverage formerly made in England as a warm gruel from wine or ale, with eggs, bread or oatmeal, sugar, and spices and usually intended for invalids).

cave à liqueurs. The French term for a LIQUEUR STAND.

caviar bowl. A large BOWL on a stemmed FOOT into which fits a smaller bowl for serving caviar; the space between the bowls is for shaved ice. Similar double bowls of smaller size are used for serving chilled fruit or seafood cocktails. The French term is *rafraîchissoir*.

celery dish. A rectangular or oval dish with low vertical sides, intended for serving stalks of celery laid in it.

celery glass (or vase). A tall glass receptacle for serving stalks of celery standing upright; usually cylindrical in shape, with the BOWL narrowing toward its flat bottom or widening to a bulbous 'thistle-like' bottom. The bowl rests on a short knop STEM, attached to a spreading flat base, often a COG-WHEEL BASE. A few have two side-handles. They have been made in England from the late 18th century, and are usually decorated with deep FACETING in overall patterns. Some Irish examples have a square MOULDED foot and a TURN-OVER RIM.

celery handle. A type of HANDLE that is decorated by laying on it threads of glass alongside each other so that it is RIBBED lengthwise (resembling celery) and widens where it is attached to the BODY. This style is found on ROMAN GLASS, 1st–3rd centuries AD.

Celtic glassware. Glassware made in Central Europe (including the region around Hallstatt in Austria) during the period of the Celtic culture, from *c.* 1200 to *c.* 400 BC. *See* HALLSTATT CUPS.

Cenedese, Gino (1907–73). The founder of a glassworks, bearing his name, at MURANO in 1945. It makes a wide variety of ornamental glass and tableware, and has created some special surface effects. The chief designer from 1965 has been Nini da Ros, and the glassware has been executed and signed by Gino Fort and Angelo Tosi. *See* SCAVO, VETRO; GRAFFITO, VETRO.

centre-piece. An ornamental object intended to decorate the centre of the dining-table. They were made, in a wide variety of forms, of porcelain or glass. Some extravagant examples were embellished with various small ornaments, miniature vases and flower pots, and a central fountain, all enclosed within a decorative railing. Also called a 'table ornament'. The Italian terms are *centro da tavola* and TRIONFO. *See* ÉPERGNE; BIRD FOUNTAIN; ALZATA; SALVER; SWEETMEAT STAND.

cesendello (Italian). A type of HANGING LAMP in the form of a tall vertical cylinder with a slightly everted rim and, at the centre of the rounded bottom, a decorative FINIAL. Such lamps were suspended, usually as SANCTUARY LAMPS, from the ceiling by three metal chains and hung just below a glass smoke shade of equal diameter. Any example in a museum seen resting on a metal stand is shown thus merely for display purposes. They were made at MURANO for local use, and also for export to Turkey and the East, in the late 16th century. They are usually decorated with coloured enamelling, some have *filigrana* decoration with white threads, of the *vetro a retorti*, or *latticino*, style. Those made for export are of coloured glass or are decorated with *lattimo* stripes. The term originally applied to such lamps made of bronze or silver.

chain handle. A form of HANDLE occurring on ROMAN GLASS of the 3rd century AD, made of two parallel bands of glass pincered into small tangent loops.

chain trailing. A decorative pattern, formed from two intersecting THREADS, resembling the links of a chain; it is often used as a border. *See* GUILLOCHE.

chair. (1) The bench used by a GAFFER while forming a glass object. It is a wide bench with flat, extended, and slightly sloping arms, and on these arms he rests his BLOW-PIPE with its attached PARAISON of MOLTEN GLASS and constantly rotates it back and forth with his left hand so that the paraison keeps its symmetrical shape and does not collapse while he works it with his right hand. The French term is *banc à bardelles*. (2) The team that assists the gaffer; it is also sometimes called a 'shop'. It usually includes about 5 to 6 craftsmen and apprentices. *See* WINE GLASS, *process of making*.

chalcedony. A variety of quartz, of various colours. It was widely imitated in coloured glassware, with several colours blending into each other. *See* AGATE GLASS.

chalice. A GOBLET or other form of stemmed bowl used in the sacrament. Some were made of glass, sometimes with a conforming PATEN. From the 9th century chalices of glass were forbidden for the Mass (for fear of their being broken) by the Council of Rheims (803) and the Council of Tribur (895), but the prohibition was not strictly enforced and some glass chalices are said to have been used. In later

centre-piece (*trionfo*). Small ornaments in transparent and opaque glass, Venice, 18th century. W. 48 cm. Museo Vetrario, Murano.

chalice. Made *c.* 1924 and signed 'Schneider', at Cristallerie Schneider, Epinay-sur-Seine. Ht 19·5 cm. Kunstmuseum, Düsseldorf.

champagne glass. Glass with inverted baluster stem and tear, *c.* 1720–30. Courtesy, Delomosne & Son, London.

chandelier (1). Small early English, with branches in one piece, flat cut, *c*.1740. Ht 74 cm. Courtesy, Delomosne & Son, London.

chandelier (2). English, with globes and bowl supporting arms, *c*.1785. Ht 1·30 cm. Courtesy, Delomosne & Son, London.

periods glassware in the form of chalices, with a bowl on a footed STEM, have been made. The German term is *Kelch*. *See* KYLIX; SCALLOP-SHELL CHALICE; AMIENS CHALICE; CHURCH GLASSWARE.

Chamäleon-Glas (German). Literally, chameleon glass. A type of LITHYALIN that was introduced in 1835 by FRIEDRICH EGERMANN.

chamber candlestick. A type of CANDLESTICK mounted in the centre of a saucer (sometimes resting on a stemmed FOOT) with a single HANDLE; sometimes called a 'hand candlestick'. Some were made in the form of a FIGURE CANDLESTICK. The French term is *bougeoir*.

chamber-pot. A bed-chamber vessel for urine. Although many porcelain or pottery examples of fine quality exist, often well decorated, only a few glass examples are known. Miniature examples made in England are known. The French term is *pot-de-chambre*. *See* BOURDALOU; URINAL.

chamfered corner. A corner formed by trimming away the angle made by two adjacent flat faces, as on some square or rectangular glass tea canisters or cruet bottles. Also called 'canted corner'.

champagne glass. A drinking glass for champagne. It is a matter of controversy if any special type of drinking glass was used in England for champagne in the 17th and early 18th centuries. FLUTES for champagne were advertised from 1773 until after 1850. One view is that although the wine was popular, it was drunk from any available suitable glass. Others have erroneously stated that glasses similar to a SWEETMEAT GLASS, but with a smooth RIM, were for champagne. Some such glasses having a flared rim are not suitable for drinking purposes. Only from *c*.1830 are found the glasses (4 to 6 oz. capacity), with a shallow saucer-like bowl resting on a stemmed FOOT (saucer champagne glasses), of the type associated now with champagne. But the preference today among champagne connoisseurs is a FLUTE, whose deep bowl and narrow mouth prevent the bubbles from escaping as quickly.

Chance, Robert Lucas. Manager from 1810 to 1815 of the NAILSEA GLASS HOUSE, the factory founded in 1788 by William Chance, John Robert Lucas, and Edward Homer. In 1824 he acquired the SPON LANE GLASSWORKS at Birmingham. He was concerned with glass for industrial purposes, but fostered experiments in COLOURED GLASS. At Spon Lane he developed from 1830 improved methods for making BROAD GLASS. The factory later became known as Chance Brothers, Ltd., and was acquired in 1945 by PILKINGTON BROTHERS, LTD.

chandelier. A lighting fixture made to be suspended from the ceiling and equipped with CANDLE-NOZZLES and DRIP-PANS. Early Venetian examples, from the 18th century, had a metal frame, decorated with clear and coloured glass flowers, fruit, and leaves (the Venetian SODA GLASS not being suitable for FACETING). In England, early metal frames hung with CRYSTAL ornaments were superseded in the early 18th century by chandeliers with a central shaft decorated with glass globes and terminating in a glass bowl from which extended glass arms for the nozzles. In the mid-18th century glass PENDANTS were introduced, and later in the ROCOCO period numerous faceted DROPS and SPIRES were attached. In the NEO-CLASSICAL period there were urn-shaped glass ornaments and festoons of prismatic drops, in the prevailing ADAM STYLE. In the ensuing REGENCY period, chandeliers appeared to be entirely of glass, with the metal parts concealed within glass and the central shaft hidden by a curtain of strings of drops; the nozzles and drip-pans were ornately formed and had deeply cut decoration. When gas-lighting superseded candles in England, candle chandeliers lost popularity, but examples in ANGLO-VENETIAN style, with coloured glass ornaments, were made in the mid-19th century. In Spain one known chandelier was made entirely of OPAQUE WHITE GLASS. Very large and ornate glass chandeliers are still made for auditoriums and public rooms in France, Venice, and Czechoslovakia, and by LOBMEYR in Vienna. The Italian terms are *lampadario* and, for early ones decorated with many coloured glass ornaments, CIOCCHE (bouquet of flowers). *See* DISH LIGHT.

charge. The act of putting the BATCH into the POT.

Charles Edward Stuart glassware. Goblets and other glassware decorated with an engraved portrait of Prince Charles Edward Stuart ('Bonnie Prince Charlie'), the Young Pretender, and usually also the motto '*Audentior Ibo*'. *See* JACOBITE GLASSWARE.

Charpentier (fl. 1813–19). A Parisian glass-engraver who decorated LEAD GLASS made at a glassworks at Vonêche, Belgium, that in 1816 became a part of the BACCARAT FACTORY. His work was in fine detail, the motifs being goddesses and *amorini*. His widow, Madame Desarnaud-Charpentier, carried on his work, winning a Gold Medal in Paris in 1819. *See* ESCALIER DE CRISTAL.

cheese dish. A utensil for storing and serving cheese. It is usually a flat tray with a high-domed cover. Generally made of porcelain or pottery, examples have also been made of glass.

chequered diamond. A type of DIAMOND FACETING having four small diamonds cut in chequerboard arrangement on the surface of each large diamond.

chequered spiral-trail decoration. A style of relief decoration on BEAKERS (and some TANKARD and STANGENGLAS examples) in the form of an applied heavy and broad spiral trail encircling the entire piece, broken vertically at short regular intervals by flat or slightly concave areas. The process involved applying the trail and then, after reheating the piece, inserting and blowing it in a vertically-ribbed mould so that the ribs made the indentations that broke the trailing, thus creating the chequered pattern that resembles cut flat diamonds spiralling upwards. The spiral trailing sometimes continues on the base to the centre of the PONTIL MARK or the KICK, and towards the rim it tends to become horizontal. A BASE RING, sometimes MILLED, is occasionally applied to pieces so decorated, their rim being usually slightly everted; and sometimes a few PRUNTS are applied to the body partway up the glass. Some Stangengläser and tankards so decorated are sometimes also given enamel and gilded decoration. For many years since 1884 these glasses have been improperly designated as *Spechter* and attributed to the Spessart region of HESSE (*see* SPECHTER), but now they are considered to have originated at Antwerp and to have been made in the lower Rhineland and southern Netherlands, *c*. 1550–*c*. 1650. For a full discussion of this ware, *see* Hugh Tait, 'Glass with Chequered Spiral-Trail Decoration', *Journal of Glass Studies* (Corning), IX (1967), p. 94.

chessmen. A set of chessmen, made in many different materials and styles, and in most countries and most periods. Some have been made of glass, with black and white or coloured pieces.

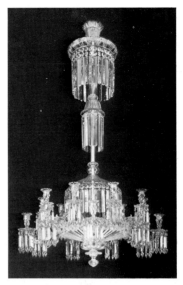

chandelier (3). English, early Victorian, with 18 lights in two tiers, and crystal glass pendants. Ht 1·90 m. Courtesy, Delomosne & Son, London.

chandelier (4). English, Regency, period, with 8 lights, ormolu rim and arms, and crystal glass drops, pendants, and nozzles, *c*. 1810–20. Ht 1·06 m. Courtesy, Delomosne & Son, London.

chessmen. Dark green and clear glass with cut decoration, English, mid-19th century. Ht 9/12 cm. Victoria and Albert Museum, London.

chequered spiral-trail decoration. Beaker with prunts and milled base-ring, probably South Netherlands, c. 1550–1650. Ht 21·5 cm. Museum Willet Holthuysen, Amsterdam.

Chinese snuff bottle. Bottle with stopper with attached spoon, single overlay decoration, 19th century. Ht 5·3 cm. Courtesy, Hugh Moss, Ltd, London.

Chesterfield Flute. The same as the SCUDAMORE FLUTE.

chestnut flask. A type of flat FLASK of circular or ovoid shape, the body of which has an overall pattern of curved RIBBING extending from the base, spreading then converging around the neck, giving the appearance of a shelled chestnut. They were made in Germany in the 18th–19th centuries. Bottles of similar shape but without the ribbing were made in the United States at many glasshouses and were called LUDLOW BOTTLES; those of small size were sometimes termed 'chestnut bottles'.

cheval-glass. A long framed dressing MIRROR, made in England from the end of the 18th century. Some have candle-holders attached to the frame. The mirrors usually swivel midway so as to be adjustable to the height of the user.

chevron. A decorative motif in the form of an inverted 'V', usually in a repeating series, resembling a zigzag pattern. It is found as a COMBED thread pattern on examples of CORE GLASS and later as enamelled decoration.

Chiddingfold. *See* WEALDEN GLASS.

chimney glass. A large framed MIRROR made to be suspended on the wall over a mantelpiece. They were introduced in France in the early 18th century, with many varied designs for the frames. Early English examples are usually rectangular, hung horizontally, with three panes of mirrors, of which the central one is about twice the width of each outer one; a few have four panes. Later examples have a single square or vertical mirror. Sometimes there is a painting within the frame just above the mirror or across the upper portion of the mirror. Some have attached branches for candle-nozzles. *See* PIER GLASS; SCONCE.

Chinese glassware. Glassware made in China from, according to tradition, 435 AD, when glass-making was introduced from the West. However, specimens of Chinese glass (identified by its containing BARIUM, not then used elsewhere) have been found in China from the late Chou Dynasty (c. 400–260 BC) and the Han Dynasty (206 BC–AD 220) along with glass imported from Syria in the Roman period. Early glassware made in China, or fashioned in China from imported glass material (*see* GLASS 'CAKES'), was mainly in imitation of carved hardstones, but some glass objects and animal figures, found in T'ang tombs, were possibly imported. After an apparent lapse of glass-making during the Sung Dynasty (960–1279) and the Ming Dynasty (1368–1644), the PEKING IMPERIAL PALACE glassworks was founded in 1680, in the Ch'ing Dynasty, within the Imperial Palaces by the emperor K'ang Hsi (1662–1722); it was the first Chinese factory to produce glass in quantity, making (under Jesuit guidance) BLOWN GLASS in the styles of Venice and the Netherlands, much of which today shows CRISSELLING. Glass of this and later reigns (some bearing reign marks) includes bowls and vases, with carved and enamelled decoration (*see* KU-YÜEH HSÜAN). The process of CAMEO decoration was apparently independently discovered, and such work was introduced in the reigns of Yung Chêng (1723–35) and Ch'ien Lung (1736–95) and continued later, especially vases and CHINESE SNUFF BOTTLES. In the 18th century coloured glass was made, particularly Imperial yellow, as well as MARBLED GLASS. In modern times glass in the form of rectangular blocks, is sent in large quantities from Po-shan (Shantung Province) to Peking for processing into amulets and toys. The Chinese terms for glass are *liu-li* (opaque glass) and *p'o li* (clear glass). *See* WATER POT; BRUSH-POT; WINE CUP; DOUBLE-GOURD SHAPE; KU; HU; KUAN LIAO; KUAN-YIN; CHINESE REIGN-MARK.

Chinese-Islamic glassware. A type of ALEPPO GLASSWARE and DAMASCUS GLASSWARE, made c. 1285–1402, with CHINOISERIE-style decoration introduced into Syria as a result of the Mongol conquest of

Mesopotamia and the ensuing Syro-Chinese commerce after the
establishment of the Yüan (Mongol) Dynasty in China (1280–1368). The
Chinese influence is shown in the motifs, such as plant life, cloud scrolls,
lotuses and peonies, and dragons and phoenixes, and in depictions of
densely packed birds and flowers within thin red outlines. This group of
glassware includes the MOSQUE LAMPS. *See* ENAMELLING, ISLAMIC;
SYRIAN GLASS.

Chinese reign-mark. The character-mark that is on some Chinese
porcelain and also some glassware, usually indicating the period of
production. It consists of four or six characters or ideograms, the first
two denoting the dynasty, the second two the name of the emperor, and
the last two represent *nien hao*, meaning 'in the reign of'. Some PEKING
IMPERIAL PALACE GLASSWARE bears the four-character mark of the reign
of Ch'ien-Lung (1736–95).

Chinese snuff bottles. SNUFF BOTTLES made in China, first in the
second half of the 17th century, and used at the Manchu Court at Peking
from *c.*1650 to *c.*1750; they were in common use from *c.*1780/1800.
They are made of various substances (e.g., hardstones, porcelain,
enamelled copper, organic materials, and also glass), all in a wide variety
of shapes and styles. Glass bottles were generally made by blowing but
sometimes by moulding or by carving from a glass block. Most are flat,
of round or ovoid shape, but some examples are of fantasy shapes. They
usually have a small spoon attached to the cork in the stopper. The glass
bottles are: (1) monochrome (often Imperial yellow, sometimes with
carved or engraved decoration); (2) enamelled on opaque white glass
(with decoration of floral patterns or landscapes, occasionally a portrait);
(3) interior painted (see below); (4) of mottled colours or imitative of
aventurine, jade, REALGAR, turquoise, etc.; (5) CASED (OVERLAY), with
a single overlay (usually ruby red or turquoise blue, sometimes hues of
green) or multiple overlays (two to four layers of differently coloured
glass, usually ruby red, but also including green or blue); (6) variegated,
with single overlays of up to eight colours applied to different areas of the
same ground. Early examples of types (5) and (6) usually have a ground
layer of colourless glass or white glass (i.e., SNOWFLAKE GLASS, glass
with embedded bubbles, or plain white); in later years coloured grounds
were introduced and these almost always have carved decoration, type
(5) in CAMEO style showing the different colours. The stoppers are often
of contrasting colour, usually green or coral. Miniature bottles (about
2·5 to 5 cm. high) have been said to have been made for ladies, but there
is no evidence of their having used snuff. Sometimes two bottles were
joined to form a double bottle. The earliest glass bottles, made at the
Imperial Factory at Peking, *c.*1700, are monochrome. Some enamelled
bottles, made from the second quarter of the 18th century, are decorated
in the KU-YÜEH HSÜAN style. The best bottles (other than some interior-
painted bottles made *c.*1900) were made *c.*1730–*c.*1850, after which
there was a period of inferior ware (with occasional exceptions) until the
art was revived *c.*1950–60, when some high-quality pieces were made
(but with no attempt to deceive as to date). Some superior bottles made
in the late 19th and early 20th century were painted by Yeh Chung-san
(fl. 1892–1920), his two sons, their pupil Wang Hsi-san, and their
followers. Most bottles were made by groups in workshops, so that few
(other than those with interior painting) bear the name of the individual
artist; and sometimes the artist signed, as a compliment, the name of an
illustrious predecessor. Some carved glass bottles (known as 'seal
bottles') after *c.*1850 bear the seal of the artist or a workshop. It is
difficult usually to make attributions to specific artists and more so to
particular dates. For a full discussion and many coloured illustrations,
see Hugh M. Moss, *Snuff Bottles of China* (1971); Bob C. Stevens, *The
Collector's Book of Snuff Bottles* (1976).   [Plate IV]
——, *interior painted.* Chinese snuff bottles (usually of glass, but
occasionally of crystal or a translucent substance, e.g., agate) decorated
by Chinese artists by painting in reverse (i.e., the foreground painted
first) on the interior of the bottle with a fine-pointed angled bamboo pen

Chinese snuff bottle. Single overlay, red,
green, and yellow, on snowflake
ground, carved lotus design,
*c.*1760–1880. Ht 9·5 cm. Courtesy,
Hugh Moss, Ltd, London.

Chinese snuff bottle: interior painting.
Scene with magpies, 1907. Ht 5·6 cm.
Courtesy, Hugh Moss, Ltd, London.

Chinese snuff bottle. Double gourd shape, semi-opaque white glass with enamel decoration in *Ku-yüeh Hsüan* style, Ch'ien Lung mark, 1736–95. Ht 4 cm. Percival David Foundation of Chinese Art, London.

inserted through the narrow neck (less than 1 cm. in diameter). The medium was watercolour, applied on early examples on the rough interior surface (resulting from hollowing and lightly polishing), but later on a superior surface treated with an acid to cause the paint to adhere and to deter wearing off and peeling. The decoration is often a landscape on the front and an inscription on the back; a few artists, e.g., Ma Shao-hsüan (fl. 1894–1925), painted portraits from photographs.

Interior-painted bottles were made during three periods: (1) The Early School, *c.* 1800–82, in Peking, represented by Kan Huan-wên (a skilled artist who possibly signed with several art names), and a later inferior group who painted on squared bottles with a polished interior; (2) The Middle School, *c.* 1882–1949, of which the leading and most copied artist was Chou Lo-yüan (fl. 1880–97), followed in Peking until *c.* 1920 by some skilled painters identified by their marks; and (3) The Modern School, *c.* 1949 and onward, after the revival of the art under the Communists, there being (a) the Yeh group (see above), and (b) a group in Shantung province, mainly (until recently) inferior artists using cheap, poor-quality bottles made at Po-shan. From the Early School until the present, interior painting has been applied to many bottles not made to be decorated, but some bottles were made expressly for such decoration, especially those of the Middle School which have a short neck and broad foot. In the Early School the painting was in sombre colours, mainly brown, black, and dark green or blue; in the Middle School, brighter colours were used, *c.* 1900–20.

chinoiserie (French). European decoration inspired by oriental sources, particularly Chinese. *Chinoiseries* are pseudo-Chinese figures, pagodas, monsters, and landscapes, with imaginative fantasy elements. Introduced at the end of the 17th century in the designs of such *ornemanistes* as Jean Bérain I, the style was developed in painted porcelain decoration at Meissen soon after 1721. It was widely used in decoration of all kinds until *c.* 1760. In England it was also popular in the REGENCY period. It was used to decorate plates of Venetian or Bohemian OPAQUE WHITE GLASS in the 18th century, imitating ENAMELLED porcelain plates so decorated, and also as ENGRAVED DECORATION on some glasses.

chintz glassware. A type of ART GLASS, made by a complicated and expensive process, having a pattern of multi-coloured vertical stripes creating a chintz-like appearance. It was made in many colours and used for bowls, vases, etc. It was developed by ARTHUR W. NASH at the Tiffany Furnaces (*see* LOUIS C. TIFFANY), but never produced there. Some pieces were later made by A. DOUGLAS NASH, but were not patented until Nash was working at the LIBBEY GLASS CO.; that company made only a few experimental pieces.

chipping. A technique for decorating glassware, developed by WILHELM VON EIFF in the 1930s.

chocolate-pot. A covered pot for serving hot chocolate, usually pear-shaped or cylindrical, and often with a handle placed at right angles to the spout, although this feature is not, as is sometimes supposed, proof that a vessel is a chocolate pot. The pot usually has a long spout extending upward from near the bottom of the body. They were usually made of porcelain or silver but some have been made of glass in the 19th century.

Choisy-le-Roi Glassworks. A glassworks at Choisy-le-Roi, in Paris, founded in 1821 by M. Grimbolt. The technical director from 1823 to 1848 was GEORGES BONTEMPS who produced URANIUM GLASS here *c.* 1838. The factory closed in 1851.

chope (French). A drinking glass, usually for beer, in the shape of a tall straight-sided tumbler (HUMPEN) or a MUG.

Christian glassware. Glass objects decorated with New Testament motifs and made in the Eastern Mediterranean region, especially in Syria, and in the Roman Empire, principally from the 3rd to the 5th centuries AD. Some Roman glassware was engraved with Christian motifs; one known early example is a round bottle of greenish glass decorated with a Christian and a Byzantine cross, with an inscription. Some Syrian square or hexagonal bottles of amber-coloured glass have on the alternate panels relief decoration of a Latin or a Maltese cross (*see* MALTESE CROSS BOTTLE) and a stylized sun-flower or palm motif. Some Roman gold-leaf SANDWICH GLASSWARE was made in the 3rd century with the engraved monogram IHS, and other such pieces with New Testament motifs were made until after the mid-4th century. *See* BIBLICAL SUBJECTS; RELIGIOUS SUBJECTS; HOPE GOBLET.

chromium. A metallic agent used in modern times in colouring glass, producing yellowish-green. When excessive amounts are used, crystals of chrome oxide separate from the glass while cooling, the tiny flakes remaining suspended to make green AVENTURINE GLASS.

church glassware. Glass objects used for sacred purposes, such as CHALICES, PATENS, RELIQUARIES, ASPERSORIA, OSCULATORIES, and oil dishes for church lights. The making of glass objects expressly for church use was restricted when the use of glass chalices for the Mass was prohibited by the Council of Rheims in 803, by Pope Leo IV in 847–55, and by the Council of Tribur in 895. Glass objects were often used for reliquaries from the time of the medieval papal edict requiring all altars to have a relic. In France during the Revolution it is said that glass altar vessels were substituted for those made of silver. *See* CHRISTIAN GLASSWARE; FLAGON.

ciborium. A cup-shaped receptacle with two vertical loop handles, a stemmed FOOT, and an everted rim; the form was derived from the Egyptian water-lily. They were sometimes decorated with circular flat PRUNTS or with JEWELLED DECORATION. The form is similar to that of a SKYPHOS except that the handles are midway on the bowl. Examples are known from the 3rd–5th centuries, generally for liturgical use.

cicada. A glass funerary object made in China in imitation of a jade cicada. They were placed on the tongue of the dead. They were made during the Han Dynasty (206 BC–AD 220).

cider glass. A type of drinking glass for cider, having an engraved motif of a spray of apple-blossoms, apple-tree branches or a fruiting apple tree.

cigar holder. A cigar-smoker's accessory made of many different materials, occasionally of glass. Examples made in the 19th century are generally in traditional shape but some have an upturned end and some are curved. They were made of clear glass, opaque white glass, and coloured glass, sometimes with cut overlay and sometimes in so-called NAILSEA GLASSWARE.

cinerary urn. A vessel to contain the ashes of the dead. Some were inscribed with the name of the deceased. Urns of this kind, varying in shape and style, were made in pottery from the earliest times, but many were made of ROMAN GLASS from the 1st to the 4th century AD. One type of glass urn has an ovoid body, with a flat BASE and a wide MOUTH, and usually two massive loop-handles extending up from the shoulder; the COVER is conical with a small flat cap. *See* KALPIS.

Cintra glass. A type of ART GLASS developed by FREDERICK CARDER at STEUBEN GLASS WORKS before 1917. It was made by picking up (from a MARVER) finely sifted powdered coloured glass on a molten matrix (PARAISON) and then, when the object had been formed into the desired shape, covering it with very thin (but occasionally thick) clear glass to

Christian glassware. Mould-blown Maltese Cross bottle, Syrian, 4th/5th century. Ht 14·0 cm. Römisch-Germanisches Museum, Cologne.

cinerary urn. Roman glass. Museo Civico, Adria.

Cintra glass. Vase designed by Frederick Carder for Steuben Glass Works, *c.* 1917, with blue and white stripes and black applied decoration. Ht 24 cm. Corning Museum of Glass, Corning, N.Y.

cistern. Vessel with gilded applied masks, rim, and twisted handles. Venetian, first half of 16th century. Ht 18 cm. Victoria and Albert Museum, London.

embed the partly fused particles. The thin specimens are usually without any air bubbles, but thicker ones have controlled or random bubbles. Such objects were made in monochrome or shaded glass, sometimes with wide vertical stripes of alternating colours (usually blue and pink or blue and yellow), produced by embedding the stripes of powdered glass in consecutive operations. Sometimes there is additional decoration in the form of a black- or blue-glass ring applied around the rim, pulled feathery borders, or applied coloured decoration on the neck and shoulder of a vase. The name was derived from the verb 'sinter', meaning to cause to become a coherent mass by heating without completely melting. A type of glass recorded by Carder *c.* 1921 and called 'Or Verre' may have been a variety of striped Cintra glass. *See* CLUTHRA GLASS; QUARTZ GLASS.

ciocche (Italian). Literally, bouquet of flowers. A Venetian CHANDELIER, from the 18th century, made on a metal frame and elaborately decorated with clear and coloured soda-glass flowers, fruits, leaves, etc. made AT-THE-LAMP. The chandeliers are usually large, 2·00–2·50 m. in diameter. The leading maker was GIUSEPPE BRIATI of Murano.

circus bowl. A glass BOWL (sometimes a JAR or BEAKER) with encircling decoration depicting scenes of a Roman circus, with gladiators, quadrigae, etc. The MOULD-BLOWN bowls of ROMAN GLASS were made in the 1st century AD. On some the scenes are mould-blown, with added inscriptions. Others, from the 3rd/4th centuries, have the scene in WHEEL-ENGRAVING, despite the difficulty of engraving on such thin glass. The German term is *Zirkusschale.*

cire perdue (French). Literally, lost wax. A process originally used in casting bronze pieces, and later for glassware. The model was carved in wax, then encased in a mould, and heat was applied to cause the wax to melt and run out, after which the mould was filled with molten glass or reheated powdered glass. It was used for pieces too detailed to be effectively produced by the usual process of making in a mould. The basic method was later used for glass by RENÉ LALIQUE and more recently by EDRIS ECKHARDT. The German term is *Wachsausschmelzverfahren.*

A more complicated process was developed by FREDERICK CARDER in 1933/34 and used by him until 1959; it required 8 steps: (1) an object was made in clay as a model; (2) a replica was made in plaster of Paris; (3) a sectional gelatin mould was made from the plaster replica; (4) a wax replica was made from the gelatin mould; (5) a ceramic mould was made over the wax replica; (6) this was dipped into boiling water, so that the wax melted out; (7) small pieces of glass rods were put into the mould which was then fired so that the glass entered each hollow of the mould; and (8) the mould and glass were annealed and the mould then broken away, leaving the glass casting of the original clay model which was a separate ornament, or a figure or a handle attached to a glass bowl or vase, or an ornamental piece to be attached to an object of DIATRETA GLASS.

cista (Italian). *See* FRUIT BOWL, *Venetian.*

cistern. A large receptacle for water, sometimes in the form of a deep basin. Generally made of maiolica, some were made of glass in Venice in the 16th century.

clapper. A glass-maker's tool with two wood arms joined at one end by a curved leather hinge, with a small aperture on one arm to fit around the STEM of the glass that was being made. It was used to squeeze a blob of molten glass in making the FOOT of the glass. *See* WOODS; TOOLS, GLASS-MAKERS'.

claret glass. A medium-size wine glass on a stemmed FOOT, intended for drinking Bordeaux red wine.

claw-beaker. A conically shaped BEAKER of green or blue coloured glass that is decorated with two or three encircling rows of hollow

PRUNTS extended to form claws or probosces, the tails of each upper row falling between the heads of the next lower row. Each claw was applied as a blob of hot glass attached to the cooled body of the piece, still kept on the blow-pipe. The blob, which warmed the adjacent part of the wall, was blown (it has been suggested that a separate special blow-pipe was used) from the inside so as to inflate the bulb, and this was then, by use of a hook or pincers, brought outward and downward, and its tip then pressed against the body. The claws usually have further decoration in the form of notched trails of coloured glass, and they are hollow so that the liquid contents flows into them. The beaker tapers downward to a small circular base with a FOOT-RING, and just below the wide everted mouth and above the foot-ring are wide bands of encircling glass trailing. Such beakers were made in Germany of FRANKISH GLASS from the 5th to the 8th centuries; they have been found in England and Scandinavia as well as in western Germany. The early examples are low (15 cm. high) and later ones are taller (25 cm.). They are also called 'trunk beakers'; the German term is *Rüsselbecher*. *See* CASTLE EDEN CLAW-BEAKER; DOLPHIN BEAKER.

claw-beaker. Green glass with hollow claws, Frankish glass, 5th century. Ht 19 cm. Römisch-Germanisches Museum, Cologne.

clear glass. Colourless TRANSPARENT glass. Ancient glass usually termed 'clear glass' is generally tinged with green or yellow, due to impurities in the materials.

Clichy Glassworks. A glasshouse founded by M. Rouyer and G. Maës, at Billancourt (Pont-de-Sèvres) in 1837 and moved to Clichy-la-Garenne (a suburb of Paris) in 1839 (or possibly at Sèvres, *c.*1838, and moved to Clichy-la-Garenne *c.*1844). By 1884 M. Clemendot was associated with the factory. From 1846 to 1857 it made PAPER-WEIGHTS of the highest quality, and also used MILLEFIORI CANES on objects other than paper-weights. It was taken over in 1885 by the Lander family that had made coloured glass, and became known as the Cristalleries de Sèvres et Clichy. *See* PAPER-WEIGHTS, *Clichy*; CLICHY ROSE.

Clichy rose. The slice (set-up) of a CANE made at the CLICHY GLASSWORKS which resembles an open rose. It is a FLORET cane depicting a stylized rose with a central cylindrical motif surrounded by flattened tubes of glass. The colours range from shades of pink to white, and also include yellow, blue, green, wine-red, and amethyst. It was perhaps adapted from a Venetian plaque dated 1846 or Bohemian PAPER-WEIGHTS. Examples are often used in Clichy paper-weights as a central motif or scattered among the canes of a MILLEFIORI weight.

cloche. A hollow dome-shaped or bell-shaped object used as an airtight cover. It usually has a FINIAL at the top to serve as a handle. They were made of NAILSEA GLASS. *See* DOME.

clock case. An ornamental piece to accommodate a clock, usually made of faience or porcelain at most of the important European factories, but also occasionally made of glass.

clock lamp. A device for telling the time in the dark by observing the amount of oil consumed from a graduated glass container or tube. It has a brass stand supporting the oil container; from the bottom of the container a holder for a wick projects horizontally. The light from the wick permits the 'hour-scale' on the oil container to be read. Examples were made in England *c.*1730.

cire perdue. Figure of a ram, colourless glass cast by lost-wax process, designed and made by Frederick Carder for Steuben Glass Works, dated 1940. Ht 24·5 cm. Corning Museum of Glass, Corning, N.Y.

club-shaped. The form of a DRINKING GLASS, especially a HUMPEN, that is tall and circular, sloping inward slightly from halfway below the MOUTH to near the bottom and then slightly outward to the BASE. *See* KEULENGLAS; IGELGLAS.

Cluny Goblet. A glass GOBLET generally attributed to JACOPO VERZELINI. It bears the initials *A : MM : DLP*, and the date 1578, and is engraved with the French *fleur-de-lis*. It was found at Poitiers, and is so-called because it is in the Cluny Museum, Paris. It has been suggested that it may have been engraved by ANTHONY DE LYSLE, while still in France, or brought later from England to France with others taken to be French, or made by another in France.

Clutha glass. A type of glass that preceded the ART NOUVEAU style, mainly greenish but also sometimes turquoise, yellow, brown-green or smoky black, with embedded air bubbles, streaks of pink and white, and speckles of AVENTURINE. It was patented in the 1890s by JAMES COUPER & SONS, Glasgow, and some pieces were made from designs by CHRISTOPHER DRESSER and GEORGE WALTON, who were influenced by Japanese motifs. The name 'Clutha' has been said, incorrectly, to have been derived from the word 'cloudy'; it is the Gaelic name for the River Clyde. The glass has sometimes been confused with the 'Old Roman Glass' made by THOMAS WEBB & SONS from 1888, but it is distinguishable by the speckles of aventurine in the METAL. *See* WEST LOTHIAN GLASSWORKS.

Cluthra glass. A type of ART GLASS developed before 1930 by FREDERICK CARDER at STEUBEN GLASS WORKS. It was made in the same manner as CINTRA GLASS and was basically similar, but characteristically it has larger air bubbles of various sizes and also it was made with larger unsifted glass particles and CASED with heavier clear glass. The random bubbles were produced by mixing with the powdered glass a chemical (probably potassium nitrate) which reacted to the heat by expanding to form the bubbles, and hence requiring a promptly applied heavier casing to control their size and spacing. The glass body is monochrome or of shaded colours. Such glass was often used for lamp bases with acid-etched designs. The origin of the name is not known, but was probably not related to 'cloudy' (*see* CLUTHA GLASS), as Carder surmised. A glass simulating Cluthra glass was made by KIMBLE GLASS CO.

Cluthra glass. Pilgrim flask, green shaded to white, with bubbles and clear casing, designed by Frederick Carder for Steuben Glass Works, *c.* 1930. Ht 25 cm. Rockwell Collection, Corning Museum of Glass, Corning, N.Y.

coach-horn. A glass ornament, up to one metre long, made in the form and size of a coach-horn. They were made of NAILSEA or BRISTOL GLASS, usually of clear or deep-blue colour. *See* BUGLE; POST-HORN.

coaching glass. A type of drinking glass with a BOWL of inverted conical shape (like a FLUTE GLASS), no HANDLE, and, instead of a FOOT, a KNOP or ball at the end of the short STEM. They were intended to be emptied by the drinker and then stood inverted on the rim, like a STIRRUP CUP. It has been suggested that they were used for service to a coach at a wayside inn or were carried in a coach.

coaster. A small circular tray on which to rest a BOTTLE, DECANTER, or a DRINKING GLASS. They were used in England from the mid-18th century, when the table-cloth was removed from the dinner-table before dessert (i.e. fruit and nuts) was served and the bottle or decanter of port was circulated by sliding it along the table to successive diners; the coaster was placed under it to prevent its scratching the table, hence it was made originally with a wooden or *papier-mâché* bottom. Modern glass coasters are small, intended to be used under a drinking glass.

cobalt. A metallic agent used in colouring glass, producing dark and light hues of blue, including the very dark blue generally referred to in England as BRISTOL BLUE. It is used as oxide of cobalt (ZAFFER) and when mixed into molten glass that is then cooled and ground, the product is known as SMALT; this is used as a colouring agent for glass and ENAMEL, and for pottery and porcelain glazes and underglazes. Glass coloured with cobalt is very dense, transmitting little light. It was used as a DECOLOURIZING AGENT at the end of the 18th century.

cocktail glass. A small drinking glass for cocktails; it has a stemmed FOOT, with a BOWL, the shape and size of which varies, but is usually funnel-shaped.

coffee-pot. A covered receptacle, made in varying forms, sizes, and styles, for serving coffee. Examples usually have one loop HANDLE and on the opposite side a SPOUT. They are generally pear-shaped or in the form of a truncated cone. Generally made of porcelain or pottery, some have been made of glass.

coffin. A small model funerary object in the form of a coffin (sometimes anthropoid) such as were made in Egypt in the XVIIIth Dynasty (*c.* 1475 BC) to be buried with the deceased, comparable to a USHABTI. One example was made on a solid core, with white trailing and applied figures of funerary deities on blue glass. Sometimes such coffins held a miniature figure with the inscribed name and titles of the deceased. See *British Museum Quarterly*, No. 30 (1966), p. 105.

cogwheel base. A type of BASE, mainly seen on low BEAKERS or TUMBLERS of the form known as RANFTBECHER, on the edge of which there is encircling deep cutting in the form of a cogwheel. It is also found on some English CELERY DISHES.

coil-base. A type of base on a vessel made by forming a ring (on which the vessel stands) around the bottom by TRAILING molten glass. *See* BASE-RING.

coin glass. A drinking glass, tankard or jug having a coin (or several, as many as seven being recorded) embedded in the hollow KNOP of the STEM or in the FOOT. The date of the coin could be earlier than the date when the object was made, but it establishes the earliest possible date of manufacture. The dates on coins in known examples of English glass range from 1679 (made later) to 1776 and, in an Italian glass, 1647. Such coin glasses are still made to commemorate some important public event, e.g., the Coronation in 1953 of Elizabeth II. The process for inserting the coin was to put it into the knop before the knop was attached to the stem. Some early Roman TEAR BOTTLES contain coins from 27 BC to AD 68, but they may have been inserted later.

Colbert, Jean-Baptiste (1619–93). The French statesman and financier who in 1665 established a glass factory, the Manufacture Royale des Glaces de Miroirs, in the Faubourg St-Antoine in Paris to make, under a monopoly, the MIRRORS to be installed in the Salle des Glaces at Versailles. The company was in 1667 united with a glassworks at Tourlaville, in Normandy, set up by RICHARD LUCAS DE NEHOU, and in 1695 that company was combined with others to form the predecessor of the present ST-GOBAIN GLASS COMPANY.

cold colours. Lacquer colours or oil paint applied as decoration to glassware, as to porcelain, without firing. Often much of this type of decoration has worn off old pieces. Such colours are more effectively used when applied to the back of the surface to be viewed and protected by a layer of varnish or metal foil, or by another sheet of glass. The process was sometimes used on HUMPEN and other large glasses of WALDGLAS, possibly because the glasses were too large for convenient firing in the MUFFLE KILN, and also because the METAL was not sufficiently durable to withstand a second firing, and the painting was done by peasants who had no furnace available. It was also used at HALL-IN-TYROL. The German term is *Kaltmalerei*, the Italian is *dipinto freddo*.

cold cutting. The process of cutting a glass vessel or object from a raw block of glass which has not previously been cast into a shape. The technique was used from early times but became more common from the 8th century BC. Evidence of the use of such a process is the existence of internal spiral grooving made by the tool. *See* EGYPTIAN GLASS; SARGON ALABASTRON; STAFF HEAD.

cogwheel base. Ranftbecher enamelled by Anton Kothgasser, Vienna, *c.* 1830. Ht 10·7 cm. Kunstmuseum, Düsseldorf.

Colinet (Colnet) family. The owners of a glasshouse at Beauwelz, near Mons (Belgium), *c*. 1506–75. The glasshouse made wares in the *façon de Venise*, and Catalogues of the period 1550–55 and the *Journal d'Amand Colimet* (1574) contain sketches of a NEF EWER and several types of drinking glass, such as were made at Murano in the 16th century, and of numerous other Venetian, French, and German shapes.

collar. A circular ring around the STEM of a glass. It is especially to be found on certain English WINE-GLASSES where it is added just under the BOWL to disguise the weld-mark of a three-piece glass whose stem was made separately and joined to the bowl; also, it sometimes occurs at the bottom of the stem where it joins the FOOT, or between larger KNOPS (with sometimes several, up to six, on a single stem). *See* MERESE.

cologne bottle. A type of bottle for holding eau de cologne (toilet water), to be kept on a dressing table. It is larger than a SCENT BOTTLE and has a glass STOPPER rather than a metal cap. They are usually of circular, oval or square section. Many made for expensive scents are ornamental; some made in England, in the second half of the 18th century, have SULPHIDES embedded in one side, and some made by FREDERICK CARDER for STEUBEN GLASS WORKS, *c*. 1928, are of CINTRA GLASS with heavy casing or are of uncoloured crystal glass with engraved decoration. Simpler examples made in the United States in the 19th century were (1) tall with panelled sides and a long tapering neck, or (2) MOULD-BLOWN with a wide range of decorative patterns and shapes; they were of coloured or colourless glass, and produced by the BOSTON & SANDWICH GLASS CO. and other factories. *See* SMELLING BOTTLE; TOILET-WATER BOTTLE.

colour etching. The process of decorating glassware by applying a thin FLASHING of coloured glass over clear glass and then ETCHING the surface glass so that a pattern appears in clear glass. It was used on BOHEMIAN GLASSWARE.

colourband glassware. A glass BOTTLE or other receptacle decorated with encircling horizontal loops of wide stripes or bands of contrasting colour.

coloured glass. Glass that is coloured by (a) impurities in the basic ingredients in the BATCH (e.g., BOTTLE GLASS), or (b) techniques of colouring TRANSPARENT clear glass by three main processes: (1) by use of a METALLIC OXIDE in the solution in the BATCH to impart a colour throughout; (2) by use of a substance in a colloidal state, e.g., GOLD RUBY GLASS with microscopic particles of gold; and (3) by use of embedded particles of coloured material, e.g., AVENTURINE GLASS. A different category of coloured glass is CASED (FLASHED) GLASS which is coloured by an OVERLAY of coloured glass or glasses over uncoloured or differently coloured glass, or vice versa. Coloured glass was made from earliest times in Egypt, and was preferred there to uncoloured glass and in all regions making ROMAN GLASS until the development of CRISTALLO GLASS in Venice in the 15th century and LEAD GLASS by GEORGE RAVENSCROFT in England, *c*. 1676. Its use was revived from the early 19th century onwards, and especially by makers of glass in ART NOUVEAU style.

colouring agents. Various METALLIC OXIDES used to impart a colour throughout TRANSPARENT or OPAQUE glass. *See* COLOURED GLASS.

column candlestick. A type of CANDLESTICK which has its nozzle on a support in the form of a column. *See* FIGURE CANDLESTICK.

columnar vase. A small vase in the form of a column, cylindrical or almost so, and sometimes with a decorative ring projecting around the neck representing the capital. They were made of CORE GLASS, usually about 10 cm. high and 2·5 cm. in diameter, with a slightly splayed base.

combed decoration. Alabastron, dark-blue core glass with combed decoration, Syrian, 4th/3rd century B C. Ht 13·7 cm. Kunstmuseum, Düsseldorf.

Examples from Egypt, *c.* 1450–1350 BC, were decorated with encircling or COMBED coloured TRAILING. Later examples, from the 4th/3rd centuries BC, are in rare cases decorated with white and coloured trails placed alternately horizontally and vertically.

combed decoration. A wavy, festooned, feathery, or zigzag pattern of decoration in two or more colours which was produced on glass objects, by applying to the molten object threads of opaque glass of a colour different from that of the body which, by MARVERING, were rolled into the body until level, after which the surface would be combed or dragged to produce the desired decorative effect. Sometimes encircling threads were combed alternately upward and downward, producing a 'palm-leaf' pattern. (The process is similar to combing a pattern on pottery, where a comb is pulled through the glaze.) The technique was used on core-made vessels made in Egypt, *c.* 1450–1350 BC, and elsewhere, *c.* 400–100 BC, in the form of small Greek vases such as the ALABASTRON, OENOCHOË, AMPHORISKOS, and ARYBALLOS. The process of dragging the design was done by pulling a toothed metal tool in the molten glass, and it is said that sometimes metal pins were inserted to hold in place the parts not being dragged. The Italian terms are *vetro a pettine* (or *a piume*). Such combing is found on some 16th-century Venetian glassware referred to at MURANO as VETRO FENICI, but often erroneously called LATTICINO.

combed fluting. A decorative pattern on CUT GLASS in the form of closely spaced vertical flutings (SPLITS) level at the bottom and having a point at the top of each split, thus resembling a comb pointed upward. It was used encircling the base of English DECANTERS, JUGS, etc. in the 1760s.

comfit glass. A small SWEETMEAT GLASS, usually used for serving so-called 'dry sweetmeats' or 'comfits', e.g., chocolates, nuts, cachous. They were about 7 to 10 cm. high. often with a rudimentary STEM or no stem, and a plain or ornamented BOWL.

Commedia dell'Arte figures. A series of FIGURES, up to sixteen in number, depicting characters from the Italian Comedy, or *Commedia dell'Arte.* The principal characters were often masked. The Comedy was played by travelling troupes, and existed only as a scenario, the dialogue and stage-business being largely improvised. It was at the height of its popularity from the 17th century to *c.* 1760. The figures were often taken, when depicted on porcelain or glassware, from Luigi Riccoboni's *Histoire du Théâtre Italien,* 1728. They have been used as ENAMELLED decoration on some Venetian glassware of the late 16th century.

commemorative glassware. Glass objects, of great variety, that are decorated to commemorate some event, person or cause, by a portrait, scene or inscription. Roman glassware shows gladiatorial events and the names of contestants. Islamic Glass bears the name of the Sultan (*see* MOSQUE LAMP). From the 16th century it became especially popular in England, Holland, and Germany to record on glass objects historical, political, and other events. The methods of decoration include ENGRAVING, ETCHING, STIPPLING, ENAMELLING, and GILDING. Among the subjects depicted are: English and foreign sovereigns and lesser rulers; statesmen and military leaders; battles; treaties; inaugural events (e.g., opening of Sunderland Bridge, 1796); battleships and regiments; guilds; clubs (e.g., MASONIC WARE) Parliamentary and local elections; family and personal events; sporting events. Sometimes the significance is concealed in symbolism. *See* ARMORIAL GLASSWARE; COIN GLASS; MARRIAGE BOWL; MARRIAGE GOBLET; CANADA, GREAT RING OF.

———, *family events.* Glass objects with engraved or enamelled designs commemorating births, betrothals, marriages, anniversaries, deaths. They were made in England and on the Continent. For English examples, *see* DIER GLASS; BUGGIN BOWLS; for Italian, BAROVIER CUP; BOLOGNA GOBLET; for German, FAMILIENHUMPEN.

columnar vases. Dark and light blue, yellow, and white with blue interior. Pre-Roman 4th–3rd centuries BC. Ht 10 cm. Museo Civico, Adria.

Commedia dell'Arte figure. 'Il Capitano', Murano, second half of 16th century. Kunsthistorisches Museum, Vienna.

concave beaker. Opaque black Roman glass, 1st century AD. Ht 10·5 cm. Römisch-Germanisches Museum, Cologne.

——, *military and naval.* Glass objects with engraved designs commemorating victories (e.g., Waterloo, Trafalgar), fighting ships (*see* PRIVATEER GLASSWARE), regiments, Peace Treaties (e.g., Utrecht, 1713; Paris, 1815). *See* WESTPHALIA TREATY HUMPEN; PRESENTATION GLASSWARE.

——, *portraits.* Glass objects bearing the engraved portrait of a prominent person, e.g., Lord Nelson, the Duke of Wellington, Napoleon III, Disraeli, George Washington, Lincoln; *see also* Royalty, below, and PORTRAIT GLASSWARE; JACOBITE GLASSWARE; WILLIAMITE GLASSWARE.

——, *Royalty.* Glass objects commemorating English and Continental royal events, e.g., births, marriages, coronations, jubilees, and deaths. For an early Egyptian example, *see* TUTHMOSIS JUG. For English examples, *see* CORONATION GLASSWARE; EXETER FLUTE; ROYAL OAK GOBLET.

companion piece. One of a PAIR, SET or GROUP of related or matching objects.

compatibility. The quality of two or more PARAISONS of glass that have similar coefficients of expansion, usually achieved by use of paraisons from the same basic BATCH. It is important in making CASED GLASSWARE or glassware with enclosed ornamentation (e.g. PAPER-WEIGHTS), as otherwise the piece will crack when subjected to even a slight change of temperature or mechanical shock. *See* PULL TEST.

comport. (1) A stand or SALVER on a heavy flared FOOT, to be placed on a dinner-table to hold fruit or JELLY GLASSES. They were made in England in the 18th century. (2) The term used in the United States for a *compôtier.*

composto, vetro (Italian). A type of glassware developed by VENINI made by a thin flashing of coloured glass over a thin glass of a different colour.

compôtier (French). A bowl, usually on a stemmed foot, for serving *compôte* (stewed fruit); often loosely called a *compôte*. Such receptacles usually have a cover. Called in the United States a comport.

concave beaker. A type of BEAKER with its side curved inward from the RIM to the bottom, and which rests on a low base about half the diameter of the rim. They were made of dark green, almost black, ROMAN GLASS in the 1st century AD.

concentric. A style of decoration of circles having a common centre, such as sometimes used with CANES around a central motif in a PAPER-WEIGHT.

cone-beaker. A drinking vessel of inverted conical shape. Examples were made of ROMAN GLASS in the 4th century AD, and similar taller vessels were made later of FRANKISH GLASS. They could be rested only on the rim, but did not necessarily have to be drained at one draught but possibly could have been used by the drinker while standing or reclining and then handed to a servant. Most examples taper to a rounded point (sometimes flattened) but some have a rounded bottom. They are sometimes decorated with glass trailing, often horizontally near the rim and with continuous vertical WOLF'S TOOTH loops towards the point. Some have JEWELLED DECORATION. Some examples have a thin attenuated tip, and may have been used as oil lamps suspended in a POLYCANDELON, but one such bears a drinking inscription. The German term is *Trichterbecher. See* KEMPSTON BEAKER; SPITZBECHER.

Conradsminde Glassworks. A small Danish glassworks founded in 1834 near Aalborg, in Jutland, and operated until 1857. It made glass of a cobalt-blue tint that is now rare, and ware of colourless glass.

cone beaker. Lamp or drinking vessel, yellowish glass with blue blobs, Roman glass, 4th/5th century AD. Ht 16 cm. Wheaton College, Norton, Mass.

constable glass. A very large English goblet-shaped glass with a funnel-shaped bowl, ranging up to 38 cm. high. It was used as a LOVING CUP on ceremonial occasions or for serving spirits to be transferred into small, often conforming, wine glasses. They were made from *c*. 1760 but lost popularity when the RUMMER came into use in the late 18th century. Also termed a 'serving rummer'.

conterie (Italian). Small beads of glass, globular or oval, smooth or irregular, of varied colours, that are sold on strings and in mass quantity, sometimes to be remelted and used to make many types of glass objects. They are made from a thin glass tube cut into short segments and then rounded by heating. The ancient (and now obsolete) Venetian term for such beads was *margarite*.

contracted neck. The NECK of a vessel that is narrow in proportion to the BODY.

cooler jug. A type of DECANTER JUG with an indented well for ice to cool, without diluting, the surrounding wine. They were made in Germany in the 18th century. The German term is *Kühlflasche*.

Copier, Andries Dirk (b. 1901). An outstanding Dutch glass-designer who became associated with the LEERDAM glassworks in 1914, creating tableware. In 1923 he became the Director of Leerdam's Unica Studio, and extended the development of modern design in household ware and also the production of single-piece ornamental glassware.

copper. A metallic agent used in colouring glass, producing turquoise blue or green, depending on the manner of use. In a REDUCING ATMOSPHERE (predominantly carbon monoxide), it produces copper-red. The blue is a TRANSMUTATION COLOUR that shows as red when seen by transmitted light; *see* FAIRFAX CUP. Copper was used to produce an opaque red glass in Egypt, and a red glass in medieval times. Tiny flakes of copper are suspended in gold AVENTURINE GLASS.

Coralene. A style of decoration on glass that has the appearance of sprays of coral extending over the surface of an object. The process involved painting in enamel the coral-like forms, and then applying to the enamelled lines tiny glass beads (colourless, coloured or opalescent), which were then fused on to the glass. Similarly produced glassware had the design in a pattern of fleurs-de-lis, herringbone lines, sheaves of wheat, etc. It was used on glassware of many forms and colours by the MT. WASHINGTON GLASS CO., which so named it, and by many other factories of the United States and Europe.

cord. A striation or line that is a defect in glassware. It may be caused in several ways: by a variation in the temperature of the furnace; by the unequal density of the materials used; or by some silica in the composition of the pot breaking off and fusing with the BATCH. When making high-quality CRYSTAL glass, it is avoided by stirring the batch.

cordial glass. A small drinking glass (about 1 to $1\frac{1}{2}$ oz. capacity) for cordials, liqueurs, and *eaux-de-vie*. They were usually in the form of wine glasses but smaller, with a BOWL (bell-shaped, conical or bucket-shaped) on a tall stem (plain, baluster, air-twist, opaque twist, colour twist, cut or Silesian) resting on a plain, domed or folded foot. They were used in homes or in taverns, from *c*. 1663 to *c*. 1883. A similar glass was made of enamelled OPAQUE WHITE GLASS in Bohemia in the 18th century. The type is often called a 'liqueur glass'.

core. (1) A glass CANE placed vertically in the centre of the STEM of a drinking glass and then enclosed within a spiral TWIST. There are many variations, e.g., the core may be a single coloured or opaque thread or a gauze twist, corkscrew twist, gauze corkscrew twist, spiral twist, ribbon twist, etc., and the surrounding twist may be of one or more ribbon spirals,

cooler jug. Jug with well for ice, Germany, 18th century. Ht 24 cm. Kunstsammlungen der Veste Coburg.

cordial glasses. English, *c*. 1765. Ht 16/17 cm. Courtesy, Delomosne & Son, London.

one or more gauze twists, or a 2-ply (or up to 15-ply) spiral. (2) The substance upon which CORE GLASS is made.

core glass. A type of glass object that was made from *c*. 1500 to *c*. 1200 BC in Mesopotamia and in Egypt, and thereafter in Mesopotamia, Egypt, and the Eastern Mediterranean region. Production declined between *c*. 1200 and *c*. 800, but revived from *c*. 800 to the 1st century BC. The process (called 'core winding') involved first making a CORE, early examples probably being of mud and straw (D. B. Harden, of the British Museum, in 1968 rejected – as now 'certain' – the possible use of sand, as had formerly been stated), but later possibly of clay, and then moulding it into the form of the desired piece. The core was probably then rotated while MOLTEN GLASS was TRAILED, by another metal rod (DIPSTICK), closely around the core until it was covered (except for small gaps left for expansion); then the glass was MARVERED and more glass was trailed onto it to obtain the desired thickness. The exterior was smooth, with no evidence of MOULDING. When the glass had been ANNEALED, the core could be readily removed (leaving on some examples a rough interior surface). It has been said that an alternative method was merely to dip the core into molten glass until it was sufficiently covered, then remove the core when the glass cooled, and polish the object. Sometimes glass of more than one colour was used, often two or three hues, and decoration was usually enhanced by COMBING. The pieces are generally in the form of small JUGS, FLASKS or BOTTLES for perfume or ointments (*see* UNGUENTARIUM). Some were FISH BOTTLES, but many were in the form of a COLUMNAR VASE or of GREEK VASES, e.g. the ALABASTRON, OENOCHOË, AMPHORISKOS, ARYBALLOS or AMPHORA. With the introduction of glass BLOWING by the Syrians, *c*. 100 BC, the method ceased to be used. It has been questioned how the core could be removed from vessels with a small mouth, but as sand was not, according to recent opinion, used for the core, it would seem that the core could have been picked out. Such pieces are sometimes still called 'sand-core glass', from the formerly held view that the core was made of sand. Recent experiments by DOMINICK LABINO in 1962/63 have suggested the method of production employed. *See Journal of Glass Studies* (Corning), VIII (1966), p. 125.

Cork glassware. Glassware made at three glassworks in Cork, Ireland: (1) Cork Glass Co., established in 1783 by Atwell Hayes, Thomas Burnett, and Francis Rowe, and closed in 1818. Numerous vessels, mostly DECANTERS, survive, marked on the bottom 'Cork Glass Co', with either a wheel-engraved or cut VESICA or other motif. Cutting is generally flat, unlike deep-cut WATERFORD GLASSWARE, and the METAL is generally good but not superior. A blue tint, often erroneously attributed to Waterford glassware, is frequently found. (2) Waterloo Glass House, established by Daniel Foley *c*. 1815 and operated by him until he retired and was succeeded by Geoffrey O'Connell in 1830; it closed in 1835. The products were considered to be imitative of those of Cork Glass Co., but some pieces bear a poorly engraved bowknot mark. (3) Terrace Glass Works, established by Edward and Richard Ronayne in 1818 and closed in 1841. The products are not identifiable. *See* IRISH GLASSWARE.

Cork glassware. Jug with moulded flutes and engraved decoration, mark 'Cork glass Co.', 1820. Ht 16·5 cm. Courtesy, Delomosne & Son, London.

Corning Glass Works. A multinational glassworks and a leading producer of speciality, industrial, technical, and domestic glassware: in 1918 it acquired the STEUBEN GLASS WORKS (now Steuben Glass, Inc.). It grew from a glassworks at Cambridge, Massachusetts, in which Amory Houghton purchased an interest in 1851. In 1864 Houghton purchased the Brooklyn Flint Glass Co. which was moved in 1868 to Corning, New York (the present headquarters), renamed it the Corning Flint Glass Co., and reincorporated it in 1875 as the present Corning Glass Works. It established in 1951 the Corning Glass Center, housing the Steuben Glass Works, the Corning Museum of Glass (ancient to modern specimens), and the Hall of Science and Industry (displays of the manufacturing processes and uses of glass). *See* PYREX; PYROCERAM; PHOTOCHROMIC GLASS; HOUGHTON, ARTHUR AMORY, JR.

cornucopia. A receptacle shaped like a horn, to be used as an ornament or sometimes as a wall-pocket (a flower vase to be attached to a wall). Some were made of OPAQUE WHITE GLASS, *c.* 1740. The French term is *corne (d'abondance)*. *See* CORNUCOPIA VASE.

cornucopia vase. A glass vase, made in the shape of a CORNUCOPIA, attached at its tip to a gilded bronze or marble stand, with the tip of the glass sometimes mounted within a gilded bronze miniature ram's head. The vase is usually FACETED and has a scalloped rim. Examples have been made by the BACCARAT GLASS FACTORY since the 19th century.

coronation glassware. A type of COMMEMORATIVE GLASSWARE to commemorate a coronation in England. The earliest example was perhaps the EXETER FLUTE to commemorate the coronation of Charles II (1660). After a lapse for James II such glasses were resumed for William and Mary, William III and thereafter for every Sovereign except Edward VIII, the most recent being for Elizabeth II in 1953. They were made in the styles of the various periods, and usually bore the royal arms or cypher and date, with a crown. Many have been made by STEVENS & WILLIAMS LTD. In addition to drinking glasses, other varieties of glassware were made.

cornucopia vase. Vase with gilded bronze mount, Baccarat, mid-19th century. Ht 31 cm. Kunstgewerbemuseum, Cologne.

corroso, vetro (Italian). Literally, corroded glass. A type of ICE GLASS made by the use of HYDROFLUORIC ACID, applying it to the glass after the surface has been covered with a type of resin that crackles when dry, so that only the crevices are affected by the acid and the covered area left smooth. Such ware has irregular mat patches of lines and rough spots, sometimes covering the entire surface. It is a modern technique developed at Murano by VENINI, *c.* 1933.

Costa, Da (fl. 1673). An Italian glass-maker who was engaged by GEORGE RAVENSCROFT to assist him in making glass in the style of Venetian glass and in developing an improved substitute.

country-market glassware. A type of peasant glassware made for country and artisan markets, probably for sale at fairs as gifts or souvenirs. It included pieces made of MILK-AND-WATER GLASS, and early glass from NAILSEA, e.g., ROLLING PINS, DOOR-STOPS and simple PAPER-WEIGHTS. Some was painted with such inscriptions as 'A trifle from . . .'. The decoration was usually in unfired COLD COLOURS and unfired GILDING, much of which has since rubbed off.

Couper, James, & Sons. *See* WEST LOTHIAN GLASSWORKS; CLUTHA GLASS.

cover. An unattached top (as distinguished from a LID, which is usually hinged) for closing the mouth of a jar, vase, pot, bowl, or other open vessel. It varies in shape, being flat, conical, or dome-shaped, and is usually surmounted by a decorative HANDLE, termed a FINIAL (sometimes loosely a KNOP). A cover usually fits within the RIM, but rare examples fit over the rim and rest on an applied ring. The French term is *couvercle*. *See* DOME COVER; CAP COVER; SCREW CAP; STOPPER.

cracking-off. The process of removing a glass object, such as the BOWL of a WINE-GLASS, from the BLOW-PIPE after it has cooled, as distinguished from the process of shearing (*see* SHEARED GLASS). Originally this process was done manually, by circumscribing the object (with a diamond or wet file) with a line and then either giving a sharp tap so that it breaks off cleanly or (at a much later period) heating the glass with a gas flame so that the tension breaks off the glass along the scratched line. The glass left on the blow-pipe is called the 'moil'. In

cover. Bowl and cover with acorn finial, English, *c.* 1700–10. Courtesy, Delomosne & Son, London.

recent years a machine has been developed for the purpose; some cut off the bowl of a number of glasses simultaneously and accurately. The process is sometimes called 'wetting-off'.

crackle glass. The same as ICE GLASS.

Crama, Elizabeth. A Dutch glass-engraver, of the last quarter of the 17th century, who decorated GOBLETS with DIAMOND-POINT ENGRAVING and CALLIGRAPHY. Her signature sometimes appears on the cover of the goblet rather than in the usual position on the foot. Only five examples have been identified. Some unsigned English glassware so decorated has been attributed alternatively to her or to WILLEM JACOBSZ VAN HEEMSKERK. See *Journal of Glass Studies* (Corning), XII (1970), p. 136.

cranberry glassware. Transparent glass objects of a reddish-pink colour made, in a great variety of forms and styles, in both England (especially at Stourbridge) and the United States from the mid-19th century, and originally so-called in the United States. Early examples are generally FRIGGERS, drinking glasses, and other useful ware, made with little decoration, but later ornamental pieces, e.g., vases and bowls, were made, decorated with TRAILING or ENAMELLING, and some with white OVERLAY (or clear glass with cranberry overlay) cut to reveal the inner layer. Often colourless glass handles and stems were attached to coloured bodies. The colour is lighter than that of GOLD RUBY GLASS, and the ware is to be distinguished from Bohemian clear glass FLASHED with ruby glass. *See* GREGORY, MARY, GLASSWARE.

cream-jug. Moulded jug, with three feet and lion's-mask bosses, *c.* 1750. Ht 8·5 cm. Courtesy, Delomosne & Son, London.

cresting. Jug with cresting along handle, Venice, 19th century. Ht 17·8 cm. Kunstgewerbemuseum, Cologne.

cream-boat. A small boat-shaped jug used for serving cream, usually having a HANDLE at one end and a pouring LIP at the other, and often resting on a SPLAYED BASE. It is similar, in general, to a SAUCE-BOAT but smaller. Generally made of porcelain or pottery, some have been made of glass.

cream-bowl. A type of circular bowl with steeply sloping sides used for settling milk so that the cream would rise and could be poured off the surface. Such bowls, some made in glass, possibly at Nailsea, *c.* 1790 and later, have a pouring lip on the rim. *See* DAIRY BOWL.

cream-jug. A small JUG or PITCHER for serving cream. It is often pear-shaped, with a single loop handle and a pouring lip, and sometimes has a cover. It is smaller than a MILK-JUG. Generally made of porcelain or pottery, some have been made of glass; rare examples have three feet. See YACHT JUG.

crenellated. In the form of battlements, sometimes called 'castellated' e.g., in the case of the RIM on some glassware with equally spaced perpendicular straight-sided projections. The French term is *crénelé*.

cresting. A decorative ornament usually found projecting from the outer edge of the wings of some WINGED GLASSES, or similar stem ornament of comparable glasses, from Venice or elsewhere of the 17th century. It is in the form of small closely spaced tufts, points or thin semi-circles, PINCERED with close parallel ridges. It has occasionally been used on the handle or the body of a Venetian or Spanish VASE or EWER.

Creusot, Le, Glass Factory. The successor to the ST.-CLOUD GLASSWORKS after it moved in 1787 from Paris to Mont-Cénis near Le Creusot, and which was acquired jointly by the BACCARAT GLASS FACTORY and the ST-LOUIS GLASS FACTORY. It specialized in making cameo incrustations (SULPHIDES), which were later made by Baccarat for PAPER-WEIGHTS. It was sold in 1837 to Schneider & Cie.

Creutzburg, Caspar (fl. 1689). A glass-engraver who worked at Gotha, in Thuringia, c. 1689, at the Court of Dukes Friedrich I and Friedrich II of Saxe-Gotha. One example signed by him is known.

Cricklight. A later adaptation by Samuel Clarke of his patented FAIRY NIGHT LIGHT (trade-mark registered in 1889), popular until c. 1930. It was the same except that the shade was of clear uncut glass so as to emit more light from the enclosed candle than the coloured shades. They were placed on dining-room tables. Some were fitted on CHANDELIERS and CANDELABRA, and on some Worcester porcelain vases and figures, by means of a glass peg that fitted into the candle-nozzle.

crimper. A glass-maker's tool used for decorating the RIM of bowls and other objects to produce a crimped effect.

crimping. (1) The process of decorating glassware by PINCERING to form a wavy EDGE. (2) The process of drawing in molten glass to form a NECK of a vessel. (3) The process of shaping a glass ornament into a decorative design.

crisselling (or **crizzling**). A basic defect in glass caused by an imperfect proportion of the ingredients in the BATCH, particularly an excess of ALKALI, resulting in a network of fine internal cracks and drops of a sour-smelling moisture that forms on the surface. Such glass eventually deteriorates, decomposes, and crumbles. This so-called 'diseased glass' is most commonly found in glass made in the late 17th century in England, Holland, Germany, and China, where there was much experimentation to find a substitute for VENETIAN GLASS. In England it was largely remedied or ameliorated by 1676 by GEORGE RAVENSCROFT, whose remedy was the addition of oxide of lead; in Germany a remedy, developed by JOHANN KUNCKEL, was the addition of chalk. It has been asserted that the defect is more prevalent in heavy glass. In China it occurred on glass from the reign of K'ang Hsi (1662–1722); its absence on glass from the reigns of Yung Chêng (1723–35) and Ch'ien Lung (1736–95) indicates that a remedy had been found.

Creutzburg, Caspar. Detail of bowl of *Pokal* (with mereses) engraved and signed 'Casp. Creutzburg fecit', Thuringia, 1689. Kunstsammlungen der Veste Coburg.

Crisselling must be distinguished from decay of glass resulting from centuries of burial in damp earth; in the latter case only the surface has been affected, due to dissolved carbonic acid in the soil which has drawn the ALKALI from the glass, often producing IRIDESCENCE.

cristallo (Italian). A type of SODA GLASS developed in Venice, perhaps before the 15th century. It was made with sea-plant ASHES or NATRON, and had a pale-yellow, straw-like or smoky-grey colour. By the use of MANGANESE as a decolourizing agent, it was made colourless and thus resembled ROCK-CRYSTAL. It was thin and fragile so that it could not be CARVED, CUT, or STIPPLED, but it was suitable for DIAMOND-POINT ENGRAVING and was so used by artists in Italy, the Netherlands, England, Germany, and Bohemia. It became the standard METAL used at Venice for its glassware decorated with elaborate TRAILING, with enamelling and gilding, and with FILIGRANA decoration.

cristallo. Tazza with lion's head masks, Venice mid-16th century. Ht 15·2 cm. Museum für Kunsthandwerk, Frankfurt.

Cros, Henri (1840–1907). A French sculptor who, with his son Jean, revived the process of making objects, especially large panels in RELIEF, in PÂTE-DE-VERRE toward the end of the 19th century. He carried on many years of research at the Sèvres porcelain factory, and produced a series of coloured reliefs from 1892 to 1903.

cross-cut diamonds. A decorative pattern on CUT GLASS, that is a variation of the RAISED DIAMOND PATTERN, having four short grooves at right angles across each diamond apex. It dates from the 19th century and, being very difficult to produce, is rare.

crown. (1) A fanciful ornamental glass piece made in the form of a crown. Examples were made in Germany or southern Netherlands in

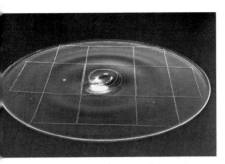

crown glass. Glass marked for cutting into panes, model of 16th/18th century method. Science Museum, London; Crown copyright.

the 17th century. (2) The clear glass dome above the decorative motif of a PAPER-WEIGHT. (3) A particular type of paper-weight; *see* PAPER-WEIGHT, *crown*.

crown glass. Flat PANE GLASS made by the process of blowing a bubble of glass, transferring it from the BLOW-PIPE to a rod, cutting it open, then rapidly rotating it, with repeated reheatings, until, by centrifugal force, it was spread into a large flat disk, up to 4 ft (1·20 m.) in diameter. The glass was then annealed and cut into rectangular or diamond-shaped pieces that were fairly thin but showed slight convexity and concentric wavy lines caused by rotating, leaving the centre boss or 'bull's-eye', where the rod had been attached. The process was known to the Romans in many parts of the Empire, and fragments exist from the Middle East from the 4th century AD, and examples occur in medieval cathedral windows (*c*. 1270–80). The process was later called the 'Normandy method', its invention having been uncertainly attributed to Philippe de Caqueray of Normandy in the 14th century and the process then brought to England – notwithstanding the fact that the process had been brought to England earlier, before 1100, possibly by Syrian glass-makers. The French terms are *verre de France* and *plats de verre*, and the German *Butzencheibe*. See BROAD GLASS; BULL'S-EYE PANE.

Crown Milano. A type of ART GLASS made of OPAL GLASS that was acid finished and had various ornate designs embedded in the glass in shades of brown and beige, and was also decorated on the exterior with enamelled motifs. It was developed by the MT. WASHINGTON GLASS CO. *c*. 1890, and first called 'Albertine Glass'. It is sometimes marked with a C over an M, above which is a 5-pointed crown. Some examples of white opal glass have a glossy surface and are termed 'Shiny Crown Milano'; these usually are decorated with floral motifs in gold, and have on the bottom an enamelled red laurel wreath.

crown pot. A type of POT (crucible) in which the ingredients for making glass were fused. It was a covered pot with a domed top and a side opening ('mouth') at the shoulder, as distinguished from the former SKITTLE POT. The crown pot is mentioned by CHRISTOPHER MERRET in 1662. The invention accelerated the banning of wood as a fuel in English glass-making. The crown pot is especially preferable in making LEAD GLASS as it prevents the gases from the combustion chamber of the FURNACE from coming into contact with and discolouring the molten glass.

cruche. A type of JUG with one large spout and an overhead handle. The same as a Spanish BOTIJO.

crucible. The same as a POT.

Crucifixion candlestick. A type of CANDLESTICK with a hexagonal base and nozzle, having a moulded glass Crucifix applied to the SHAFT or column. They were made at the NEW ENGLAND GLASS CO. until 1888, of many types of glass and in various colours.

cruciform. Shaped like a cross or, more usually, in quatrefoil section with four equal lobes or foils. Some English WINE BOTTLES and DECANTERS are so shaped.

cruet (or **cruet bottle**). A small jug-shaped receptacle, usually with a lip or spout, a handle, and a stopper or lid, used for serving liquid condiments (oil, vinegar, also lemon juice, garlic juice, etc.) at the table. Some are inscribed with the name of the contents. (When they have a pierced COVER and are for pepper, dry mustard, or sugar, and are sometimes so inscribed, the proper term is a CASTOR.) Early English forms are mallet-shaped, then pyriform, cylindrical, and urn-shaped. They are of clear glass, decorated with engraving, cutting, or gilding, or of OPAQUE WHITE GLASS with enamelling; many have a silver MOUNT.

They are often made as a CRUET SET, and accompanied by a CRUET-STAND. The PONTIL-MARK was usually removed so that the cruet could sit firmly in the stand. The French term is *burette*. *See* DOUBLE CRUET; SETRILL; VINAIGRETTE.

cruet set. A SET of two to five CRUETS, usually each inscribed with the name of the contents. They were generally accompanied by a CRUET-STAND of silver or glass.

cruet-stand. A receptacle, often part of a dinner service, and usually made of silver but sometimes of glass, for holding two or more CRUETS. They are sometimes called a 'bottle-stand'. *See* PORTE-HUILIER.

crumb bead. A BEAD having tiny pieces of glass of different colours affixed to the surface. A 'flush crumb bead' has the pieces marvered flush into the surface. They were first made in the Aegean Islands, *c.* 1300 BC, and in Egypt several centuries later.

Crutched Friars Glasshouse. A glasshouse near Aldgate, in London, established in 1568 by JEAN CARRÉ to make CRISTALLO. It was operated by several Italians whom he had brought with him to England, including JACOPO VERZELINI, who managed it after Carré's death. In 1576, after the glasshouse had burned down, Verzelini rebuilt it and a home for himself within its walls.

crystal. (1) The same as ROCK-CRYSTAL. (2) The colourless TRANSPARENT glass, such as Bohemian potash-lime glass, Venetian CRISTALLO, English LEAD GLASS, and Continental crystal glass, that resembles rock-crystal. It is a term heretofore used loosely for all fine glass, especially glass made for the dinner-table decorated with FACETING and faceted DROPS on CHANDELIERS and CANDELABRA. Today glass so-called and made with LEAD must contain at least 24% lead oxide; it is sometimes called 'half-lead' and that with 30% lead is called 'full lead' or *'cristal supérieur'*. Recent regulations of the European Economic Community (EEC) forbid any glass to be called or labelled 'crystal' unless meeting such standards, and a special label is provided for each class.

Crystal Palace. The vast structure erected in Hyde Park, London, to house the Great Exhibition, from 1 May to 11 October 1851; it was visited by over six million persons. Later moved to Sydenham, it was a great iron-framed building enclosed with about 300,000 panes of glass, whence its name. It was destroyed by fire in 1936. Among the exhibits was a vast amount of contemporary glassware, including a large FOUNTAIN. See *Journal of Glass Studies* (Corning), XVI (1974), p. 95.

crystalline. (1) Consisting of crystals, unlike glass which is amorphous. (2) A frosty texture on the surface of some glass produced by the use of hydrofluoric acid, in a process developed at Zwiesel, in Bavaria. (3) Glass approaching CRYSTAL (2) in quality, called in Italian *cristallino*.

crystallo-ceramie. A term used in England for a SULPHIDE. Also called 'encrusted cameo' and 'cameo incrustation'.

cucumber trainer. A glass tube used as an horticultural aid to keep a growing cucumber straight. It is a glass tube, about 37 cm. high, with a rounded upper end having a small orifice. They were made of NAILSEA GLASS.

cucurbit. A glass bottle, used with an ALEMBIC, to contain a liquid that is to be distilled; it is supported over a flame, and upon it rests the alembic. Sometimes the two parts were made in one integrated piece.

cullet. Broken pieces (FRAGMENTS) from glass objects discarded during production as imperfect or broken by the workmen, and remnants

cruet set. Cruet-stand with five inscribed blue glass cruets, *c.* 1785. Courtesy, Maureen Thompson, London.

cucurbit. Joined to alembic, clear glass, possibly 19th century. Science Museum, London; Crown copyright.

(MOIL) left on the BLOW-PIPE or the PONTIL, which are all melted together with the fresh ingredients of a new BATCH. As the cullet melts faster, this effects a saving in fuel cost, and also in the re-use of expensive ingredients such as red lead. Sometimes out-of-fashion glassware is broken and used as cullet, e.g., in England after 1745 when the GLASS EXCISE ACT taxed new glass-making ingredients but exempted a certain amount of cullet. Cullet usually constitutes about one-quarter to one-half of a batch; it can be the sole ingredient but seldom is. The French term is *verre brisé*.

cup. (1) A small bowl-shaped receptacle employed for drinking. Some have no HANDLE (like a Chinese tea-bowl); but most have one side-handle (sometimes two), and some have a COVER. A cup may be on a STEMMED FOOT or on a FOOT-RING. Generally it is accompanied by a matching SAUCER similarly decorated. They are usually made of porcelain or pottery, but some (despite occasional fears of breakage caused by the hot contents) have been made of glass, including OPAQUE WHITE GLASS. The French term is *tasse*; the German is *Tasse* or *Kelch*; the Italian is *tazza*. *See* TEACUP. (2) Any small bowl-shaped hollow object used as an ornament. (3) A term sometimes applied to a large bowl-like receptacle. *See* WEDDING CUP; STANDING CUP; PROCESSION CUP.

cup plate. A plate (about 7.5 to 9 cm. in diameter) that was used as a rest for a cup (perhaps dripping) while the tea (or sometimes coffee) was being drunk from its saucer (whence the expression 'to have a dish of tea'). A great number were produced in MOULD-PRESSED GLASS (especially LACY GLASS) by factories in Massachusetts, Pennsylvania, and Ohio from the 1820s to the 1860s. The custom was strictly American, and such plates were rarely made in Europe except in Staffordshire printed pottery produced for the United States market. The decoration on the plates included a great variety (over 1,000) of historical subjects or conventional patterns; usually circular, often with a scalloped or serrated rim, but occasionally octagonal, the plates are of clear or coloured glass (occasionally opaque white or opalescent).

cupping glass. A glass cup used for cupping, i.e., placing it over a part of a person's body to draw blood to the surface, either by dry cupping or, if an incision has been made, by wet cupping. A partial vacuum was first created in the cup by burning in it some small substance. Such objects were used from Roman times and in modern times from the 16th century.

Curling creamer. A type of CREAM-JUG of PRESSED GLASS inscribed 'R B Curling & Sons, Fort Pitt' (that being a glasshouse at Pittsburgh, Pennsylvania, founded in 1827 by Robert B. Curling and William Priest, and known from 1834 until the 1840s as Fort Pitt Glass Co.). It is of conventional shape, with a heavy handle. Its origin has been debated between the named company and the BOSTON & SANDWICH GLASS CO.

curtain tie-back. A glass ornament affixed to a wall-bracket for tying back a curtain. Some were made of LACY GLASS.

curved cutting. A decorative pattern on CUT GLASS made by deflecting the glass object out of line with the edge of the wheel, rather than the usual cut patterns that are made in straight lines with the object held firmly in line. Various motifs occur, but are difficult to make. The German term is *Kegelschliff*. *See* KUGELGRAVÜREN; PRINTIE.

cushion pad. A hollow ornament of flattened globular (oblate) shape placed, singly or in a group, between the bowl and the stem of a drinking glass; if more than one is used, they usually diminish in size as they descend. They generally have a diameter smaller than that of the bowl, but some extend beyond the bowl, as in the Prince of Wales service (*see* SERVICE). *See* KNOP, *cushion*.

cupping glass. Mould blown. Ht 5.3 cm. Collection of Dr John W. Scott, Toronto.

cut glass. Vase with cut decoration, French; Empire period, *c.* 1810. Ht 33 cm. Courtesy, Delomosne & Son, London.

cusp. (1) A pointed end. (2) A pointed KNOP on the STEM of a glass, usually centrally placed.

custard glass. A small glass bowl for serving an individual portion of custard. Some are on a stemmed FOOT and some have a single HANDLE. They were made in England in the early 19th century and later. *See* JELLY GLASS.

cut glass. Glassware decorated with FACETS, grooves, and depressions made by the process of CUTTING into its surface by the use of a rotating wheel of iron or stone, usually of large diameter, as distinguished from glassware decorated by WHEEL ENGRAVING and from SCULPTURED GLASS. The glass (usually LEAD GLASS) is ground to form facets which have a high degree of light refraction.

———, *cutting.* The process of cutting glass to make FACETS, grooves, and depressions by GRINDING, using rotating disks of various materials, sizes, and shapes and a stream of water with an abrasive. The process involves four stages: (1) marking, i.e., painting on the glass the outline of the pattern or design, usually with a mixture of red lead and turpentine; (2) roughing, i.e., cutting the basic pattern or design, using an iron (or, in recent times, carborundum) wheel, fed with wet sand, the cutter pressing the object against the fixed revolving wheel, the edge of which is flat, convex, or mitred (V-shaped), to make the type of cut desired; (3) smoothing, i.e., refining the cut pattern or design with a wheel made of copper or sandstone fed with pulverized sand or emery; and (4) polishing, i.e., removing the dull mat surface either (a) by the use of wheels made of wood fed with putty (calcined tin) powder or of felt fed with rouge (colcothar), or (b) by a chemical method, involving dipping the object momentarily into a mixture of HYDROFLUORIC ACID and sulphuric acid. The wheel has been turned from early times by foot-pedal, water-power or other mechanical means; steam-power was used at Stourbridge from *c.*1807, and modern factories use electricity. As glass cutting is difficult and costly, some pieces, especially large ones, are often first partially decorated by MOULDING and then refined by grinding and polishing; and since the 19th century the process has been aided by removing part of the glass by the use of acids.

———, *history.* The art and methods of decorating glass by cutting its surface were known before the 8th century BC (*see* SARGON ALABASTRON). In the 1st century BC the Romans decorated (a) glass objects with interlacing grooves by wheel-cutting in the manner of the lapidary craftsmen cutting semi-precious stones and ROCK CRYSTAL, (b) glass objects with relief cutting, and (c) CASED GLASS with sculptured pictorial subjects (*see* PORTLAND VASE). Cutting was done at Cologne in the 3rd and 4th centuries. In the 5th and 6th centuries pre-Islamic glass from Persia and Mesopotamia was cut with diamond and circular concave FACETING, and Islamic glass from the 9th and 10th centuries was cut with facets and arabesque designs. In the 9th to 12th centuries cutting was done in Egypt. After a lapse of several centuries glass cutting was revived in Bohemia and Germany from *c.*1588. In England cutting, in the modern sense, appeared from *c.*1719 (*see* JOHN AKERMAN), and from the development in 1676 of LEAD GLASS, with its inherent softness and refractive qualities, cutting became the prevalent method of decorating English and Irish glassware, during the second half of the 18th and the first half of the 19th century.

———, *patterns.* Designs made on cut glassware by the use of three types of cutting disks: (1) flat edge, for making flat FACETING; (2) mitred (V-shaped) edge, for making grooves in straight line patterns, and (3) convex edge, for making depressions or hollow patterns, e.g., PRINTIES. Designs are made with one type or a combination of two or three types, usually depending on the shape of the object being decorated and the area to be cut. The mitred patterns are usually of straight lines, with the object held firmly in line with the edge of the wheel; curved patterns are produced by deflecting the object out of line. Straight-line cut patterns include: ALTERNATE PANELS; BLAZES; FANS; FLAT DIAMONDS;

cut glass. Decanter with hob-nail cutting, fluted neck, lapidary stopper, and star base. Royal Brierley (Stevens & Williams), Brierley Hills, Staffordshire.

cutlery. Various styles of *filigrana* decoration, Venetian glass. L. 19·5 and 25·5 cm. Royal Collection, Rosenborg Castle, Copenhagen.

Czechoslovakian glassware. Furnace-shaped, garnet-coloured ornament, designed by Pavel Hlava, 1971. Museum of Decorative Arts, Prague.

FLUTES; HEXAGONS; HOBNAILS; prisms (*see* PRISM CUTTING); RAISED DIAMONDS; SPLITS; STAR CUTTING; STRAWBERRY DIAMONDS.

cutlery. A set of knife, fork, and spoon, which was sometimes made entirely of glass, e.g., Venetian glass with opaque white thread decoration. They were probably made more for ornament than use. *See* KNIFE; KNIFE AND FORK HANDLES; SPOON.

cut-velvet glassware. A type of glassware made with two layers of glass, the inner layer mould-blown in various patterns to simulate patterns found on cut-velvet and the outer layer acid-finished to simulate the velvet texture. It is said to have been made by many United States glassworks.

cyathus. The Latin form of KYATHOS.

cylinder glass. The same as BROAD GLASS.

Cyprian glass. A type of ART GLASS developed by FREDERICK CARDER at Steuben Glass Works, made from *c.*1915 until the 1920s. It is similar to AQUA MARINE GLASS except that the pieces have a celeste-blue glass ring applied around the rim. It has been erroneously referred to as VERRE DE SOIE or Aqua Marine Glass. It does not in any way resemble CYPRIOTE GLASS.

Cypriote glass. A type of opaque ART GLASS developed by ARTHUR J. NASH for LOUIS COMFORT TIFFANY, having its finely pitted nacreous surface wholly or partially roughened. It was inspired by the corroded IRIDESCENT surface texture of long-buried ROMAN GLASS, and was made in several colours. The rough surface resulted from rolling the heated body on a MARVER spread with powdered glass. The term was loosely applied by Tiffany to other types of glass with an antique finish; and the Cypriote glass was sometimes labelled by Tiffany merely as 'Antique glass'.

Cyrén, Gunnar (b. 1931). A glass-designer since 1959 at ORREFORS GLASBRUK. He has created a style called 'Pop Age Glasses' which uses an opaque white body or stem decorated with numerous bands of different bright colours.

Czechoslovakian glassware. Glassware made in the former Bohemian glassworks after 1918 when Czechoslovakia became independent and in new ones established since then, all of which were nationalized in 1948. The traditional heavy CUT GLASS is still made, but new techniques and designs have been introduced, especially by those who have attended the many glass schools. The Studio Lobmeyr (Stefan Rath), at Kamenický Šenov (formerly Steinschönau) still operates but is no longer connected with J. & L. LOBMEYR, Vienna; JAROSLAV HOREJC has been a designer there. At the former HARRACHOV GLASSWORKS, Miloš Pŭlpital developed HARRTIL GLASSWARE. LADISLAV OLIVA used SAND-BLASTING, 1959–60, to produce original designs with deeply cut patterns. Many new glassworks have been established in towns throughout the country, but mainly at Karlovy Vary (formerly Karlsbad), Novy Bor (formerly Haida), Zelesny Brod, and recently Skrdlovice. *See* BOHEMIAN GLASSWARE. For names of makers and decorators of modern Czechoslovakian glassware with many illustrations, *see* Josef Raban, *Le Verre Moderne de Bohème* (1960).

D

Dagnia family. A family of Italian glass-makers who were brought to England by SIR ROBERT MANSELL from L'ALTARE, *c.* 1630. One branch of the family founded, *c.* 1651, the first recorded glasshouse in Bristol. Three brothers, Onesiphorous, Edward, and John, moved to Newcastle-upon-Tyne and established their first glasshouse there in 1648, later introducing to the region the making of LEAD GLASS based on the formula of GEORGE RAVENSCROFT and developing the style of glasses still known as 'Newcastle'. They operated, *c.* 1684, the Closegate Glasshouse. In 1731 the widow of one of the Dagnias married John Williams, of Stourbridge, who entered the business, and it became the Dagnia-Williams Glassworks, for which at one time WILLIAM and MARY BEILBY probably worked. The factory made brilliant CUT GLASS from *c.* 1725, and later made glasses for JOHN GREENWOOD.

Dahlskog, Ewald (1894–1950). A Swedish designer with KOSTA GLASSWORKS from 1926 until 1929. He is noted for his work in CUT GLASS.

dairy bowl. A low circular bowl for separating cream. Examples in NAILSEA GLASSWARE have a hollow everted rim with a pouring lip. *See* CREAM BOWL.

daisy glassware. A type of MOULDED glassware decorated with a stylized replica of a daisy, enclosed within a square, diamond or hexagon. It was a popular pattern in the United States, called 'daisy-in-square', etc. It has been attributed to the glasshouse of HENRY WILLIAM STIEGEL.

damasco, vetro. A type of ornamental glass developed in Italy for BAROVIER & TOSO by Ercole Barovier, *c.* 1948; the ware is characterized by broad vertical bands, one translucent (of clear amethyst colour) and the other of a group of white FILIGRANA threads set in gold glass.

Damascus glassware. A type of Syrian Islamic glassware made *c.* 1250–1402; it is identified with, but was not necessarily produced at, Damascus (it was possibly also made at Raqqa or Aleppo) until 1402 when Damascus was sacked by the Mongols under Tamerlane and the local glass industry was moved to Samarkand in Transoxiana. The ware includes flasks, beakers, candlesticks, and bowls decorated with figures in bright colours, and MOSQUE LAMPS decorated with enamelled and gilded designs. The ware was sometimes referred to as being '*à la façon de Damas*'. *See* ENAMELLING, ISLAMIC; SYRIAN GLASS.

Dammouse, Albert Louis (1848–1926). A potter at the Sèvres porcelain factory who from *c.* 1898 began making MOULDED small vessels in a translucent material similar to PÂTE-DE-VERRE, composed of a mixture of soft porcelain paste and glass. He also made objects in PÂTE D'ÉMAIL, which was similar but instead of being made in a mould it was dried and then fired without external support; small coloured, decorative, floral pieces so made were applied sometimes to a piece made of *pâte-de-verre*.

Danish glassware. Glassware made in Denmark, of which the earliest recorded was from a glassworks located in Jutland from *c.* 1550 until *c.* 1650, when it closed due to the shortage of wood fuel. Later the idea was conceived of using peat for glassworks fuel, and the Holmegaards Glassvaerk was founded in 1825. The Kastrup Glassworks was started in 1847 to make green-glass beer bottles (*see* KASTRUP & HOLMEGAARDS

daisy glassware. Flask, deep amethyst colour. Possibly glasshouse of Henry William Stiegel, Manheim, Pennsylvania, 3rd quarter 18th century. Corning Museum of Glass, Corning, N.Y.

Damascus glassware. Flask, blown glass, decorated with enamel and gold, Syria, 13th/14th century. Ht 8·3 cm. Freer Gallery of Art, Washington, D.C.

Daum Frères. Vase in Art Nouveau style, *c.* 1900, Cristallerie de Nancy.

Daumenglas. Covered glass with four finger-holes and trailed decoration, Germany, 17th century. Ht 26 cm. Kunstgewerbemuseum, Cologne.

GLASSWORKS). Production of other types of glass in the 19th century was imitative of German and English styles and forms, but from 1925, when JACOB E. BANG began designing for Holmegaards, modern Danish glassware has been made, of good quality but simple form and decoration, principally by Holmegaards and Kastrup, now merged since 1965 and controlling almost the total production of Danish service and industrial glassware. There are a few artist-craftsmen making ornamental engraved STUDIO GLASSWARE. *See* CONRADSMINDE GLASSWORKS.

Daphne Ewer. An opaque white glass EWER that has been stated to be the most remarkable example of ancient Roman enamelling and gilding. It depicts the myth of Daphne turned into a laurel tree to save her from the pursuing Apollo. The ewer, which was probably made at Antioch in Syria, is in the Corning Museum of Glass.

date flask. *See* GRAPE-CLUSTER FLASK.

Daum Frères. A glasshouse, for many years known as Daum Freres but recently called the Cristalleries de Nancy, at Nancy (France) that has been operated by the Daum family since a glassworks there was taken over in 1875 by Jean Daum (1825–85) when he fled from Alsace. It was continued by his two sons, Jean-Louis Auguste (1853–1909) and Jean-Antonin (1864–1930), who made glassware in ART NOUVEAU style as followers of EMILE GALLÉ. They made coloured glass with enamelled floral decoration, glass etched with hydrofluoric acid, CASED GLASS, and glass of many other techniques. Among the new forms created by them is a vase, called a BERLUZE, with a tall thin neck. Paul Daum (1890–1944) and Henri Daum (1894–1966), sons of Auguste, departed from the traditional production styles in 1925, and with Michel Daum (b. 1900), son of Antonin, began the making of clear crystal glass in free forms. Since 1966 art glass made of PÂTE-DE-VERRE has been reintroduced, designed by outside artists, including Salvador Dali. Since 1965 the president has been Jacques Daum (b. 1919), grandson of Auguste. All ware made by the factory bears the engraved Daum signature; the signature has varied in form many times over the years. Antonin Daum, who was a vice-president of the School of Nancy when it was founded in 1903 by Emile Gallé, has contributed to the development of new types of glass and the creation of new styles of form and decoration.

Daumenglas or **Daumenhumpen** (German). Literally, thumb glass. A type of BEAKER that is barrel-shaped, but tapering inward slightly toward the mouth, decorated with circular indentations, for gripping with the thumb and fingers, and usually also with bands of encircling TRAILING (simulating barrel hoops) above and below the indentations. They were made, in Germany or the Netherlands, of greenish WALDGLAS in the 16th and 17th centuries. Some have a (domed) COVER with a FINIAL.

Davenport's Patent Glass. Glassware decorated with the intention to imitate ENGRAVING or ETCHING by a process patented in 1806 by John Davenport, an English potter who was also a glass-maker. The process involved covering the outer surface of glassware with a paste containing powdered glass, then removing the surplus paste so as to leave the intended design and quickly firing at low temperature to fuse the glass powder on to the surface without melting it. Such glassware is usually inscribed 'Patent' on a label made and affixed by the same method. *See* POWDER-DECORATED GLASS.

decalcomania. The process of transferring pictures or designs, as from specially prepared paper, to glass, ceramic ware, etc., and permanently affixing them thereto, e.g., as on WITCHBALLS made of NAILSEA GLASS.

decanter. A type of decorative BOTTLE used for serving wine at the table, after being decanted from the bottle to leave the dregs. When the binning of wine was introduced in England in the mid-18th century, decanters were used for serving wine taken from the bin in the original bottle or the DECANTER BOTTLE. Some were used for sherry or port, to be served with matching glasses; and later some were for spirits, cordials, and even ale. Early English decanters, *c*. 1725–50, were similar in form to the decanter bottles, but had a projecting STRING RIM for tying the cork; or after *c*. 1730 glass STOPPERS came into use. From *c*. 1775 the early club-shaped bottles became slender, and some were barrel-shaped or mallet-shaped, with a disk stopper. Towards 1800 broad types were made, with a bulbous body and with glass rings around the NECK; they usually had a mushroom stopper. In the second half of the 18th century, decanters with a square cross-section were introduced; they are used today mainly for whisky. Decanters are usually of clear uncoloured glass, but some are of coloured glass (e.g. blue or green); and they are often decorated with FACETING, RAISED DIAMONDS or HOBNAIL CUTTING. Some decanters, from *c*. 1755, bore – engraved (if of uncoloured glass) or in gilt or enamel (if coloured) – a BOTTLE LABEL with the name of the contents. A few were provided with silver MOUNTS. The capacity varied from half-pint to quart size with larger examples being called a magnum decanter (*see* below). *See* also DECANTER BOTTLE; DECANTER JUG; DECANTER SET; WINE URN; DECANTER RINGS.

decanter, bell. Decanter with mushroom stopper, English, *c*. 1710. Ht 26·5 cm. Courtesy, W. G. T. Burne Ltd., London.

TYPES OF DECANTER

——, *bell*. A type of decanter, the body of which is bell-shaped.

——, *bottle-shaped*. A type of decanter with a circular body having a sloping shoulder extending into its neck.

——, *claret*. A type of decanter or JUG with a globular or ovoid BODY, a thin tall neck having a pouring LIP, and a curved HANDLE attached at the shoulder and near the top of the neck. They have a flat base or a low conical foot. Most have a glass STOPPER. They were made in the 18th century (e.g. at La Granja de San Ildefonso in Spain and Val-Saint-Lambert in Belgium). They are sometimes called a 'DECANTER JUG' or a 'claret jug'. *See* ELGIN JUG.

——, *club-shaped*. A decanter with straight sloping sides and shoulder extending into the neck.

decanter, club-shaped. Decanter with engraved vines and label, English, *c*. 1765. Courtesy, Delomosne & Son, London.

——, *Cluck-Cluck*. The onomatopœic name for a decanter made for over 100 years by Holmegaards Glassworks (*see* KASTRUP & HOL-MEGAARDS GLASSWORKS), in the general form of the KUTTROLF, with four vertical tubes connecting the lower and upper parts of the body. The Danish term is *Klukflaske*.

——, *cruciform*. A type of decanter of which the body has a section resembling a Greek cross. Examples are known dating from *c*. 1740–60.

——, *lipped pan*. A type of decanter of which the upper part above the NECK is wider than the neck and shaped somewhat like a shallow pan, and has pouring lip. *See* PAN-TOPPED.

——, *guest*. A small decanter (with no STOPPER) to be filled with wine or spirit and placed at a guest's dinner place or on his bedside table.

——, *magnum*. A decanter of large capacity, usually about 2 quarts.

——, *mallet*. A type of decanter, cylindrical or hexagonal, shaped like a stone-mason's mallet, with almost vertical sides and a shouldered body. They were made of English glass, *c*. 1740–80, and are somewhat in the form of the Chinese porcelain mallet vases of the Sung dynasty (960–1280).

——, *onion-shaped*. A decanter of which the body is almost globular, with a sloping shoulder.

——, *piggy-dog*. A type of fanciful horizontal decanter in the form of a standing hollow dog or pig, with curved tail forming the handle and the snout forming the spout, sometimes with a glass stopper. They were made in the 19th century.

——, *Rodney*. The same as a ship's decanter (see below); the name refers to Admiral Lord Rodney, commemorating his victory over the French in 1782; such decanters, however, usually have no special Rodney decoration or inscription.

——, *shaft-and-globe*. An early type of English decanter from *c*. 1725 made of SODA GLASS before the development of LEAD GLASS. It has a globular body with sloping shoulder and a tall cylindrical neck. Some are decorated with GADROON ribbing on the lower half, and others with NIPT-DIAMOND-WAIES. *See* WINE BOTTLE, *English.*

——, *ship's.* A decanter of circular section, having an especially broad bottom, intended to ensure stability at sea, for use in the cabin of a ship's officer. The sides of the body sometimes extend in a straight line from the neck to the bottom, but others have gracefully curved concave sides. Most are highly ornamented with FACETING and CUTTING, and usually have several applied rings on the neck spaced to fit the fingers so as to afford a firmer grasp; the rings require skill to make a perfect join. Sometimes termed a 'Rodney decanter' (see above).

——, *shouldered.* A decanter with sides sloping slightly inward toward the bottom and with a shoulder sloping in and narrowing to form the neck.

——, *spirit.* A decanter for spirits (e.g. Whisky – see below) rather than wine. They usually bore, engraved or painted in gilt, the name of the contents (Gin, Brandy, Rum), with sometimes the initial on the STOPPER.

——, *stirrup.* A type of decanter with a pointed bottom, and sometimes with a pouring lip and handle; it was carried by an attendant to the horsemen, set into a hole in a tray from which it was taken, used, and replaced. The tray also provided STIRRUP CUPS or glasses.

——, *tantalus.* A set of three or four decanters in a locked rack. The decanters were square and of CUT GLASS, usually decorated with the ALTERNATE PANEL PATTERN. *See* DECANTER SET.

——, *tapered.* A decanter of which the sides are almost vertical, then sloping into the neck.

——, *toilet-water.* A decanter of small size, for various toilet waters, and sometimes bearing a label designating the contents. They usually occur in SETS of three. *See* COLOGNE BOTTLE; TOILET-WATER BOTTLE.

——, *whisky.* A decanter of square section, introduced in England in the second half of the 18th century and used usually for serving whisky rather than wines or other spirits. They are often of elaborate CUT GLASS and are made in many factories in England and elsewhere. Some were engraved with the name of the distillery. The shape was convenient for keeping a set of 3 or 4 of them in a fitted box, sometimes also provided with glasses EN SUITE. They are sometimes called 'squares' or 'case bottles'. *See also* tantalus decanter (above).

decanter bottle. A type of BOTTLE that preceded the DECANTER and was used for bringing wine to the table from the cask, hence sometimes called a 'serving bottle'. The earliest form, from the mid-17th century, has a bulbous or wide-shouldered BODY, a long NECK, and a KICK BASE. The form changed *c.*1680 to a short neck and a broad body, and again *c.*1735 to a longer body and neck; from *c.*1750 the bottles resembled a modern wine bottle, with a cylindrical body and straight sides. Early ones were made and imported by the GLASS SELLERS' COMPANY. Such bottles usually bore, on a glass pad on the shoulder, the date, the initials of the glass-maker, or his device. They were made originally of blackish-green or dark-brown BOTTLE GLASS. An example in the British Museum, of LEAD GLASS and having a flattened globular body, is attributed to the glasshouse of GEORGE RAVENSCROFT, *c.*1677–80.

decanter jug. A type of JUG for use as a DECANTER, usually having a shouldered body sloping inward towards the base, a thin neck with a FUNNEL MOUTH or a lipped mouth, and characterized by a loop-handle extending from the rim to the shoulder. Some have a glass STOPPER, some a metal hinged LID. Usually there is a frilled collar at the base of the neck. Some examples were made of English glass engraved in the Netherlands, and some were made of clear LEAD GLASS, possibly by GEORGE RAVENSCROFT, *c.*1680–85.

——, *octagonal.* A type of decanter jug of which the lower part of the body is of octagonal section, and having a sloping shoulder, a long thin neck, a loop-handle, and a stopper.

decanter, Cluck-Cluck. Decanter in the form of a *Kuttrolf*, Holmegaards Glasvaerk, Copenhagen.

decanter, ship's. Decanter with raised diamond cutting and three neck rings, with conforming mushroom stopper, Georgian, *c.*1810. Courtesy, Delomosne & Son, London.

decanter rings. The rings placed around the neck of most English and Irish DECANTERS to afford the user a firmer grip. The number varies from one to four, and the styles of the rings vary with the period and maker. Some rings are (in section): (1) triangular, projecting from the body to form an acute angle; (2) square, projecting in a square form; (3) rounded, projecting in a semi-circular form. Some are decorated with cutting or a feathered or notched pattern. Some such rings are known as 'double rings', meaning not two rings but one ring slightly divided to appear as two; and similarly there are 'triple rings'.

decanter set. A SET, usually consisting of three or four DECANTERS, each often marked with the name of the contents, e.g., Rum, Gin (or Hollands), Brandy, Whisky. They were made in England in the 18th and 19th centuries. Some were provided with a leather case for travelling or a silver or Sheffield-plate stand for the sideboard; the latter sometimes have a central silver post surmounted by a SWEETMEAT DISH and an overhead HANDLE, and with rings attached to the stand to hold the STOPPERS when removed. *See* DECANTER, *tantalus.*

decanter-wag(g)on. A rectangular stand (about 33 cm. long), with recesses to hold two DECANTERS, mounted on four wheels and having a small pulling handle; it is used to pass decanters at the dinner-table. The origin is said to be from a suggestion of George IV to avoid the need for a guest sitting beside the King to have to rise to pass the port to the guest on his other side. The prototypes, a pair in silver gilt, were made by Sir Edward Thomason, of Birmingham, and later presented by the King to the Duke of Wellington; they are now at Stratfield Saye, near Reading. Modern examples are of many styles.

deceptive glassware. Various glass drinking vessels (e.g., wine glasses, sherry glasses, TOAST-MASTER GLASSES) made with a bowl having a thick base and side so as to contain a lesser quantity than the size of the glass would suggest. They are sometimes called 'disguised glasses'.

Deckel. The German term for a COVER.

decolourizing agent. A mineral used to counteract in glass the dark greenish or brownish colour resulting from iron particles in the SILICA (sand) ingredient or imparted to the BATCH by iron or other impurities in the POT or elsewhere in the production process. The decolourizing agent does not remove the colour but acts as a chromatic neutralizer, it and the iron absorbing the light that the other transmits (if in equal quantity); if too much of both is present, little light can be transmitted, and the resultant glass, if thick, becomes grey or blackish. The method of using MANGANESE oxide (glass-maker's soap) was known to the Romans in the 2nd century. Later NICKEL alone or in blue CULLET) was used, and the colour became bluish. At the end of the 18th century, COBALT in minute quantities was used.

Décorchement, François Emile (1880–1971). A French painter and potter at Conches (Eure), who in 1902 started his glass-factory there, making principally circular bowls and vases of PÂTE-DE-VERRE in subdued colours and with cloudy translucency. His early work included statuettes and small bowls with enamel decoration, but he soon developed a new style with a rough surface, and from 1905 to 1910 decorated with relief floral motifs and insects. From 1912 he developed the use of PÂTE-DE-CRISTAL, and made pieces in massive style with marbled colours. He used the CIRE PERDUE technique, with relief decoration of floral, feathery, and abstract designs. His work shows continuous artistic evolution. After World War II he resumed his work, making ware of opaque and marbled glass and also stained-glass windows.

déjeuner service. A small breakfast service, including a platter, cups and saucers, teapot or coffee-pot, sugar-bowl, cream-jug or milk-jug,

decanter jug. Octagonal moulded body and stopper with ball finial, *c.* 1715. Courtesy, Delomosne & Son, London.

deceptive glass. Wine glass with ogee bowl, double-series opaque twist stem, and conical foot. Ht 13·7 cm. Courtesy, Sotheby's, London.

Décorchemont, François-Emile. Bowl in
pâte-de-verre in Art Nouveau style,
Conches, France, *c.* 1919, signed 'F.
Décorchemont'. Ht 7·8 cm.
Kunstmuseum, Düsseldorf.

jam-pot. Such services were occasionally made in glass as luxury items,
such as one made for Louis XVI.

Dennis Glassworks. An English glasshouse, at Amblecote, near
Stourbridge, built in 1855 by Thomas Webb I (1804–69) and operated
as THOMAS WEBB & SONS. On Webb's retirement in 1863 his sons,
Thomas Wilkes Webb II (1837–91) and Charles Webb, took over, and
upon their father's death in 1869 they were joined by another brother,
Walter Wilkes Webb. After Thomas II died, the business was continued
by Walter and Charles until Walter's death in 1919. The business was
then sold to Webb's Crystal Glass Co. Ltd, and it is now a part of Dema
Glass, Ltd, having been acquired by the Crown House Group in 1964.

desk set. A set for use on a writing desk, usually consisting of two
PAPER-WEIGHTS and one INK-BOTTLE; they were made at Stourbridge.

Despret, Georges (1862–1952). A French glass-maker at Jeumont
(Nord) who made some vases of OVERLAY glass with entrapped air
bubbles and abstract patterns in various colours, but from 1890
specialized in making objects in PÂTE-DE-VERRE, particularly female
TANAGRA-STYLE FIGURES and fish and animals in the round. He signed
pieces with his engraved signature 'Despret'. His factory was destroyed
in World War I but was reopened in 1920; it closed *c.* 1937.

Desprez, Barthélemy (fl. 1773–1819). A Parisian sculptor who
introduced the process of making SULPHIDES. He was a sculptor at
Sèvres from 1773 to 1792, when he started his own factory for making
porcelain and stoneware in Paris in the Rue des Récollets du Temple
and began making sulphides. His son succeeded him *c.* 1819 and
continued until *c.* 1830. The surname and address were usually
impressed on the reliefs. The sulphides were made from moulds from
coins and medals bearing portraits of famous persons, and the moulds
were sometimes used more than once.

dessert glass. A large glass receptacle for sweetmeats which in the early
18th century superseded the SWEETMEAT GLASS and which was also
used for serving fruits. The early term was often 'desart glass'.

deutsche Blumen (German). Literally, German flowers. Naturalistically painted flowers, either separate or loosely tied in bouquets,
introduced as decoration on porcelain by J. G. Höroldt at Meissen,
c. 1740. They followed the style known as INDIANISCHE BLUMEN, and
were extensively copied at many factories between 1750 and 1765.
Although primarily a porcelain decoration, it was also used on some
opaque glassware in Germany in the 18th century.

devitrification. The process of converting glass into a crystalline
substance. By heating the materials of which glass is made (the BATCH)
above the temperature needed for VITRIFICATION (about
1,300–1,550° C.), the glass liquefies; by permitting it then to cool too
slowly or too long, it devitrifies and becomes crystalline, with a milky
appearance and is no longer capable of use as glass. *See* MOLTEN GLASS.

dew-drop glass. A type of ART GLASS that has an overall surface
decoration of 'hobnail' protuberances, made by first MOULD-PRESSING,
then expanding the object by blowing, and completing it by tool
manipulation. It was developed in 1886 by William Leighton, Jr. and
William F. Russell, at HOBBS, BROCKUNIER & CO. Sometimes it has an
OPALESCENT surface, and is then called 'opalescent dew-drop'.

Dewar flask. A double-walled glass vessel for holding liquid air,
having a vacuum in the space between the walls so as to prevent
conduction of heat, and sometimes having the glass silvered to prevent
absorption of radiation. They are called, depending on the shape, a

'Dewar bulb' or 'Dewar tube'. They were named after Sir James Dewar (1842–1923), a British chemist.

diafano, vetro (Italian). A type of ornamental glass designed by Angelo Barovier, for BAROVIER & TOSO, *c*. 1975. It is clear transparent glass decorated with wide engraved festoons whose translucency creates a diaphanous effect.

diamond faceting. A style of FACETING forming a DIAPER pattern of vertically placed diamonds; it is used on the stems and bowls of English drinking glasses, and on decanters, etc. *See* CHEQUERED DIAMOND.

diamond-moulded pattern. A moulded decorative pattern on American glassware in the form of diamonds of varous sizes, varying according to the mould and the degree of expansion after removal from the mould. Small 'diamonds', usually circular or oval, were sometimes called in England 'pearl ornaments'. There were three varieties of the pattern: (1) overall; (2) rows above vertical fluting; (3) CHEQUERED DIAMONDS. It is found mainly on SALTS.

diamond-point engraving. The technique of decorating glass by ENGRAVING, using a diamond point to scratch the surface. Early examples date possibly from Roman times, although it has been said that these may have been done by use of flints. It was used on ISLAMIC GLASS, and in the 16th century by glass-decorators in Venice and in HALL-IN-TYROL on fragile Venetian glass, and also by German engravers who copied the Venetian style. It was also used by Dutch and English engravers. The process first occurs in England in the late 16th century on glassware made at the glasshouse of JACOPO VERZELINI. In Holland it was used mainly by amateur decorators in the 17th century, especially those who used the method for decorating in CALLIGRAPHY. Diamond-point engraving was superseded in the 18th century by WHEEL ENGRAVING and by decoration by ENAMELLING, but the diamond point continued to be used for STIPPLING from the 1720s to the late 18th century, and also in England for JACOBITE GLASSWARE and COMMEMORATIVE GLASSWARE. The Italian term is *inciso a punta di diamante*.

diamond-studded. A style of decoration on some English glassware, e.g. the Silesian stems (*see* STEM) of some SWEETMEAT GLASSES or some domed feet (*see* FOOT) on TAPERSTICKS and wine glasses, in the form of raised diamonds.

diamonding. A form of decoration on bevelled MIRRORS made by cutting a series of intersecting hollows along the bevelled edges, as a result of which a diamond pattern was formed at the points of intersection. It was the earliest English form of CUTTING. A patent for a mechanical process of diamonding was granted in 1678 to John Roberts.

diamonds. Various decorative patterns on CUT GLASS, e.g., hollow (or shallow) diamonds, raised diamonds, strawberry diamonds. *See* CUT GLASS, *patterns*.

diaper. A repetitive pattern with the units of the design being similar and connected, or nearly so, to each other. *See* SCALE FACETING; HEXAGONAL FACETING; DIAMOND FACETING; CARPET GROUND; DOT AND CROSS DIAPER.

Diatreta glass. A type of glass created by FREDERICK CARDER, *c*. 1952, at the STEUBEN GLASS WORKS, to simulate the ROMAN GLASS of the VASA DIATRETA type. The ornamental pieces were produced by an adaptation of the CIRE PERDUE process and fused to the body of the object by tiny glass struts. A decoration of comparable appearance was produced on some vases of TIFFANY GLASSWARE by similar applied ornamentation. *See* PSEUDO VASA DIATRETA.

diafano, vetro. Bowl and vase, designed and signed by Angelo Barovier, *c*. 1975. Barovier & Toso, Murano.

Diatreta glass. Light-blue and green bubbly vase, cast by *cire perdue* process, designed and made by Frederick Carder for Steuben Glass Works, dated 1952. Corning Museum of Glass, Corning, N.Y.

Dier Goblet. Diamond-point engraved, Jacopo Verzelini glasshouse, London, dated 1581. Ht 21 cm. Victoria and Albert Museum, London.

diminishing lens. Covered beaker of flashed glass (ruby over clear glass), Conrath und Liebsch, Steinschönau, Bohemia, *c.* 1912. Ht 18 cm. Kunstgewerbemuseum, Cologne.

diatretarii (Latin). The decorators of glass of ancient Rome, as distinguished from the VITREARII, who were the glass-makers and glass-blowers. They did all types of decorating by abrasion, and not only such deep cutting as on the CAGE CUPS, as was once supposed and which led to such cups being called VASA DIATRETA.

dice glass. A TUMBLER, about 7·5 cm. high, with a double-walled bottom in which are enclosed dice which can be shaken and played as a game.

dichromatic glass. Glass showing different colours depending on the angle of light falling on it, as when seen by reflected or transmitted light. This is sometimes due to the addition to the BATCH of a tiny percentage of colloidal GOLD (*see* LYCURGUS CUP; AMBERINA GLASS) or URANIUM, or the use of COPPER OXIDE in a REDUCING ATMOSPHERE (*see* FAIRFAX CUP). *See* TRANSMUTATION COLOUR.

Dier Goblet. A glass GOBLET generally regarded as having been made by JACOPO VERZELINI. It has three engraved panels with the names 'John-Jone' and 'Dier 1581', and the Royal (Elizabethan) Arms. It is presumed to be a MARRIAGE GLASS. Encircling the bowl of the goblet, above the names, are depicted a stag, a unicorn, and two hounds. The bowl is of rounded-funnel shape, and the stem has a large hollow fluted bulb between two small KNOPS. *See* VERZELINI GLASSWARE.

diminishing lens. A polished glass concave lens that concentrates the light and reduces the image, being the opposite of a MAGNIFYING LENS. It was first used, before 1646, as a decorative feature on a piece of glassware by GEORG SCHWANHARDT. They have been used singly and in an overall pattern on German and Bohemian glasses. Sometimes a portrait or other design is placed exactly opposite such a lens and shows through it in reduced size. *See* CONVEX MIRROR.

diota (Greek). A JAR, generally of Greek pottery, for storing water, wine or oil. It is somewhat similar in form to an AMPHORA, but has a pointed bottom, and is characterized by two side HANDLES which gave rise to the name (meaning 'two-eared') of the vessel. Some small examples were made of ROMAN GLASS, from the 2nd century AD, with the loop handles extending from midway up the neck (attached to an encircling thread) downward to the shoulder.

dip-mould. A one-piece MOULD open at the top and used to make glassware for shape only or to make PATTERN-MOULDED GLASSWARE which is later expanded by BLOWING. *See* PIECE-MOULD.

dipper. *See* KYATHOS.

dipstick. A rod of iron or bronze used for removing a small GATHER of MOLTEN GLASS from the POT and then for trailing it onto an object being decorated by such trailing, e.g., CORE GLASS. The rod usually has one end bent at a right angle so as more readily to apply the trailing. *See* GATHERING IRON.

diseased glass. *See* CRISSELLING.

disguised glasses. The same as DECEPTIVE GLASSWARE.

dish. A shallow utensil on which food is served or from which it is eaten. The dish is usually circular or oval, with a ledge and a well. The term is usually reserved for utensils 30 cm. or more in diameter, those smaller being termed a PLATE. Although generally made of porcelain or pottery, a number have been made of glass. Examples in ROMAN GLASS, cast in two-piece MOULDS, are decorated with CUTTING. The French term is *plat*; the German is *Schüssel* if there is a ledge, *Schale* if none. *See* BOWL.

dish light. A type of CHANDELIER in the form of a wide shallow circular piece of glass to be suspended from the ceiling by three chains attached to the encircling metal rim. Above the glass, arms extend from the rim to a central metal frame in which to place a FONT for oil. The pieces were made in many sizes, for use in narrow vestibules or large rooms, and sometimes c͏͏ ͏ᴉe in pairs. The chains, rim, finial below the glass, and ceiling attachment were often made of decorative ORMOLU. They were used in England in the Regency period, *c*. 1810. *See* LAMPADA PENSILE.

disk. A small circular disk of glass with a circular hole in the centre. They were made in Egypt *c*. 200 BC, the purpose not now known, and in China (*see* KU).

dispensary container. A glass storage-jar for liquids, used in the dispensary of a pharmacy. They were of various sizes, known as 'pottles', 'Winchester quarts', and 'Corbyns', but these terms have been loosely used for varying sizes. *See* CARBOY.

distaff. *See* SPINDLE.

documentary specimen. Any piece which throws light on glassware history, including those: signed by the glass-maker or decorator; bearing a rare mark or informative inscription; discovered during a particular glasshouse excavation (*see* WASTER); descended in a family connected with an early glasshouse; decorated with armorial bearings; described in old documents.

Dolce Relievo glassware. A type of glassware made of opaque white or ivory-coloured glass FLASHED with clear glass; a design then made on the surface was etched away, leaving the design in shallow relief on the ground colour. It was also called 'soft relief', and was made by STEVENS & WILLIAMS, LTD, in the 1880s.

dolphin beaker. A type of BEAKER with two encircling rows of hollow 'elephant-trunk' PRUNTS (as on a CLAW-BEAKER, except that those on the upper row terminate in a fish-like tail). The prunts of the upper row have ornamentation in dark-blue glass; the upper piece at the top of the prunt is in a form suggestive of a dolphin's head, and the lower piece is pincered to suggest a pair of fins. They were made in Cologne, Germany, in the 4th century. Other beakers of the same origin and date have decoration of different shellfish, e.g. shrimp. *See* SHELLFISH BEAKER; DOLPHIN BOTTLE.

dolphin bottle. A small globular receptacle, similar to a Greek ARYBALLOS, resting on a flat or rounded bottom and with a disk or funnel mouth. They have two thick curled handles on the shoulder, sometimes reaching to the disk encircling the mouth, and each handle has a pierced hole for a bronze chain or handle to be suspended from the user's wrist. They are so-called because the form of the handles of some is said to resemble a dolphin. They have been attributed to Roman origin, from the 1st to the 4th centuries AD. *See* DOLPHIN BEAKER.

dome. A tall cylindrical glass object whose sides curve inward at the top to form a globular dome. They were used in Victorian days to cover displays of reproductions of fruit or bouquets of artificial flowers or a BIRD FOUNTAIN. They were made by the same process as for making BROAD GLASS, by blowing a large bubble and allowing it to form a cylinder and then trimming the end opposite the dome. *See* SHADE.

dome cover. A type of COVER that is circular and domed in shape, and fits on the upright collar of a vessel. Some are 'double-domed', with a small dome resting on top of a larger one. *See* CAP COVER; SCREW CAP.

Donovan, James (fl. 1770–1829). An enameller on glassware, porcelain, and pottery. He owned a glasshouse in Dublin, but no signed specimen of his glassware is known.

dish light. Glass dish with diamond-cutting and chased ormolu mounts, English, *c*. 1810. Ht 1·40 cm. Courtesy, Delomosne & Son, London.

dolphin beaker. Greenish glass with dark-blue ornamentation, Cologne, 4th century. Ht 13 cm. Römisch-Germanisches Museum, Cologne.

dolphin bottle. Thick-walled vessel with circular and oval concavities, probably Egyptian, 3rd/4th century. Ht 8·5 cm. Victoria and Albert Museum, London.

door-stop. Green and white cased glass, English, early 19th century. Ht 12·6 cm. Victoria and Albert Museum, London.

Doppelsturzbecher. Double upside-down glass, Georg Schwanhardt, Nuremberg, mid-17th century. Ht 22·8 cm. Kunstsammlungen der Veste Coburg.

door knob. A knob, sometimes made of glass, for releasing the door-latch. Examples are found decorated with MILLEFIORI designs. A patent for a door knob in PRESSED GLASS was granted in the United States to John Robinson, of the STOURBRIDGE FLINT GLASS WORKS, Pittsburgh. *See* PAPER-WEIGHT, *crown.*

door plate. A flat decorative glass piece, usually rectangular, to be set into the panel of a door, secured by external framing of metal or wood or by screws. It serves to prevent marring the door. Examples with engraved decoration were made in England by APSLEY PELLATT.

door-stop. A solid glass object of irregular global or ovoid shape, with a flat bottom. They are from 7·5 to 15 cm. high, and weigh up to 6 lb. (2·75 kilos). All have a PONTIL MARK. They were made of greenish METAL such as was used for making bottles, often the dregs in the melting POT at the end of a day which would otherwise be discarded or 'dumped', hence the colloquial term 'dumps' sometimes applied to them. They are usually larger and heavier than PAPER-WEIGHTS; sometimes small ones are termed 'bottle-green paper-weights'. Some are decorated with a multitude of interior tiny silvery air-bubbles, usually in some sort of pattern. The bubbles were produced by (1) use of a SPOT MOULD or (2) pressing the PARAISON into chalk spread on the MARVER, then adding more molten glass and repeating the process, perhaps several times, the heat causing the chalk to emit gas that formed the bubbles in the pattern of the chalk. Some were decorated with an internal pot of stylized flowers or several flowers in tiers; the process of production involves making a PARAISON with three or more layers of CASED GLASS, after which a pointed instrument is pushed into the top of the mass so as to depress the layers and form trumpet-like shapes in the piece, with sometimes a cavity at the top, and when another layer of glass is added, an air bubble or TEAR is trapped in the cavity. Some large examples have the bubbles in pyriform shape with a silvery thread at the narrow end. In England such pieces were made from the late 19th century in Yorkshire, and some have, over the pontil mark, the seal of J. Kilner, of Wakefield. They continued to be made until a much later date at Stourbridge and elsewhere in England by many glassworks making bottles.

Doppelpokal (German). A POKAL that has as its cover another *Pokal.*

Doppelscheuer (German). A type of covered CUP of which the BOWL is in the basic form of a SCHEUER, with a COVER that repeats the form of the bowl (without the handle) and has a flat top. Some were made in glass, attributed to Venice (or perhaps southern Germany), dated 1518, but examples are known in northern Europe, made in the 15th–16th centuries, of silver, crystal or wood (called a 'covered mazer').

Doppelsturzbecher (German). A type of drinking glass in two parts, each of which is a STURZBECHER, with one inverted so that its rim rests on the rim of the other. Sometimes the stems are of different length, that of the upper glass being shorter and having the appearance of a FINIAL. Examples with WHEEL-ENGRAVING were decorated by GEORG SCHWANHARDT at Nuremberg in the mid-17th century.

Doppelter Krautstrunk (German). Literally, double cabbage-stem glass. A type of KRAUTSTRUNK in the form of one such glass superimposed on the rim of another. They were made in Germany in the 16th century.

Doppelwandglas (German). Literally, double-wall glass. The same as ZWISCHENGOLDGLAS.

Dorflinger Glass Works. A glasshouse founded at White Mills, Pennsylvania, by Christian Dorflinger (1828–1915), a native of Alsace who trained as a glass-maker in France and emigrated to the United States in 1846, acquiring in 1852 the Long Island Flint Glass Works in Brooklyn, and by 1860 two others, all of which made flint glass. At the Dorflinger Glass Works, founded in 1865, he made blanks for other factories, and from 1867 did glass cutting and engraving. His four sons and his grandsons carried on after his death as Dorflinger & Sons until the factory closed in 1921. The factory made superior glassware, including tableware for the White House, and also a few PAPER-WEIGHTS made probably as gifts.

Dorsch, Christoph (1676–1732). A glass-engraver of Nuremberg, *c.* 1712, the son of Erhard Dorsch (1649–1712), also a glass-engraver of Nuremberg.

dot and cross diaper. A DIAPER pattern made of parallel crossing lines decorated with two dots on the lines between each intersection and one dot within each resulting square or diamond.

double bottle. A BOTTLE that is divided internally into two separate compartments, each with its own MOUTH or SPOUT. Some are globular shaped, some rectangular. *See* MULTIPLE BOTTLE; DOUBLE CRUET.

double-cone bottle. A glass vessel in the form of a truncated inverted cone with a flat base, on the RIM of which appears to be superimposed another conical-shaped vessel tapering upward to form a slender neck. The rim of the upper cone projects slightly over the lower cone, in an overhanging fashion. They were made by reheating the long glass bubble near the middle and then pressing the lower half into the upper half, making a fold. They were made in Germany in the 14th–15th centuries.

double cruet. A bottle with an interior division so as to make two separate containers, each with its own neck, mouth, and stopper. They were made as separate bottles blown and then fused together. They had a flat bottom for table use, and were usually intended for oil and vinegar. They were made in England (attributed usually to NAILSEA GLASS), Germany, and Spain in the 17th to 19th centuries, and in the United States in the 19th century. *See* DOUBLE BOTTLE; MULTIPLE BOTTLE; GEMEL FLASK.

double-eagle glass. A type of WINGED GLASS that has an elaborate stem, constructed of THREADS, having the appearance of a double-headed eagle, made in the FAÇON DE VENISE in Germany and the Netherlands in the late 16th and early 17th centuries. Sometimes the twisted rods were further embellished by clear or coloured pincered CRESTING along the outer edge. The German term is *Doppeladlerglas*.

double-ended gin glass. A type of drinking glass consisting of two FUNNEL BOWLS joined at their bases. Sometimes they are DECEPTIVE GLASSES, with thick sides.

double etching. A decorative technique by which a design is etched on an object of CASED GLASS and then an added design is produced by a second dipping in the acid while the first design is covered, creating two depths of etching. When done on such glass as that developed by FREDERICK CARDER at the STEUBEN GLASS WORKS, with a layer of ALABASTER GLASS cased between two layers of coloured glass, the pattern on the outer layer appears in three shades of colour.

double flashing. FLASHING with two or more layers of glass of different colours. It occurs in German glass of the 1830s. The German term is *Doppelplattierung*.

double flint glass. *See* FLINT GLASS.

double glass. *See* ZWISCHENGOLDGLAS.

double-cone bottle. Green iridescent glass, German, 15th century. Ht 14 cm. Kunstgewerbemuseum, Cologne.

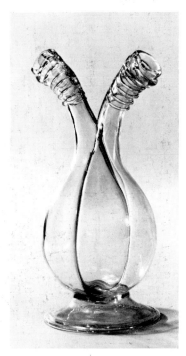

double cruet. Cruet for oil and vinegar, German, 17th century. Ht 22 cm. Kunstsammlungen der Veste Coburg.

double wine glass. White enamelled decoration, with pincered threads, Germany, 17th/18th century. Ht 24 cm. Museum für Kunsthandwerk, Frankfurt.

double gourd shape. The shape of a vase or other object with the lower part in globular form and the upper part similar in shape but slightly smaller, the space between being a shorter contracted waist. Such pieces were made in Chinese porcelain (and imitated in European faience) but some CHINESE SNUFF BOTTLES were also made in glass in this shape.

double ogee. A shape or ornamentation in the form of an extended OGEE moulding, resembling two connected S's. The BOWL on some VASES or drinking glasses were so shaped.

double unguentarium. A receptacle in the form of two UNGUENTARIA joined together with one loop handle on each side and an overhead loop handle extending over both vessels. Sometimes called a DOUBLE BALSAMARIUM.

double wine glass. A WINE GLASS with a bowl at each end of the STEM. The glass rests on the RIM of one bowl while the other bowl is upright. One bowl is usually larger than the other bowl, for red and white wine, respectively. Such glasses are used mainly for drinking Rhine wines. They are usually of light-green coloured glass, sometimes with gilt decoration. Some ornate examples have an elaborate stem, occasionally in the form of a figure, so that when the piece rests on the larger bowl, it resembles a woman carrying a basket on her head.

dove flask. A fanciful FLASK for perfume, made in the form of a dove. Some were made of ROMAN GLASS, with smooth sides and an elongated tail as the spout (the tip having been broken off to make the orifice). They were made in variegated colours in the 1st century AD. Later examples were made of OPAQUE WHITE GLASS, modelled in greater detail and enamelled with feather markings in various colours; these were made in Europe in the 18th century. The German term is *Taubenflasche*.

dragon glass. A type of WINGED GLASS, of which the stem, constructed of THREADS, is convoluted into the form of a dragon, made in the FAÇON DE VENISE in Germany and the Netherlands in the late 16th and early 17th centuries. Sometimes the twisted rods were further embellished by clear or coloured pincered CRESTING along the outer edge. The German term is *Drachenglas*.

Drahonovsky, Josef (b. 1877). A Czechoslovakian sculptor and glass-engraver who, as a professor at the Prague School of Industrial Art, has been very influential. He has designed glassware in classical style. Among his pupils was JAROSLAV HOREJC.

Drahtglas (German). Literally, wire glass. The German term for NETTED GLASSWARE. *See* HECKERT, FRITZ.

dram glass. A small drinking glass for spirits (capacity, about 2 to 3 ounces), with a funnel or conical bowl and a short or RUDIMENTARY STEM resting on a heavy foot. They were sometimes made as 'double-dram glasses' with bowls of different sizes.

drawn shank. The STEM of a drinking glass that is made by drawing the METAL from the bottom of the BOWL, in contrast to a STUCK SHANK. In this method the glass is made in two parts, the bowl with the stem attached and the FOOT. Also called a 'drawn stem' or a 'straw stem'. The glass so made is called a 'shank-stem glass'.

Dresser, Christopher (1834–1904). An English designer in many fields of applied art who was influenced by the ART NOUVEAU style. He designed bowls, vases, and goblets to be made in CLUTHA GLASS by James Couper & Sons of Glasgow, the pieces often being marked 'Clutha designed by C D'.

drinking bowl. Turquoise glass, Islamic, 9th century, with gilded silver jewelled Byzantine-style rim, 10th/11th century. Treasury of St Mark's, Venice.

dressing. The various glass ornaments on a CHANDELIER or CANDELABRUM, including various types of DROP, SNAKES, arms, and PINNACLES.

dressing-table set. A group of glass objects for feminine use at a dressing-table, including such pieces as perfume bottles, powder bowl, ring stand, and jars, sometimes with a tray.

drinking bowl. A bowl, usually circular and low, such as was used from the 5th to 9th century as a communal drinking vessel. Some, made of Islamic glass (possibly in Persia) in the 9th century, had the appearance of being made of another material, e.g. turquoise, and were provided with a gilded silver rim and mount in the Byzantine style of the 10th and 11th centuries.

drinking glass. A hollow glass vessel intended for an individual to use for drinking wine, water or other beverage. They are almost invariably symmetrical and may have a flat bottom or a stemmed FOOT. Some are made in two parts (*see* DRAWN SHANK), and in three parts (*see* STUCK SHANK). They have been made in many different shapes and sizes, with styles varying as to the BOWL, STEM, and BASE and as to the styles of decoration (CUT, ENGRAVED, ENAMELLED, STIPPLED, ETCHED). *See* BEAKER; GOBLET; TUMBLER; MUG; STEIN; TANKARD; WINE GLASS; CHAMPAGNE GLASS; ALE GLASS; WHISKY GLASS; GIN GLASS; TODDY GLASS; RUMMER; PUNCH GLASS; JUICE GLASS; DRINKING HORN; THISTLE GLASS.

———, *English.* Many types of glasses used for particular beverages (e.g., wine, champagne, ale, ratafia, toddy, etc.) and of many shapes and sizes, the styles of the BOWLS, STEMS, KNOPS, and FEET changing from time to time and thus identifying the date of production of many examples. *See* L. H. Bickerton, *Eighteenth-Century English Drinking Glasses* (1971), for illustrations of many types and styles.

———, *German.* Glasses made in many shapes and sizes, decorated with enamelling, engraving or PRUNTS, and often with long inscriptions. *See* HUMPEN; POKAL; RÖMER; BERKEMEYER; MUG; TANKARD; STANGENGLAS; NUPPENBECHER. *See* Axel von Saldern, *German Enameled Glass* (1965), for illustrations of many shapes and styles of decoration.

drinking horn. A drinking vessel in the form of a crescent-shaped horn. Some examples are of ROMAN GLASS from the 1st century AD, but most are of FRANKISH GLASS, *c.* 400–700, blown as a CONE-BEAKER and bent while warm. The latter are usually decorated with applied threads in horizontal spirals, continuous vertical loops, or zigzag WOLF'S-TOOTH patterns, sometimes combined. The tip is pointed or flattened. Unlike other contemporary drinking vessels, the mouth usually does not have a smooth rounded edge, but is knocked off and sharp (*see* GROZING). A few such horns have been found in England, but are thought to have been imported rather than made locally. They were later made in the Netherlands and Germany until the 17th century, with DIAMOND-POINT ENGRAVING and with two small suspensory rings. Some examples of ROMAN GLASS are decorated by being shaped at the point in the form of a faun's head. *See* RHYTON. For Germanic glass drinking horns, see *Journal of Glass Studies* (Corning), XVII (1975), p. 74.

drip pan. A fixed or removable small tray immediately beneath and surrounding the NOZZLE of a CANDLESTICK, for the purpose of catching the wax drippings. Also called a 'grease-pan'. The French term is *bobêche*. *See* SAVE-ALL; SPREADER.

driven glass. The same as SHEARED GLASS.

dromedary flask. A type of flask in the form of a dromedary bearing a globular vase. Examples have been attributed to Syria, in the 6th to 8th

dragon glass. Winged goblet with dragon stem, Netherlands, second half of 16th century. Ht 23·7 cm. Kunstmuseum, Düsseldorf.

drinking horn. Faun's head, Roman glass, 1st century AD. L. 16 cm. Museo Civico, Adria.

centuries AD. The vase is decorated with zigzag threads of glass of different colours.

drop. A glass ornament, smooth or FACETED, that is suspended ornamentally from a CHANDELIER, CANDELABRUM, GIRANDOLE or CANDLESTICK. They were made in a variety of shapes and sizes, and are usually faceted. Some have a small hole for direct suspension, or are hung from an intervening glass button.

——, *Albert.* A form which is triangular in section, having a head shaped like an arrow, and faceted edges. They are found on English LUSTRES of the Victorian period.

——, *button.* A small circular drop which usually hangs above a larger drop.

——, *icicle.* A long thin faceted drop resembling an icicle. They are cut with three or more sides.

——, *mirror.* A drop with, at the back, two facets cut at an angle that permits no light to penetrate, so that light is reflected back as from a mirror.

——, *pear.* A drop that is pyriform and usually is FACETED. Some are long and tapering, some flat.

——, *prism.* A solid glass drop, triangular in section, with plane surfaces which by refraction of light reveal the colours of the spectrum; they are often affixed to a CHANDELIER or, especially, a LUSTRE.

——, *rule.* A long drop with perpendicular sides.

——, *spear.* A long drop with a shaped point like the head of a spear.

dropper flask. A type of FLASK or BOTTLE with a globular body and a wide tall neck having a funnel mouth. The characteristic feature is a small attenuated tube, projecting outward from the middle of the body, which serves as a dropper. Some have a loop handle at right angles to the dropper. Examples were made of SYRIAN and ROMAN GLASS in the 1st/4th centuries. They have been called 'feeding bottles', but such use seems doubtful, especially as the tip of the tube is broken off and rough, and moreover the handle is more serviceable for pouring than for baby feeding. The German term is *Tropfflasche.*

drug jar. A JAR for holding wet or dry drugs in a pharmacy. Generally made of tin-glazed earthenware, some have been made of glass in the form of an ALBARELLO.

dropper flask. Roman glass, 1st century AD. Ht 9 cm. Museo Vetrario, Murano.

drum-stick. A type of FRIGGER in the form of a stick for beating a drum. It is usually a straight stick, about 20 cm. long, at one end of which is a globular beater. They were made of NAILSEA GLASSWARE.

Dublin glassware. Glassware identifiable as made in Dublin. (1) The term is usually considered limited to a few moulded pieces marked on the bottom 'C M & Co' for Charles Mulvaney & Co., established in 1785 and closed by successors in 1846. Some retail glass-dealers in Dublin had their names placed on glassware, including Francis Collins, Armstrong Ormond Quay, and Mary Carter & Son; the ware was possibly made by the Mulvaney company, but does not have any style identifiable as from a particular glasshouse or from Dublin. (2) Flint glass made by the Pugh family in Dublin, first at the Liffey Flint Glass Works by T. & R. Pugh, 1855/63 and later at the Potter's Alley Glasshouse, 1863/93; the ware included COMMEMORATIVE GLASSWARE, some engraved by Bohemian artisans. (3) Ware made by RICHARD WILLIAMS & CO. *See* IRISH GLASSWARE; AYCKBOWM GLASSWORKS.

ductile. Susceptible to being drawn out into a desired shape, e.g. in the case of molten glass. *See* PLASTIC; GLYPTIC; MOSAIC.

dump. *See* DOOR-STOP.

Dutch glass. *See* NETHERLANDS GLASS.

Dutch, Royal, Glassworks. A glass factory established in 1765 at Leerdam, near Rotterdam, making ornamental ware and table glass. It still operates today. *See* LEERDAM GLASSWARE.

dwarf ale glass. A small short-stemmed drinking glass, the BOWL usually of inverted-conical shape, made in England from the early 17th century (when ale was stronger and drunk in smaller draughts) until the early 19th century. The decoration on the bowl is never cut, and the STEM and FOOT occur in various styles. The capacity was about 3 to 4 ounces. *See* ALE GLASS; GEORGIAN ALE GLASS; JELLY GLASS.

Dyottville Glass Works. A glasshouse that was originally the Philadelphia & Kensington Glass Works founded in 1771 in the Kensington district of Philadelphia. Thomas W. Dyott (1771–1861) acquired, *c.*1825, an interest in it and became its owner by 1831, changing its name *c.*1833/34. He was an Englishman who migrated to the United States in 1795, sold patent medicines, styled himself Dr Dyott, and in 1817 became agent for the Olive Glass Works, Glassboro, New Jersey. The factory under him made bottles and HISTORICAL FLASKS, and from 1831 it made FLINT GLASS. He went bankrupt in 1838 from financial speculation, and the factory was taken over by his brother Michael and then many successive owners until after 1926. It is best known for the historical flasks that depict Presidents of the United States.

drug jar. *Albarello*, enamelled clear glass, in the *façon de Venise*. Ht 11 cm. Kunstsammlungen der Veste Coburg.

E

ear trumpet. A hearing aid, some examples being of NAILSEA GLASS in the form of a tube, about 36 cm. long, with a rounded upper end having a small orifice with a flattened everted rim.

Eckhardt, Edris (1910–   ). An American studio artist and teacher of ceramics and glass, from Cleveland, Ohio. She has researched in new methods of making ZWISCHENGOLDGLAS with many laminated layers of glass embedding gold foil, and in making glass sculpture by the CIRE PERDUE method.

Eda Glassworks. A Swedish glasshouse that operated from 1833 until 1953. It was directed by EDVARD STRÖMBERG from 1928 until 1933.

Eder, Paulus (fl. 1685–1709). A glass-engraver of Nuremberg. He decorated glassware, some of it imported from Bohemia, engraving in TIEFSCHNITT in great detail. Eight pieces signed by him are known.

edge. (1) The narrow band between the upper and lower RIMS of a PLATE or the inner and outer rims of a VASE, BOWL, or other hollow vessel. The edge of some English glass vessels is rounded, while that of Continental ones is flat. On English glass of before *c*. 1830, the edge of bowls is uneven and shows shearing marks; later ones are smooth, due to the use of a wooden smoothing tool (burned by the heat of the molten glass) which eliminated these flaws. *See* ROPE EDGE. (2) The thin surface of an object where two plane surfaces meet, such as the edge of a box.

Edinburgh Crystal Glass Co. A large maker of ornamental cut glass, now located at Penicuik, near Edinburgh. It had grown from glassworks operated in Leith, Scotland, from 1864, particularly the Norton Park works operated by Jenkinson & Co. from 1866, and was later known as the Edinburgh & Leith Flint Glass Co., and from 1955 as the Edinburgh Crystal Glass Co. It was acquired in 1919 by Webb's Crystal Glass Co., which then also acquired THOMAS WEBB & SONS, and in 1964 they all became units, with Dema Glass Ltd, Chesterfield, of the group owned by Crown House Ltd. In 1974 the old Norton Park factory was closed and the whole operation moved to Penicuik.

Edkins, Michael (1734–1811). A decorator of delftware and glassware in Bristol who, as a freelance, worked from 1762 for various Bristol firms, including LAZARUS JACOBS, *c*. 1785–87. He kept detailed ledgers from 1762 that have survived and show the nature of his work, which was mainly painting on OPAQUE WHITE GLASS and blue glass, and also some gilt decorating. Motifs that have been attributed to him are mainly floral designs, birds, insects, and CHINOISERIES. However, only a few works are now tentatively attributed to him (none definitely on glass from Bristol), and some pieces of opaque white glass, formerly attributed to him, have recently been said to be more probably by an unidentified London or Midlands workshop. *See* R. J. Charleston, *Journal of Society of Glass Technology*, XXXVIII (1954), p. 13. *See* P F GLASSWARE.

Edwards, Benjamin, Senior (d. 1812). A glass-maker from Bristol who in 1771 set up a glassworks at Drumrea, in County Tyrone, Ireland, and later, *c*. 1776, moved it to the Long Bridge in Belfast. His sons, John, Hugh, and Benjamin, Jr., became partners in 1800; after his death they were associated with several other glasshouses, but with little success. *See* BELFAST GLASSWARE.

Edkins type. Opaque white glass jar, enamelled in the style formerly attributed to Michael Edkins, English, *c*. 1760. Ht 20·3 cm. British Museum, London.

églomisé. Panel, 'Annunciation to the Virgin', Italian, 16th/17th century. Ht 22·5 cm. Walters Art Gallery, Baltimore, Md.

Egermann, Anton Ambros (d. 1888). A glass-enameller of Novy Bor (Haida), In Bohemia, who managed his father's factory and who also worked for J. & L. LOBMEYR, *c*.1878.

Egermann, Friedrich (1777–1864). A native of Blottendorf, near Haida (Bohemia), who was at first an enameller of glassware in the style of ANTON KOTHGASSER. He later turned to developing new styles of coloured glass and new techniques, and his factory at Haida became very prosperous. He developed and produced LITHYALIN GLASS from 1828 until *c*.1840. He also discovered a method of yellow STAINING (*c*.1820) and of red staining (*c*.1840); the colour stain was applied with a brush to give the effect of thin FLASHING. *See* CHAMÄLEON-GLAS.

egg. An ovoid object made of glass, serving as an ornament or a HAND-COOLER.

egg and star border. An ornamental border pattern consisting of alternate ovals and stars. It was used as an engraved motif on some English glasses.

egg-cup. A small semi-ovoid cup on a stemmed foot, for serving a boiled egg in the shell. Glass examples were made in the 18th century, some of coloured or opaque white glass, and usually with a folded rim for added durability. Some were made in sets accompanied by a silver stand from which they were suspended, along with small silver spoons, and having at the top a glass salt-cellar.

églomisé (French). A style of decorating glass by applying gold leaf (occasionally silver leaf, or both) and then engraving it with a fine needle point, but without firing the piece further. Since such work (similar to unfired painting) would readily rub off and was not durable, it was usually applied on the reverse of the surface to be viewed and then protected by a layer of varnish or metal foil or another sheet of glass. Sometimes there is supplemental background decoration in black or colour where the foil has been removed. The term is derived from the name of the 18th-century French writer, artist, and art dealer Jean-Baptiste Glomy (d.1786), who used the process extensively. Such engraving of gold leaf had been done in ancient times by the Egyptians and Romans (*see* FONDI D'ORO) and in Bohemia (*see* ZWISCHEN-GOLDGLAS; MEDAILLONBECHER). Examples were made in Italy, at Venice, Bologna, and Siena, and especially fine medallions were executed in the second half of the 14th century, possibly by the Sienese(?) painter Paolo di Giovanni Fei. Later, ZEUNER decorated glass objects with Dutch and English landscapes in *églomisé* style. Modern examples have been made in England and elsewhere. *See* GOLD ENGRAVING.

Egyptian faience. A ware made in Egypt, from before 3000 BC, of ground quartz fused by the use of an ALKALI and covered with a glaze made of a similarly produced material, but coloured with copper and finely pulverized. The term 'faience' is incorrect, since the body is not pottery and the glaze contains no tin oxide as in the case of glazed faience. Examples of such ware are the mummiform USHABTI tomb figures. The glaze employed was an early forerunner of similar objects made of glass (*see* EGYPTIAN GLASS).

Egyptian glass. Apart from a vitreous glaze on stone and ceramic objects (e.g., BEADS) and on so-called EGYPTIAN FAIENCE, both made before 3000 BC, the earliest surviving glass vessels said to have been made in Egypt are three small vessels with a cartouche of Tuthmosis (Thothmes) III (1504–1450 BC), which were made *c*.1470 BC (*see* TUTHMOSIS JUG), later than some glass made in Mesopotamia *c*.2000 BC. Glass was made in Egypt, especially CORE GLASS, until *c*.1150 BC when the glass-making industry declined until the founding of Alexandria, 332 BC, and the beginnings of its important glass-making and exporting industry. During the following Hellenistic period until *c*.AD 100, core

Eight Priests, Goblet of the. Standing cup, Islamic, from Raqqa, 13th/14th century, with silver-gilt foot. Ht 20 cm. Douai Museum, France.

glass was again made, including small UNGUENTARIA in various GREEK VASE forms, as well as MOSAIC GLASS and glass made by COLD CUTTING and CASTING. *See* GLASS: *pre-Roman history.*

Ehrenfeld. The site, near Cologne, of a glassworks that, *c.*1886–92, specialized in making WINGED GLASSES, characterized by a STEM made of SNAKE THREADING (often with a snake in the form of a figure 8 within a loop) and with CRESTING. It also made copies of 16th/17th century German glassware. Some examples bear on the FOOT a mark of the Rheinische Glashütte, Ehrenfeld (the initials R G, with 'Ehrenfeld' and an outline of a goblet).

Eiff, Wilhelm von (1890–1943). A German glass-designer and engraver who trained with several glass firms, including LALIQUE and LOBMEYR. He taught glass-decoration at Stuttgart from 1922, and his pupils have achieved reputations in many countries and spread his influence. He himself was proficient at all phases of engraving and cutting glass, especially in HOCHSCHNITT, and designed his own pieces, making miniature engraved portraits and figure subjects in connection with developing new techniques of carving in HIGH RELIEF and for chipping the surface (*see* GESCHNITZT). He invented in the 1930s a new flexible engraving tool. He has been regarded as the finest glass-engraver of the 20th century. Some of his pieces are marked 'W.v.E.'

Eight Priests, Goblet of the. An Islamic GOBLET of colourless glass with enamelled decoration of strapwork on a white beaded ground, with a European silver-gilt foot of the 13th–14th century. It is tentatively attributed to Raqqa (*see* RAQQA GLASSWARE) in the late 12th or early 13th century. It was part of a bequest recorded in 1329 which included an endowment for the poor clergy of Douai, France – hence the name. It was preserved in the Douai Museum until destroyed in World War I, along with a French leather case, the arms on which suggest that it was brought from Palestine by a Crusader in 1251. A similar goblet is in the Museum at Chartres, and a comparable later one is in the British Museum. Fragments of similar goblets are at the Victoria and Albert Museum, London.

Eisch, Erwin (1927–). A Bavarian glass-maker with a studio at Frauenau. He specializes in glassware in original forms made by free blowing.

electioneering glassware. Plates, jugs, and other glassware inscribed with names or slogans of electoral candidates, e.g., in Parliamentary elections. Examples are found in English and other glassware, with the design done by DIAMOND-POINT ENGRAVING, WHEEL ENGRAVING or ENAMELLING.

Elector's Beaker. *See* KURFÜRSTENHUMPEN.

Elgin Jug. An ovoid claret decanter (*see* DECANTER, *claret*) with a tall neck, a mouth with a pouring lip, one curved handle, and a stopper. It was WHEEL-ENGRAVED by FREDERICK E. KNY, a Bohemian artist working for the DENNIS GLASSWORKS of THOMAS WEBB & SONS. It was exhibited uncompleted at the Paris Exhibition of 1878 and finished in 1879. Its decoration is an encircling frieze in RELIEF depicting an adaptation of a section of the Elgin Marbles, featuring a group of horsemen from the Parthenon Frieze. Additional encircling decoration includes bands of ACANTHUS and REEDING. *See* ELGIN VASE.

Elgin Vase. An urn-shaped VASE engraved by JOHN NORTHWOOD, having an encircling frieze with an adaptation of a section of the Elgin Marbles, featuring a group of horsemen from the Parthenon Frieze. Additional encircling decoration includes bands of acanthus, Vitruvian scrolls, and ivy leaves. The vase, begun in 1864 and completed in 1873, was commissioned by Benjamin Stone and made of clear glass at his factory, Stone, Fawdray, & Stone, in Birmingham.

Elgin Jug. Claret decanter, engraved by F. E. Kny for Dennis Hall Glasshouse of Thomas Webb & Sons, Stourbridge, 1879. Ht 36·2 cm. Victoria and Albert Museum, London.

Elixir of Life Bottle. A flattened globular BOTTLE, flat-bottomed and having an attenuated neck tapering upward to a narrow mouth. The neck is almost entirely covered with thick spiral trailing. Such bottles have an inscription (meaning 'Elixir of Life') in gilt lettering. They were made as wine containers in Persia, possibly at Qum, probably in the 18th/19th centuries.

Empire style. A version of the NEO-CLASSICAL STYLE, current in France *c*. 1800–20, which followed the Directoire style. It is more or less equivalent to the REGENCY STYLE in England, which is often called 'English Empire'. It is noted for the use of florid classic motifs, to which were added Egyptian motifs and others stemming from the Napoleonic campaigns. Mounting in GILT BRONZE continued, and new decorative themes were introduced.

enamel. A pigment of a vitreous nature coloured with METALLIC OXIDES and applied to glassware, as to porcelain, as surface decoration by low-temperature firing. (*See* ENAMEL FIRING.) It differs from COLD COLOURS in that (1) it is fused into the glass surface and so is not so readily affected by wear; (2) it results in a smooth surface only slightly palpable to the fingers; and (3) its final colour is often not apparent upon application but only after firing. Enamel colours were OPAQUE or TRANSPARENT. (*See* TRANSPARENT ENAMEL.) Enamels were usually mixed with a FLUX to facilitate melting at a low temperature; *see* MUFFLE KILN. The process is different from decorating glass by LUSTRE PAINTING. *See* STAINING.

enamel colours. Colours applied in ENAMELLING glassware and fixed at a low temperature in the MUFFLE KILN. Enamel colours are METALLIC OXIDES mixed with a glassy FRIT of finely powdered glass and suspended in an oily medium for ease of application with a brush, the medium burning out during firing. The actual colours develop in the kiln. In early enamelling, the colours were usually opacified with tin oxide, and on thin English glass this resulted in more than the usual palpability, because it was too risky to heat the soft glass sufficiently for the colours to sink in; but later TRANSPARENT ENAMEL was developed, *c*. 1810 (although JOHANN SCHAPER had used coloured stains in the 17th century). The enamels must fuse at a lower temperature than the glass (*see* ENAMEL FIRING), and they and the glass must have about the same ratio of contraction. *See* COLD COLOURS.

enamel firing. A low-temperature FIRING of glassware in the MUFFLE KILN, to affix permanently ENAMEL COLOURS; in the 16th century both Venetian and German glassware was fired in the ordinary furnace. The method is similar to the *petit feu* firing of decoration on porcelain, and is at a temperature in the range of 700–900° C. The enamel colours are mixed with a FLUX (e.g., BORAX) to lower the fusing point to below that of the glass object. Objects to be fired are enclosed in a MUFFLE. The low temperature is necessary to fuse the enamelling to the glass object without damaging its form. Sometimes several firings are needed for different colours.

enamel glass. A term sometimes misleadingly and incorrectly applied in the 18th century (and sometimes later) to OPAQUE WHITE GLASS, because its appearance and composition (tin oxide) are similar to those of enamelled or tin-glazed pottery. (*See* NEVERS FIGURE.) The term 'enamel' with respect to glassware should be restricted to the vitreous fusible pigment used for surface decoration; *see* ENAMEL; ENAMEL COLOURS; SCHMELZGLAS.

enamelling. The process of decorating glassware by the use of ENAMEL on the surface to paint scenes, figures, and inscriptions. Enamelling was done on clear or coloured glass or OPAQUE WHITE GLASS. The process of using ENAMEL COLOURS by ENAMEL FIRING was known in Roman times and during the Islamic period in Egypt and Syria, and in China in the

Elgin Vase. Vase by John Northwood, completed 1873. Ht 40 cm Birmingham City Museum & Art Gallery.

enamelling, German. *Ranftbecher* with enamelled view of St Stephen's Cathedral, Vienna, by Anton Kothgasser; Vienna 1812–30. Ht 11·7 cm. Museum für Kunsthandwerk, Frankfurt.

18th century; it was used extensively in Venice from the 15th century and elsewhere in Europe from the 16th century. Opaque enamels were used generally but TRANSPARENT ENAMEL was developed *c*.1810.

——, *English*. The earliest mention of enamelling on glassware in England is a notice by a Mr Grillet in 1696, but no examples are extant. Early enamelling was introduced by artists from Germany and the Low Countries, and was of the type known as 'thin' or 'wash' enamel, more suitable for the softer English LEAD GLASS, and was executed within outlines previously etched on the glass. Later 'dense' enamel became the usual medium, first with designs of festoons and flowers, but later landscapes, figure subjects, and coats-of-arms. The leading enamellers in the 18th century were WILLIAM and MARY BEILBY and possibly MICHAEL EDKINS; they decorated on clear glass and OPAQUE WHITE GLASS, respectively. Some ware was enamelled by glasshouses in Stourbridge in the 19th century, and possibly by London jewellers and silversmiths doing work on glass boxes and SCENT BOTTLES, as well as by artists who decorated porcelain in imitation of Sèvres porcelain and who did like work on glassware in the mid-19th century (*see* SÈVRES-TYPE VASE).

——, *German*. The earliest German ENAMELLING on glass was in the second half of the 16th century, copying the Venetian enamelled armorial glasses that had been previously imported. The German enamelling is found on the typical German HUMPEN and STANGEN-GLAS, where the style evolved to cover almost the entire surface with painting, subordinating the glass itself. Bright colours created decorative effects, but the painting was not of superior quality. The early motifs were coats-of-arms, followed by a great variety of subjects, e.g., paintings of a religious, allegorical, historical or scenic nature, or showing artisan or guild activities, or satirical or family scenes, usually with dates or long datable inscriptions added. (*See* REICHS-ADLERHUMPEN, KURFÜRSTENHUMPEN, APOSTELGLAS, WILLKOMM, FAMILIENHUMPEN, WESTPHALIA TREATY HUMPEN.) The place of production is usually unidentifiable, except in some special types such as the OCHSENKOPFGLAS (Franconia), HALLORENGLAS (Halle), and HOFKELLEREIGLAS (Saxony or Thuringia). For a full discussion of German opaque-enamelled glassware, with factories, forms, and decoration, and many illustrations, see Axel von Saldern, *German Enameled Glass* (1965). In the 17th century enamelling became more restrained and skilfully done, with the artists usually signing the pieces. A new style of enamelled decoration, SCHWARZLOT, was introduced by JOHANN SCHAPER and his followers, JOHANN LUDWIG FABER, ABRAHAM HELMHACK, and HERMANN BENCKERTT. The use of opaque enamel was superseded by transparent enamel, introduced *c*.1810 by SAMUEL MOHN and also used by his son GOTTLOB SAMUEL MOHN and FRANZ ANTON SIEBEL. In the 1750s some enamelling was done on OPAQUE WHITE GLASS (*see* BASDORF GLASS).

——, *Islamic*. Opaque vitreous enamelling (with gilding) as decoration on glassware made between 1170 and 1402 – when Tamerlane (Timur) sacked Damascus and the local glass industry was moved to Samarkand – and some glassware associated with Fostat (Egypt) said to date from *c*.1270–1340. It is of five tentatively identifiable groups based on the criteria of style and date rather than any proven connection with the places of production identified with the groups, i.e. RAQQA GLASSWARE, SYRO-FRANKISH GLASSWARE, ALEPPO GLASSWARE, DAMASCUS GLASSWARE, CHINESE-ISLAMIC GLASSWARE, and SYRIAN GLASSWARE.

——, *Venetian*. Enamelling done at MURANO on clear coloured glass and OPAQUE WHITE GLASS. Although enamelling on glassware has recently been said to have been done in Venice at the end of the 13th century, no example is reported. From the late 15th century enamelling was done in a manner similar to that of Islamic glassware of the 13th/14th centuries, the technique being said to have been re-invented by Anzolo Barovier. It was first used on sapphire-blue glassware (*see* BAROVIER CUP) and other coloured transparent glassware, but towards the end of the 15th century coloured glass tended to be superseded for

enamelling, Venetian. Beaker with decoration of *amorini* and crabs, Venice, *c*.1500. Museum of Decorative Arts, Prague.

this purpose by colourless glass in Venice. By the end of the 16th century enamelling was little used there except on OPAQUE WHITE GLASS (*see* LATTIMO), which again became popular in the 18th century (*see* VENETIAN SCENIC PLATES).

——, *Viennese*. Decoration on glassware at Vienna, especially by GOTTLOB SAMUEL MOHN and ANTON KOTHGASSER.

encased glassware. Objects of coloured glass or OVERLAY GLASS that are covered with a layer of clear glass, such as PAPER-WEIGHTS. The process of encasing an overlay paper-weight requires that, after applying the coloured overlays, the weight must be cooled and ANNEALED so that the WINDOWS can be ground, and then placed again on a PONTIL for reheating and fusing the outer overlay of clear glass.

enclosed ornamentation. A type of decoration in the interior of solid glassware, such as is found in many forms of PAPER-WEIGHTS (e.g. flower, etc.) and some DOOR-STOPS. *See* J. Stanley Brothers Jr, 'Enclosed Ornamentation', in *Journal of Glass Studies* (Corning), IV (1962), p. 117.

encrusted cameo. A term used in England for a SULPHIDE. Also called 'cameo incrustation' or 'crystallo-ceramie'.

encrusted glassware. Glass objects which have attached to the body, in a haphazard manner, numerous very small fragments of glass, projecting from the surface as a result of having been applied after the piece was MARVERED and shaped. Sometimes the bits are of opaque white glass affixed to a body of coloured glass; sometimes they are very tiny and coloured. The decoration was used on ROMAN GLASS of the 1st to 4th century. *See* FLECKED GLASSWARE; BLOBBED GLASSWARE; JEWELLED GLASSWARE; BRÖCKCHEN.

end-of-day-glass. An American term for FRIGGERS, also called 'off-hand-glass'. The term is sometimes applied to SLAG GLASS, but incorrectly, as such glass is made for specific purposes, while friggers are made from remnants of various types of molten glass left in the POTS.

engraved designs. Engravings of pictorial subjects and decorative patterns employed by glassware decorators as inspiration for their work. *See* CALLOT FIGURES; COMMEDIA DELL'ARTE FIGURES; SCENIC DECORATION.

engraving. The process of decorating glass by cutting the design into the surface of the glass by a diamond, metal needle or other sharp implement or a rotating wheel. There were various techniques: DIAMOND-POINT ENGRAVING; WHEEL-ENGRAVING; STIPPLING. Related processes are ETCHING and SAND-BLASTING. Some of the techniques, similar to the work of the lapidary, were used in earliest times, in Egypt and Rome, and later in Venice, Germany, Bohemia-Silesia, Holland, France, and England. Engraving was best done on glass that had a hard brilliance when cut and was of sufficient thickness; hence the process developed after the introduction in Bohemia and Silesia of glass made with chalk and in England of glass made with lead, superseding the fragile Venetian LIME-SODA GLASS. *See* INCISED.

——, *Bohemian-Silesian*. WHEEL-ENGRAVED decoration on BOHEMIAN and SILESIAN GLASSWARE. It was introduced in Central Europe by CASPAR LEHMAN(N), probably before 1600, at Prague, and developed by GEORG SCHWANHARDT at Nuremberg. It was continued by many unidentified artists in various glass centres, but especially at localities where glassworks were under the patronage of the local ruler, e.g., Petersdorf, under the Counts von Schaffgotsch (*see* FRIEDRICH WINTER). Several glass engravers worked in the mid-18th and early-19th centuries, in the BIEDERMEIER period, at the spas during their season, engraving portraits, e.g., at WARMBRUNN (*see* CHRISTIAN GOTTFRIED SCHNEIDER), and later at Franzensbad (*see* DOMINIK

encrusted glassware. Vase, violet-coloured with white flecks, Roman glass, 1st century AD. Museo Civico, Adria.

engraving. 'Wild Strawberry' bowl, designed by Sigurd Persson, engraved by Lisa Bauer, 1974. Ht 23 cm. Courtesy, Kosta-Boda Glassworks, Sweden.

engraving, German. Covered *Pokal* with decoration in *Tiefschnitt*, with arms of Friedrich II of Gotha-Altenburg, by Georg Ernst Kunckel, Gotha, *c.* 1730. Ht 24·8 cm. Kunstsammlungen der Veste Coburg.

BIEMANN). Much of the finest work of the early period was in elaborate BAROQUE style, featuring HOCHSCHNITT and TIEFSCHNITT, and later in ROCOCO STYLE. A large number of glassworks were founded throughout the region, especially in Bohemia in the late 17th century and in Silesia from *c.* 1725; a school for engravers was opened at Karlsbad in the early 19th century by ANDREAS MATTONI.

——, *English.* The earliest engraving on English glassware is the DIAMOND-POINT ENGRAVING generally attributed to ANTHONY DE LYSLE or craftsmen at the glassworks of JACOPO VERZELINI after the latter's death. (*See* DIER GLASS; BARBARA POTTERS GOBLET.) Later diamond-point engraving is found on JACOBITE GLASSWARE and other COMMEMORATIVE GLASSWARE, as well as being used for inscriptions of contents of vessels or convivial subjects. There are no well-known artists, and the work may have been done by silver engravers. After LEAD GLASS was introduced, the thickness and hardness of the METAL were more conducive to WHEEL-ENGRAVING, as seen on some FLOWERED GLASSES, and to decoration by CUTTING.

——, *French.* Glassware was not decorated by ENGRAVING in France before the mid-19th century but by then the glass industry there was further developed and some superior engraved pieces were made at the CLICHY GLASSWORKS at Clichy-la-Garenne, *c.* 1862.

——, *German.* After the introduction in Central Europe by CASPAR LEHMAN(N) of WHEEL-ENGRAVING to decorate glassware, probably before 1600, the art was greatly developed in the 17th and 18th centuries in the German glass centres, such as Nuremberg (by GEORG SCHWANHARDT, HERMANN SCHWINGER, GEORG KILLINGER, JOHANN HEEL, ANTON WILHELM MÄUERL) and Potsdam (Berlin) especially at the glassworks established by JOHANN KUNCKEL (by GOTTFRIED SPILLER, MARTIN WINTER, HEINRICH JÄGER, ELIAS ROSBACH), as well as in other cities, e.g. Magdeburg (HEINRICH FRIEDRICH HALTER), Frankfurt-am-Main (JOHANN BENEDIKT HESS), Weimar (ANDREAS FRIEDRICH SANG), and Brunswick (JOHANN HEINRICH BALTHASAR SANG). In several localities glasshouses were set up under the patronage of the local rulers, e.g. Hesse-Cassel (FRANZ GONDELACH), Gotha (GEORG ERNST KUNCKEL), Brandenburg (MARTIN WINTER), and Arnstadt (SAMUEL SCHWARTZ). The work in the 17th and 18th centuries was in the BAROQUE and ROCOCO styles, with elaborate wheel-engraving in HOCHSCHNITT and TIEFSCHNITT.

——, *Islamic. See* under ISLAMIC GLASSWARE.

——, *Italian.* Decoration on glass by DIAMOND-POINT ENGRAVING was done in Italy from the mid-16th century to the 17th century but was soon discontinued in favour of other styles, although it was used elsewhere on glassware *à la façon de Venise.* Wheel-engraving was not suitable, as the Venetian LIME-SODA GLASS was too thin and fragile to be a medium for that style, but it was probably introduced in the 18th century.

Ennion. A celebrated glass-maker who, in the first quarter of the 1st century AD, made MOULD-BLOWN glass in Syria (*see* SIDONIAN GLASS), sometimes marking on a pad on the handles of the vessels his name (in Greek and Latin letters) and sometimes the word 'Sidon'. Occasionally his name is moulded on the piece. As some pieces marked 'Ennion' have been found in Italy, it has been conjectured that he later moved there. His marked pieces include cups, jugs, and two-handled bowls, decorated with stylized plant forms and REEDING. Twenty-one signed pieces were known in 1962. *See* GLASS-MAKER'S MARK.

epergne (French). A centre-piece for a dinner-table, which includes numerous small dishes, CRUETS, and SALT-CELLARS supported by branching arms; it is also sometimes provided with candle nozzles. English versions are often based in design on contemporary silver, and were made in ADAM STYLE in the last quarter of the 18th century. One English example has small baskets for sweetmeats suspended from arms extending from a central column.

Ennion. Greenish glass bowl, Syrian, 1st century AD. Ht 12 cm. Museo Civico, Adria.

Ericsson, Henry (1898–1933). The artistic adviser to the RIIHIMÄKI GLASSWORKS from 1928 when the making of ornamental glass was begun in Finland.

Escalier de Cristal. A glass-decorating and selling company featuring GLASS FURNITURE. It was established in 1802 in the Palais Royal in Paris by CHARPENTIER. By 1816 his widow, Madame Desarnaud-Charpentier, had taken over the business, making tables and metal-mounted ornaments. The glass was made at the VONÊCHE GLASSWORKS and decorated in Paris.

Estonian glassware. Glassware produced in Estonia after the territory was ceded by Sweden to Russia in 1721 after the 1700–21 War and when the export of timber from Estonia was banned by Russia. Its first recorded glassworks was the Pirsalche Glase Fabrik at Piirsalu founded before 1740, which, after making mainly window glass and some hollow ware, closed c. 1790–98. The largest factory in the 18th century was at Rekka; it and other factories made bottles, mirrors, and windows, and during the 19th century some household glassware, but none made artistic glass. Of more than forty glassworks since the 17th century, only four were known to remain in 1969. *See Journal of Glass Studies* (Corning), XI (1969), p. 70.

etching. Decoration on the surface of glass by the use of HYDRO-FLUORIC ACID, varying from a satin MAT finish to a deeper rough effect. The process involves covering the glass with an acid-resistant wax, varnish or drying linseed oil, and then scratching with a sharp tool to form a design; then a mixture of hydrofluoric acid, with potassium fluoride and water, is applied to corrode (etch) the exposed design into the body of the piece (hence sometimes called 'acid etching' or 'acid engraving'). The depth of the etching varies with the time of exposure. Hydrofluoric acid was not generally used until the mid-19th century, although it was discovered in 1771; an etching on a bowl dated 1686 attributed to HEINRICH SCHWANHARDT may have resulted from a different acid (sometimes called *aqua fortis*) that reacted on the glass. In the 19th century the process was used to decorate extremely thin tableware that was not susceptible to wheel-engraving. It was later used by EMILE GALLÉ and his followers to facilitate engraving in HIGH RELIEF, and at Zwiesel in Bavaria it was used for frosted decoration. It was used in England in the 1830s by Thomas Hawkes of Dudley and in the 1860s and 1870s by J. & J. Northwood and by Guest Brothers. Acid etching is also used on modern ARCHITECTURAL GLASS to produce textures and patterns. The process is being replaced by SAND BLASTING. *See* SATIN GLASS; COLOUR ETCHING; DOUBLE ETCHING.

Etruscan-style decoration. A style of decoration that imitated that on ancient Etruscan pottery, employed by W. H., B. & J. RICHARDSON, of Stourbridge, England, c. 1850. It depicted figures and floral designs in black enamelling.

étui (French). A small case fitted with miniature implements, such as scissors, bodkins, needles, a tiny knife, etc. They were variously shaped, delicately decorated, and made of porcelain or glass, richly mounted with gilt hinges and collars.

everted. Turned outwards in shape, such as the LIP of a JUG, PITCHER or SAUCE-BOAT, or the RIM of some vessels.

ewer. A type of JUG, usually tall with globular or ovoid body resting on a flaring BASE and having one large loop handle extending from the shoulder to or above the rim. They have a tall neck and a lipped pouring rim or occasionally a spout upward from the body and connected to it by a bridge. They are sometimes accompanied by a basin EN SUITE. Generally made of porcelain or pottery, some examples have been made of glass, e.g., marbled or opaline. The French term is *aiguière. See* AMPULLA.

épergne. Coloured glass flowers, *reticello* base and bowls, Venice. Ht 51·5 cm. Royal Collection, Rosenborg Castle, Copenhagen.

ewer. Agate glass with gilt Arabic script, Venice, early 16th century. Museum für Kunsthandwerk, Frankfurt.

Excise Acts, Glass. Revenue measures enacted to raise funds, usually for financing foreign wars. The earliest in England was imposed in 1645 by Oliver Cromwell, at the rate of 1 shilling for each £1 of value (i.e. 5%). That of May 1695, imposed for five years by William III, levied a tax on FLINT GLASS and PLATE GLASS of 20% and of 1 shilling a dozen on common glass bottles; due to the public outcry, this was halved in 1698 and repealed in 1699. The Act of Parliament of 1745 taxed glass (but excluding OPAQUE WHITE GLASS and some CULLET) on the weight of materials used (9s. 4d. per hundredweight, later 21s. 5½d.); the result was a reduction in the size of glassware (e.g., the substitution of lighter BALUSTER wine glasses for the heavy baluster), and the elimination of embellishments such as KNOPS and the folded FOOT, and greater use of surface decoration (e.g., ENGRAVING, ENAMELLING, and GILDING), of coloured, opaque white, and air TWISTS, and also of FACETING, as well as greater use of cullet. This tendency was increased by the higher duties of the Acts of 1777, 1781, and 1787, and the granting of free trade with Ireland in 1780 (where there was no excise duty until 1825). The Excise Acts were repealed in 1845, resulting in a great increase of glass production of all types.

Exeter Flute. A FLUTE GLASS made to commemorate the Coronation of Charles II (1660) and probably engraved for a banquet held at Dorchester in 1665 for a visit by him. It is DIAMOND-POINT ENGRAVED with a portrait of Charles II, the inscription 'God Bless King Charles II', and a truncated oak-tree with a sprouting branch. (The oak stump is said to represent the beheading of Charles I and the sprout the Stuart Restoration under Charles II.) The flute was formerly attributed to the Savoy Glasshouse of the DUKE OF BUCKINGHAM, but W. B. Honey has written that it was both made and engraved in the Netherlands, similar to the SCUDAMORE FLUTE and the ROYAL OAK GOBLET. The Exeter Museum (the owner) records an attribution of the engraving to Christof le Jansz, a Dutch engraver. The Museum has no record of its use ever at Exeter and suggests that the name is a misnomer. A very similar flute, but without the sprouting oak-tree, is in the Toledo (Ohio) Art Museum, attributed to the Netherlands, c. 1660–80; and still another, but with only the engraved coat-of-arms of Charles II, is known.

exotic birds. A decorative motif depicting birds of exotic types. It was originally introduced as decoration on Meissen porcelain and called *Fantasie Vögel*; it was later used at Vincennes and Sèvres, and in England at Bow, Chelsea, and Worcester. Such decoration is also found on some English glassware with gilded decoration attributed to the workshop of JAMES GILES.

expanded glassware. Glassware which, after being shaped in a MOULD, was enlarged by BLOWING, before any further decorating was done by manipulation. Examples are American glassware with 'expanded vertical ribbing' or 'expanded diamonds', especially STIEGEL-TYPE GLASSWARE.

extrinsic decoration. Various styles of decoration on glassware after it has been shaped and ANNEALED, e.g., CUT GLASS, *patterns*; ENAMELLING; ENGRAVING; GILDING.

eye. (1) The centre of the grate in the SIEGE of the glass FURNACE, the hottest part. (2) The hole ('*occhio*') in the floor of a LEHR through which heat is transmitted. (3) The coloured tip (PEARL) on the PRUNTS of some NUPPENBECHER and other 15th/16th century Antwerp and German glassware. *See* GLASS EYE; LITTLETON, HARVEY K.

eye-bath. A small cup-like utensil having a rim made to conform to the shape of the eye, and usually resting on a stemmed FOOT. Some examples have a globular reservoir between the cup and the base. Examples have been made in clear or coloured glass, and decorated with FLUTING.

Exeter Flute. Diamond-point engraved portrait of Charles II, 1660. Ht 44–5 cm. Royal Albert Memorial Museum, Exeter.

eye-bead. A BEAD, of several varieties, all somewhat resembling an eye, generally made of glass, and almost always talismans. Chief varieties are: (1) 'spot beads', each with a single spot of glass of one colour stuck on or pressed into the surface of the matrix; (2) 'stratified eye beads', each with a spot of coloured glass pressed into the surface, and another spot affixed on it, and so on; (3) 'ring eye beads', each with a ring or rings of glass of a different colour pressed into the surface; and (4) 'cane eye beads', each with a slice of a CANE pressed into the surface. The first two types were made as early as 1300 BC, the third *c*. 9th century BC, and the fourth *c*. 6th century BC, and all continued to be made during Roman times. Such beads were also made in China in the 4th and 3rd centuries BC, being the earliest form of Chinese manufactured glass.

eye-paint phial. A small glass receptacle for containing eye paint (kohl), sometimes accompanied by a glass rod for application. They were made in Egypt in the XVIIIth Dynasty (*c*. 1380 BC). They are in the form of a thin vertical tube, about 7·5 cm. high.

exotic birds. Scent bottle with gilt birds and inscription, studio of James Giles, London. Ht 7·4 cm. Courtesy, Sotheby's, London.

F

façon de Venise. Vase with double-ogee body overlaid in blue, handles blue with clear cresting, blue prunts, made in *façon de Venise*, 19th century. Ht 25 cm. New Orleans Museum of Art (Billups Collection).

Fairfax Cup. Tumbler of opaque glass with enamelling, Murano, *c.* 1480. Ht 9·2 cm. Victoria and Albert Museum, London.

Faber, Johann Ludwig (fl. 1678–97). A HAUSMALER in Nuremberg who was a follower of JOHANN SCHAPER as a painter of STAINED GLASS, faience, and glass vessels, and decorated glass in SCHWARZLOT. He invented some new TRANSPARENT ENAMELS. Only one signed piece is recorded.

fable glassware. Glassware with enamelled decoration depicting animals and figures from fables, such as *Aesop's Fables*. Such decoration is found on HUMPEN, executed in Bohemia in the 16th and 17th centuries.

faceted glass. Glass (other than flat glass) that is decorated by grinding to give it a partial surface area that is flat or nearly so. Glass so decorated is a type of CUT GLASS. SODA GLASS is not suitable for such cutting, while POTASH GLASS has been often so decorated in Bohemia, and LEAD GLASS, being heavy and having a high refractive index, is generally so decorated in England and Ireland. The STEMS of many drinking glasses are faceted by being given a series of encircling vertical connected plane surfaces (called a 'FLAT CUT STEM') or by SCALE FACETING or HEXAGONAL FACETING. Some glassware is decorated with shallow circular concavities (PRINTIES). Faceting also is found on HANDLES of glass vessels and on KNOPS of stems. This form of decoration did not appear on English glasses until towards the middle of the 18th century, by which time it was highly developed in Germany; it had appeared on Middle Eastern glass as early as the 8th century and also occurs on ROMAN GLASS. *See* BEVELLED GLASS.

faceting, circular. A style of decoration made by FACETING in the form of small circular concavities, sometimes in an overall pattern or sometimes in an encircling band, as PRINTIES.

façon de Venise, à la. In the style of Venice, as applied to high-quality glassware made throughout Europe, often by emigrant Venetian glassworkers, especially the thin SODA GLASS and glassware decorated in FILIGRANA or with ornate embellishments such as the WINGED GLASSES. It was developed in the mid-16th century and flourished throughout the 17th century.

factory-mark. A mark or inscription on glassware to indicate the place of production. Such marks on glassware are usually impressed, scratched, etched or engraved on the bottom. Factory marks were used less often on glassware than on ceramic ware, but have been used by Irish glasshouses (*see* IRISH GLASSWARE, *decanters*), some French factories (e.g., BACCARAT), and some makers of English SLAG GLASS. *See* GLASS-MAKER'S MARK; RAVEN'S HEAD.

Fairfax Cup. A TUMBLER with sides sloping slightly inward toward the bottom. It is of opaque turquoise glass, the colour turning to red when viewed by transmitted light. It is decorated with an encircling scene of a fountain and figures depicting the legend of Pyramus and Thisbe, in blue, green, and yellow enamelling, with two encircling gilt bands. It was made at MURANO, *c.* 1480, and was in the Fairfax family from before 1643. *See* TRANSMUTATION COLOUR.

Fairy Night Light. A small glass dome-shaped shade, with a vent hole in the top, that was placed over a special type of candle made by Samuel Clarke, *c.* 1886–92 in London, a maker of candles and tapers, not a glassmaker. He patented the shade in several countries and licensed glass-makers in England (e.g. SOWERBY and THOMAS WEBB & SONS) and on the Continent and in the United States to make them. The shades were made of glass of various colours (e.g., BURMESE GLASS) and in many models, some with SATIN or QUILTED surface, and some in novelty form (e.g. as an animal). The candles were of wax with a double wick of twisted rush; they burned for 4 to 11 hours. The shades were used on tables, but also affixed to CHANDELIERS and CANDELABRA or placed under food-warmers. They were also called 'Clarke's Patented Pyramid Night Lights'. Some unlicensed glass-makers made imitations. *See* CRICKLIGHT.

fake. A piece of genuinely old glassware that has been altered or added to, in its body or decoration, for the purpose of deceptively enhancing its value, e.g. one that has had added ENGRAVING or ENAMELLING. *See* FORGERY; REPRODUCTION.

Falcon Glassworks. Originally two glasshouses in Southwark, London, said to have been founded in 1693 by the glass-makers Francis Jackson (d. *c.*1701) and John Straw, and later operated by William Jackson until 1752. After 1752 one of these glasshouses, the 'Cockpit', is said to have been taken over by Hughes & Winch (during whose operation JEROM JOHNSON possibly was a partner, *c.*1757), and then in 1760 was operated by Hughes, Hall & Co. and in 1765–80 by Stephen Hall & Co., after which it was probably abandoned. The other, the 'Falcon', probably operated from 1752 to 1803 under William Barnes & Co. and successors and from 1803 by Green & Pellatt (*see* APSLEY PELLATT); in 1831 Apsley Pellatt & Co. ceased business. From 1820 the glasshouse made SULPHIDES. It supplied glass to JAMES GILES.

Familienhumpen (German). Literally, family beaker. A type of HUMPEN (or STANGENGLAS) decorated with the enamelled portrayal of a couple and their children lined alongside by ages, and usually with inscriptions showing their names and ages. Sometimes there is, instead of figures, a family tree. They were made in the 17th and 18th centuries, mainly in Franconia. *See* ENAMELLING, *German.*

fan cutting. A decorative pattern on CUT GLASS usually found on the scalloped rim of a bowl. The 'fan' is a series of mitred grooves cut within the scallop and radiating to the upper edge, and usually with the extremity of each mitred groove notched. This pattern is also called the 'shell border' or the 'fan escallop'.

fan vase. A type of vase (ranging from 4 cm. to 20 cm. high) having a narrow bowl with parallel sides and shaped like a triangular fan with a slightly convex rim; the bowl is supported above the base by one or more knops. They are of plain glass or decorated with bubbles or engraved pattern. They were made by FREDERICK CARDER in the 1920s.

fans. *See* CUT GLASS, *patterns.*

Farquharson, Clyne. An English glass-designer who worked in the 1930s for John Walsh Walsh at Birmingham, and thereafter for STEVENS & WILLIAMS, LTD.

Favrile glass. A type of glass developed *c.*1892 by LOUIS COMFORT TIFFANY in ART NOUVEAU style, having an IRIDESCENT surface simulating somewhat that of ancient Roman glass excavated after being buried for centuries. The bluish-green, golden and other surface colours, in a nacreous effect, were produced by spraying the surface, while hot, with metallic salts which were absorbed into the glass. The lustre effect depended on the colour and density of the METAL. The multi-coloured decoration, sometimes embedded or applied, features naturalistic motifs. The glass was used for a great variety of objects, but mainly vases of various sizes and shapes. Examples are signed with Tiffany's name or initials. The name 'Favrile' (registered in 1894),

Familienhumpen. Beaker with enamel figures and inscriptions, Germany, 1671. Ht 23·5 cm. Kunstgewerbemuseum, Cologne.

feeding bottle. German, 18th century.
L. 23 cm. Kunstsammlungen der Veste
Coburg.

festooned glassware. Jug, Nailsea-type
greenish glass with opaque white
festoons, early 19th century. Ht
15·8 cm. Victoria and Albert Museum,
London.

added on the glass or a label, is derived from *fabrile* (an old English word meaning 'belonging to a craftsman or his craft) which was first used before inventing a name that could be registered. The glass was imitated by the LÖTZ GLASSWARE factory and other Bohemian glass-makers. *See* TIFFANY GLASSWARE: *Favrile fabrique.*

feather holder. An object made to hold a feather as a hat ornament. They were made of glass in China in the Ch'ing Dynasty from *c.* 1662.

feathered decoration. A style of decoration on ancient CORE GLASS in the manner of COMBED DECORATION, making a zigzag pattern. It is also found on glassware of post-Roman date, on some ISLAMIC GLASSWARE, and on some Venetian and Spanish glassware.

feeding bottle. (1) A boat-shaped vessel completely enclosed except for a hole in the top for filling (and for controlling the flow by use of the thumb) and a small projecting tube at one end to which can be attached a teat. Such bottles were used for feeding infants and invalids. They were usually made of porcelain or glazed earthenware, but also, *c.* 1840, of colourless glass; they are usually of undecorated form, but one ornate Continental example is in the form of a large bird. Some made in England, *c.* 1890, have a built-in thermometer. (2) An upright bottle, the forerunner of the modern baby-feeding bottle; it is usually cylindrical or pear-shaped. Some examples have a glass feeding tube extending through the stopper and down to the bottom of the bottle, to prevent sucking air.

feeding cup. A type of CUP, partially covered to prevent spilling, for feeding children and invalids. It has a SPOUT (usually straight on early examples, curved on later ones) at the front and generally a loop HANDLE at right angles to the spout. Some made in Venice in the 16th and 17th centuries were in fantastic form, e.g., shaped like a fish. *See* HALF-DECKER GLASSWARE; POSSET POT.

fenici, vetro (Italian). Literally, phoenix glass. A type of glassware, so-called at Murano, with decoration of bands of opaque white glass (LATTIMO) manipulated with a special tool (*maneretta*) so as to create a pattern of white waves, festoons, feathers, or ferns. It is a modern adaptation of ancient glassware with COMBED DECORATION which was made with coloured bands on glass of contrasting colour (termed *vetro a pettine* or *vetro a piume*).

fern-ash. Ash obtained by burning ferns (bracken). It was used in making VERRE DE FOUGÈRE, to provide the POTASH that contained the ALKALI ingredient for such glass; the alkali acted as a FLUX. *See* SODA.

fern-decorated glassware. Glassware decorated with engraved depictions of ferns. It was mainly produced in Scotland, by John Ford at the Holyrood Glassworks, and was exhibited in 1862. It was produced until the end of the century, being engraved by Emanuel Lerche and Johan Millar, and other Bohemian immigrants, and became very popular at the time.

festoon. A decorative pattern in the form of garlands of flowers, leaves, fruit, ribbons or drapery, hanging in a natural curve and suspended from the two ends. *See* SWAG.

festooned glassware. A type of glassware formerly attributed generally to the NAILSEA GLASSHOUSE, but now ascribed to other unidentified factories. The METAL is brownish-green BOTTLE GLASS, and the decoration is bold white festoons covering the entire surface of the vessel. Such festoons are sometimes called 'quillings'.

fibre glass. *See* GLASS FIBRE.

fibula. An ancient brooch consisting of a bronze pin with a coiled spring and catch (the most common form of which resembled the modern safety pin), but some have a curved bow joined to the pin. The bow is sometimes enclosed within a glass leech-shaped tube (or a 'runner' in the form of a half-section of such a tube). The glass is sometimes opaque white, decorated with marvered coloured combed trailing in a feather pattern. Examples have been found at Etruscan sites in Italy and attributed to the 9th/7th centuries BC. It has been suggested that some were probably made locally, but that some were made in the Illyrian region (present-day Yugoslavia), and that the glass for the making of such objects may have been imported from elsewhere in the form of ingots and worked locally. *Fibulae* have been said to distinguish the Greek historic period from the prehistoric, and may have been introduced by the Dorians, *c.* 1100 BC, or made from fragments of bracelets of CELTIC GLASS. *See* LEECH BEAD.

Fichtelgebirgeglas (German). *See* OCHSENKOPF.

figure. An object representing a human being or an animal, alone or in a GROUP, or as an ornament attached to a vessel such as a VASE. The French term is *statuette*. *See* NEVERS FIGURE.

figure candlestick. A CANDLESTICK having its SHAFT in the form of a figure supporting the candle NOZZLE. Some are in the form of a caryatid or dolphin.

figured beaker. A type of BEAKER decorated with relief figures. A well-known group consists of MOULD-BLOWN cylindrical beakers which are slightly tapering inward towards the bottom, and characterized by an encircling decoration of four standing figures in relief; the figures face to the right or in opposite directions, and are separated by columns, usually with a pediment or a festooned garland above them. In this group the figures represent various mythological characters (not all yet identified). The beakers are said to have been made in Syria and probably other places, in the 1st and 2nd centuries AD, and are sometimes referred to by the German term *Götterbecher*. Over twenty of the group have been identified, ranging in height generally from about 12·0 to 12·6 cm. See *Journal of Glass Studies* (Corning), XIV (1972), p. 26.

figured engraving. A style of decoration done by hardstone-point engraving, or by linear WHEEL-ENGRAVING, depicting usually mythological scenes and figures encircling shallow colourless bowls. Early examples are from Egypt, and later ones are known from Rhenish workshops. Some pieces have added faceting and also shading of ABRADED DECORATION. The early examples so decorated have usually an encircling engraved band just below the rim. Other examples exist from the 17th and 18th centuries.

figurine. (1) A small individual FIGURE, the term being employed especially of ancient examples. They were made of EGYPTIAN GLASS, in the 14th–13th centuries BC, by fusing powdered glass in the CASTING process. Those with a flat back were made in an open PRESS-MOULD, but those in-the-round were perhaps cast in two-piece moulds. (2) A common expression in the United States for any figure.

filigrana or **vetro filigranato** (Italian). Literally, thread grained. A term that has been applied to glassware (with various styles of decoration) of clear glass, made originally at MURANO, *c.* 1527–49, by the use of opaque white or of coloured glass threads (or even sometimes a single white thread), but which term is preferably used generally (as proposed by Astone Gasparetto, of Murano, in 1958, and other later writers) to refer to all styles of decoration on clear glass that are made with a pattern formed by embedded threads of glass, including (1) VETRO A RETORTI (glass with embedded twisted threads forming various lace-like patterns), (2) VETRO A RETICELLO (glass with embedded criss-cross

figures. Opaque white glass with polychrome enamelling. Venice, 18th century(?). Museo Vetrario, Murano.

figured beaker. Mould-blown Roman glass beaker with four columned niches, probably Syrian, 1st century AD. Ht 12·5 cm. Cinzano Glass Collection, London.

filigrana (1). Detail of plate with opaque white glass threads in *retorti* pattern. Museo Vetrario, Murano.

filigrana (2). Detail of plate with opaque white glass threads in *reticello* (network) pattern. Museo Vetrario, Murano.

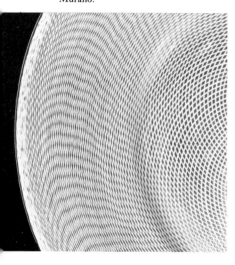

filigrana (3). Detail of plate with opaque white glass threads in spiral pattern (*vetro a fili*). Venini, Murano.

threads forming a regular fine network, usually with tiny air bubbles trapped between the crossed threads), and (3) *vetro a fili* (glass with embedded threads in a spiral or helix pattern, e.g. on plates, and sometimes in a spiral or volute pattern, e.g. on vases). This general usage of *filigrana* includes all three such patterns made with opaque white or with coloured threads, or both, and is favoured as relegating that oft abused word LATTICINO (or *latticinio*) to designate only such decoration made exclusively with white threads. Thus used, it would also include glassware heretofore sometimes loosely designated as VETRO DI TRINA (lace-glass). The *filigrana* style originated with the Venetians in the 16th century, but was perhaps inspired by the twisted opaque white threads in the ROPE EDGE of some bowls of ROMAN GLASS, *c.* 100 BC–AD 100), although the process of making the latter was different; it was perfected in Venice in the 16th and continued in use till the 18th century (a specialist was Giuseppe Briati, d. 1772), and it has been used in Bohemia, Silesia, Germany, Netherlands, Belgium, France, Spain, the United States, and even China. It has been revived at Murano in the 20th century. The usual English term is 'filigree', and other terms are 'threaded glass', 'lace glass', 'muslin glass', and (in the U.S.A.) 'cotton glass'; the French term is *verre filigrane*, and the German is *Fadenglas* (thread glass).

filigree. The English term for FILIGRANA. This term, as applied to glassware, is to be distinguished from the same word used to designate metalwork made with wires.

Findlay glassware. A type of ART GLASS moulded in various patterns and made in many colours that were shaded by controlling the repeated firing; the usual colour was a 'syrup' brown. It was made by Dalzell, Gilmore, Leighton Co., Findlay, Ohio, *c.* 1889.

finger-bowl. A small bowl, usually of glass, for holding water to rinse the fingers at the dinner-table. Early English examples were cylindrical, like a TUMBLER, and sometimes called a 'water-glass'; and it is said that they were then used to rinse the mouth after a meal. Later the name became 'wash-hand glass', and then 'finger cup', and finally 'finger-bowl', and the form changed to a cup-shaped or widened cylindrical bowl. They were made in coloured or clear glass, or OPAQUE WHITE GLASS with ENAMELLED decoration; some had a glass cover with a knopped FINIAL. Finger-bowls are usually circular, but some are three-lobed or of other ornamental shape. Some examples are provided with a matching saucer.

——, *lipped.* A cylindrical bowl with a flat bottom and having one or two lips in the rim. They are today often called a FINGER-BOWL, but it is also said that they served as a WINE-GLASS COOLER, and it may be that they were used for both purposes.

finger-glass. The same as THUMB-GLASS.

finial. The terminal ornament of any object, but particularly that on the COVER of any receptacle, where it also serves as a HANDLE; sometimes termed a KNOP, but that term is better reserved, especially when dealing with glassware, for bulbous decorations of a STEM. Finials are made in a great variety of forms and styles, such as ACORNS, figures, animals, flowers, fruit, loops, mushrooms, pine-cones, shells, spheres, etc. Certain finials are characteristic of particular periods, e.g., flowers and fruit for the ROCOCO style, pine-cones and acorns for the NEO-CLASSICAL STYLE, and sphinxes for the EMPIRE STYLE. *See* SWAN FINIAL.

finned bowl. A bowl decorated with encircling almond-shaped projecting bosses (pointed fins). One example, with twenty such fins and the body decorated with INTAGLIO engraved decoration of petals grooved lengthwise, was moulded, then finished by cutting and polishing. It is from the last quarter of the 3rd century BC and was found at Canosa, in Apulia, Italy. *See* D. B. Harden, *Journal of Glass Studies* (Corning), X (1968), p. 21.

Finnish glassware. Glassware made in Finland, where the earliest recorded glasshouse was founded by Gustaf Jung at Nystad in 1681 and closed in 1685. It was soon followed by one at Avik from 1748 to 1833, another at Mariedal from 1781 to 1824, and others also short-lived. They worked mainly with Swedish craftsmen, making simple utilitarian glassware. The first modern artistic glass was made from *c.* 1928. The leading glasshouses now are RIIHIMÄKI, IITTALA, and NOTSJÖ. Recent wares by Finnish designers combine functional qualities with artistic styles in the modern Scandinavian mood, and some household and ornamental glassware is created in very imaginative forms.

fire-clay. Clay capable of being subjected to a high temperature without fusing, and hence used for making CRUCIBLES in which formerly a BATCH of glass was fused. The clay contains much SILICA, and only small amounts of lime, iron, and alkali.

Fireglow glass. A type of ART GLASS with a reddish-brown colour intensified by transmitted light. It was made by the BOSTON & SANDWICH GLASS CO. in a tan colour and also by European factories, all usually with acid finish and elaborate designs of various harmonizing colours.

fireman's horn. A glass replica of a fireman's horn, made in the 19th century as a presentation piece. *See* HORN.

fire-polish. The brilliant surface condition given to glass by heating at the mouth of the furnace (GLORY-HOLE) after the manufacture of a glass object. The process removes the dull surface sometimes imparted to glass by iron in a MOULD. *See* POLISHING; AT-THE-FLAME.

firing. The process of heating the BATCH in order to fuse it into glass by exposing it to the requisite degree of heat in the CRUCIBLE or of reheating unfinished glassware while it is being worked or of reheating glassware in order to fix ENAMELLING decoration or GILDING. The melting of the batch may require a temperature of 1300–1500° C and the MUFFLE KILN a temperature in the range of 700–900°.

firing glass. A low drinking glass (about 10 cm. high), usually a DRAM GLASS, with a short thick STEM, a thick FOOT, and a BOWL of varying shape. They were used at a ceremonial occasion, after a toast was offered, by being rapped loudly on the table. They were made in England during most of the 18th century. Some bear MASONIC or JACOBITE emblems or insignia, but most are undecorated. Sometimes called a 'bumping glass'.

fish. A shape used from ancient times in glassware. The FISH-BOTTLE was made of EGYPTIAN GLASS and ROMAN GLASS. Ornamental pieces in such form were made of Venetian glass in the 17th century, and it is a popular form for present-day Venetian glass ornaments, made of solid glass in many shapes and colours. A modern group has been made for the Monaco Aquarium; *see* JEAN SALA.

fish-bottle. A type of BOTTLE of which the BODY is in the form of the head and front half of a scaly fish, extending into a long tapering NECK but not the tail of the fish. They were made of ROMAN GLASS, MOULD-BLOWN, in the 1st–3rd centuries A D. Earlier forms were made of opaque EGYPTIAN GLASS *c.* 1504–1320 BC (XVIIIth Dynasty); these were of CORE-GLASS, with the orifice in the open mouth of the fish, the BODY having COMBED DECORATION dragged forward to simulate the scales.

fisherman's float. A large glass ball used as a float for a fisherman's net, along with a duplicate ball to be hung in the window of his cottage as a spiritual contact with the fisherman at sea. The float was inscribed with a name and number, and sometimes a scriptural text; it was expected to

fireman's horn. Engraved presentation piece, Pittsburgh, Pa., 1850–80. Ht 50 cm. Corning Museum of Glass, Corning, N.Y.

finned bowl. Greenish glass with 20 encircling almond-shaped bosses (fins), pre-Roman glass, late 3rd century BC. Ht 9·2 cm. British Museum.

fish bottle. Blown olive-green glass, Roman, 2nd/3rd century. L. 27·8 cm. British Museum, London.

flagon. Double-handled with cresting, Roman glass, 1st/2nd century. Ht 23·2 cm. British Museum, London.

Flask. Clear glass with high kick-base, Persian, 18th/19th century. Ht 23·5 cm. Kunstgewerbemuseum, Cologne.

float ashore in case of an accident at sea. Some were plain green, some decorated with FILIGRANA stripes. They differ in appearance and purpose from a WITCH BALL.

flacon. A small bottle with a STOPPER, e.g. a SCENT BOTTLE or a SMELLING BOTTLE.

flagon. A pouring vessel, similar to a TANKARD. Early forms were like a PILGRIM BOTTLE with a screw cap, but later examples have one loop-handle, sometimes a long spout, and generally a hinged metal lid with a thumb-rest for opening. They were used at the table for serving wine. The term has sometimes been applied to a vessel to hold the wine at the Eucharist. They were made of ROMAN GLASS from the 1st century, probably as sacramental vessels, but later ones were for drinking beer, comparable to many made of ceramic ware in Germany.

flammiform. A decoration in the form of a flame. It is found as enamelled decoration on some glassware, or as decoration on the upper edge of WRYTHEN moulding.

flash. A thin covering of translucent coloured glass that is sometimes added to the CROWN of a PAPER-WEIGHT or across its bottom so as to show a colour through the crown.

flashing. The superimposition of a thin layer of glass on the body of a glass object; it is done by dipping the object into molten glass (unlike CASED GLASS). The flashing, if of a contrasting colour, could be ground through to produce a pattern, as in the case of CAMEO GLASS produced by the Romans. All medieval ruby window-glass consists of a thin flashing of red glass on colourless glass. The process was used on ISLAMIC GLASS, and fragments of glass found at Samarra (Mesopotamia) show such flashing. The process was used in Bohemia in the BIEDERMEIER period, and also in England in the 19th century, especially by GEORGE BACCHUS & SON. It was used in more complicated form by EMILE GALLÉ in his VERRE DOUBLÉ. A piece (e.g., PAPER-WEIGHT) covered with clear glass is said to be 'encased'. *See* DOUBLE FLASHING.

flask. A small receptacle for holding spirits. There are two types: (a) Those for standing on a table are usually globular or pear-shaped, with a short neck and a small mouth, and provided with a stopper. Some made in Germany are eight-sided (really four-sided with CHAMFERED CORNERS) and decorated with SCHWARZLOT or painted in ENAMEL or COLD COLOURS. Some have been made in fantastic shapes, e.g., in the form of a dog. (2) Those for carrying on the person usually are of elliptical or oval cross-section, with the sides flat or convex, and rise to a shoulder or taper directly into a short neck. These have a capacity (always only approximate) of one pint; some of ½-pint capacity are sometimes called 'pocket bottles'. They are often, especially those made in the United States, MOULDED with various forms of decoration on the sides. *See* BARREL FLASK; BELLOWS FLASK; DROPPER FLASK; FLOWER-ENCLOSING FLASK; GEMEL FLASK; GRAPE-CLUSTER FLASK; HEAD FLASK; HELMET FLASK; HEXAGONAL FLASK; HISTORICAL FLASK; HOLY-WATER FLASK; INDENTED FLASK; LENTIL FLASK; LENTOID FLASK; PICTORIAL FLASK; PISTOL FLASK; PITKIN FLASK; SHELL FLASK.

flask within flask. A glass FLASK or BOTTLE made with a small flask on the inside, attached to the bottom. The outer flasks are variously shaped, but often have SNAKE-TRAILED decoration. They were made contemporaneously with Syro-Palestinian glass and independently in the Rhineland, 3rd/4th century. Their purpose is not known.

flat bouquet. A decorative design in a bouquet PAPER-WEIGHT in which the replica of the bouquet of flowers and leaves is placed flat parallel to the base. *See* STANDING BOUQUET.

flat diamond cutting. A variation of the RAISED DIAMOND CUTTING, made by leaving an uncut area between the criss-crossed diagonal grooves so that a flat area results instead of pointed diamonds. This area is left uncut, in contrast to the STRAWBERRY DIAMOND CUTTING.

flat flutes. *See* FLUTING.

flecked glassware. A type of glassware of which the surface is mottled with flecks of opaque glass, usually white, of various shapes and sizes (some sizable, some tiny specks) scattered at random. Broken chips of glass were spread on a MARVER, after which a GATHER of molten glass was rolled over them; when the mass was reheated and blown and possibly further marvered, the chips became embedded in the surface of the vessel and were no longer palpable. The process was used on ROMAN GLASS made at Alexandria in the 1st century AD, on French and German glassware in the 17th and 18th centuries, and was copied in England in the 19th century (sometimes using chips of different composition from that of the body, probably derived from broken discarded pieces); such English ware has been attributed to NAILSEA, but is now ascribed to other unidentified factories. In England the tax on glass (*see* EXCISE ACTS), which was lower on bottle glass than on FLINT GLASS, led to the making of this inexpensive decorated ware. *See* ENCRUSTED GLASS-WARE; BLOBBED GLASSWARE.

flint. An impure variety of quartz which, after being heated to $c.400°$, can be easily crumbled and powdered. It was used in England from $c.1647$ in making FLINT GLASS.

flint glass. A misnomer for English LEAD GLASS, probably given currency because the evolution of lead glass occurred at a time when calcined or ground flint was substituted for Venetian pebbles as the source of SILICA for making the glass. Later, sand replaced the flint, but the name 'flint glass' has persisted and is still sometimes used to refer to all English lead glass. Flint glass has been classified since 1682 as (1) 'single flint' or 'thin flint', and (2) 'double flint' or 'thick flint'. It has been said that both types were of the same composition, with equal lead content, but that the 'double flint', made to remedy the lightness and fragility of 'single flint', was produced by a double GATHERING of METAL, resulting in a sturdier and more popular (though costlier) glass. A modern name for 'flint glass' is 'lead crystal'.

flip. A type of BEAKER, so-called in the United States, that is 14–17 cm. high, taller than the usual 10 cm. tumbler. Examples are usually cylindrical or slightly spreading toward the rim. The larger ones usually had a cover with a dome having a hollow pointed globular FINIAL. They were made in the 18th and 19th centuries.

float glass. SHEET GLASS made by the process developed by PILKINGTON BROTHERS LTD and introduced in 1959. It is now the world's principal method of flat glass manufacture. By this process a continuous ribbon of molten glass, up to 3·30 metres wide, moves out of a melting furnace and floats along the surface of a bath of molten tin. Control of atmosphere and temperature cause irregularities to melt out, and the glass becomes flat, to a thickness of from 2·5 to 25 mm. The ribbon is fire-polished and annealed, without grinding or polishing. By a 1967 development, called the 'electrofloat process', the glass can be tinted as it passes through the float bath.

Florentia Glass. A type of ART GLASS developed by FREDERICK CARDER at STEUBEN GLASS WORKS and made in the 1920s and early 1930s. It is decorated with an encircling relief pattern of adjacent leaves with the points extending upward from the base. The leaves were made of powdered coloured glass (usually pink or green) and were fused onto the body of the vase or other object; after annealing, the entire surface was sometimes given a MAT finish.

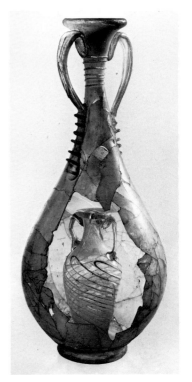

flask within flask. Roman glass, made in the Rhineland, 3rd/4th century. Ht 23·5 cm. Römisch-Germanisches Museum, Cologne.

flower-enclosing flask. Clear glass with polychrome flowers, Persia, 19th century. Ht 25 cm. Kunstgewerbemuseum, Cologne.

flower-holder. Germany, 17th century. Ht 17·5 cm. Kunstgewerbemuseum, Cologne.

Florentine glassware. Glassware made in Florence from the 14th century (mainly bottles and beakers) and later, during the 16th and 17th centuries, under the patronage of the Medicis, luxury glassware as well as objects for medical and scientific purposes, e.g. RETORTS, ALEMBICS, LENSES, BAROMETERS, and THERMOMETERS, made AT-THE-LAMP by immigrant Venetian artisans. In 1567 Luigi Bertolo of Murano fruitlessly negotiated with Cosimo de'Medici for a glass-making privilege, but later a 14-year contract was made, and the furnace was at work by 1579, making glassware in the Venetian style. In the 17th century a glassworks was established under Don Antonio de'Medici where experimental work was done by ANTONIO NERI. *See* TUSCAN GLASSWARE.

floret(te). A slice (set-up) of a large CANE composed of a group of varied coloured RODS arranged, usually concentrically, so as to form a stylized floral pattern. It sometimes refers to a cane whose cross-section resembles a single natural or stylized flower. They are used especially in PAPER-WEIGHTS of the millefiori, circlet, concentric or garland types.

flower block. A glass receptacle to hold standing cut flowers. It is made with a vertical column of glass supporting glass disks at two levels, each having holes for the sprays. Some made by FREDERICK CARDER at STEUBEN GLASS WORKS are embellished with a crystal glass figure or animal, usually with MAT finish, affixed by a peg in a hole in the base. The basic form was patented by Carder in 1921, but a similar receptacle had been patented by ARTHUR NASH in 1917 for Tiffany Furnaces.

flower-enclosing flask. A type of Persian glass flask (sometimes a jug or a *kalian*), usually pear-shaped with a tall thin neck and a foot-ring, the characteristic feature of which is a coloured glass floral (or fruit) ornament within the bowl and attached to the bottom. The ornament was made first AT-THE-LAMP and then attached within the flask while it was being blown. They are said to have been made in Venice in the 18th/19th century for the Eastern market, and then copied in Persia in the early 19th century. The Italian term is *pipe con frutto entro*. *See* VASE WITH ENCLOSED BOUQUET.

flower stand. An ornamental glass receptacle for holding cut flowers, having several vertical tubes (3 to 6) for individual flowers. They were made in England (probably Stourbridge, *c.* 1895) and also in Germany in the 17th century. The German term is *Blumengefäss*. *See* BOUQUETIÈRE.

flowered glassware. A type of WHEEL-ENGRAVED glassware made *c.* 1735–85 in England with shallow engraving depicting flowers or floral festoons, or motifs of grapes and vines (for WINE GLASSES or DECANTERS), hops and/or barley (for ALE GLASSES), or apple branches (for CIDER GLASSES).

Flügelglas (German). Literally, wing glass. The German term for a WINGED GLASS.

fluorescence. *See* ULTRA-VIOLET RADIATION.

flute cordial glass. The same as a RATAFIA GLASS.

flute cutting. A pattern on CUT GLASS consisting of a series of parallel grooves, either mitred (V-shaped), called 'pillar flutes', or of concave semi-circular or semi-elliptical section, made with a convex wheel and called 'round flutes' or 'hollow flutes'. (The term 'hollow flutes' has sometimes been applied to convex PILLAR MOULDING.) Sometimes the grooves are separated by a common arris (as on a Doric column). Fluting on glass objects may be in straight lines, curved (spiralling) or decoratively shaped (e.g., petal-shaped). On some English drinking glasses, the fluting on the bowl extends from midway down past the junction of the stem, making a CRESTING on the bottom part of the bowl, with the upper ends of the flutes terminating in a point; it is called 'bridge fluting'. A FACETED stem is sometimes said to have 'flat flutes' or 'hollow flutes'. The ridges of some fluting is notched. *See* PILLAR CUTTING.

flute glass. A tall and extremely narrow drinking glass of which the BOWL is of inverted conical shape and rests on a very short STEM (often merely a KNOP or a small BALUSTER with MERESES). In the 17th century Dutch and German examples, for wine drinking, were from 40 to 50 cm. high. Similar tall examples were made in Venice, some FILIGRANA examples have a COVER. They usually have engraved decoration, sometimes commemorative portraits. Modern flutes, about 12 cm. high, are used as a type of CHAMPAGNE GLASS. The Italian term is *flauto*. *See* EXETER FLUTE; SCUDAMORE FLUTE; TULIP GLASS.

fluting. RELIEF ornamentation in the form of a series of parallel concave grooves, being the converse of REEDING or RIBBING. *See* CUT GLASS: *patterns* (*fluting*).

flux. A substance added to ENAMEL COLOURS so as to lower their fusion point during FIRING to below that of the glass BODY to which they are to be applied, although some softening of the glass is essential to key the colours to it. A flux was also added to the BATCH in making glass in order to facilitate the fusing of the SILICA. Fluxes include BORAX, POTASH, and SODA. *See* ASHES.

fly-trap. A vessel with a small orifice, through which flying insects can enter but not readily escape. Some were made of glass. One example rests on three feet and has an opening in the bottom whose sides extend upwards and into the body of the vessel so as to form an interior trough into which ale is placed to attract and drown flies (or wasps). Also termed a 'wasp-catcher'.

flying-bird flask. A type of ornamental FLASK (ZIERGLAS) with four apertures in the body through each of which a glass pigeon, attached to the hole, appears to be flying. The pigeons are of white glass with blue heads, three flying in one direction, the fourth opposite. They were made of ROMAN GLASS in the Rhineland in the 3rd/4th century. The German term is *Vögelchenflasche*.

foam glass. A lightweight cellular form of glass produced by the chemical foaming of molten glass. It is made by packing crushed grains of glass into moulds mixed with a chemical substance which, when heated, emits bubbles of gas; the resultant slabs are sawn and used for walls and insulation. It is light enough to float on water.

fondi d'oro (Italian). The bottoms of glass cups and bowls, with a small central medallion of engraved gold leaf (sometimes additionally painted) protected by a sheet of fused glass. Fragments have been found in Roman catacombs where they had been embedded in the walls of *loculi* (cells), presumably by relatives of a deceased to mark a grave. The subjects depicted are from Christian and Jewish symbolism and Biblical history, and also from classical mythology and games, with sometimes inscriptions or dedications to saints and heroes. The outer layer of glass is sometimes coloured but usually both layers are colourless glass. The earliest period of their making was probably the 3rd century, and the last was *c.* 410 when catacomb burial was discontinued. These were sometimes forged in the 18th century. *See* GOLD ENGRAVING.

font. (1) A HOLY-WATER STOUP. (2) The oil reservoir for an oil lamp.

foot. The part of a vessel or object (other than its BASE) on which it stands. The term applies either to the part which broadens out from the STEM or to the lower part of individual legs attached to the BODY. On English drinking glasses, the foot sometimes had a sharp EDGE until

flutes. Carafe with cut wide and narrow flutes, English, *c.* 1810. Ht 12·5 cm. Courtesy, Delomosne & Son, London.

flying-bird flask. Clear glass, birds white with blue heads, Rhineland, 3rd/4th century. Ht 24·5 cm. Römisch-Germanisches Museum, Cologne.

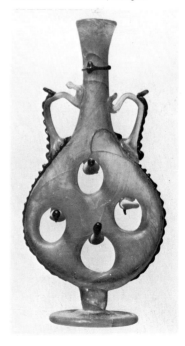

fondo d'oro. Medallion with gold-leaf engraving on back (same persons occur on another such piece in British Museum), Roman glass, 4th century. W. 7 cm. Museo Sacro, Vatican.

Forest Spires. Crystal glass triangular prism, designed by Golda Fishbein for Steuben Glass, Inc., 1973. Ht 33·5 cm. Steuben Glass, New York.

c. 1730–80 but thereafter a rounded or folded edge. The foot is of various shapes (circular, square, hexagonal, scalloped or ornamental) and styles that serve to identify the period of production. Generally, the diameter of the foot is equal to that of the bowl. The edge of a foot should be examined to ascertain if it has been trimmed to remove chips. *See* BUN FOOT.

TYPES OF FOOT

——, *annulated.* A foot that is composed of a series of equal or unequal rings, vertically placed.

——, *beehive.* A foot that is domed and has external horizontal ridges, simulating a beehive. It is a development of the terraced foot (see below).

——, *conical.* A foot shaped like a low cone, either solid or hollow. The latter type was used during most of the 18th century so that the PONTIL MARK would not rest on the table and cause instability and wear.

——, *corrugated.* A foot that has a series of small steps so as to offer a corrugated effect; an adaptation of the terraced foot (see below).

——, *domed.* A foot that is domed, high or low, and that has several variations: domed and terraced (either solid or hollow), terraced-domed, beehive domed (either solid or hollow), double-domed, with one dome above another. They are circular but some have a square BASE. The domed foot was made originally so that the mark of the PONTIL could be pushed upward in order that the glass would stand on a level bottom. It is found on drinking glasses, candlesticks, and other glass objects.

——, *firing.* A foot that is thick and heavy, such as that found on a FIRING GLASS; it often slopes slightly outward and has a squared edge.

——, *flanged.* A foot that is hollow on the bottom and rests on an incurved flanged rim.

——, *flat.* A foot that has a flat bottom. They were made in the 19th century, after there was no longer need for a PONTIL MARK.

——, *folded.* A foot that has its RIM slightly turned under, making a double layer of glass to lend stability and lessen the risk of breakage. Variations are conical or domed (see above). They are found on drinking glasses, candlesticks, and other glass objects. The fold was from 0·5 cm. to more than 1 cm. on some large glasses. Venetian glasses of the 16th and 17th centuries, with an intricate STEM, often had a folded foot, as did some English glasses by JACOPO VERZELINI. The folded foot was rarely used in England after the Glass Excise Act (see EXCISE ACTS) of 1745 made it uneconomical, but it was briefly revived *c.* 1780 in a modified form; it never went entirely out of use, particularly on objects other than drinking glasses, surviving well into the 19th century. On some Continental glasses the fold is not under but is over the foot.

——, *helmet.* An exceptionally high domed foot (see above).

——, *honeycomb.* A domed foot (see above) that is decorated with a moulded honeycomb pattern.

——, *lemon-squeezer.* A foot, found usually on an Irish fruit bowl (*see* IRISH GLASSWARE) or CANDELABRUM, moulded with deep radial ribs on the underside of a dome and usually set above a solid square foot (see below).

——, *overseam.* A type of foot on a drinking glass made with applied glass threads radiating outward from the base of the stem to just over the edge of the foot. Such glasses were made in England in the mid-18th century, and are usually a DRAM GLASS.

——, *overstring.* A type of foot on a drinking glass similar to an overseam foot (see above) but having thicker radial trails. Such glasses were made in England in the 18th century, and are usually a DRAM GLASS.

——, *pedestal.* (1) A foot of a drinking glass that is tall and wider toward the bottom, and has a concave sloping exterior. Variations are solid or hollow. It is found often on a PASSGLAS or a STANGENGLAS, and sometimes has a KICK or a PONTIL MARK. (2) *See* PAPER-WEIGHT, *pedestal.*

——, *square.* A foot of square shape, a type much used in England and elsewhere in the early 19th century. It was used mainly on

RUMMERS, BONNET GLASSES, BOWLS, CANDELABRA, etc. Some were
topped by a circular dome-shape or by a lemon-squeezer foot (see
above).

––––, *star*. A foot of a drinking glass with its bottom decorated with a
WHEEL-ENGRAVED star. Early English examples have 8, 12, or 16 points.
Later ones are more sharply cut and extend to the edge of the piece.
From *c*. 1830 to 1840 the stars had 24 points, and *c*. 1840–50 they had 32
points; *c*. 1850–60 the 'Victorian Glory Star' has points of three different
lengths, and *c*. 1880–1900 the German 'Jewel Star' had 16 points, with
rounded points and cross-cutting.

––––, *stemmed*. The foot of a vessel that is attached to the BODY by a
stem, particularly on drinking glasses. The foot may be circular, square,
or ornamentally shaped, and of various styles.

––––, *terraced*. A foot that has a series of steps extending one beyond
the other. Variations (see above) are: domed and terraced, with a domed
section above the terracing; and terrace-domed, with terracing on the
domed section.

foot-bath. A deep oval basin, usually with a HANDLE at each end of the
long axis. Generally made of glazed earthenware, some have been made
of glass, e.g., in Venice in the 16th century. Sometimes called a 'cistern'.

foot-maker. A member of the team (CHAIR or SHOP) assisting the
GAFFER in making glass objects such as wine-glasses.

foot-ring. A slightly projecting rim on the underside of the bottom of a
vessel or object, on which it stands. It may be made by (a) applying a
glass coil, (b) pinching out a small fold, or (c) fusing on to the bottom of
the object a thin layer of glass with its rim bent downward. It is also
called a 'basal rim' or a 'foot-rim'. *See* BASE-RING.

footed beaker. A type of BEAKER that is supported by a stemmed FOOT
or sometimes by three bun feet (*see* SCHAPER GLASS).

footed bowl. *See* STANDING BOWL.

Forest Spires. A unique ornamental glass piece made by STEUBEN
GLASS, INC., in 1973. It is a tall prism with its three sides engraved with a
group of interlacing fir trees. The fringed branches overlap so that their
reflections appear and disappear as the viewer's position changes. It was
presented to Emperor Hirohito by President Gerald R. Ford on his visit
to Japan in 1975. It was designed by Golda Fishbein.

forgery. A close copy of valuable old glassware, made with the intention
to deceive prospective buyers and offered for sale as genuine at a high
price. The types of ware subject to frequent forgery are: German
ENAMELLED ware (especially ARMORIAL WARE); English twist-stemmed
WINE GLASSES; JACOBITE GLASSWARE; WILLIAMITE GLASSWARE;
PAPER-WEIGHTS; NAILSEA GLASSWARE; WATERFORD GLASSWARE;
YARD-OF-ALE; ZWISCHENGOLDGLAS. The detection of a forgery is
aided by such matters as: the presence or absence of a PONTIL MARK
(forgeries may have one, but the absence of one is not conclusive
evidence; *see* CRUET BOTTLE); relationship between weight and
resonance (not only LEAD GLASS emits a ring, but also BARIUM GLASS,
and even heavy lead-glass DECANTERS will not ring); weight (some
vessels of SODA GLASS are thickened to simulate the weight of lead-
glass); colour (e.g., the alleged WATERFORD 'pale blue tinge' has been
imitated); presence of lead (*see* LEAD TEST); enamel (differences between
thick and thin); and, most important, faults of form, style or
workmanship. *See* FAKE; REPRODUCTION.

Fostoria Glass Co. The largest factory producing hand-made glass in
the United States. It was founded in 1887 at Fostoria, Ohio, but when
local gas supplies became insufficient it was moved to its present location
at Moundsville, West Virginia. Its principal production is blown and
moulded stemmed drinking glasses.

foot, domed. Detail of candlestick,
English, *c*. 1765. Ht 25·7 cm. Courtesy,
Sotheby's, London.

foot, folded. Detail of wine glass,
English, *c*. 1720. Courtesy, Derek C.
Davis, London.

foot, pedestal. Detail of *Stangenglas*,
Venice, dated 1596. Ht 32·5 cm.
Museum für Kunsthandwerk,
Frankfurt.

foot, terraced. detail of liqueur glass,
English. Courtesy, Maureen
Thompson, London.

Four Continents. Opaque white glass tumbler with decoration depicting 'Europe', Bohemia, *c.* 1770–80. Museum of Decorative Arts, Prague.

found. The amount of time needed for melting the BATCH, from the CHARGE until the MOLTEN GLASS is ready for working.

founding. The initial phase of heating the BATCH in the FURNACE, when the materials must be heated to a temperature of about 1400° C. before a maturing period of about 12 hours during which the MOLTEN GLASS cools down to a working temperature of about 1100° C.

fountain. An elaborate glass fantasy made to function as a fountain; one well-known example contained four tons of glass and measured 27 ft (8·30 m.) in height. It was made by E. and C. Osler, of Birmingham, for the 1851 Great Exhibition at the CRYSTAL PALACE, London, and received much acclaim as well as much adverse criticism. *See* BIRD-FOUNTAIN.

Four Continents, The. An allegorical motif depicting the four continents, Europe, Asia, Africa, and America. Often depicted in porcelain as figures, the subject was used as enamelled decoration on some OPAQUE WHITE GLASSWARE in Bohemia in the second half of the 18th century.

Four Elements, The. An allegorical motif depicting the four elements of antiquity – Fire, Water, Air, Earth. It is found in porcelain, as figures or as painted or transfer-printed decoration, made in the 18th century. It also appears as decoration on glassware, e.g., enamelled opaque white glass made probably in Bohemia in the second half of the 18th century.

Four Seasons, The. An allegorical motif depicting the four seasons, usually with their attributes. It is found more often on porcelain, but examples are known in glass, either a set of four such figures (e.g., one made at Nevers) or an engraved or enamelled design on some GOBLETS or HUMPEN. *See* NEVERS FIGURE.

fox-head. A STIRRUP-CUP made in the form of a fox's head. Often made of porcelain or pottery, some were made of glass, toward the end of the 18th century.

fragment. A piece of a broken glass object. They are found less often than ceramic shards at excavation sites and waster heaps because broken pieces of glassware were re-used as CULLET in making new glass. Shards and fragments usually serve as a reliable means of identifying similar surviving ware, but it is important to distinguish between those found at a factory site, which are likely (but not necessarily) to be pieces made at that factory and those discovered at dumps which may be from other sources, not even nearby. *See* WASTER.

Franck, Kaj (1911–). A Finnish glass-designer who became Art Director at the NOTSJÖ GLASSWORKS in 1951. He has designed household glassware in modern techniques and styles, as well as ART GLASS.

Franconian glassware. Glassware made in Franconia, in middle Germany. The best-known products are the OCHSENKOPF beakers, made at Bischofsgrün in the Fichtelgebirge ('Fir Mountain') region, where there were glasshouses from 1561. From the Kreussen district in Upper Franconia come enamelled glassware having dotted borders with cresting and interlaced arcs, predominantly in white, as also found on Kreussen stoneware. Generally, light enamel colours were used – white, yellow, and pale blue.

Frankish glassware. Glassware made from *c.* 400 to *c.* 700 (when the custom of burying vessels with the deceased was discontinued), found throughout central Europe but principally in Belgium and the Rhineland, hence the surmise that most of it was made there. Such glassware, made during the so-called 'Dark Ages', after the decline of

Four Elements. Beaker or ice pail, opaque white glass with enamelled scene depicting 'Das Feuer' (Fire), probably Bohemia, *c.* 1750–70. Ht 22·9 cm. British Museum, London.

the Roman Empire in the West, was still SODA-LIME GLASS, before the use of POTASH for WALDGLAS, *c.* 1000 AD; but the forms and decoration of the glassware changed to suit the Teutonic taste. The examples consist mainly of drinking vessels without a handle, foot or base, such as the CONE BEAKER and DRINKING HORN. They are usually of impure greenish glass, with decoration only of trailed threads, and without any painting, cutting or engraving. During this period are found early examples of the NUPPENBECHER, with rows of various coloured drops ('eyes') of glass applied to the bowl, and of the RÜSSELBECHER (*see* CLAW-BEAKER); there are also some early examples of the KUTTROLF. The ware is sometimes termed 'Merovingian glassware' or 'Teutonic glassware'.

Franklin Flint Glass Co. A glasshouse at Philadelphia, Pa., established in 1860 by WILLIAM T. GILLINDER. In 1867 it was taken over by Gillinder's sons, James and Frederick, and its name was changed to Gillinder & Sons. It made CUT and PRESSED GLASSWARE, and *c.* 1890 moved the pressed glass department to Greensburg, Pa. It closed in the 1930s.

free-blown glass. Glassware shaped solely by the process of BLOWING with the use of a BLOW-PIPE and shaped by manipulation with tools. *See* FREE-FORMED GLASS.

free-formed glass. Glassware shaped by hand-processes in asymmetrical form, often in the ART NOUVEAU style. It was sometimes made of FLASHED GLASS or CASED GLASS, sometimes with glass enclosing threads of white glass. This glassware has been made in Venice (*see* HANDKERCHIEF BOWL) and also by EMILE GALLÉ, DAUM FRÈRES, and the BACCARAT FACTORY, as well as by FREDERICK CARDER at the STEUBEN GLASS WORKS. *See* GROTESQUE GLASSWARE.

French glassware. Glassware of various types made in France, from VERRE DE FOUGÈRE and the renowned medieval STAINED-GLASS WINDOWS to 17th-century CROWN GLASS, BROAD GLASS, MIRRORS, CHANDELIERS, and NEVERS FIGURES, and table-glass in the FAÇON DE VENISE, the *façon de Bohème*, and the *façon d' Angleterre*, and later from the 18th century, the CUT GLASS, engraved glass, CAMEO GLASS, and PAPER-WEIGHTS of BACCARAT, ST-LOUIS, and CLICHY, and the modern glass of EMILE GALLÉ, RENÉ LALIQUE, DAUM FRÈRES, and MAURICE MARINOT. The early French glass-makers of the Renaissance period were itinerant artisans from L'ALTARE and some immigrants from Venice. Although glassworks existed throughout France, the principal centres were Nevers, Rouen, Orléans, and Nantes, apart from the big factories of Baccarat, St-Louis, Clichy and ST-GOBAIN. For a discussion of French glass *see* J. Barrelet, *La Verrerie en France* (1953), and (as to the 18th century) R. J. Charleston, 'The Art of Glass in France', *Antiques Magazine* (1952), pp. 156, 253.

fret. A decorative border pattern in continuous repetitive form made of short lines of equal length meeting at 90 degrees, or (less often) at a slightly larger or smaller angle. Fret patterns have been used to decorate glassware, e.g., PANELS of ROMAN GLASS, *c.* 100 BC–AD 100, and glassware decorated by ISAAC JACOBS at Bristol, *c.* 1805–10. *See* KEY-FRET.

frigger (or **friggar**). A glass object, of various forms, made by a glass-maker in his own time and for his amusement and home decoration or for sale by him. They were usually made from the molten glass remaining in the POT at the end of the day, considered as the workman's perquisite. In some regions, they were made on Saturdays when the glasshouse was not working, and on Sunday each factory group paraded with its accomplishments (e.g., from Stourbridge to Wolverhampton), stopping at each public house en route to have the pieces voted on, and the most popular received a prize and the assurance of factory

Four Seasons. 'Winter', detail from diamond-point engraved bowl, Netherlands(?) in *façon de Venise*. Kunstsammlungen der Veste Coburg.

Frankish glassware. Tumbler of greenish glass with white threads, 6th/7th century. Römisch-Germanisches Museum, Cologne.

production. They were made in many English regions, especially at Nailsea, in the late 18th and 19th centuries, and included such objects as jugs, candlesticks, ship models (NEFS), animal figures, TOBACCO PIPES, ROLLING PINS, BELLS, HATS, WITCH BALLS, MUSICAL INSTRUMENTS, and WALKING-STICKS. Such pieces have been extensively copied in recent years. They have been sometimes called in England 'end-of-day-glass' and in the United States 'off-hand-glass'.

frit. Some of the ingredients used in making glass, such as SAND and ALKALI, pre-heated in a CALCAR but not completely melted or fused, and when cooled ground into a powder, to be added to the final ingredients that went into the POT (CRUCIBLE) to be melted into glass. *See* ENAMEL COLOURS.

Fritzsche, Elias (fl. 1630). A glass-engraver who was employed at the glassworks at Tambach by Duke Bernhard von Sachsen-Weimar. *See* THURINGIAN GLASSWARE.

Fritsche, William. A Bohemian engraver who had a workshop at the DENNIS GLASSWORKS of THOMAS WEBB & SONS. He introduced the style of decoration known as ROCK-CRYSTAL ENGRAVING, often engraving so deep that the effect had the appearance of glass sculpture. His work was exhibited at the 1878 Paris Exhibition. An example of his work is the 'Fritsche Ewer', with a mask of Neptune under the lip and waves and fish on the body and shells on the foot. He signed much of his work.

Frontinus bottle. Mould-blown pale-green Roman glass, crisselled, 3rd/4th century. Ht 12·7 cm. Pilkington Museum, St Helens, Lancs.

Frontinus bottle. A type of BOTTLE of cylindrical or barrel shape, made of MOULD-BLOWN green glass, with a thin vertical neck having a disk mouth and with a bent handle (two on later examples) attached at the shoulder and the rim of the mouth. The characteristic feature is a series of encircling horizontal bands or hoops covering the upper third and lower third of the body. They bear on the base the moulded mark of the maker, e.g., 'Frontin O' (hence the name) or 'Daccius F' or 'Felix Fecit'. Frontinus had a glassworks said to have been in Gaul, at Boulogne or Amiens, in the 3rd–4th centuries. Such bottles have not been known to be found in the East, but like vessels without the hoops have been discovered in Nubia and Egypt.

frost. Finely powdered glass made by crushing glass bubbles that have been blown very thin. It was used by FREDERICK CARDER at STEUBEN GLASS WORKS for surface decoration on CINTRA GLASS, CLUTHRA GLASS, and QUARTZ GLASS.

frosting. (1) A network of small cracks on the surface of a glass object, due to WEATHERING. (2) A MAT effect obtained by the use of HYDROFLUORIC ACID on the surface of a glass object. *See* MAUDER, BRUNO.

fruit. A fanciful glass ornament in the form of a piece of fruit, e.g., an apple or a lemon. Some were made of Venetian glass in the 17th century and later. *See* PAPER-WEIGHT, *fruit.*

fruit bowl, Irish. *See* IRISH GLASSWARE.

fruit bowl, Venetian. A large cylindrical receptacle, about as high as it is wide, resting on three small bun feet and having two vertical scroll HANDLES and a domed COVER with a high FINIAL. They were made at MURANO in the 16th century, for use in wealthy households. Some are decorated in RETICELLO style and some with DIAMOND-POINT ENGRAVING. The Italian terms are *cisto con coperchio* (covered bowl) and *cisto biansato* (two-handled bowl).

fruit bowl. Venetian *reticello*, mid-16th century. Ht 33·5 cm. Museo Vetrario, Murano.

Fry, H. C. (1840–1929). Founder of the H. C. Fry Glass Co. at Rochester, Pennsylvania, formed in 1901 to succeed the Rochester

Tumbler Co. established by Fry in 1872. It made cut glass in the 1920s and, from 1922, milky-white opalescent glassware with handles, finials, and other attached parts of coloured glass. It also made vases, jugs, etc. of opalescent ART GLASS called 'Foval'; some pieces have applied sterling silver mounts made and hallmarked by Rockwell Silver Co., of Meriden, Connecticut. Some of its ware resembled the opalescent glassware of DORFLINGER & SONS and the JADE GLASS of STEUBEN GLASS WORKS; some pieces made by Dorflinger have been attributed to Fry. The factory closed in 1934.

full-size piece-mould. A PIECE-MOULD that makes glassware in approximately the finished size, without later BLOWING, making it in the final shape and style of decoration, as distinct from glassware that has been PATTERN-MOULDED and later expanded. *See* PART-SIZE PIECE-MOULD.

funnel. An object with a thin tube for pouring liquid into a narrow-neck receptacle, and usually having above the tube a hemi-spherical or cone-shaped mouth. They were made of ROMAN GLASS, adapted from bronze prototypes, in the 1st century and later. The Italian term is *imbuto*.

funnel-mouth. A MOUTH (of a receptacle) shaped like a funnel. *See* SPITTOON.

furnace, pot. The source of heat in a glassworks in which the POTS are placed for fusing the glass ingredients to make household and ornamental glassware. The heating formerly was done with wood or peat, and great quantities were used, so that the glass-makers frequently moved from place to place in forest areas as the fuel supplies became exhausted. In England from *c.*1615 pit-coal was used instead of wood, which had been banned to glass-makers with the invention of coal-burning furnaces (*see* CROWN POT). The use of coal spread to other countries, and it was used until recently when oil, gas, and electricity have been used. A pot furnace normally held up to twenty pots, arranged in a circle above the source of heat. The direct application of heat was superseded by a process using hot gas through 'recuperators'. The melting of the ingredients may take as long as forty-eight hours, and requires a temperature of 1300–1500° C. *See* FURNACE, TANK; REVERBERATORY FURNACE.

furnace, tank. A type of furnace used for the making of such articles as bottles and window-glass, where the glass is melted in a shallow tank, sometimes holding up to several hundred tons of molten glass. *See* FURNACE, POT.

furnace-shaped. Formed by heat from the furnace rather than by moulding, blowing or pressing.

furniture. Pieces of furniture decorated with glass. Some examples were entirely covered with glass, such as was made in Russia in the 18th/19th centuries, e.g., a card-table designed by Charles Cameron (fl. 1779–1811) and a divan at the ST PETERSBURG IMPERIAL GLASS FACTORY, *c.*1822, for the Shah of Persia, with head-boards, side-boards, and bedposts entirely covered with blue and colourless crystal glass. At Murano some bureaux were made with the drawer fronts decorated with plaques of coloured glass. Similar mirror-decorated furniture was made in Germany at Brunswick. Glass-decorated furniture was particularly sought-after in France; *see* ESCALIER DE CRISTAL.

fusiform. Shaped like a spindle or a cigar, tapering inward at both ends. The shape is found in some examples of the ALABASTRON made of CORE GLASS, *c.*200–100 BC, and on the STEM of some Venetian TAZZE.

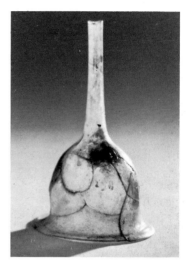

funnel. Roman glass, 1st century AD. Ht 11 cm. Museo Vetrario, Murano.

G

gadroon. Detail of goblet showing gadrooning on bowl, English, *c.* 1690. British Museum, London.

G S Goblet. A GOBLET attributed to the glasshouse of JACOPO VERZELINI, bearing below the rim the engraved monogram 'G S' tied with a lover's knot, and the date 1586. The engraving has been attributed to ANTHONY DE LYSLE. The semi-ovoid bowl is encircled by two applied bands of colourless glass and two opaque white threads, between which is engraved 'IN : GOD : IS : AL : MI : TRUST' which was the motto of the Pewterers' Company. The goblet is 13 cm. high, with a hollow ribbed KNOP in the stem and a plain flat foot engraved with petals. The rim and the knop have GRANULAR GILDING; it is the only so-called Verzelini goblet that combines gilding with engraving. Another so-called Verzelini goblet, with the initials 'K Y' and dated 1583, bears the identical inscriptions as do some Dutch and German diamond-engraved glasses. *See* VERZELINI GLASSWARE.

gadget. A metal rod, used since *c.* 1760–1800, that has a circular spring-clip that grips the foot of a glass while the GAFFER trims its RIM by SHEARING, and hence avoids leaving a PONTIL MARK; it does, however, leave its own pressure mark on the foot. After *c.* 1900 the clip was wrapped with asbestos threads to eliminate the pressure mark. Later, when CLAPPERS were used, there were no marks of any kind on the foot. A 'plunger' closes the gadget to attach the clip to the glass and opens it to release the finished glass. *See* TOOLS, GLASS-MAKERS'.

gadroon. An ornamental decoration in a continuous pattern of short, repetitive sections of REEDING, set vertically, diagonally, or twisted. Originally it was inspired in the late 17th century by contemporary silverware. It has been used in England on the lower part of the bowl of some drinking glasses or JELLY GLASSES or of some BOWLS. Some gadrooning (called 'spiked gadrooning') has tiny protruding points at the upper end of the row of gadroons. When gadrooning, instead of being in parallel ribs, radiates from a central point, it is called 'radial gadrooning'. The French term is *godron*.

gaffer. The master-worker of the team (CHAIR or SHOP) engaged in making glass objects such as wine-glasses. He sits at the CHAIR (bench) and on its sloping arms he rests the BLOW-PIPE or the PONTIL holding the embryonic glass object (the PARAISON) and constantly rotates it so that the attached paraison retains its symmetrical form and does not collapse. He does the most skilled and intricate work and controls the procedure of the team until the object is completed and sent to the LEHR for annealing.

Gallé, Emile. Vase, signed 'Gallé', Nancy, *c.* 1895. Ht 18 cm. Kunstmuseum, Düsseldorf.

galanterie (French). Various small trinkets given in the 18th century by gentlemen to ladies. They were intended to be carried in the pocket or a reticule, kept in a dressing-table, or displayed in a cabinet. Included are ÉTUIS, SNUFF-BOTTLE, SCENT-BOTTLES, SMELLING BOTTLES, PATCH-BOXES, PATCH-STANDS, NEEDLE-CASES. Generally made of porcelain, some examples were made of glass, especially OPAQUE WHITE GLASS with enamelled decoration. The German term is *Galanteriewaren.*

gall. The scum that comes to the surface of a BATCH of glass in the process of being melted, and which must be skimmed off.

Gallé, Emile (1846–1904). A glass-maker and designer from Nancy, France, who was a leading exponent of ART NOUVEAU. The son of Charles Gallé-Reinemer, a glass-maker in Lorraine, he learned glass-making and later studied art at Weimar, and made visits to Paris and London to study glass. He established his own glasshouse in 1867, and in

1874 he and his father began production of ART GLASS at Nancy; the factory later expanded, having many employees and creating a vast amount of ware, including vases, lamps, and tableware. Gallé soon began to develop his own original techniques for making and decorating glass, including CAMEO GLASS, being influenced by Japanese styles and hence featuring naturalistic decoration with engraved floral and plant designs. He developed complicated cut CASED GLASS, but the factory also made, from the early 1890s, much 'standard Gallé' with coloured floral relief patterns on opaque white backgrounds. He exhibited at the Paris Expositions of 1878 and 1884, and also, with great acclaim, of 1889 and 1900. The 'School of Nancy' grew up around him. The pieces made by him, especially those after 1889, are signed conspicuously, but also some made by his collaborators. His factory remained in operation after his death, under the direction of his son-in-law Pédrizet-Gallé, and until 1913 it was under the artistic direction of his lifelong friend and assistant VICTOR PROUVÉ; it closed in 1931. *See* VERRERIES PARLANTES; VERRE DE TRISTESSE; MARQUETERIE DE VERRE. *See* Philippe Garner, *Emile Gallé* (1976).

galleried rim. The RIM (of a bowl or other receptacle) which extends horizontally a short distance outward from the body and then vertically upward, the rim thus being greater in diameter than the body of the vessel. It is found on some glass sugar-bowls of the 19th century made in the United States. The cover rests within the rim, unlike the usual form of cover that is flanged and rests on the rim of the vessel.

game pieces. Coloured 'eyes' of marvered cane sections, late 1st century AD. W. 2·5/2·8 cm. British Museum, London.

game pieces. Sets of glass objects made for playing games. One example, late 1st century BC, consists of twenty-four domed pieces, differently coloured and decorated in four groups of six each, decorated with vari-coloured sections of CANES. *See* ASTRAGAL.

Gampe, Gottfried (fl. 1668). An engraver and enameller who worked at the glasshouse at Marienwalde, in Brandenburg, and was granted the privilege to engrave in 1668. He came from Bohemia with his brothers Daniel and Hans Gregor. Later, in 1750, Samuel Gottlieb Gampe worked at the same factory.

garniture. A set of decorative ornaments, such as VASES or CANDELABRA, for display on the mantelpiece or table. It usually consists of three pieces, sometimes five, with the centre piece somewhat larger. They were usually made of porcelain, but some were of glass, e.g., some vases of OPAQUE WHITE GLASS enamelled with a blue ground and white reserves with flowers, to simulate Worcester porcelain. Some garnitures of candelabra consisted of three such pieces (one with two arms and two outer pieces with one arm each). The French terms are *garniture de cheminée* and *garniture de table*.

gasolier. A type of CHANDELIER whose source of illumination is gas, rather than candles. An important English example, in the Banqueting Room at the Royal Pavilion, Brighton, was designed in 1828 by Robert Jones, who worked for Frederick Grace, but the maker is unknown; the jets are enclosed within shields of opaque white glass in the shape of lotus flowers.

Gatchell, Jonathan (d. 1823). A clerk at the Waterford Glass Co. who learned glass-making from JOHN HILL and succeeded him as Director, *c.* 1786. He operated it successfully, becoming a partner in 1799 and acquiring the company in 1810, managing it until his death. *See* WATERFORD GLASSWARE.

Gate, Simon. The Bacchus Festival Bowl, designed by Simon Gate, 1923. Orrefors Glasbruk, Orrefors, Sweden.

Gate, Simon (1883–1945). A glass-designer who joined in 1915 the ORREFORS GLASBRUK. He was previously a trained artist but not acquainted with glass techniques. He first made tall tinted vases, but soon experimented with CASED GLASS and was instrumental in the development of GRAAL GLASS. While at Orrefors he collaborated with EDWARD HALD.

Gates, John Monteith (1905−). An American architect who was appointed by AMORY A. HOUGHTON in 1933 as design leader at STEUBEN GLASS, INC. He designed the engraved decoration on important pieces, e.g., the VALOR CUP (1940). His meeting with Henri Matisse in Paris in 1937 inspired the series of pieces designed by eminent European and American painters and sculptors, with a view to developing collaboration between artists and artisans. When he retired in 1970 he was succeeded by PAUL SCHULZE.

gather(ing). A blob of molten glass attached to the end of a GATHERING IRON, BLOW-PIPE or PONTIL, preparatory to forming an object. It may be enlarged by collecting additional molten glass as work progresses, and the later gathering of the same colour does not appear as a layer but is fused into one mass. A glass object that is thick and heavy is said to have been made from a 'heavy gather'.

gathering iron. A long thin rod of iron used to dip into a POT (CRUCIBLE) of molten glass and on to which adhered a GATHER of the molten glass. Sometimes a BLOW-PIPE was so used when it was intended to shape the glass at once by blowing. *See* DIPSTICK; PONTIL.

gauze. *See* TWIST, *gauze.*

Gazelle Bowl, The. A crystal glass BOWL made by STEUBEN GLASS WORKS in 1935, the first engraved piece designed for the factory by the sculptor SIDNEY WAUGH. It is of blown, clear crystal glass and rests on a solid base cut with four flanges. The encircling engraved design depicts twelve leaping gazelles. It was first engraved by Joseph Libisch, who was brought to Steuben from Europe by FREDERICK CARDER and who was the master engraver there for many years. The piece has been commissioned more than fifty times, and replicas are in many museums.

Geares, Wenyjfrid, Goblet. A GOBLET attributed to the glasshouse of JACOPO VERZELINI with gilding possibly by ANTHONY DE LYSLE; it bears the name 'Wenyjfrid Geares' and the arms of the Vintners Company, with the date '1590'. It has elaborate gilded decoration of heraldic devices and inscriptions, the only so-called Verzelini goblet having decoration entirely in gilding. *See* VERZELINI GLASSWARE.

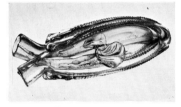

gemel flask. Pale green glass with rigaree trailing and applied leaf, Tansboro Glass Co., New Jersey, 1848–85. Ht 20 cm. Corning Museum of Glass, Corning, N.Y.

gemel flask. A FLASK with an interior division so as to make two separate containers, each with its own neck and mouth. They were made as separate bottles blown and then fused together. The term is derived from the Zodiacal sign of Gemini, the Twins. Some English examples have been attributed to so-called NAILSEA GLASS; these usually have a body of clear glass with opaque festoons as decoration or of opaque glass with coloured festoons. They were popular in the United States in the first half of the 19th century; they were often decorated with pincered RIGAREE trailing and other applied ornament or were of coloured or white glass with loopings. Those with a flat base were probably used as a DOUBLE CRUET for oil and vinegar. Sometimes called a 'gimmel flask'.

gemmato, vetro (Italian). Literally, gem-like glass. A type of glass developed by Ercole Barovier, *c.*1936–38, which has a slightly roughened overall surface somewhat resembling a stone.

gemmaux (French). Pieces of glassware made in France by a technique involving fusing small pieces of vari-coloured glass into a relief pattern.

George III glassware. Glassware whose decorative motif is related to George III, either an engraved portrait or cipher. As George I and successive Kings of England were also the Electors (later kings) of Hanover, Germany, from 1714 to 1837, the portrait of George III − sometimes with that of his consort, Queen Charlotte (of Mecklenburg-Strelitz) − also appears on some contemporary glasses made in

Germany. The royal cipher, strangely, is engraved on a wine glass made by the NEW BREMEN GLASSWORKS, Maryland, as late as *c.* 1785.

Georgian ale glass. A type of ALE GLASS made from *c.* 1770 that has a BOWL slightly more globular than conical and a stemmed FOOT, some stems having a MERESE under the bowl.

German glassware. Apart from some early ROMAN GLASS and FRANKISH GLASS, the earliest wares were made from *c.* 1400 in WALDGLAS, principally in the wooded regions of the Rhineland, Hesse, Thuringia, etc. This early *Waldglas* of the 15th century included the MAIGELEIN, the RÖMER, the NUPPENBECHER, and the KUTTROLF. In the 16th, 17th, and 18th centuries new forms and styles developed, including the STANGENGLAS and the HUMPEN, and also thin glassware in the FAÇON DE VENISE decorated with DIAMOND-POINT ENGRAVING and later thicker glass with WHEEL-ENGRAVING in HOCHSCHNITT and TIEFSCHNITT (*see* GONDELACH, FRANZ; SPILLER, GOTTFRIED; WINTER, MARTIN; SCHWANHARDT, GEORG; SCHWINGER, HERMANN; KILLINGER, GEORG). ENAMELLED GLASSWARE was also introduced, including that in SCHWARZLOT (*see* SCHAPER, JOHANN; HELMHACK, ABRAHAM; BENCKERTT, HERMANN) and TRANSLUCENT ENAMEL (*see* MOHN, SAMUEL; MOHN, GOTTLOB SAMUEL; SIEBEL, FRANZ ANTON). Later in the 18th century CUT GLASS was made in English style, and in the late 19th century and early 20th century German glassmakers were leading exponents of the *Jugendstil* (ART NOUVEAU). *See* BAVARIAN GLASSWARE; BRANDENBURG GLASSWARE; FRANCONIAN GLASSWARE; HESSE GLASSWARE; NUREMBERG GLASSWARE; SAXON GLASSWARE; THURINGIAN GLASSWARE; EHRENFELD.

geschnitzt (German). Literally, chipped. Decoration on glassware, done by chipping; a technique developed by WILHELM VON EIFF.

Gesù, Al, glassworks. A glass factory at MURANO that was first recorded in 1537 when it was owned by Nicolò Andrigo. It was operated from 1542 by Antonio Miotti and thereafter by his descendants until the late 18th century. It made OPAQUE WHITE GLASS with polychrome enamel decoration, examples of which exist with the factory name as its mark and some of which are dated from 1731 until 1747. It is said that the factory introduced AVENTURINE GLASS. *See* MIOTTI FAMILY; ANTONIO MIOTTO; VENETIAN SCENIC PLATES.

gilding. The process of decorating glassware, on the surface or on the back of the glass, by the use of gold leaf, gold paint or gold dust. Gold leaf with engraved designs was sandwiched between layers of glass from ancient times, e.g., Hellenistic glass (*see* CANOSA BOWL), ROMAN GLASS (*see* FONDI D'ORO), and later ZWISCHENGOLDGLAS. Some Roman glass gilding was done by applying gold leaf to a hot bubble of glass which, when blown, would break the leaf into speckles. Another method, used on Roman glass and later on Venetian glass, was sprinkling granular gold dust on molten glass. Islamic glass in the 12th to 14th centuries was very rarely decorated only with gilding, applied by use of colloidal gold and then fired; but mainly such glass combined gilding with enamelling (*see* MOSQUE LAMPS). Some Venetian glassware of the 16th century has gilding as the sole decoration, but normally enamelling was combined with it. In some German DIAMOND-POINT ENGRAVING the decoration is supplemented with gilding. By the 17th century decoration on the surface of glass was done by FIRING with gold leaf, using one of several methods (see below: *honey*; *mercuric*) or, especially in Venice, HALL-IN-TYROL, and the Netherlands, by unfired gold painting (see below: *unfired*); such methods of decoration were for gold borders on engraved glassware and in Germany on some Court glassware with gilt in relief. Gilding on the surface of English glassware was done in the 18th–19th centuries by LAZARUS JACOBS, the BEILBYS, MICHAEL EDKINS, and JAMES GILES. In Spain gilding was applied to WHEEL-ENGRAVED glassware at LA GRANJA DE SAN ILDEFONSO in the 18th century, and

Gazelle Bowl. Blown engraved bowl on cut base, designed by Sidney Waugh for Steuben Glass Works, 1935. D. 16·5 cm. Metropolitan Museum of Art, New York. Courtesy, Steuben Glass.

similar work was done in Germany. *See* GOLD ENGRAVING; GRANULAR GILDING.

——, *fired. See* below under *mercuric*; *honey*.

——, *honey*. The process of applying gold to the outer surface of glass (as to porcelain) in a manner to assure reasonable permanency (as distinguished from unfired gilding – see below) by using gold-leaf pulverized in honey, or powdered gold, and affixing it in a low-temperature FIRING. The resulting appearance is dull, with a rich and sumptuous effect, but it can be burnished to brightness.

——, *mercuric*. The process of applying gold to the outer surface of glass (as to porcelain) in a manner to assure permanency (as distinguished from unfired gilding – see below) by brushing on an AMALGAM of gold and mercury. Low-temperature FIRING caused the mercury to vaporize, leaving a gold deposit forming the design which could then be burnished to brightness. Mercuric gold has a thin, metallic, brassy appearance, quite unlike the dull rich colour of honey gilding (see above); it is much cheaper and easier to apply and fix.

——, *unfired*. The process of applying gold to the outer surface of glass (as to some porcelain) without FIRING, resulting in a lack of permanency. It was a primitive method by which a preparation of linseed oil was applied to the glass with a brush, after which gold leaf was affixed and allowed to dry without firing; so that subsequently it readily rubbed off. It could not be burnished. This method was used in England on some OPAQUE WHITE GLASS and on COUNTRY-MARKET GLASS of the 19th century. It was also used on some German, Bohemian, and Spanish glass, but has lasted better than that done in England, perhaps due to the fixative employed. Where such gilding has been rubbed off, the design may still be observed by the pattern on the glass left by the fixative. This method is also called 'cold gilding' or 'oil gilding'. The process is similar to applying COLD COLOURS (KALTMALEREI).

Giles, James (1718–80). An English independent decorator of porcelain by enamelling. His studio in London also, according to Robert J. Charleston, decorated glassware in similar style, but no example of glassware is known to have been decorated by him personally. The products so attributed are of opaque white glass decorated with enamelling, also blue and green glassware with gilding, often with pheasants and EXOTIC BIRDS, as well as neo-classical motifs and flowers, as found on Worcester porcelain. The pieces included mainly SCENT BOTTLES, DECANTERS, CRUET BOTTLES, FINGER-BOWLS, and VASES; the glass was obtained from the FALCON GLASSWORKS and PARKER CUT GLASS MANUFACTORY. *See* Robert J. Charleston, 'James Giles as a Decorator of Glass', *The Connoisseur*, June 1966, p. 96, and July 1966, p. 176.

Gilliland, John L. A glass-maker who worked at the NEW ENGLAND GLASS CO. and in 1820 founded, with John and Richard Fisher, the Bloomingdale Flint Glass Works in Manhattan, New York (it closed in 1840). In 1822 Gilliland withdrew and formed John L. Gilliland & Co. in Brooklyn, which operated the Brooklyn Flint Glass Works that made high-quality cut glass. In 1864 a controlling interest was acquired by Amory Houghton, Sr. and Jr., and they moved the factory to Corning, New York, where they established the CORNING GLASS WORKS. Gilliland has been said to have made PAPER-WEIGHTS (possibly due to confusion with the name of the glasshouse of WILLIAM T. GILLINDER), but no example has been identified nor any record found.

Gillinder, William T. (1823–71). A glass-maker born at Gateshead, near Newcastle-upon-Tyne, England, who worked in Birmingham, learning to make PAPER-WEIGHTS at GEORGE BACCHUS & SON there. He migrated to the United States in 1853 and, after a brief time at the NEW ENGLAND GLASS CO. (where he made some paper-weights), moved west, working at various glassworks in Pittsburgh, St Louis, and elsewhere, until he returned to Philadelphia in 1861. There he bought a glassworks that he renamed the Franklin Flint Glass Works, and later he added to it a new glassworks. He was joined in 1863 by Edwin Bennett, and they formed Gillinder & Bennett which in 1867 (after Gillinder's two sons, James (b. 1844) and Frederick (b. 1845), had joined the firm) became Gillinder & Sons. They soon acquired the Philadelphia Flint Glass Works. After their father's death the sons took over, but later, in 1912, the three sons of James Gillinder withdrew from the family concern to start their own company, Gillinder Brothers, Inc., at Port Jervis, New York. The Philadelphia factory closed in 1930. Charles Challinor (1841–1932), who was an apprentice at the factory from c. 1867, made paper-weights. *See* GILLILAND, JOHN L.

gilt bronze. Bronze mercurially gilded, employed in France as a MOUNT for glass and porcelain ware, especially during the period of the EMPIRE STYLE and after. It is often loosely called 'ORMOLU'.

gimmel flask. *See* GEMEL FLASK.

gin glass. A drinking glass, varying in shape, for serving gin; it is similar to a WINE GLASS but smaller, usually about 9 to 11 cm. high.

girandole. (1) An English term for a CANELABRUM, from 1766 until c. 1792. They were also sometimes called a LUSTRE or TABLE LUSTRE. (2) The term is sometimes used loosely, especially in the United States, to include a SCONCE, a convex wall mirror with two branched candlesticks. (3) The term is also occasionally used for the nozzle and drip-pan of a CANDLESTICK.

girasol (Italian). An Italian term that means 'sunflower' and also a milky semi-precious stone that in sunlight is reddened and resembles an opal. Glass resembling such stones was made in Italy from the beginning of the 18th century and was so-called.

Gjövik Glasverk. *See* NORWEGIAN GLASSWARE.

glass. An amorphous, artificial, non-crystalline substance, usually TRANSPARENT but often TRANSLUCENT or OPAQUE, made by fusing some form of SILICA (e.g., SAND) and an ALKALI (e.g., POTASH or SODA), and sometimes another base (e.g., LIME or LEAD OXIDE). It is PLASTIC when molten, DUCTILE, e.g., after having cooled from a molten state or after having been reheated, and GLYPTIC when cold. It can be formed into various shapes when in a molten state by BLOWING, MOULDING, PRESSING, or CASTING, coloured by use of various METALLIC OXIDES, and decorated by ENGRAVING, ETCHING, CUTTING, CASING, SAND-BLASTING, ENAMELLING, PAINTING, GILDING, TRANSFER PRINTING, or applied ornament. It has been used for decorative purposes (e.g., STAINED GLASS WINDOWS and MIRRORS), utilitarian purposes (e.g., BOTTLES, JUGS, GLASS BULBS, PHARMACY WARE), as tableware (e.g., DRINKING GLASSES, BOWLS, DECANTERS), ornamentally (e.g., VASES, FIGURES, PAPER-WEIGHTS), as jewellery (e.g., BEADS, BANGLES), and industrially (e.g., PLATE GLASS, LAMINATED GLASS, GLASS FIBRE). The ingredients and quality of the METAL and the forms and decoration of glass objects have changed from earliest times to modern times with contemporary techniques and styles. The hardness of average glass is 5·5 on the MOHS' SCALE. *See* DEVITRIFICATION.

———, *pre-Roman history*. Glass originated, according to recent opinion, in Mesopotamia in western Asia, c. 3500–3000 BC where it was used as a vitreous glaze on stone and ceramic beads. It was first used independently for objects made wholly of glass c. 2500 BC, mainly for glass beads and amulets. Evidence has been found of glass making in Mesopotamia by 2000 BC. The earliest evidence of glass vessels are fragments, probably from Mesopotamia, c. 1525–1475 BC found in western Asian sites near Nineveh. After the conquests in Asia by the Egyptian Pharaoh Tuthmosis (Thothmes) III (c. 1504–1450 BC), a glass-vessel industry is said to have been started, c. 1475–1450 BC, in Egypt by Asiatic workers. The earliest-known surviving glass vessels are three small Egyptian vessels that may have been made, c. 1470 BC, to honour

Giles. James. Blue-glass vase, with gilded decoration of exotic birds, studio of James Giles, 1770. Ht 27·7 cm. Courtesy, Maureen Thompson, London.

him. (*See* TUTHMOSIS JUG.) CORE GLASS was made in Egypt from the second half of the 15th century (the late XVIIIth Dynasty) until the end of the Bronze Age, *c.* 1150 BC; there was then a glass-making lapse there until the 4th century BC. In Syria SIDONIAN GLASSWARE was made from before 1500 BC, but from *c.* 1150 BC until the 9th century little glass was produced there except beads and trinkets, but then the glass centres of Sidon and Tyre and of Mesopotamia were revived, and the Phoenician traders spread the products widely. By the 7th and 6th centuries BC glass-making had started in Cyprus and Rhodes, and from the 9th century BC in Italy, spreading north to Adria by the 5th century. From the 4th century less Mesopotamian glass was produced, but Syrian centres were revived, and also Egyptian glass upon the founding of Alexandria and its glasshouses in 332 BC. In the Hellenistic period, 320–100 BC, much glass from Syria and Egypt was exported to Greece and Italy. Glass-making in Italy really started about 100 BC, brought by the Alexandrians, who were followed by Syrian craftsmen in northern Italy after BLOWING was invented.

Glass made in the pre-Roman period was produced by (1) the CORE GLASS process, (2) fusing coloured glass rods to make MOSAIC GLASS, (3) COLD CUTTING from a raw block of glass (*see* SARGON ALABASTRON), or (4) CASTING by the use of a MOULD, either fusing powdered glass in a mould (*see* PÂTE-DE-VERRE) or using the CIRE PERDUE method.

glass 'cakes'. Flat tablets, usually of opaque glass (sometimes loosely called 'enamel glass'), made in various shapes and sizes, often bearing the impressed mark of the MURANO glass-maker who in the 18th century made them to be sold commercially (sometimes as far afield as China), probably for making coloured enamels for decorative purposes. They were usually of brilliant colours (e.g., sealing-wax red or turquoise blue), but sometimes white, and some were of transparent glass. Ingots of similar material made in Egypt seem to have been their precursors. The Italian term is *focaccie*. *See* SMALTO. For a discussion of such pieces, *see* Robert J. Charleston, 'Glass Cakes', *Journal of Glass Studies* (Corning), V (1963), p. 54.

glass eye. An artificial removable glass eye made and decorated in the form of an eye-ball. They were first made *c.* 1579 by Ambrose Paré (1510–90) in France, although artificial eyes of stone and other materials (for humans, statues, and mummies) had been made in Egypt *c.* 500 BC. In Germany Ludwig Müller-Uri, a glass-blower in Lauscha, Thuringia, made improved glass eyes for humans, but previously glass eyes had been made in Venice, Bohemia, England, and France. Glass models were further developed in Germany in 1898 by Snellen (b. 1862). Most glass eyes were made in Germany until World War I, and even after 1918 much of the special glass came from Germany; but in 1932 suitable glass was developed by technicians in Chicago. Due to the fragility of glass eyes, substitute materials have been developed since the 19th century, and today are in general use (e.g. acrylic eyes). The globe of a glass eye is made of blown opaque glass, the cornea of clear glass, and the colours are produced by metallic oxides. *See* J. H. Price, *Ocular Prosthesis* (1946); for a very old explanation for making a glass eye, *see* H. Bléricourt, *The Art of Glass* (English translation, London 1699), p. 353.

glass fibre. A generic term to designate any fibre drawn as a thread from molten glass, regardless of its form, appearance, method of manufacture or ultimate purpose. Such fibre can be a continuous fine filament (about 0·02 mm. thick) that can be drawn out and wound on a spool. Some ancient examples of glass fibre are recorded from Egypt and Rome. Fibre sufficiently fine to be woven or knitted was developed from *c.* 1840, and used to make imitations of silk brocade, and even to make a neck-tie or a dress. It was also used in making some boat-shaped FRIGGERS (some rested on a sea of glass fibre) or a BIRD FOUNTAIN. From 1893 it was woven with silk thread to make fabrics which, however, were too coarse to be folded. Later the process was refined and wider uses developed,

glass hunt. Coloured glass riders, hounds, stag and tree; mirror background, first half of 19th century, signed 'Aubin'. W. 38 cm. Courtesy, Delomosne & Son, London.

e.g. insulation, sound absorption, boat hulls, fabrics, etc. Obsolete terms are 'glass silk' and 'spun glass'.

glass flowers. *See* HARVARD GLASS FLOWERS.

glass-grinder. A glass-worker who was a grinder and shaper of MIRRORS as distinguished from a cutter of ornamental glass. His work included grinding the mirror to a true surface and also BEVELLING the edges and making them into ornamental forms by diamond FACETING (*see* DIAMONDING) and shallow CUTTING. *See* BETTS, THOMAS.

glass hunt. A three-dimensional tableau in a metal frame with clear glass at the front and mirrors at rear and sides, depicting with miniature coloured glass figures and animals a hunting scene, with trees, riders, stag, and hounds in full chase. They were made AT-THE-LAMP in England in the first half of the 19th century, some signed 'Aubin'.

glass-maker's mark. A mark to identify his work, affixed by a glass-maker from early times until today, in the form of a signature, inscription or device. Such marks have been affixed by MOULDING, scratching or ENGRAVING. In the 1st century AD some glass objects were marked to indicate the workshop of ENNION, a famous maker who worked at Sidon and later in Italy; other early marks, in Greek and Latin, are those of ARISTEAS, Ariston, Artas, JASON, MEGES, and NEIKAIS. Some German and Bohemian decorators in the 16th and 17th centuries signed their work, e.g., CASPAR LEHMAN(N), HEINRICH JÄGER, JACOB SANG, ELIAS ROSBACH; also some Dutch decorators, e.g., DAVID WOLFF and FRANS GREENWOOD, and Austrian, e.g. ANTON KOTH-GASSER. In later times, those who have signed their work include: in France, EMILE GALLÉ and MAURICE MARINOT; in England, the BEILBYS and ISAAC JACOBS; in the United States, LOUIS COMFORT TIFFANY and FREDERICK CARDER.

glass-maker's mark. Detail of foot of *Pokal* engraved 'Rosbach Fecit Berlin', Elias Rosbach, *c*. 1735. Kunstgewerbemuseum, Cologne.

glass-makers' soap. Oxide of MANGANESE used as a DECOLOURIZING AGENT for glassware.

glass-makers' tools. *See* TOOLS, GLASS-MAKERS'.

glass-of-lead. An early English term for LEAD GLASS.

glass paste. *See* PÂTE-DE-VERRE.

glass picture. A picture executed on flat glass, by any of several processes. *See* PANEL; MIRROR PAINTING; BACK PAINTING; PLAQUE; TRANSFER ENGRAVING.

glass porcelain. *See* BASDORF GLASS.

Glass Sellers' Company. A corporation, initially composed of mirror-makers, hour-glass makers and merchants; it was granted a charter in 1635 by Charles I and reincorporated in 1664 as 'The Worshipful

Company of Glass Sellers of London'. A purpose after 1670 was to promote research in order to make English glass-makers independent of Continental materials. By 1675 it had 88 members. Initially it imported MURANO glassware; from 1667 to 1673 it sent to Murano designs for glassware in more simplified English styles, made by JOHN GREENE and Michael Measey. In 1674 it contracted with GEORGE RAVENSCROFT to start research toward a new type of glass, and by 1674 he was the company's official glass-maker, with his glasshouse founded in 1673 in the Savoy in London and another that the company provided for his further experiments in 1674 at Henley-on-Thames. By 1676 Ravenscroft had developed the process of making LEAD GLASS, having found a way to reduce the problem of CRISSELLING; in recognition he was granted permission to use the RAVEN'S HEAD seal on the Company's glassware. From 1676 the Henley glasshouse was directed by HAWLEY BISHOPP, who took over also the Savoy in 1681. By 1685 the making of lead glass was successfully established. *See* JOHN AKERMAN.

glass silk. An obsolete term to designate relatively long but not continuous monofilaments of glass produced by a number of early processes for mechanical drawing of GLASS FIBRE.

glass wool. A fluffy mass of relatively short GLASS FIBRES collected together or felted together into a thick blanket or lightweight slab. It is generally produced by a blowing or centrifugal process. The threads have a much lower density than so-called SPUN GLASS. It is used for heat insulation, filtration of acids, etc.

Glassboro Glassworks. The second glasshouse started in New Jersey, founded at Glassboro, Gloucester County, 1780–81, by Jacob Stanger and two brothers, former employees of CASPAR WISTAR. It had many changes of ownership until it became the Whitney Glass Co. that was acquired in 1918 by the Owens Bottle Co. Its wares in the 18th century were of the style of SOUTH JERSEY TYPE GLASSWARE, but no piece has been definitely attributed to this factory.

glassware. Objects and ware that are made of glass. These fall into several categories, which are distinguished by the French thus: (1) *verrerie*, vessels, vases, and useful ware; (2) *verroterie*, beads, toys, and jewellery; (3) *vitrail*, window glass, and (4) *verre plat*, plate glass.

glasswort. A salt-marsh plant, *Salicornia herbacea*, the ash of which was used to supply ALKALI (SODA) in making glass from Renaissance times or earlier. It was known in Spain as BARILLA and that coming from Egypt as ROQUETTA.

Glastenbury Glass Factory. A glasshouse founded in 1812 on the Hartford-New London Turnpike, near Hartford, Connecticut. It made principally bottles. After several successive changes of ownership, it closed *c.* 1824. Some pieces probably made by it have heretofore been attributed to the PITKIN GLASS WORKS, in East Hartford, as products from this factory were not known until recently excavated. *See Journal of Glass Studies* (Corning), V (1963), p. 117.

glazier. A specialized artisan who from early times made, fitted, and painted flat glass for WINDOWS. They were formed into guilds from *c.* 1328 in England and in the 14th century in Continental countries.

glazing bars. Wooden framework into which glass panes were fitted in windows and doors of 18th-century cabinets.

globe. A terrestrial globe, such as one made entirely of crystal glass for Louis XVIII in 1818 by the ST-LOUIS GLASS FACTORY.

globe. Crystal glass globe made for Louis XVIII, 1818, St Louis Glass Factory.

globular unguentarium. A type of UNGUENTARIUM made in the form of a globe, said to have been blown with a small closed neck and then broken to make a tiny opening for filling, the contents having been said to be probably a powder. They were made of ROMAN GLASS, probably in northern Italy, in the 1st century. *See* ROUGE POT.

glory-hole. A hole in the side of a glass FURNACE, used when reheating glass which has already been molten and is in the process of being fashioned or decorated, or ware which, having been MOULD-BLOWN, is FIRE-POLISHED to remove imperfections remaining from the mould. *See* BOCCA; BYE-HOLE; SPY-HOLE.

glue chip decoration. A type of decoration made by covering the surface of a PARAISON (e.g., of JADE GLASS) with a glue which shrank when it dried and pulled away flakes of glass from the surface; then an acid bath was used to remove sharp edges and provide a satin-finished texture. It was used by FREDERICK CARDER at STEUBEN GLASS WORKS in the 1920s.

glyptic. Susceptible to carving, especially hardstones and gems, but also applicable to certain types of glass, as evidenced by incised relief cutting such as HOCHSCHNITT, and CAMEO cutting (as on the PORTLAND VASE). *See* PLASTIC; DUCTILE.

goblet. A drinking glass with a large BOWL, of various shapes, resting on a stemmed FOOT. The usual height is from 20 cm. to 25 cm., but some have been made as tall as 45 cm. The stem is of various forms and styles, some very elaborate and occasionally in the form of a figure or an animal. Most 16th–17th century examples, particularly from the Continent, have a COVER with an ornate FINIAL. The German term is POKAL. *See* COVERED GOBLET; WINGED GLASS; GOBLET, TWO-TIERED; BELL GOBLET; STANDING CUP; ROYAL OAK GOBLET; DIER GLASS.

goblet, armorial. A GOBLET decorated with a coat-of-arms. They were made in Germany and elsewhere, but especially highly regarded are those with the Royal Arms of George III enamelled by WILLIAM BEILBY.

goblet, two-tiered. A double GOBLET with one resting on top of the other. Examples were made in Bohemia.

goblet vase. A type of VASE in the shape of an over-size GOBLET, with large bowl and footed stem, and sometimes with a mouth narrower than the bowl to provide a collar on which to rest a cover.

Gol, José Maria. A glass-designer from Barcelona who, during the period between the two World Wars, designed some glassware in traditional style as well as pieces in fanciful shapes with enamelled decoration in the form of floral and abstract motifs.

gold. (1) A metallic agent used in colouring glass; *see* GOLD RUBY GLASS and PURPLE OF CASSIUS. (2) The metal used in various forms to decorate glassware, e.g. as gold-leaf (*see* ZWISCHENGOLDGLAS; FONDI D'ORO; GOLD ENGRAVING), or in GRANULAR GILDING, as well as gilt painting (both fired and unfired) – *see* GILDING.

gold-band glassware. Glassware decorated completely with vertical wavy bands of vari-coloured and colourless glass, some of the bands being formed of two layers of clear glass enclosing a layer of gold foil. Examples exist in the form of small *alabastra* made in Egypt in the 2nd/1st century B.C.

gold engraving. A style of decoration done on gold-leaf applied to the back of glass and engraved with a point. The design is protected by the glass itself or sometimes additionally by a layer of glass fused or

gold-engraved glass. Goblet with cover, with gold-engraved decoration on interior, and ribbing on exterior of bowl, Bohemia, second quarter of 18th century. Ht 25 cm. Museum für Kunst und Gewerbe, Hamburg.

cemented over it. The technique was used in the 3rd century BC in Alexandria. The former type includes some miniature portraits, *c.* 3rd century (*see* PLACIDIA, GALLA, CROSS), the French glass known as ÉGLOMISÉ, some Italian PANELS of the 14th century (the best examples being made at Padua and Bologna), and the Netherlands glass decorated by ZEUNER. The latter type includes the FONDI D'ORO, the 18th-century German ZWISCHENGOLDGLAS, some Bohemian and German pieces, *c.* 1725–42, backed with coloured marbling, and the MEDAILLONBECHER decorated by JOHANN JOSEF MILDNER.

gold ruby glass. Beaker with cover, decorated in *Hochschnitt* by Gottfried Spiller, Berlin, late 17th century. Ht 15 cm. Museum für Kunst und Gewerbe, Hamburg.

gold ruby glass. A type of glass of deep-ruby colour, coloured by means of a precipitate of colloidal gold. The process for making it was hinted at by ANTONIO NERI in his book published in 1612. The method was developed by JOHANN KUNCKEL, probably before 1679, at the Drewitz glasshouse at Potsdam, near Berlin, by the use of gold chloride (applying the previous discovery by Andreas Cassius of the effect of gold chloride in making his PURPLE OF CASSIUS) and by then reheating (STRIKING) the greyish finished object, the reheating developing the ruby colour. Such glass was made thereafter in Nuremberg and elsewhere in south Germany, some examples having Nuremberg and Augsburg marked mounts and characteristic engraving, some perhaps by JOHANN HEEL. Some Bohemian glass imitations were made possibly using copper instead of gold by FLASHING ruby-coloured glass over colourless glass or by adding a ruby STAINING over colourless glass; these were embellished by cutting or engraving that showed the design contrasted against the under-layer of clear glass. In the United States solid gold ruby glass was made by the leading glassworks; a variation was made called AMBERINA. Being expensive, ruby glass was sometimes used only for parts of objects. It has long been a belief by some glass-makers that the colour was produced or ensured by throwing into the BATCH a gold coin or gold ring; this was perhaps done as a ceremony or by superstition, but such solid gold objects would only melt in the bottom of the pot and not disperse into the mix, as occurred only when gold was used in a solution. The German term is *Goldrubinglas* or *Rubinglas*. *See* LYCURGUS CUP.

gold sandwich glass. *See* ZWISCHENGOLDGLAS.

goldstone glass. The same as gold AVENTURINE GLASS; it is brownish with innumerable tiny and closely spaced spangles of gold. It has been used in decorating some CHINESE SNUFF BOTTLES and some PAPER-WEIGHTS.

Gondelach, Franz (1663–1726). A German engraver of glass who was born at Gross-Almerode, near Cassel (Kassel), in Hesse, apprenticed to a gem-cutter at Cassel, and in 1688 became glass-engraver at the Court of Landgrave Carl of Hesse-Cassel. His work included mainly HOCHSCHNITT, and INTAGLIO (with signature in DIAMOND-POINT ENGRAVING) and some cutting of ROCK CRYSTAL. He depicted allegorical, heraldic, biblical and pastoral subjects, as well as making portraits. His work is considered the supreme achievement of BAROQUE art in glass engraving. He often engraved an eight-point rosette under the base of a glass. He was succeeded by Johann Heinrich Gondelach. *See* HESSE GLASSWARE.

Gondelach, Georg. A native of Hesse, Germany, who, after operating a crystal-glass factory at Dessau, was made, in 1677, the glass-master at the new glassworks that had been established at Drewitz (near Potsdam) by the elector of Brandenburg CHRISTOPH TILLE, who also came from Dessau, followed him to Drewitz. *See* BRANDENBURG GLASSWARE.

Götterbecher. *See* FIGURED BEAKER.

Gottstein, Franz (fl. 1810–30). An Austrian glass-engraver from Gutenbrunn; he worked at the Strany glassworks in Moravia and later, from 1823 to 1830, at Gutenbrunn, signing some pieces.

Graal glass. A type of ornamental glass made by ORREFORS GLASBRUK, developed in 1916 by SIMON GATE with Albert Ahlin and Knut Bergqvist. The process was an improvement on the CASED GLASS of EMILE GALLÉ and involved first making an object in coloured glass by CUTTING and ETCHING the pattern, and then subjecting the piece to the heat of the furnace to impart fluidity and smoothness to the design, and finally FLASHING with clear glass. In early examples the METAL was thin and light, but after 1930 the glass was heavy and massive, thickly cased, and sometimes combined with the technique of ARIEL GLASS. The name was inspired by the saga of the Holy Grail. The technique has been further developed in recent years by designer Eva Englund, making pieces featuring animal and flower (especially rose) motifs.

Graffart, Charles (1893–1967). A leading Belgian designer and maker of glassware at VAL-SAINT-LAMBERT, becoming master-engraver in 1926 and designing about 300 pieces until 1929. From 1929/30 he made models for the workshops. In 1942 he succeeded Joseph Simon as the artistic director. He decorated glassware with WHEEL-ENGRAVING and he carved pieces from solid blocks of glass, and also made pieces of PÂTE-DE-VERRE.

Graal glass. Bowl by Eva Englund in Graal technique, 1974. Courtesy, Orrefors Glasbruk.

graffito, vetro (Italian). Literally, engraved glass. A type of ornamental glassware with surface decoration made by engraving applied gold leaf.

Gral Glashütte. A German glasshouse founded in 1930 in Dürnau, near Göppingen, east of Stuttgart. A subsidiary established in 1938 in Bohemia was soon discontinued and in 1945 was combined with the factory at Dürnau. Among its designers have been Konrad Habermeier (b. 1907), Josef Stadler, Günther Hofmann (b. 1940), and Karl Wiedemann (b. 1905).

Granja de San Ildefonso, La. The site, near the Spanish Royal Palace south-west of Segovia, of a royal glassworks. It was founded in 1728 by (BUENA) VENTURA SIT with artisans from glassworks in Spain and various countries of Europe, and came under the patronage of Queen Isabella (Isabel Farnese). Its early production was mainly large MIRRORS (for which it received the patronage of Philip V), later plate glass for mirrors (under an Irishman, John Dowling), and from c. 1763 CHANDELIERS and VASES (engraved with royal heraldic devices), all made for the Royal Palace. At a separate workshop, the Réal Fábrica de Cristales (under Lorenz Eder, of Sweden, and from 1768 until after 1791 Sigismund Brun, a German), blown, opaque, and coloured SODA LIME glass was made in the 18th century, with cut and engraved decoration, fired GILDING, and ENAMELLING, mainly inspired by foreign styles. After c. 1800 glass was made in the Anglo-Irish style. In 1829 the factory was leased, and thereafter made only common glass. *See Journal of Glass Studies* (Corning), III (1961), p. 119. *See* SPANISH GLASS.

granular gilding. A style of gilt decoration on glassware having a granular dust-like appearance. It was used on some Venetian CRISTALLO glass of the early 16th century, sometimes over a broad expanse of the surface or as a wide border around the rim of a bowl, and on some examples the lettering of inscriptions was erased from the gilding. The style is also found on some glass from Antwerp or southern Netherlands of the 16th century but to a lesser degree, as on rims and on applied decoration. It also was used on some of the goblets attributed to JACOPO VERZELINI; *see* G S GOBLET.

grape-cluster flask. A type of FLASK (AMPHORISKOS) decorated with a MOULD-BLOWN cluster of grapes, identical on both sides. Some have merely a short neck and small mouth, but others, with a spreading BASE or flat foot, have a tall neck, a disk mouth, two vertical side-handles, and two small loops at the rim for a carrying strap. Often the seam indicates where the two identical sides were joined. They were made of ROMAN GLASS in Syria, and probably also in Italy and Gaul, in the 1st to 3rd

grape-cluster flask. Double-moulded flask, Roman glass, c. 300 AD. Ht 17 cm. Römisch-Germanisches Museum, Cologne.

Mary Gregory glassware. Vase, cranberry colour with white enamelling. Henry Ford Museum, Dearborn, Mich.

centuries. Similar flasks were in the form of a bunch of dates. For the same technique, *see* SHELL FLASK.

grape ripener. A hollow dome-shaped object of clear glass, about 26 cm. high, having one flattened side and an orifice on the shoulder through which to insert a developing bunch of grapes to promote their growth and ripening out of doors. There is a knob at the top for suspension. They were made of NAILSEA GLASSWARE.

grapefruit-glass. A modern piece of glassware, having a hemispherical bowl resting upon an attached flat base, used at the table for serving a half grapefruit.

Gray-Stan Glassware. Glassware made at the glasshouse of Mrs Graydon-Stannus, London, *c.* 1923–32. Some pieces designed by James Manning are of clear glass with applied coloured decoration, opal glass with FLASHING, and coloured glass with marbled effect. Other ware is of clear glass engraved by Noel Billinghurst, *c.* 1932, and pieces in the style of CLUTHRA GLASS.

grease-pan. The same as a DRIP-PAN of a CANDLESTICK.

Greek fret. *See* KEY-FRET.

Greek glassware. No glass-making site on the Greek mainland or islands dating from antiquity had been discovered until recently; some specimens of inferior glass have been attributed to a factory in Crete (possibly at the city of Elyros). Covered circular examples of the PYXIS (copied from marble prototypes) were possibly made in moulds, using crushed glass that was fused by heat, and date from perhaps the 3rd/1st centuries BC. There is evidence to suggest that glass was also made on Rhodes and Cyprus, and a workshop for melting glass has been discovered on the mainland at Olympia. Glass was also made at Corinth and elsewhere in the 11th/12th centuries AD. *See Journal of Glass Studies* (Corning), I (1959), p. 11.

Greek inscription. An inscription in Greek, such as appeared on ROMAN GLASS vessels (e.g. CIRCUS-BEAKERS) made in the 1st century AD, sometimes including the name of the maker, e.g. ENNION or MEGES. *See* GLASS-MAKER'S MARK.

Greek vases. Jugs, bowls, cups, and other utentils (the Latin word *vasa* broadly means 'utensils') generally made of pottery in ancient Greece (and later in Rome). Some were copied from metal vessels, and vice versa. There are various basic forms having accepted names, but these forms are known in many variations. *See* ALABASTRON, AMPHORA, AMPHORISKOS, ARYBALLOS, ASKOS, DIOTA, KYLIX, KALPIS, KANTHAROS, KARCHESION, KRATER, KYATHOS, LAGYNOS, LEKYTHOS, OENOCHÖE, OLPE, PELIKE, PHIALE, PYXIS, PROCHOÖS, SKYPHOS, STAMNOS. For Roman names, *see* ACETABULUM, AMPULLA, MODIOLUS, POCULUM, STAMNIUM, TRULLA. Most of these types have been copied in EGYPTIAN GLASS and in ROMAN GLASS.

green glass. (1) Glass that is of a natural green or brownish colour, due to the presence of particles of IRON in the SILICA of the BATCH. Being used to make some bottles, it is sometimes called 'BOTTLE GLASS'. *See* WALDGLAS; NAILSEA GLASS. (2) Glass that is deliberately coloured green, by use of oxide of COPPER, CHROMIUM or iron, and also sometimes uranium (*see* URANIUM GLASS). *See* ANNAGRÜN.

Green Opaque glassware. A type of ART GLASS, having a yellowish-green solid colour. It is uncased, and often a blue stain is applied in a mottled pattern on the upper part of the object, below which is a gold border. It was developed by the NEW ENGLAND GLASS CO., *c.* 1887.

Greene, John (d. 1703). A founding member and master (1679) of the GLASS SELLERS' COMPANY who, with another member, Michael Measey, corresponded with Alessio Morelli, a glass-maker at Murano, complaining about the Venetian glass and sending instructions and over 400 sketches as to the types of glass that his company would import. He placed orders from 1667 to 1673. His efforts led to changes in the styles of the Venetian glass and the development of the forms of STEMS and BOWLS that were to prevail in England; the so-called 'Greene glasses' were in general of simple shape and well proportioned. General dissatisfaction with the Venetian ware and breakages in transit led eventually to the development of an English substitute glass, LEAD GLASS, encouraged by the Glass Sellers' Company.

Greenwood, Frans (1680–1761). An amateur engraver who was the first to decorate glass exclusively by STIPPLING, from *c.*1722 onward. He was a native of Rotterdam, but of English ancestry, and became a resident of Dordrecht from *c.*1726. He seldom designed but usually copied from mezzotints and prints of contemporary paintings or depicted flowers and fruits and also CALLOT FIGURES. Most of the glassware decorated by him was supplied from Newcastle, England, it being more suitable for his medium. He signed and dated many of his pieces. *See* W. Buckley, *Notes on Frans Greenwood* (1930).

Gregory, Mary, glassware. Glassware, clear or of various colours, decorated with paintings, in white or pink-white ENAMEL, of a young boy and/or girl in Victorian dress. The METAL was often of a ruby or dark pink colour, hence the use of the name 'Cranberry glass' applied to it in the United States. The ware was popular in the Victorian period, after which many crude versions, imitating CAMEO GLASS, were made. 'Mary Gregory' has not been identified, but it has been suggested that there was a worker so named at the BOSTON & SANDWICH GLASS CO., *c.*1870–80, where some glassware of this type has erroneously been said to have been made, as well as possibly at other United States glasshouses.

grenade. A type of BOTTLE, globular and short-necked, which served as a primitive fire-extinguisher by being filled with water and hurled into the flames. Examples (about 8·5 cm. in diameter) have been found at the Arsenal at Rhodes. *See* BOMB.

grey cutting. The frosted surface on glassware when decorated by cutting and before any polishing. Sometimes it is combined for artistic effect with areas of polished surface.

Grice, Edwin (1839–1913). An English glass-engraver and glass-cutter who began working on glass in 1861 as a decorator for J. & J. Northwood, at Wordsley, and who was connected from 1879 until 1904 with Guest Bros., of Brethell Lane, where he introduced many improved processes. *See* PEGASUS VASE; NORTHWOOD, JOHN.

grisaille. (1) Decorative painting on glass, as well as on porcelain or pottery, in varying shades of grey (termed '*en grisaille*'); it was sometimes used to imitate relief sculpture. *See* MONOCHROME; CAMAÏEU, EN. (2) A brownish paint made with oxide of iron that was fused onto the surface of glass to define details in STAINED-GLASS WINDOWS.

Grotesque glassware. A type of glassware developed and so-called by FREDERICK CARDER at the STEUBEN GLASS WORKS, *c.*1929–30. The pieces (BOWLS and VASES) are of FREE-FORMED GLASS, usually of colourless glass toward the bottom and shading up into various transparent colours. The shape usually undulates with areas of thin glass between heavy corner ribs. *See* SHADED GLASS.

grotesques (from Italian *grotteschi*). A decorative motif consisting of fanciful figures and half-figures that are partly human and animal and

Grotesque glassware. Vase designed by Frederick Carder for Steuben Glassworks, 1930s, and so-called by him. Ht 22·7 cm. Rockwell Collection, Corning Museum of Glass, Corning, N.Y.

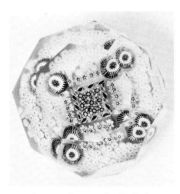

ground, carpet. Patterned millefiori faceted paper-weight with carpet of hollow white canes, St Louis. W. 6·4 cm. Courtesy, Sotheby's, London.

ground, jasper. Flower paper-weight, St Louis, W. 7·8 cm. Courtesy, Christie's, London.

ground, upset muslin. Snake paper-weight, Baccarat. W. 7·8 cm. Courtesy, Christie's, London.

partly acanthus foliage, used in conjunction with masks (based on the old classical theatrical masks) and such mythological figures as satyrs and sphinxes. They were originally inspired by mural decoration in the ruins of Nero's Golden House in Rome, and adapted by Raphael in the decoration of the *loggie* of the Vatican (hence sometimes called '*Raffaelesche*'). They should not be confused with ARABESQUES. They continued to be popular in various guises and modifications until the beginning of the ROCOCO period, *c*.1730. Although principally a decoration on ceramic ware, it also is found on some glassware.

grotto. A tableau of a grotto with miniature glass figures, made AT-THE-LAMP, of people and animals (often religious groups). The figures were made at Nevers, or in the style of NEVERS FIGURES. The tableau was usually on a stand enclosed in a glass case, with a mirror or painted landscape at the rear, or sometimes was a table decoration on a mirror tray. the stand was usually made of papier-mâché moulded to resemble rocks, covered with ground glass, and embellished with shells and dried plants. Such pieces were made at Nevers and Paris.

ground. The flat or slightly domed decorative background for the interior motif in a PAPER-WEIGHT. Sometimes the ground is glass of a single opaque or translucent colour, but usually grounds are very detailed and varied, made of many CANES or of FILIGRANA or LATTICINO glass, and they often contribute largely to the value of highly prized weights.

TYPES OF GROUND

———, *basket*. A ground that is funnel-shaped and made of double-swirl canes of *filigrana* glass. It is found on some ST-LOUIS weights, and sometimes the basket appears to have an arched overhead handle.

———, *carpet*. A ground made of many small canes of the same type and colour, tightly packed so as to form a 'carpet' and into which are usually set a few interspersed contrasting canes, e.g. SILHOUETTES, FLORETS. *See also*, below: star-dust ground; *choux-fleur* ground.

———, *chequer*. A ground composed of regularly spaced canes in a chequerboard pattern, separated by a grid of FILIGREE canes. Called in the United States a 'checker ground'.

———, *choux-fleur*. A ground in the style of a carpet ground (see above) but with white canes shaped somewhat irregularly and loosely set, so as to give the appearance of a cauliflower.

———, *cushion*. A ground that is somewhat arched and elevated above the basic ground so that some parts of the motif can be sometimes seen in the intervening space. It is usually made of double-spiral opaque-white or coloured FILIGRANA threads.

———, *jasper*. A ground that is mottled with particles of glass of two colours. *See also* pebble ground, below.

———, *latticino*. A ground composed of threads of white FILIGRANA (LATTICINO) glass in a double swirling pattern, one layer superimposed over the other in the opposite direction. Some such grounds are made with coloured threads. They were made by ST-LOUIS.

———, *moss*. A ground made of green canes creating a moss-like appearance, sometimes including tiny white stars. It was used only at CLICHY.

———, *mottled*. A ground made of fused chips of vari-coloured glass.

———, *muslin*. A ground composed of lengths of white FILIGRANA (LATTICINO) glass rods laid side by side, so as to appear when viewed from the top of the weight as floating gauze or muslin. Sometimes there is an upper layer jumbled together to give the appearance of twisted floating gauze; these are referred to as 'upset muslin'. Such a ground is used especially at CLICHY (*see* PAPER-WEIGHTS: *millefiori*; *flower*). Also called 'gauze', 'lace' or 'filigree'.

———, *pebble*. A ground similar to a jasper ground (see above), but with larger and coarser canes.

———, *rock*. A ground that has the appearance of quartz-like rockwork, being made of various elements, e.g. bits of glass or

SULPHIDES, sand, mica, etc. Sometimes it is the sole decoration of a paper-weight, but usually it is the ground for the 'snake' variety made by BACCARAT (*see* PAPER-WEIGHT : *snake*).

——, *sodden snow*. A ground consisting of textured OPAQUE WHITE GLASS.

——, *star-dust*. A carpet ground (see above) that is composed of a great number of tiny white canes (star-shaped or corrugated), sometimes with a coloured dot or star in the centre. Usually larger canes with FLORET or SILHOUETTE patterns are interspersed.

groviglio, vetro (Italian). A type of ornamental glassware developed by VENINI *c.* 1950, the characteristic feature of which is a mass of irregularly twisted thin copper wire embedded in the glass.

grozing. Snipping off the raw edge of a glass vessel by nipping with a tool, leaving a slightly jagged edge; sometimes termed 'knocking off'.

gugglet (or **guggler**). A term usually referring to a long-necked porous earthenware vessel for holding water, but as applied to glassware in some inventories of the 18th century it is (in the opinion of Robert J. Charleston) more likely to be a type of globular bottle with a tall thin neck and onion-shaped mouth.

guild glassware. Glassware made for the many German and Bohemian guilds, with enamelled decoration depicting members, emblems, and tools. *See* HALLORENGLAS.

guilloche. A decorative pattern, often used as a border, formed by two or more continuous closely interlaced bands twisted over each other in a manner usually resulting in a small circle-like space between the intersections (sometimes filled in with a small ornament). It is found on mosaics from the 2nd century AD, and on porcelain from the second half of the 18th century, especially from Sèvres. In glassware, the term is usually applied loosely to a pattern of CHAIN TRAILING, with a large space within the loops between the intersections of the pincered trailing.

Gumley, John (fl. 1694–1714). The founder, in 1705, of a glasshouse in Lambeth, London, that made good-quality MIRRORS, which he had been making since the late 17th century, and CHANDELIERS.

Gundelach. *See* GONDELACH, FRANZ; GONDELACH, GEORG.

guttular jar. A type of JAR whose decoration is a series, usually in two rows, of small decorative S-shaped hooks, attached to the body by a drop of glass, and looping downward and then outward. Such jars, usually of clear glass, may have the appendages of clear or coloured glass. They are said to come from the neighbourhood of Olbia, in Russia, from the 3rd/4th centuries. The French term is '*bol orné de guttules appendiculées*'.

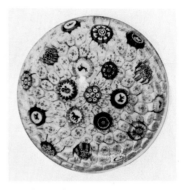

ground, star-dust carpet. Paper-weight with white corrugated canes enclosing stars and silhouette canes, Baccarat, dated 'B 1848'. W. 7·6 cm. Courtesy, Sotheby's, London.

guttular jar. Colourless glass body with emerald-green hooks, attributed to Olbia, Russia, 3rd/4th century. Ht 11·9 cm. British Museum, London.

H

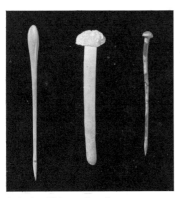

hair pins. Chinese glass. L.
12·5–15·8 cm. British Museum,
London.

Hallstatt Cup. Green transparent glass
with ribbing, *c.* 500–400 BC. Ht 4·9 cm.
Natural History Museum, Vienna.

Hadelands Glasverk. A Norwegian glasshouse founded in 1762, first to make bottles and after 1850 to make tableware and later crystal glass. After 1845 the factory extended its production to include coloured glass, and engaged designers who created new forms and styles of decoration. It has been operated since the 1850s by six generations of the Berg family, the latest director being Jens W. Berg. It employed the designer SVERRE PETTERSEN from 1928. Among its recent designers have been Benny Motzfeldt (b. 1909), and Gro Sommerfelt (b. 1940).

Haedy, Christopher (fl. 1769–85). A glass-maker in London who advertised CUT and ENGRAVED glass between 1769 and 1785. He is thought to be the son of a Bohemian glass-cutter named Haedy who was employed by JOHN AKERMAN from 1719 and who was referred to in advertisements in newspapers in Birmingham and Bath. Other members of the Haedy family advertised as glass-cutters in London until 1810.

haematinum (or **haematinon**). An opaque blood-red glass in which, under a magnifying glass, the cuprous oxide crystals become visible against a colourless background. It was made in the early Egyptian period and the Roman period, and was used in enamels and mosaics.

hair pin. Thin pins, varying in length, from *c.* 12 cm. to 16 cm., and in various decorative styles. They were made of glass in China during the T'ang Dynasty (618–906). Some were hollow, with heads in the form of a lotus bud.

Hald, Edward (1883–   ). A glass-designer who in 1917 joined the ORREFORS GLASBRUK. He was a painter from Stockholm, and had designed pottery. He collaborated with SIMON GATE in the development of the CASED GLASS for which Orrefors became renowned. He gave an exhibition of his latest creations in 1973.

half-decker glassware. Glass receptacles in the form of an oval sauce-boat but with the front half of the opening covered, leaving only a small orifice for pouring. It is uncertain if their purpose was to serve as a sauce-boat, a cream-boat, or a baby's feeding vessel.

half-post method. A process of moulding an object by which a GATHER ('post') is dipped again into the POT so as to be covered half-way ('half-post'), and thus is thicker on the lower half. It is so-called in the United States, although the process was introduced by Continental glass-blowers. It was used in making some PITKIN FLASKS.

Hall-in-Tyrol. The site, in Austria, east of Innsbruck, of an important glassworks founded in 1534 by Wolfgang Vitl, of Augsburg, Germany, and which prospered *c.* 1540–69 under the direction of Sebastian Hochstetter, also from Augsburg, and under the patronage of Archduke Ferdinand I of Tyrol. It continued until 1602 under others from Augsburg, always employing German craftsmen but influenced by Venetian forms and styles of decoration, adapted to local ideas. When Archduke Ferdinand founded a rival glassworks at Innsbruck in 1570 with Venetian workers, Hall, under Chrysostomos Hochstetter, adapted its style to that of Innsbruck. In 1635, after 10 years of disuse, Hall was given to the Franciscan Order and in 1640 it closed. Hall produced mainly goblets and ewers, in colourless glass and also blue and green glass. The decoration was in DIAMOND-POINT ENGRAVING, unfired GILDING, and ENAMEL PAINTING in COLD COLOURS (red, green),

V *Harvard glass flowers.* Tea-plant (*Camellia sinensis*), from the botanical series made by Leopold and Rudolf Blaschka, Dresden. Ware Collection, Botanical Museum, Harvard University, Cambridge, Mass.

VI *Lycurgus Cup*. Two views
showing dichroic effect: opaque
pea-green when viewed in
reflected light, and wine-red
when viewed in transmitted light;
Roman glass, 4th century AD, ht
16·5 cm. (6½ in.). British
Museum, London.

featuring heraldic devices. The engraved designs covered almost the entire surface of the piece. *See* INNSBRUCK GLASSWARE.

Hallorenglas (German). A HUMPEN or STANGENGLAS (sometimes having a cover) decorated in coloured enamelling depicting the biennial Pentecost ceremony of the guild of salt-workers of the town of Halle-an-der-Saale, in Saxony. They were decorated in four styles, but all showing basically a panorama of the town, the guild-house, the procession of the marchers, and the arms of the guild. It was customary to carry such a beaker in the procession. They were made in Saxony from 1679 to *c.*1732.

Hallstatt cups. Small CUPS of CELTIC GLASSWARE (found at Hallstatt, Austria), from the 6th to the 5th century BC, and said to have been among the earliest glass vessels not made over a CORE. The cups taper inward to a small base and are usually decorated with vertical RIBBING.

Halter, Heinrich Friedrich (fl. 1700–20). A glass-engraver from Magdeburg, Germany, who decorated GOBLETS (probably of Potsdam glass) with landscapes and portraits. Some were signed with his initials.

Hammonton Glassworks. A glasshouse at Hammonton, New Jersey, founded by William Coffin, Sr, with Jonathan Haines, in 1817 and operated until 1836 as Coffin & Haines. Thereafter it was operated as Coffin & Hay by Coffin's son Bodine and a son-in-law Andrew K. Hay. It made SOUTH JERSEY TYPE GLASSWARE.

hanap. A large 17th-century drinking cup of goblet shape. Some were made of glass.

hand-bell. A type of bell with a vertical handle, usually from 18 to 20 cm. high. They were made of glass in many varieties and styles, and at many factories. Such bells were made in Germany and in England at Bristol, in the 18th and 19th centuries. Some handles are in the form of a leg or an arm. Some Victorian hand-bells were made in sets of three, each varying in size.

hand-blown. The same as FREE-BLOWN GLASS, made without a MOULD.

hand cooler. A solid ovoid object, said to have been used (1) by ladies in ancient Rome to cool the hands, (2) in later times by young ladies when being wooed, or (3) to serve as a darning egg. Some examples in glass have been made by ST-LOUIS with MILLEFIORI decoration, including RETORTI canes of alternating colours.

Handel Glass Co. A glasshouse founded in 1885 at Meriden, Connecticut, by Philip J. Handel (d. 1914), and continued by his widow and then from 1919 by his cousin, William Handel, until it closed in 1936. It made mainly decorative lamps with glass shades, of many types and sizes, generally in the style of TIFFANY LAMPS. The shades were decorated by 'chipping', a patented process by which the design was SAND-BLASTED and then filled in with glue which, upon reheating, contracted and fell away with some glass particles, leaving a frosted surface. The ware was usually marked 'Handel'.

handkerchief bowl. A modern decorative circular BOWL without a foot; its side has the appearance of the loose folds of a cupped handkerchief. They were made of VETRO A RETORTI (ZANFIRICO) or of OPALINE by VENINI at Murano, *c.*1952–59. The Italian term is *vaso fazzoletto* (literally, handkerchief vase).

handle. That part of a CUP, MUG, JUG, VASE or other vessel by which it is held by the hand when used or carried, usually with a space for the fingers or hand to be passed through, as distinguished from a LUG.

Hall-in-Tyrol. Pokal with diamond-point engraved decoration, *c.*1572–80. Ht 21·5 cm. Kunstsammlungen der Veste Coburg.

handkerchief bowl (fazzoletto). Vase in opaline, Venini, Murano.

Although usually single, double handles are not uncommon, while pieces with three and more handles have been recorded. They can be horizontal or vertical, simple or interlaced, at the side of the vessel or overhead (i.e. bail) HANDLES, and of many shapes and styles, e.g. INTERTWINED HANDLE; CELERY HANDLE. The French term is *anse*, the German *Henkel*.

hanging lamp. A LAMP that is suspended by chains from the ceiling, usually in a church. There are several types made of glass. The simplest form is a shallow bowl, usually with a ball finial at the bottom, that is suspended by chains attached to the rim; these were made of PERSIAN GLASS. *See* MOSQUE LAMP; LAMPADA PENSILE; CESENDELLO; SANCTUARY LAMP.

hanging unguentarium. A type of UNGUENTARIUM with loops for a suspensory cord. Some have a high overhead loop, and some DOUBLE UNGUENTARIA have two such loops, above which is a third loop. Others have a series of small loops along the sides.

Harcourt. A style of crystal glass tableware designed and made by the BACCARAT FACTORY since 1842. It has six ovoid facets on the bowl of goblets, and a hexagonal foot supporting a hexagonal stem with a faceted knop. It sometimes has gilding, and then is termed 'Empire'. Services in this style have been acquired by many heads of state.

harmonica, glass. *See* MUSICAL GLASSES.

Harrachov glassworks. A large glasshouse, founded in 1712 at Novy Svet (Neuwelt) in northern Bohemia, which greatly expanded until 1808 when it was taken over by Johan Pohl (1796–1850) who further enlarged its facilities and made high-quality glassware. In 1887 it was acquired by Josef Riedel, who had previously operated another glasshouse, and he combined these and other glassworks into a group that produced a large quantity of luxury glass. *See* HARRTIL GLASSWARE.

Harrtil glassware. A type of glassware with intricate patterns of inlaid coloured threads. Made since 1955 by the former (now nationalized) HARRACHOV GLASSWORKS in Czechoslovakia. It was developed by the company's manager, Miloš Pŭlpitel.

Harvard Glass Flowers. A collection of naturalistic life-size models and anatomical sections made in glass, all with botanical accuracy. The collection consists of 784 American and other flowers and 3,218 enlarged flower and anatomical sections, representing more than 780 species and varieties of 164 families of flowering plants. They were made AT-THE-LAMP by LEOPOLD BLASCHKA and his son RUDOLF BLASCHKA, near Dresden, Germany, from 1887 to 1939, and were donated, in memory of Charles Eliot Ware, to the Botanical Museum of Harvard University, Cambridge, Mass., by Elizabeth C. Ware and Mary Ware, his widow and daughter, who together financed the production by the Blaschkas, and for whom a unique BOUQUET was made. The collection is viewed by over 200,000 visitors annually but serves mainly for study purposes, due to the scientific accuracy with which the actual flowers were copied by skilled naturalists who were also artistic glass-makers. [*Plate V*]

hat. (1) A fanciful form of bowl made in the form of a hat, e.g., one in the shape of a tricorne made in Catalonia, Spain, in the 18th/19th century. (2) A type of FRIGGER, made in various hat styles, mainly of so-called NAILSEA GLASS.

hatched. Decorated with fine lines to give the effect of shading; sometimes two sets of equidistant parallel lines crossing each other at an oblique or right angle ('cross-hatching'). During the 16th century, this style was used at Hall-in-Tyrol as a decoration on glass by the process of DIAMOND-POINT ENGRAVING, applied to thin Venetian METALS and

Harvard Glass Flowers. Mountain laurel, from the botanical replicas by Leopold and Rudolph Blaschka, Dresden, 1887–1939. Ware Collection, Botanical Museum, Harvard University.

hat. Bowl in form of a tricorne hat, Catalonia, Spain(?), 18th/19th century. W. 16 cm. Kunstgewerbemuseum, Cologne.

producing a smooth graduated tone. It was also used in the Netherlands and possibly in England, the work being that of Dutch engravers, although some has been questionably attributed to England. It is particularly associated with WILLEM MOOLEYSER. It was also used by an unidentified engraver using the initials 'C F M' and whose earliest dated work is of 1644; he used close hatching and cross-hatching to produce a chiaroscuro effect.

hatpin. A long pin for securing a lady's hat, used especially from Victorian days until *c.*1940. The head has been decorated with ornaments of various materials and in many styles, but some have a small glass ornament, such as a tiny bird or a coloured BEAD.

Hausmaler (German). Literally, home painter. A decorator of glass (as well as of porcelain or faience) during the 17th and 18th centuries, who bought undecorated ware from the factory and painted it with enamels either at home or in his own studio. The best of the *Hausmaler* were often superior to the factory's own painters, and some maintained studios with a number of assistants and apprentices. The *Hausmaler* operated chiefly in Germany. The French term is *chambrelan*, the Austrian is *Winckelmann*, and the English is 'independent' (or 'outside') 'decorator'. *See* JOHANN SCHAPER; DANIEL PREISSLER; JOHANN FABER.

Hawkes, Thomas G., & Co. A glassworks, established in 1880 at Corning, New York, at which decoration was done by cutting and engraving, using BLANKS supplied by CORNING GLASS WORKS. In 1903 Thomas G. Hawkes joined with FREDERICK CARDER to establish the STEUBEN GLASS WORKS.

Hay, Sir George (1572–1634). A Gentleman-of-the Bedchamber to James I, who, in 1610, procured from the King a glass-making monopoly for Scotland. He established a glassworks at Wemyss, in Fifeshire, which he leased to a Scotsman named Crawford and later to James Ord. Ord resisted the attempts of SIR ROBERT MANSELL to attack his monopoly. In 1621 Hay obtained a licence to import into England glass from Scotland. In 1627 Hay sold his rights to Thomas Robinson in ignorance of the fact that Robinson was an agent of Mansell. *See* SCOTTISH GLASSWARE.

Hazard, Charles-François (1758–1812). A glass-maker in Paris who made glass figures of important persons, e.g. Henry IV, of the 16th and 17th centuries in the style of the NEVERS FIGURES.

head. A full-relief three-dimensional representation (usually a portrait) of a person, but only from the neck up. A head of the Emperor Augustus, in ROMAN GLASS from the 1st century, exists. The German terms are *Kopf* or *Köpfchen*. *See* BUST.

head flask. A type of FLASK having a bowl in the form of a human head (e.g., Medusa, Bacchus) and a tall thin neck (sometimes a low cup-like mouth) rising above the centre of the head; sometimes there is a loop handle extending from the rear of the head to the middle of the neck. Some examples are janiform (*see* JANUS FLASK). They were MOULD-BLOWN, of ROMAN GLASS from the 1st to the 4th centuries. Rare examples have a small hole in the bottom so that the piece can serve as a sprinkler. The German term is *Kopfflasche*. *See* SCHUSTERGLAS.

Heaton, Maurice (1900–   ). A glass-maker who went from England to the United States in 1914. From 1931 he made modern glass murals and lighting fixtures. In 1947 he invented a process for firing and shaping in the furnace glassware enamelled and decorated with scratched figures, and from 1961 he covered the enamel with several layers of glass. His studio is at Nyack, New York.

head. Emperor Augustus, green glass, Roman, 1st century AD. Ht 4·7 cm. Römisch-Germanisches Museum, Cologne.

head flask. Light-green Roman glass, made in the Rhineland, 3rd century AD. Ht 18·8 cm. Römisch-Germanisches Museum, Cologne.

Hedwig Glass. Decoration in *Hochschnitt* and *Tiefschnitt*; said to have belonged to St Elizabeth (1207–31), niece of St Hedwig, and later owned by Martin Luther. Ht 10·3 cm. Kunstsammlungen der Veste Coburg.

Heckert, Fritz. A glass-enameller in Petersdorf, in Bohemia, who established a glass-decorating works in 1866 and a glass factory in 1889. They operated mainly from 1866 until *c.* 1890, making a large variety of German-type enamelled HUMPEN, etc., often copying woodcuts and engravings. He may have made the glassware wrapped in wire netting and bearing the enamelled date '1616', and certainly made glassware blown into a wire mesh. *See* DRAHTGLAS.

Hedwig Glasses. Thick-walled BEAKERS deeply cut in relief decoration, with stylized figures of a lion, griffon, and eagle, and palm-leaf motifs. They were so-called because two of the surviving examples supposedly belonged to St Hedwig (1174–1243), patron saint of Silesia (and wife of the Duke of Silesia) who is said to have witnessed the miracle of water turning into wine in one of the glasses. They were formerly thought to have been made in Egypt or in a Western country by an Egyptian craftsman, or in Persia, in the 11th to 12th century, but B. A. Shelkovnikov, of the Hermitage Museum, Leningrad, proposed in 1966 that they are the work of a 12th-century, White Russian workshop in NOVOGRUDOK, which once belonged to Poles and was under Byzantine influence and where similar fragments have been found. The origin is still regarded as unknown. Some have been used as reliquaries and have 13th-century metal mounts.

Heel, Johann(es) (1637–1709). A glass-worker in Nuremberg who was a follower of GEORG SCHWANHARDT; he cast glass figures and decorated by cutting. He was also a painter of faience, to whom some pieces decorated with birds and fruit have been attributed; based on stylistic similarity, some vessels in GOLD RUBY GLASS have been ascribed to him.

Heemskerk, Willem Jacobsz van (1613–92). An amateur decorator of glass, of Leiden, Holland, who did DIAMOND-POINT ENGRAVING, often on green bottles, but who is known mainly for his work in CALLIGRAPHY on a variety of objects, e.g. BOTTLES, DISHES and RÖMERS. Much of his work bears his signature and the date. *See* CRAMA, ELIZABETH.

Heesen, Willem (1925–). A glass-designer at the Leerdam Glassworks (*see* LEERDAM GLASSWARE). He has revived the decoration of glass by DIAMOND-POINT ENGRAVING as well as developing decoration of ornamental ware with rough surfaces.

Heisey, A. H., Glass Co. A glasshouse at Newark, Ohio, making table glassware and ornamental glassware in modern styles. It was founded in the mid-1860s by George Duncan with Daniel C. Ripley and others, to make cut and engraved crystal glassware and pressed glassware. In 1874, after financial difficulties, it was reorganized by Duncan as George Duncan & Sons. The business was carried on by his sons and by Augustus H. Heisey, his son-in-law, and later by the latter's son, T. Clarence Heisey.

helmet flask. A type of small FLASK of blown ROMAN GLASS which, when inverted and resting on the rim of the FUNNEL-NECK, is in the form of a head wearing a crested helmet, the features of the face being of trailed or flattened threads. The funnel neck is cut in deeply internally at its base, where it joins the bowl (the head), leaving a tiny opening so that the piece functions as a perfume dropper. They were made, probably in the Rhineland, in the 3rd/4th century AD. Only three examples are recorded. The German term is *Gladiatorenhelm.*

helmet jug. A tall JUG shaped like an inverted helmet. It rests on a spreading base and has a scroll handle. Generally made of porcelain or pottery, but some examples were made of glass. Two known examples of these come from the Savoy Glasshouse in London, *c.* 1675–80, the form being derived perhaps from France where helmet jugs were a frequent form in silver or faience. They were also made

helmet flask. Blown Roman glass, 3rd century AD. Ht 10·1 cm. Römisch-Germanisches Museum, Cologne.

*c.*1800 in Ireland of clear glass, usually without much cutting or engraving.

Helmhack, Abraham (1654–1724). A HAUSMALER of Nuremberg who decorated faience and glass in SCHWARZLOT. He signed many faience jugs, but there is no recorded signed example in glassware. He was a follower of JOHANN SCHAPER, but in a broader style. His work included painting slides or transparencies for use with the newly-invented 'magic lantern'; these were in unfired colours, each one being set in sequence on the periphery of a large disk.

Herman, Samuel J. (1936–   ). An American glass-maker with a studio in England. His recent pieces show originality in form, texture, and colour. Many examples are of CASED GLASS, with embedded multi-coloured streaks and chips. He founded in 1969 the Glasshouse, a studio-workshop in Covent Garden, London. He was invited in 1974 to start a glass workshop at Adelaide, South Australia. He has designed glass in ART NOUVEAU style for the VAL-SAINT-LAMBERT FACTORY.

Herman, Sam. Vase mainly in green and purple, Sam Herman, 1971. Ht 25 cm. Courtesy, The Fine Art Society, Ltd, London.

Hess, Johannes Benedikt, The Elder (1590?–1674). A glass-engraver of Frankfurt whose family engraved thin glass resembling wares decorated at Nuremberg. He had been a follower of CASPAR LEHMAN(N) at Prague before migrating to Frankfurt. His son Johannes Benedikt Hess the Younger also did glass-engraving at Frankfurt; and his sons were lapidaries.

Hesse glassware. Glassware made in the *Land* of Hesse, Germany, an important glass centre in the Middle Ages, especially in the region of the Spessart Mountains where the workers in 1406 formed, under the protection of Landgrave Philipp, a guild which was located in Gross Almerode in 1525 and had great influence until the 17th century. The METAL was pale green or golden-yellow, and the glassware was decorated with enamelled coats-of-arms of local families. The style was like that of Bohemia, and featured gilding and borders of engraved scale patterns and white dots. The main centre was Almerode (famous for its fire-resistant pot clay), in the forest region of the Kaufungerwald. An early piece bears the initials 'AG' for Adam Götze (fl. 1658); the added initials 'WL' and 'C' under a crown are those of the Landgraves Wilhelm and Carl. In 1680 Landgrave Carl brought to his Court at Cassel (Kassel) CHRISTOPH LABHARDT who excelled in cutting rock-crystal and engraving glassware. He was succeeded by FRANZ GONDELACH who worked at Cassel *c.*1689–1717, engraving glassware in HOCHSCHNITT in Baroque style, perhaps using glass from Potsdam until *c.*1690 when fine crystal was produced at the Hesse factory at Altmünden, founded in 1594. His work was carried on by Johann Heinrich Gondelach, who became Court Engraver, and later by JOHANN FRANZ TRÜMPER and by CARL LUDWIG TRÜMPER at Cassel. *See* SPECHTER.

Hewes, Robert (1751–1830). A tanner of Boston, Mass., who became interested in making glassware before 1780, and tried unsuccessfully to found a glass factory at Temple, New Hampshire, *c.*1780–81. He had financial difficulties and early attempts failed. A factory was completed *c.*1781 but survived only about two months.

hexagon cutting. A decorative pattern on CUT GLASS, occurring on English glassware; it is similar to a HOLLOW DIAMOND pattern, but with six-sided motifs, to create a honeycomb effect. It is found on the STEMS of some drinking glasses, mainly from the third quarter of the 18th century.

hexagonal bottle. A six-sided MOULD-BLOWN bottle of coloured glass, the body of which is formed by six vertical, or slightly sloping, panels, with a cylindrical neck that extends upward from the shoulder and has a disk, funnel-shaped or cup-shaped mouth. The panels of the Syrian examples usually have relief decoration, sometimes of religious symbols; *see* MALTESE CROSS BOTTLE. Similar smaller hexagonal flasks are presumably perfume flasks, having moulded relief decoration depicting toilet articles. Most of them are of ALEPPO GLASSWARE, 1st to 5th centuries, and of ROMAN GLASS, 1st to 2nd centuries; later examples are of German glass, 15th to 16th centuries. It has been suggested that the hexagonal shape was originally derived from architectural forms, a few from Roman times but many from the Early Christian era, and has religious significance. The German term is *Sechskantflasche*. See SQUARE BOTTLE.

hexagonal faceting. A style of FACETING forming a DIAPER pattern of vertically placed elongated hexagons, used on the STEMS of some English drinking glasses.

hieroglyph. A tiny hieroglyphic character (about 2 cm. high or wide) made in Egypt in the Ptolemaic period, *c*.100 BC, and used for decorating some object by being inlaid in it. *See* INLAY.

high relief. RELIEF decoration where the projected portion is half (or more than half) of the natural circumference of the object of the design. The French term is *haut relief*, the Italian is *alto rilievo*. The German term *Hochschnitt* embraces all degrees of engraved or cut relief work, above the surface of the object. *See* MEDIUM RELIEF; LOW RELIEF.

Highdown Hill Goblet. A GOBLET with a globular body, a tall neck, and TRUMPET MOUTH, resting on a knopped FOOT. It is WHEEL-ENGRAVED with an encircling scene of running hares and a hound, and with encircling bands of pothooks below the frieze and vertical strokes around the neck. The Greek inscription around the mouth is 'Use me and may you be happy'. It is said to be of Eastern origin, probably Egyptian, late 4th or early 5th century. So-called because found in Highdown Hill Cemetery near Worthing in Sussex, England.

Hill, John. A glass-maker from Stourbridge who was brought to Waterford, Ireland, in 1783 by George and William Penrose to establish there a glassworks financed by them; he was accompanied by workmen from Worcestershire. He left the factory after three years and was succeeded by JONATHAN GATCHELL. *See* WATERFORD GLASSWARE.

Hinterglasmalerei (German). The German term for painting on the back of glass. *See* BACK PAINTING; MIRROR PAINTING.

historical flask. A type of FLASK for whiskey, made in great numbers in the United States in the early 19th century. They are of MOULD-BLOWN coloured glass, the colours ranging from tones of BOTTLE GLASS to brilliant hues of blue, green, and purple. They depict in relief a vast variety of subjects (over 400), including portraits of all the early Presidents and of other famous persons, e.g., Lafayette, JENNY LIND, and also national emblems (e.g. flags and eagles) and contemporary slogans (AMERICAN SYSTEM), as well as transportation subjects (horse and cart scenes, locomotives), Masonic emblems, and a few decorative subjects (sunburst, cornucopia). After *c*.1850 the style changed, with fewer portraits and more scenic motifs (Pike's Peak, Baltimore Fire) and emblematic designs (clasped hands). They were made by many glassworks in various regions; some bear the factory name. They are sometimes termed a 'pictorial flask'. *See* Helen McKearin, *American Historical Flasks* (1953).

historical subjects. Decoration on glassware, ENGRAVED or EN-AMELLED, depicting an historical event (*see* NAVAL SUBJECTS) or commemorating such an event (e.g., WESTPHALIA TREATY HUMPEN).

Hlava, Pavel (1924–   ). A Czechoslovakian glass-maker and designer who decorated ornamental glassware in a strikingly original manner,

Highdown Hill Goblet. Probably made at Alexandria, *c*.400 AD. Ht 20·3 cm. Worthing Museum & Art Gallery, Worthing.

mainly with cut and engraved decoration. He has developed a style of hollow glassware with metals, especially silver, fused to the glass at a second heating. Some ornamental vases designed by him and made by Borské sklo glasshouse, at Nový Bor, *c.* 1959, are of thick glass with simple lines and cut surfaces.

Hobbs, Brockunier & Co. A glassworks at Wheeling, West Virginia, established by John L. Hobbs and James B. Barnes, when they left the NEW ENGLAND GLASS CO., and bought the Plunket & Miller Glasshouse at Wheeling in 1945, naming it Hobbs, Barnes & Co. They were later joined by their sons, John H. Hobbs and James F. Barnes. In 1863 Hobbs, Brockunier & Co. was founded by the two Hobbs sons and Charles Brockunier, and became the only important producer of fine glass in the region. It employed WILLIAM LEIGHTON, and made PEACH BLOW GLASS, GOLD RUBY GLASS, DEW-DROP GLASS, SPANGLED GLASS, and, under licence from NEW ENGLAND GLASS CO., a type of AMBERINA called 'Pressed Amberina'.

hobnail. A term applied to several styles of decoration on glassware: (1) cut decoration on some ANGLO-IRISH GLASSWARE; *see* HOBNAIL CUTTING; (2) small circular PRUNTS arranged in an overall pattern on some glassware; (3) a decoration made of entrapped air bubbles in a regular pattern made by the object having been shaped in a mould with projecting circular points that left indentations in the sides of the object which were covered with an overlay. *See* AIR-LOCKS.

hobnail cutting. The same basic pattern as a STRAWBERRY DIAMOND CUTTING, but with each flat area cut with a four-pointed (or eight-pointed) star instead of small diamonds. This pattern is found on ANGLO-IRISH GLASSWARE from *c.* 1800. The term is sometimes erroneously applied to a RAISED DIAMOND CUTTING, a STRAWBERRY DIAMOND CUTTING or an ALTERNATE PANELS CUTTING.

hobnail glass. The more recent name for DEW-DROP GLASS.

Hochschnitt (German). Literally, high engraving. The German term for all degrees of RELIEF decoration produced on glass objects by WHEEL-ENGRAVING (carving). It is the opposite of TIEFSCHNITT or INTAGLIO. The process involved the making of the design in relief by cutting away the ground, in the manner of CAMEO relief. It became a dominant style of glass decoration in Bohemia and Germany in the late 17th century when POTASH glass with LIME was developed and its thickness and hardness afforded a suitable medium for deep cutting. The technique had been used by the Romans (*see* PORTLAND VASE) and others (*see* HEDWIG GLASSES), and was later used by the Chinese on SCENT BOTTLES, SNUFF BOTTLES, etc., and in the 19th century by ÉMILE GALLÉ and others in making glassware in ART NOUVEAU style.

Hochstetter, Sebastian. An Augsburg merchant who directed the glasshouse at HALL-IN-TYROL in the third quarter of the 16th century.

hock glass. A DRINKING GLASS intended primarily for German white wine. (German white Rhine wines are usually called 'hock' in England, from 'Hochheimer'.) The glasses are tall with a stemmed FOOT and a medium-size bowl of various shapes (usually globular) and often of various colours, frequently light-green, occasionally with gilt decoration.

Hodgetts, Richardson & Co. A glasshouse at Wordsley, near Stourbridge, England, during the 1870s. In 1878 William J. Hodgetts was granted a patent for a mechanical apparatus for applying glass threads to decorate glass vessels; it kept a bubble or blown cylinder of glass near a GATHER of glass from which threads were drawn and which could be moved so that the threads could be applied to make a pattern. In 1870 Henry Gething Richardson patented a method for ornamenting

historical flask. Pike's Peak (Colorado) flask, probably Pittsburgh, 19th century. American Museum in Britain, Bath.

Hochschnitt. Pokal with acanthus design, Friedrich Winter, Hirschberger Tal, Silesia, *c.* 1687–91. Ht 16·7 cm. Kunstmuseum, Düsseldorf.

Hofkellereiglas. Humpen with coat-of-arms of Friedrich I of Saxe-Gotha, and emblem of the Elephant Order, Saxony, dated 1680. Ht 17 cm. Kunstsammlungen der Veste Coburg.

Hogarth glass. English wine glass, engraved by Jacob Sang, Netherlands, signed and dated 1762. Ht 21·8 cm. Victoria and Albert Museum, London.

glassware by enamelling a design on the interior of a heated wide-mouthed vessel and then inserting a PARAISON that was blown until it became attached to the vessel and embedded the design, after which the glass was shaped to the desired form. A school of CAMEO GLASS existed at the factory, the most important artists being Alphonse Lechevrel, a Frenchman who carved a well-known cameo vase, *c.* 1878, in the style of JOHN NORTHWOOD, and Lechevrel's pupil, JOSEPH LOCKE, who was employed by the factory to make his copy of the PORTLAND VASE.

Hoffmann, Josef (1870–1955). A Viennese architect, member of the *Wiener Sezession* (1897) and founder of the *Wiener Werkstätte* (Vienna Crafts Centre) in 1903, who combined architecture with handicraft designing and created the style of decorating glass known as BRONCIT GLASSWARE which was produced by J. & L. LOBMEYR.

Hofglasschneider (German). A highly skilled glass-cutter attached to the Court of one of the German states in the 18th century. Many of them signed their work, but such persons (with some exceptions, e.g. GOTTFRIED SPILLER) did not create artistic styles and new techniques like the earlier glass-engravers, and were mainly responsible for skilfully decorated goblets and other objects for the Court. *See* KIESSLING, JOHANN CHRISTOPH; SANG, ANDREAS FRIEDRICH; SANG, JOHANN; NEUMANN, JOHANN; TRÜMPER, CARL LUDWIG; KUNCKEL, GEORG ERNST; KÖHLER, HEINRICH GOTTLIEB.

Hofkellereiglas (German). Literally, Court Cellar Glass. A type of HUMPEN with enamelled armorial decoration showing the arms, and usually the initials and titles, of the owner of the castle. They were made in Saxony and Thuringia from *c.* 1610 for many of the German local ducal wine cellars. Some rare examples were double and triple, with one or two *Humpen* stacked on each other.

Hogarth bottle. A type of WINE BOTTLE with a globular body and a tapering cylindrical neck that has been so termed because it has been said to resemble a bottle shown in an engraving from the series of *The Rake's Progress* by William Hogarth (1697–1764). Other similar wine bottles, of different size and varying slightly in shape, are sometimes called 'Hogarth-type'. They were made in the early 18th century by English and United States factories.

Hogarth glass. A type of GOBLET with an inverted bell-shaped bowl, and having a short stem formed mainly by a large KNOP resting on a dome FOOT. They have a domed cover with a pointed FINIAL. Examples are known of Netherlands glass (and some probably of English glass), engraved by JACOB SANG, *c.* 1762. The name is derived from such glasses depicted in engravings by William Hogarth (1697–1764).

Hohlbalusterpokal (German). A type of POKAL, the STEM of which has one or more hollow KNOPS. Examples used at Nuremberg in the late 17th century were decorated by a number of notable engravers, including HEINRICH SCHWANHARDT, HERMANN SCHWINGER, and HANS WOLFGANG SCHMIDT.

Holden, Henry (fl. 1683). A glass-maker who had a glasshouse in the Savoy, London, and who was reported in 1683 to have been appointed a glass-maker to Charles II and to have been given permission to use as his mark the Royal Arms. No specimen so marked has been recorded.

hollow baluster stem. A type of STEM composed of one or more hollow KNOPS, of BALUSTER shape, placed vertically and separated by small cylindrical sections and MERESES.

hollow (shallow) diamond cutting. A decorative pattern on CUT GLASS, made by four-sided diamond-shaped depressions sunk into the glass with a convex cutting wheel so that the edges touch and intersect diagonally, giving the effect of a slightly raised design. It is found on the bowl or the stem of some English drinking glasses. *See* RAISED DIAMOND CUTTING; HEXAGON CUTTING.

holly-amber glassware. A type of ART GLASS moulded with sprigs of holly and with amber colour shading from dark amber to white. The same moulds were also used for pieces in clear glass. It was made by the Indiana Tumbler & Goblet Co., Greentown, Indiana.

Holmegaards Glassworks. *See* KASTRUP & HOLMEGAARDS GLASS-WORKS.

holy-water flask. A type of FLASK, sometimes rectangular or hexagonal, decorated with the enamelled or cold-painted portrait of a Saint. They were made in Venice in the 17th century.

holy-water stoup. A small receptacle for holy-water. Often made of faience, some have been made of glass in France and the Netherlands in the late 17th and early 18th centuries. They are sometimes termed a 'font'. The French term is *bénitier*.

Honesdale Decorating Co. A glasshouse at Honesdale, Pennsylvania, established in 1901 by C. DORFLINGER & SONS. In 1916 it was purchased by Carl F. Prosch (d. 1937) of New York, who operated it until it closed in 1932. Prosch was a Viennese glass-maker who had been engaged by the Dorflinger company to decorate its glassware; after he acquired the factory, it bought BLANKS of LIME GLASS on which etched decoration with gold tracing was added. Some ware was decorated with pure gold, *c.*1914–18. Most of the pieces are marked 'Honesdale' in script, or with the initials 'HDC' or an initial 'H' on a shield.

honeycomb pattern. A decorative pattern MOULDED in the form of a diaper design of connected irregularly-shaped hexagons. It was used as an overall pattern on some bowls of ROMAN GLASS of the 4th century AD. They are said to have been made in Syria or Egypt and probably widely exported to many places where they have been found. The pattern is also found on 18th and 19th century glassware.

honey-jar. A small JAR for serving honey, usually globular or cylindrical and provided with a COVER and some with a fixed saucer-like stand. Some are urn-shaped with a square foot. Examples in Irish glass are often of CUT GLASS, with RAISED DIAMOND or other pattern.

hookah base. A BOTTLE, to which is attached a long flexible tube, for smoking tobacco in the Near and Middle East. The smoke is cooled by being drawn through water in the base after it has been brought into the water by a tube from a receptacle for the burning tobacco. Sometimes there are two apertures in the base, but some types have above the bowl a metal disk with two holes through which passes a tube for inhaling and another upward to the tobacco receptacle and descending down into the water. Such hookah bases are of three styles: (1) globular or nearly so (the *chambu* style), (2) bell-shaped, or (3) small and globular, with conical bottom, for holding in the hand; they all have a ring around the middle of the neck to facilitate holding. Such pieces were generally made of porcelain or pottery, but some were made of INDIAN GLASS in the 17th century or at Shiraz from the 17th century to the early 20th century, or in England or Ireland from the early 18th century for export to India and Persia. The Indian term is *huqqa*, the Persian is *kalian*, the Turkish is *narghile*, and an English term is 'hubble-bubble'.

Hoolart, G. H. (b. 1716?). A Dutch engraver who, *c.*1775–80, did STIPPLING in the style of FRANS GREENWOOD, said to have been his uncle.

holy-water flask. Hexagonal flask with enamelled figure of a Saint, Venetian, 17th century. Ht 19 cm. Kunstsammlungen der Veste Coburg.

hookah base. Blown blue and colourless glass, with gilded reticulated decoration, India, early 18th century. Ht 17·0 cm. Pilkington Museum, St Helens, Lancs.

Hope Beaker. Colourless glass with enamelled decoration, 'St Peter and St Paul', attribution uncertain. Ht 19 cm. British Museum, London.

horseshoe-shaped objects. Glass objects of unknown purpose, possibly Islamic. Collection of Alfred Wolkenberg, New York City.

hunting flask. Greenish-brown glass with lugs and prunts, Germany or South Netherlands, 17th century. Ht 11·5 cm. Kunstgewerbemuseum, Cologne.

Hope Beaker. A well-known BEAKER with decoration in enamelling depicting the Virgin and Child, St Peter and St Paul, and two attendant angels. with a Latin inscription. It was once supposed that this and a group of about twenty-five related pieces were of Venetian workmanship, but W. B. Honey, in 1946, attributed them to a Syrian decorating workshop at a Frankish Court, made in the second half of the 13th century. The piece is so-called because it is in the Hope Collection in the British Museum; since 1968, the Museum has doubted its authenticity, suspecting the age and origin of the piece in view of the form of the inscription, the quality of the glass, and the palette of the enamelling. *See* ALDREVANDINI BEAKER; SYRO-FRANKISH GLASSWARE; ENAMELLING, ISLAMIC.

Horejc, Jaroslav (1886–   ). A Hungarian glass-designer who, from *c*. 1922, worked at the Stephen Rath Glassworks of J. & L. LOBMEYR at Steinschönau (Kamenický Senov), his best known work depicting human figures cut in HOCHSCHNITT.

horn. (1) A glass replica of various types of musical wind instruments. *See* HUNTING HORN; FIREMAN'S HORN; COACH HORN; POST HORN; BUGLE; TROMBONE. (2) *See* DRINKING HORN.

horn-of-plenty pattern. A pattern used on early PRESSED GLASS made in the United States, consisting of a series of adjacent cornucopias (horns-of-plenty). Each alternate cornucopia is decorated with a circle of diagonal raised diamonds in its mouth and a 'bull's-eye' raised pattern on the horn,. while the intermediate cornucopias are conversely decorated, with the raised diamonds along the horn and a 'bull's-eye' raised ornament in the mouth. The pattern was used on various objects, e.g. plates, butter dishes, tumblers, spoon-holders, etc. *See* BULL'S-EYE AND DIAMOND-POINT PATTERN.

horseshoe-shaped glass. A small glass object shaped in the form of a horseshoe, with its two arms becoming flattened and widened toward the extremities and marked with several parallel indented ridges. The ends appear to have been broken in antiquity, being CRISSELLED like the rest of the piece. They are probably made of SYRIAN GLASS, and are of unknown purpose.

hot printing. The technique of TRANSFER PRINTING by means of tissue paper, so-called because a mixture of oil and colouring matter was added hot to a heated copper plate before the tissue was placed on the plate. The whole was covered with flannel and passed through a roller-press, after which the printed tissue was placed on the glass object. The back of the tissue was rubbed with a boss to transfer the design evenly, and the object was subjected to a LOW-TEMPERATURE FIRING in a MUFFLE-KILN. *See* COLD PRINTING.

Houghton, Arthur Amory, Jr. The president and policy director of Steuben Glass, Inc., from its incorporation in 1933 (to take over the STEUBEN GLASS WORKS as a division of the CORNING GLASS WORKS) until his retirement in 1972. He is a great-grandson of Amory Houghton, founder of Corning. During his regime the policy of producing coloured glass with special surface effects, the creations of FREDERICK CARDER, was superseded by a policy of making highest quality clear crystal glass of superior design and workmanship, regarding all three elements as being of equal importance.

hour-glass. *See* SAND-GLASS.

Houston, James (1921–   ). A Canadian glass-designer (also an artist and writer) who joined the design department of STEUBEN GLASS, INC. in 1962. Many of his pieces of ornamental glassware reflect his interest in nature and are influenced by his twelve years spent in the Arctic. Some of his work combines glass with gold ornament; *see* TROUT AND FLY.

Hu. A non-existent glass-enameller created by legend (and perpetuated by many modern writings) on the basis that the component Chinese characters that form the words for KU-YÜEH HSÜAN ('Ancient Moon Pavilion') make 'Hu', when joined into a single character. As a consequence, many pieces of Chinese porcelain and glass bearing that mark or of PEKING IMPERIAL PALACE GLASSWARE made in that style (or even later, imitative of the style) have been erroneously attributed to Hu.

Humpen (German). A large and wide, nearly cylindrical, BEAKER frequently of tinted WALDGLAS, with straight sides and a slightly projecting BASE, and sometimes a COVER. They were made of many styles and sizes (some up to 60 cm. high) and were used for drinking beer or wine. The decoration was usually elaborate ENAMELLING, often depicting armorial bearings, figures, animals or local scenes; the painting was often in COLD COLOURS (*see* ENAMELLING, *German*). Most *Humpen* were made in Germany, Bohemia or Silesia from the mid-16th century to the 18th century, succeeding the earlier taller STANGENGLAS. Some examples, difficult to distinguish, were made in Venice for decoration in Germany. *See* REICHSADLERHUMPEN; KURFÜRSTEN-HUMPEN; WESTPHALIA TREATY HUMPEN; HOFKELLEREIGLAS; APOSTELGLAS; HALLORENGLAS; JAGDHUMPEN; FAMILIENHUMPEN; STANGENGLAS; WALZENHUMPEN; WILLKOMM; PASSGLAS; BAND-WURMGLAS.

Hungarian glassware. Glassware made from the 14th to the 19th centuries in various glassworks in Hungary, the earlier in the style of Venice and later that of Bohemia and Germany. Main articles were JUGS and BOTTLES decorated in ENAMELLING in the style of peasant art, and also SPA GLASSES made from *c.* 1820 for those taking the cure at the local spas.

hunting flask. A type of FLASK in the form of a PILGRIM-FLASK but with a hole through the centre of the BODY. They were made in Germany or south Netherlands in the 17th century, and in France in the 18th century. The German term is *Jagdflasche*.

hunting horn. A glass replica of a hunting horn (a type of bugle used in the chase), of curved tapering form with a small mouthpiece. They are occasionally engraved with hunting motifs (e.g. a fleeing stag) and the name of the hunt. *See* HORN.

Hurdals Verk. *See* NORWEGIAN GLASSWARE.

hurricane shade. A glass screen, usually cylindrical with everted rim, made to be placed around a candle, standing in a candlestick or a lustre, to protect the flame from the wind. *See* CANDLE COVER.

Hutton, John (1906–). A freelance glass-engraver, born in New Zealand, who designed glass panels for the new Guildford and Coventry Cathedrals (the former executed by SAND BLASTING). He engraved the glass panels at Coventry with a special type of movable wheel invented by him. He has adapted some of his panel designs to decoration on vases, executed by WHITEFRIARS GLASS WORKS.

hyacinth-vase. A tall glass vase, sometimes on a stemmed FOOT, with a cup-shaped bowl, for growing a bulb. Also called a 'bulb glass' or a 'flower-root glass.'

Hyalith glass. A dense opaque glass, coloured sealing-wax red or jet-black, developed (following the emergence of ruby-red and other strong colours for glass) at glasshouses owned by COUNT VON BUQUOY in southern Bohemia. The red glass, from *c.* 1803, was used in forms and patterns possibly suggested by Josiah Wedgwood's *rosso antico* pottery, and the black glass, from *c.* 1817, in forms and patterns of Wedgwood black basaltes ware. An eight-year monopoly was granted to von Buquoy

Humpen. Vogelfängerhumpen with decoration depicting family of a bird-catcher using sticks smeared with lime, Bohemian, 1594. Ht 27·7 cm. Kunstmuseum, Düsseldorf.

hurricane shade. Probably Florentine in *façon de Venise*, after 1532. Corning Museum of Glass, Corning, N.Y.

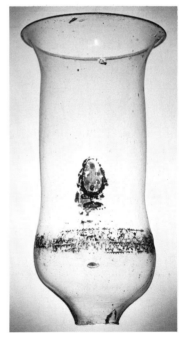

hyacinth-vase. Irish glass, probably made at Cork, *c*. 1800–10. Ht 17·5 cm. Collection of Phelps Warren, New York City.

Hyalith. Dark-red saucer with gilt *chinoiserie* decoration, glasshouse of Count von Buquoy, *c*. 1820–30. Kunstmuseum, Düsseldorf.

in 1820. The decoration was usually by fired gilding. Black glass was also made at ZECHLIN from 1804 onwards, also imitating Wedgwood's ware.

hyaloplastic (derived from Greek *hyalos*, glass). Relating to the decoration on glassware made by manipulation, e.g. by BLOWING, THREADING, TRAILING, PINCERING (rather than by MOULDING, PRESSING, CUTTING, ENGRAVING, etc.).

hydriske (Greek). A small vessel similar in shape to a Greek *hydria*, being urn-shaped and with a flat bottom, and having two horizontal loop handles and sometimes one additional vertical handle. The size is between that of a JUG and an AMPHORISKOS. They are CORE GLASS, usually decorated with a feathery or zigzag coloured pattern, made in the 4th/3rd centuries BC.

hydrofluoric acid. A colourless, volatile, highly corrosive acid. It attacks all silicates, such as glass and porcelain, and hence it is the agent used in the process of ETCHING. It was discovered by a German chemist, Karl Wilhelm Scheele (1742–86) while working in Sweden, and published in 1771. The effect of the acid varies, depending on what it may be mixed with, e.g., pure acid dissolves the glass and leaves a bright surface (*see* ACID POLISHING), but mixed with sulphuric acid, it produces on LEAD GLASS a high gloss, and mixed with ammonia, the acid is neutralized but leaves a frosted effect. *See* CORROSO, VETRO.

I

ice block. An ornamental irregularly shaped block of green-tinted glass made to simulate a block of ice with jagged exterior and with an imaginative scene or group of persons depicted by carving in the block. Such pieces were made by the KOSTA GLASSWORKS, designed by VICKE LINDSTRAND.

ice glass. A type of decorative glassware which has a rough irregular outer surface resembling cracked ice. It was produced by two processes: (1) the partially blown gathering of white-hot glass was plunged momentarily into cold water and immediately reheated lightly so as not to melt the cracks caused by the sudden cooling, and then fully blown so as to enlarge the spaces in the labyrinth of small fissures; or (2) the hot glass was rolled on an iron table (MARVER) covered with a scattering of small glass splinters (sometimes coloured) that adhered to the surface and became fused when lightly reheated, which process also removed the sharp edges. Such glass was first produced in Venice in the 16th century; and in Liège and Spain in the 17th century. It was revived, *c.* 1850, in England by APSLEY PELLATT, who called it 'Anglo-Venetian glass'. In France in the 19th century the first method produced a glass called *verre craquelé*, and the second method an effect called *broc à glaces*. A third method, developed in modern times by VENINI, involves the use of HYDROFLUORIC ACID; this type is termed *vetro corroso* (*see* CORROSO, VETRO). Ice-glass is sometimes called 'crackled glass'. It has been produced in the United States since the 19th century, called there 'overshot glass'. The general Italian term is *vetro a ghiaccio*.

ice pail. A bucket-shaped glass receptacle for holding ice, intended for chilling a wine bottle or drinking glass; examples occur with or without handles. The French terms are *seau à glace* and *jatte à glace*. Such receptacles, in varying sizes, were used to chill: a wine bottle (*seau à bouteille*); liqueur bottle (*seau à liqueur*); or wine glass (*seau à verre*). *See* WINE COOLER, WINE-GLASS COOLER.

ice plate. A circular dish, usually with a shallow well, varying in width from 15 to 17 cm.; they are usually found in pairs. The decoration is generally a cameo incrustation (*see* SULPHIDE) depicting a prominent person, but some have a classical subject, a coat-of-arms or a crest. The ground is usually frosted, and the edge and the cut star on the base are polished. Such dishes were a speciality of FALCON GLASSWORKS.

icicle glassware. A type of glassware decorated with encircling glass icicles hanging from the rim of the object. It has been said that it may have been made by the BOSTON & SANDWICH GLASS CO.

Igelglas (German). Literally, hedgehog (or urchin) glass. A type of drinking glass that is of inverted BALUSTER or somewhat CLUB-SHAPED form, with a wide bulge above the middle (or sometimes a wide cylindrical shape with such a bulge), and having a spreading KICK-BASE on a stemmed foot. Although sometimes said (e.g., by W. B. Honey) to have decoration of 'drops drawn out into a multitude of points' (perhaps suggested by the name), examples sketched by German writers show a smooth exterior. The origin of the name is not known. Some examples have been attributed to Germany in the second half of the 17th century.

Iittala Glassworks. A glassworks founded in Finland in 1881, which acquired in 1888 the art and tableware section of the Karhula Glassworks, becoming known as Karhula-Iittala Glasbruk. In 1917 it became a unit of the large industrial Ahlström Group, and is now known

ice block. Design by Vicke Lindstrom. Courtesy of Kosta Glassworks, Kosta, Sweden.

ice glass. Fruit bowl with bail handle, Venice, 16th century. Museo Vetrario, Murano.

as Iittala. It makes modern glassware of high quality, especially tableware of original designs and surface textures. It has produced modern ornamental ware designed by ALVAR AALTO. From 1947 TAPIO WIRKKALA and from 1950 TIMO SARPANEVA have been its designers. *See* FINNISH GLASSWARE.

Ikora-Kristall. A type of glassware made *c.* 1930 by the glass division of the WÜRTTEMBERGISCHE METALLWARENFABRIK (Württemberg Metal Factory), at Göppingen, Germany, under the supervision of the Director, Hugo Debach. It is of a heavy METAL, some pieces with coloured inlays and bubbles in a carefully arranged pattern and some with inlaid coloured stripes.

illuminating glassware. *See* CANDLESTICK; CANDELABRUM; SCONCE; CHANDELIER; LUSTRE; LAMP; LANTERN; LANTHORN; OIL LAMP.

imbricated. Arranged to overlap, in the manner of roof-tiles or fish-scales. The term is applied to SCALE GROUND patterns and to SCALE FACETING.

Imperial Glass Co. A glasshouse founded in 1901 at Bellaire, Ohio, by a group of local citizens; after bankruptcy in 1931, it was reorganized as the Imperial Glass Corporation. It produced mainly glassware with an IRIDESCENT surface, especially vases.

Imperial Glass Factory. *See* ST PETERSBURG IMPERIAL GLASS FACTORY.

incised. Scratched into the BODY of a vessel, either as decoration or to record a name, date or inscription. On glassware, the work was done on the molten glass usually with a blunt instrument to make a coarse indentation, in contrast to the fine lines cut by ENGRAVING on the finished hardened glass. It has been suggested that on some ROMAN GLASS a diamond rather than a flint was used.

inciso, vetro (Italian). Literally, incised glass, A type of ART GLASS developed by VENINI, the characteristic feature of which is that the entire surface is covered with closely packed thin and shallow adjacent grooves or incisings, all running in the same direction producing a slightly wavy effect.

inclusions. Tiny specks of a foreign substance within glass, e.g. the metallic particles in AVENTURINE GLASS, the mica particles in SILVERINA GLASS, and fine white flecks in SNOWFLAKE GLASS occurring in some CHINESE SNUFF BOTTLES.

indented flask. A type of FLASK (or BEAKER) of various shapes, characterized by having, as its principal decoration, pushed-in concavities, vertical or circular. They are of ROMAN GLASS, 1st–4th centuries AD, and also in the form of RING-FLASKS.

index of refraction. A measure for determining the refractive power (and therefore also the brilliance) of glass.

Indian glassware. Glassware made in India during the Mogul (Mughal) period (1526–1857), the most important being from the 17th century (apart from beads and other small objects made from as early as the 5th century BC), but not to be confused with imports from Persia in the 16th century or from Venice or Germany in the 17th century. The indigenous Indian glassware is decorated some with wheel-cutting but mainly gilded or enamelled formalized flowers (e.g., poppies), possibly the work of immigrant Persian artisans. The main objects were the HOOKAH BASE, SPRINKLER, SPITTOON, and DISH. For the history of

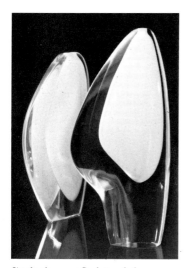

Iittala glassware. Sculptured glass ornaments designed by Timo Sarpaneva, 1953. Iittala Glassworks, Finland.

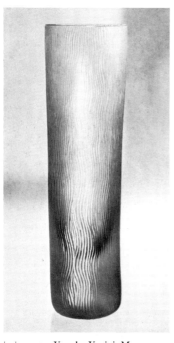

inciso, vetro. Vase by Venini, Murano.

Indian glass and the ware made, with many illustrations, *see* M. G. Dikshit, *History of Indian Glass* (1969).

Indianische Blumen (German). Literally, [East] Indies flowers. A style of enamelled floral decoration introduced by H. G. Höroldt on porcelain at Meissen soon after 1720. It was principally based on Japanese Kakiemon styles, with some inspiration from Chinese porcelain of the *famille verte* style. The name is derived from the fact that much Oriental export porcelain arrived in Europe in ships of the East India Companies. It was extensively used on porcelain until *c*. 1740, when the DEUTSCHE BLUMEN style partially replaced it. Although primarily a porcelain decoration, the style was also used on some opaque glassware in Germany in the second half of the 18th century.

Ingrand, Max (1908–). A French glass-designer who has used various techniques, including SAND-BLASTING and ETCHING, to decorate church windows, table lamps, and decorative flat glass.

ink bottle. A small receptacle for ink; some examples were made in glass with an oblate BODY, a tall NECK, and a STOPPER, but they also occur in various forms and styles. Some were decorated on the base and stopper in MILLEFIORI style; some were of OPAQUE WHITE GLASS decorated in ENAMEL. Those made at Stourbridge usually have five concentric circles of florets in the base and in the stopper. The French term is *encrier*. *See* DESK SET.

inkstand. A desk stand for writing materials, usually including an ink-well, a pounce-pot, and a pen-holder (or a drawer for pens). Often made of porcelain or pottery, some were made of enamelled OPAQUE WHITE GLASS in Bohemia in the 18th century. Also called an 'inkstandish'.

inlay. A flat piece, usually small, of decorative glass made to be inlaid as a component in a larger decorative design. They were made in the form of faces and of patterns of many shapes, in polychrome glass. Plain inlays come from ancient Egypt, mosaic-type inlays from the later Hellenistic period. *See* HIEROGLYPH.

Innsbruck glassware. Glassware made at the Hofglashütte at Innsbruck, in the Austrian Tyrol, founded by Archduke Ferdinand II (1520–95), after he moved from Prague to Innsbruck in 1563. It was operated as a hobby, employing artisans from Venice who brought their own tools and models. The ware was at first window panes but later ornate glass vessels were made and decorated with DIAMOND-POINT ENGRAVING and unfired painting, mixing German and Venetian styles. *See* AUSTRIAN GLASSWARE; HALL-IN-TYROL.

inscriptions. Words, names, and dates, and occasionally MARKS inscribed on some glassware to record the name of the maker or decorator, the date of manufacture or decoration, information about the circumstances in which it was made or the purpose for which it was intended or some sentimental expression. Some ancient GLASS-MAKER'S MARKS were accompanied by a Greek inscription, e.g. on pieces by ENNION, JASON, and NEIKAI(O)S. *See* DOCUMENTARY SPECIMENS; COMMEMORATIVE GLASSWARE; JACOBITE GLASS.

intaglio. The style of decoration created by engraving or cutting below the surface of the glass so that the apparent elevations of the design are hollowed out and an impression from the design yields an image in relief. The background is not cut away, but is left in the plane of the highest areas of the design. It is the opposite of HOCHSCHNITT. Such work was done on CRYSTAL and on glass by WHEEL ENGRAVING in medieval Rome but especially in Germany and Silesia from the 17th century. The design was left MAT or was partially polished for greater effectiveness. It is comparable to lithophane decoration on porcelain made from a wax model. It is sometimes called 'hollow relief' or 'coelanaglyphic'. The

indented beaker. Mould-blown Roman glass, 1st/4th centuries. Ht 7·9 cm. Cinzano Glass Collection, London.

intaglio. Gold ruby glass, tumbler, Potsdam, *c*. 1680–88. Ht 9·8 cm. New Orleans Museum of Art (Billups Collection).

German term is TIEFSCHNITT, the Italian is *cavo relievo*. *See* WHEEL ENGRAVING: *intaglio*.

Intarsia glass. A type of ART GLASS, developed *c*. 1920 by FREDERICK CARDER at STEUBEN GLASS WORKS, having a layer of coloured glass (usually black, blue or amethyst) etched with a design and CASED between two layers of clear glass (the total thickness usually being less than 3 mm. or $\frac{1}{8}$ in.). Experiments in the very difficult process were carried on *c*. 1916–17, involving laying on a MARVER fragments of coloured glass to form a design, then picking them up on a PARAISON of clear molten glass and covering them with another layer of clear glass. Later, *c*. 1920, an acceptable technique was developed, using three layers of glass, the two exterior layers of clear glass enclosing a layer of coloured glass etched through with a floral or arabesque design; the piece was reheated and blown, reducing the thickness and lightening the colour. Most of the examples (less than 100) are bowls or vases made in the 1930s. In Carder's early experiments, the term 'Intarsia' was applied to pieces, made *c*. 1905–06, by two wholly different processes and with different appearance, involving rubbing decorative glass threads into the matrix or into pieces of Gold Aurene (*see* AURENE GLASS) with marvered finishing, termed 'New Intarsia Ware'. The name 'Intarsia' is derived from the Italian word *intarsiatura*, meaning a 15th-century type of inlaid woodwork. All examples made for sale bear Carder's facsimile signature engraved on the lower side of the bowl of the piece, rather than on the bottom.

Intarsia glass. Vase, designed by Frederick Carder for Steuben Glass Works, *c*. 1930, with blue pattern encased in two layers of clear glass. Ht 16 cm. Private collection. Courtesy, Corning Museum of Glass, Corning, N.Y.

interior painted. *See* CHINESE SNUFF BOTTLES, *interior painted*.

intertwined handle. A HANDLE made of two loops, in plain or strap form, which overlap at least twice.

investiture.glass. A GOBLET decorated with the arms and insignia of certain Continental noble orders, and used at a banquet following the investiture of a new knight, when a toast was drunk to him from this special glass that was then presented to him. Such goblets were made in the 17th and 18th centuries in Bohemia, Silesia, Germany, and the Netherlands. During the 19th century they were made for military orders and awarded for accomplishment. Although some beakers, flutes, and goblets engraved with the crest of the British Order of the Garter are in the Royal Collection, there is no tradition for their special usage at investitures.

iridescence. The rainbow-like play of different colours, changing according to the angle of viewing or with the angle of incidence of the source of illumination. Some of the finest iridescent effects have been produced naturally on ancient Roman glass that is LAMINATED as the result of burial for a long period in damp soil. The iridescence is due to the diffraction of light reflected from the roughened surface. It differs from absorption colours by its shifting character and is less striking than the brilliant colours reflected by PRISMS. Natural iridescence must be distinguished from that of LUSTRE PAINTING and surface use of acids on glass. Iridescent is also termed 'opalescent'. Naturally produced iridescence is one form of the changes on the surface of glass that are included in the term WEATHERING. *See* IRIDESCENT GLASS.

iridescent glass (1). Square bottle, Roman glass, *c*. 100 AD. Kunstmuseum, Düsseldorf.

iridescent glass. (1) Glassware that has acquired an IRIDESCENT appearance as the result of being buried for a long period in the soil, as in the case of some ancient ROMAN GLASS, which is often paper-thin and feather-light, and some 17th and 18th-century bottles. If such glass is immersed in water, the iridescence disappears when the laminations are filled, but returns when the glass dries. (2) Glassware decorated with LUSTRE PAINTING. (3) Modern opalescent glassware that has been treated with metallic oxides and heated in a controlled atmosphere to develop an iridescent effect. It was produced commercially in 1863 by J. & L. LOBMEYR, and thereafter by many glasshouses under various

VII *Mosaic bowl*. Multi-coloured bowl, Roman glass, 1st century AD; diam. 15·5 cm. (6¼ in.).
Victoria and Albert Museum, London.

VIII *Opaline glassware.* Group of objects, all with gilt-bronze mounts:
box with lid (*gorge-de-pigeon*); basket (blue); and bell (lavender). Courtesy, Roger Imbert, Paris

patents, including one in England granted in 1877 to Thomas Wilkes Webb, of THOMAS WEBB & SONS, Stourbridge. The effect was popularized by the FAVRILE GLASSWARE made by LOUIS COMFORT TIFFANY, by some LÖTZ GLASSWARE, and by many varieties of glass developed by FREDERICK CARDER. *See* CARNIVAL GLASS.

Iris glassware. A type of ART GLASS developed in 1910 by FOSTORIA GLASS CO. and trade-marked in 1912. It is an iridescent glass which was made in many colours and decorated with trailing of coloured glass in various patterns, e.g. 'spider-web pattern', 'pulled leafy pattern', and 'heart-and-vine pattern'.

Irish glassware. Glassware (apart from some said to have been made *c.* 1585) made at several factories in Ireland that were established after *c.* 1780 when the ban on exporting glass from Ireland was removed and several Irish financiers and English glass-makers saw the opportunity to avoid the high English taxes under the EXCISE ACTS of 1745 and 1777. *See* BELFAST GLASSWARE; CORK GLASSWARE; DUBLIN GLASSWARE; WATERFORD GLASSWARE. The Irish glassware was made until soon after *c.* 1825, when the first Irish Glass Excise Duty was imposed. It initially followed English styles, but later developed an Irish character. It was made almost exclusively of heavy LEAD GLASS, and decoration was by wheel cutting, engraving, or produced by MOULD BLOWING. Most of it, except a few marked pieces, cannot be distinguished by factories, and identification as Irish glass is made chiefly by weight, form, cutting, and absence of colour except for a few bluish pieces made at Cork (the claim that Waterford glass has a blue tint having been authoritatively repudiated). For the principal glass objects made in Ireland, *see* below. For a discussion of factories and styles, and many illustrations, *see* Phelps Warren, *Irish Glass* (1970).

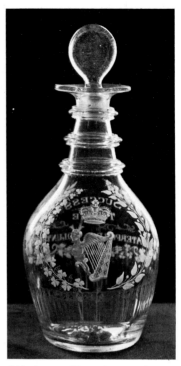

Irish glassware. Three-ring decanter with engraved decoration, 1782. Victoria and Albert Museum, London.

TYPES OF WARE

——, *bowls.* Bowls of a great variety of shapes, sizes, and styles. Some have a turned-down rim, some a scalloped rim. Some rest on a tubular stemmed base, a square solid base, a circular domed base, or 'tripod legs' (three legs). Some examples have a separate base. The type known as a 'kettle-drum bowl' is cylindrical and rests on a stemmed foot. Most types are more or less heavily cut.

——, *decanters.* DECANTERS made, *c.* 1771–1850, by several glasshouses, and sometimes identifiable by factory marks. (1) Waterford: Usually three plain rings on the neck and at the top a flat wide circular pouring lip. The lower part of the bottle has cut or moulded vertical fluting (COMBED FLUTING). The body cutting has a variety of motifs, 'pillar and arch' being most frequent. A cut or moulded mushroom STOPPER is usual. The mark is 'Penrose Waterford'. (2) Cork: Very similar in form to Waterford, but identifiable by a VESICA PATTERN, cut or engraved. The mark is 'Cork Glass Co.' or 'Waterloo Co. Cork'. (3) Belfast: Usually tall and tapering, with comb flutes around the base, a narrow flat lip, and two rings around the neck. The stoppers are circular (target stopper) or lozenge-shaped. The mark is 'B Edwards Belfast'.

——, *fruit (or salad) bowl.* A large circular or CANOE-SHAPED bowl, some circular examples having the rim curved outward and downward (called a TURN-OVER RIM), and some having a vertical rim that is scalloped, sometimes bevelled. The bowls rest on a short moulded STEM rising from a heavy FOOT which may be circular, square or oval, sometimes a lemon-squeezer foot. Some later circular examples are made in two sections (sometimes called a 'two-piece bowl'), with a solid boss or peg on the bottom of the bowl that fits into the top of the base section. All such bowls have cutting characteristic of WATERFORD GLASS CO. or CORK GLASS CO., but some may have been made at other Irish glassworks. The oval form and the turn-over rim are typically Irish.

——, *jugs and ewers.* Vessels to serve claret, milk and cream, made of luxury glass, *c.* 1780–1835, in a wide variety of forms and sizes, from low ample receptacles to tall so-called ewers with a swan-neck handle and

Irish glassware. Fruit bowl with scalloped rim and lemon-squeezer base. L. 26 cm. Courtesy, Derek C. Davis, London.

resting on a footed base. The early examples have modest cutting but later ones have rich ornamentation. The handles are of a uniform thickness and generally are undecorated, but often are curled at the lower end. The factory of origin is difficult to distinguish in individual pieces, except when they are marked.

——, *mirror*. A type of MIRROR made in Dublin, *c.*1780–1835, in oval form, to be hung vertically, and with a framing of pieces of moulded (or sometimes faceted) glass (called 'jewels') of various colours (red, blue, green, gilded, silvered) set in a single or double row within a lead channel encircling the mirror. Often the outer row is of clear glass, and the inner alternating coloured pairs. The overall heights vary upward from *c.*60 cm. Such mirrors were frequently hung on a wall to act as background reflecting a half-chandelier suspended before them; others were made with a scroll base to stand on a table.

——, *salt-cellars*. Salt-cellars made during the late-18th and early-19th centuries, usually of three different types: (1) a small version of the CANOE-SHAPED fruit or salad bowl (see above), with an oval base; (2) a circular bowl on a stemmed square base, sometimes with a TURN-OVER RIM; and (3) a ball-shaped bowl, open at the top with a flat base and having a turned in rim (making an interior flange) or a scalloped rim, and often accompanied by a saucer with a centre recess for the bowl. All types are usually heavily cut, often with STRAWBERRY DIAMONDS.

iron. A metal that occurs as a natural ingredient in minute quantities in SILICA (sand); hence it imparted a greenish or brownish colour to early glass until eliminated by the use of DECOLOURIZING AGENTS. (*See* BOTTLE GLASS.) When used deliberately to colour glass, it produces a deep green, or when mixed with MANGANESE it produces a colour varying from deep amber to brownish-black.

iron-red. An orange-red colour made from a base of ferric oxide. It was used in conjunction with decoration in SCHWARZLOT. The French term is *rouge de fer*.

Islamic glassware. Glassware made during the Islamic period, from 622 until after 1402, including that produced in Egypt, Syria, Mesopotamia, and Persia after the Islamic conquest. It embraces that made from the Umayyad dynasty (661–750), mostly glassware for common use, and the Abbasid dynasty (750–1258) at Baghdad and Samarra (836–883) and thereafter continuing into the 15th century, after the Mongol conquest under Tamerlane (Timur) in 1402. The types produced included: (1) WHEEL-ENGRAVED glass (with linear relief and CAMEO decoration and with oval and circular FACETING, from Persia, Mesopotamia, and Egypt; (2) glass with applied decoration (TRAILING and PRUNTS); (3) LUSTRE-PAINTED glass; (4) gilded glass; and (5) enamelled and gilded glass (*see* ENAMELLING, *Islamic*), as well as MOSQUE LAMPS. *See* HEDWIG GLASSES; PERSIAN GLASSWARE.

Islamic vase. *See* MOSQUE LAMP.

Islington Glass Works. A glasshouse at Birmingham, listed in 1803 as operated by Owen Johnson. It has many changes of management being known in 1849 as Rice, Harris & Son; it became in 1860 the Islington Glass Co., Ltd. It made some PAPER-WEIGHTS but only a few are recorded; some include CANES lettered 'I G W'.

Italian Comedy figures. *See* COMMEDIA DELL'ARTE FIGURES.

Ivanov, I. A. (1779–1848). A Russian architect-designer, associated with the ST PETERSBURG IMPERIAL GLASS FACTORY from 1806, who, from 1815, was entrusted with major design projects, including ware made for the Tsar, executed under his supervision. *See* RUSSIAN GLASSWARE.

Irish glassware. Salt-cellars on stemmed foot, *c.*1810–20. Courtesy, Delomosne & Son, London.

ivory-imitation glassware. A type of glassware made to imitate a piece of old carved ivory. A design was etched lightly on ivory-coloured or opaque white glass, then deepened by wheel-engraving, and finally rubbed with a coloured stain that showed in the recesses and the high points. The ware was sometimes embellished with enamelling and gilding. It was made by THOMAS WEBB & SONS, having been patented by Thomas Wilkes Webb in 1887 in England and in 1889 in the United States. GEORGE and THOMAS WOODALL made such ware, using Oriental objects as models.

Ivrene glass. A type of ART GLASS developed by FREDERICK CARDER at STEUBEN GLASS WORKS and produced in the 1920s. It is ivory-coloured and translucent with a satin surface that is slightly IRIDESCENT. It was produced by including in the BATCH feldspar and cryolite, and the iridescence resulted from spraying with stannous (tin) chloride applied AT-THE-FIRE. A variation, termed 'Ivory', is similar but without iridescence. The glass was used originally for lighting fixtures, but later ornamental objects were made of it, some engraved to show the ivory-coloured glass under the surface lustre. It has been sometimes erroneously called 'White Aurene'.

Izmailovskii glasshouse. A state-owned glasshouse at Izmailovskii, near Moscow, founded in 1660; initially it employed some European craftsmen familiar with Venetian techniques. It operated until after 1706, making colourless and green-tinted glass.

Izmailovskii glasshouse. Siphon glass, c. 1670–88. Ht 29 cm. Hermitage Museum, Leningrad.

J

Jack-in-the-pulpit. Vase in Art Nouveau style by Louis Comfort Tiffany, New York, *c.* 1900, signed 'L C Tiffany-Favrile'. Ht 52 cm. Kunstmuseum, Düsseldorf.

Jäger, Heinrich. Beaker with *Hochschnitt* engraving, monogram 'HI' probably Heinrich Jäger, Berlin, 1715–20. Ht 14 cm. Museum für Kunsthandwerk, Frankfurt.

jack-in-the-pulpit vase. A type of vase made to resemble the American woodland flower of that name, produced in coloured IRIDESCENT glass. Examples were made of Tiffany FAVRILE glass *c.* 1890–1910, and also by FREDERICK CARDER at STEUBEN GLASS WORKS.

Jacobite glassware. English glassware, usually wine glasses used for drinking toasts to loyalty and hopes for victory prior to the defeat at Culloden of Prince Charles Edward Stuart ('Bonnie Prince Charlie') in 1746. The engraved decoration includes portraits, symbols (e.g., a star rose, or thistle), and inscriptions indicating sympathies with the exiled James II (after his abdication in 1698) and his descendants (James Edward Stuart, the 'Old Pretender', and his son Charles Edward Stuart the 'Young Pretender'). The usual theme in the first period until 1745 is the English rose, perhaps representing the Crown, with one or two buds and with 6, 7 or 8 petals, and with the mottoes 'Fiat' ('Let it be done'), 'Redeat' ('May he return'), and 'Audentior Ibo' (possibly 'I will go more boldly'). In the second period, *c.* 1746–70, the decoration consisted of disguised Jacobite symbols and flowers connoting names; in this period the AMEN GLASSES were probably made. In the third period, after 1770 after the cause was lost, secrecy was gradually abandoned. Glasses with portraits (engraved or enamelled) of the Young Pretender were made after 1745 and until after 1770. The glasses have various forms of STEM (air-twist, opaque-twist, and baluster), and a variety of BOWLS. The decoration, including the portraits and symbols, is usually WHEEL ENGRAVED. The group includes some TUMBLERS and DECANTERS. There are many FORGERIES and FAKES of all types. *See* WILLIAMITE GLASSWARE; CHARLES EDWARD STUART GLASSWARE.

Jacobs, Isaac (fl. 1790–1835). A gilder of glassware at the NON-SUCH FLINT GLASS MANUFACTORY at Bristol, England. He had worked at the Temple Street Manufactory, operated by his father, Lazarus Jacob (d. 1796). Isaac decorated many blue-glass wine coolers, decanters, and finger bowls with a gilt KEY-FRET pattern and signed some pieces on the bottom 'I Jacobs – Bristol.' He was appointed 'Glass-Maker to George III'.

Jacobs, Lazarus (d. 1796). A glass-maker at Bristol, England, who founded the TEMPLE STREET GLASSWORKS. He was the father of ISAAC JACOBS. The gilt labelling on some Jacobs glasses has been attributed to MICHAEL EDKINS, who worked for Jacobs from 1785 to 1787.

Jacob's ladder. A type of FRIGGER in the form of a vertical rod resting on a flat circular base and standing within two intertwined spirals of glass threads of two colours. They were made of NAILSEA GLASS.

Jade glass. A type of ART GLASS the appearance of which is suggestive of jade stone. It is usually of pale-green colour but sometimes is of blue, rose, yellow or other colour. The body of the object is of translucent ALABASTER GLASS, coloured with metallic oxides, and sometimes has satin finish. Often applied stems or handles are of colourless glass. Sometimes Jade glass was CASED over Alabaster glass and decorated with an engraved design. It was made by STEUBEN GLASS WORKS and by STEVENS & WILLIAMS. *See* ROSALINE GLASS; PLUM JADE GLASS.

Jagdglas (German). Literally, hunting glass. Drinking glasses with enamelled or engraved decoration depicting various game being pursued by hunters and hounds, in addition to which a man and woman are sometimes shown. The ware included various types, e.g. HUMPEN, STANGENGLÄSER, GOBLETS, RÖMER, etc. It was made in Bohemia, Thuringia, and elsewhere in Germany in the 16th–19th centuries.

Jäger, Heinrich (fl. 1690–1720). An engraver of glass who was a pupil of GOTTFRIED SPILLER. He was born in Reichenberg, Bohemia, and became a citizen of Berlin in 1704 and a member of the Berlin Glass-cutters' Guild. He worked in Berlin until 1706; thereafter there is no documentary evidence of his activities, but he may have worked until 1715 somewhere in central Germany and from 1715 to 1720 at Arnstadt in Thuringia. At least eighteen glasses engraved with TIEFSCHNITT (occasionally with some HOCHSCHNITT) are known, many of them depicting human figures with the engraved monogram 'HI' or 'IH'. These are all regarded by Robert J. Charleston as done by the same hand, who states that 'HI' may be Heinrich Jäger, who is sometimes referred to as 'The Monogrammist HI'. Some of these glasses were made at Potsdam and have deteriorated due to CRISSELLING. For a detailed discussion of this topic, *see* Robert J. Charleston, *Journal of Glass Studies* (Corning), IV (1962), p. 67.

Jamburg Glasshouse. A Russian glass-factory at Jamburg (now Kingisepp), near Moscow, which operated in the second decade of the 18th century; it closed in the 1730s.

Jamestown. The site, in Virginia, of the first permanent English settlement in America, 1607, and of the first glass-making in America. The first venture, 1608–09, employed European glass-workers said to have been 'Dutch' and 'Polish' (but perhaps German and Bohemian), and a second venture, 1621–24, employed Italians. In recent excavations many bottles, seals, and fragments (over 20,000), probably English glassware dating from 1650, have been found, but no locally made glassware has been identified, except possibly some beads. *See Journal of Glass Studies* (Corning), III (1961), p. 79.

Janus flask. A type of HEAD FLASK that is janiform, i.e. with two heads addorsed. Some examples have different faces. They were MOULD-BLOWN of ROMAN GLASS from the 1st to the 4th centuries.

Japanese glassware. Some glassware dating from the 6th century but mainly objects made after AD 1750. No glassware made in Japan has been recorded or identified from the Jômon period before 200 BC, and only beads (which may have been imported from China) from the Yanoi period (200 BC–AD 300). Some MATAGAMA BEADS from the 3rd/6th centuries have been found, and also some glass vessels attributed to Japan from the Asuka/Nara periods (552–793), e.g., bottles for Buddhist relics and cinerary urns. The making of glass slackened in the 9th/12th centuries and ceased in the 13th century. It was revived *c.* 1750 after the opening of the port of Nagasaki by the Portuguese. Glass made before World War II was in traditional European engraved style, but since the war two factories in Tokyo are making glassware in contemporary style. *See* AWASHIMA GLASS CO.; KAGAMI CRYSTAL WORKS. In addition, STUDIO GLASSWARE is now being made, the most notable artist being K. Fujita.

jar. A deep wide-mouthed receptacle for holding a variety of substances, usually without HANDLES and generally cylindrical, although sometimes of baluster or other shapes. They vary greatly in size and style. Mainly utilitarian, many jars are undecorated, but small ones for such purposes as the storage of toilet preparations are often artistically decorated.

jarrita (Spanish). A type of BEAKER with a rounded body having a flat solid BASE or a pedestal FOOT, and a contracted NECK extending into a large FUNNEL-MOUTH. It is sometimes encircled by a row of vertical loop HANDLES extending from the BODY to the NECK, the handles with

Jacobite glassware. Goblet with engraved portrait of Prince Charles Edward Stuart. Ht 18 cm. Courtesy, Maureen Thompson, London.

Jacob's ladder. Two friggers in coloured glass. Collection of Mr and Mrs S. T. Painter, Somerset. Photo Keith Vincent.

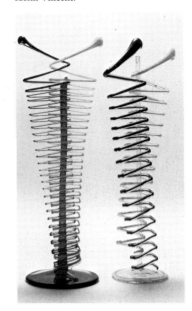

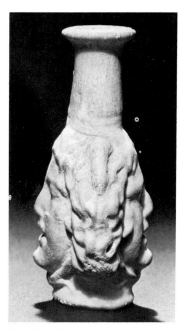

Janus flask. Opaque glass, with two Medusa heads addorsed, Roman glass, 1st/2nd centuries. Ht 7·2 cm. Römisch-Germanisches Museum, Cologne.

jarrita. Vase with eight handles and pincered edges, probably Almeria, Spain, 16th century. Ht 19·5 cm. Victoria and Albert Museum, London.

PINCERED trailing or SERRATED edge. Some examples have two rows of such handles, four large ones alternating with four smaller ones. They were made in southern Spain in the 16th and 17th centuries.

Jarves, Deming (1790–1869). A glass technologist who founded and then controlled the NEW ENGLAND GLASS CO. from 1818 until he left it in 1825 to found the BOSTON & SANDWICH GLASS CO. He withdrew from the latter in 1858 to establish the CAPE COD GLASS CO., which closed at his death. In 1837 he had founded the MOUNT WASHINGTON GLASS WORKS for his son George Jarves. He and his associate Hiram Dillaway (from 1828) perfected the process of making MOULD-PRESSED GLASS, and he obtained several patents from 1828 onwards; his companies made such ware by mass production. He also obtained in 1829 a patent for making pressed glass door knobs with a glass screw instead of a metal screw. He also developed the style of mould-pressed glass known as LACY GLASS.

Jason. A glass-maker who made MOULD-BLOWN glass in Syria in the 1st century AD, and sometimes inscribed and signed his pieces. At least six inscribed BEAKERS made by him were known in 1962; they are all alike in shape and size, apparently being from the same mould. They have straight sides sloping inward from below the inscription to the base. *See* GLASS-MAKER'S MARK.

jasper. A 17th-century English term for AGATE GLASS. The French term is *jaspe*.

Jean, A. (fl. 1878–90). A French potter who also made glassware. His early work included some iridescent glass with engraved or enamelled floral motifs and some coloured glass with gilded and enamelled decoration. Later he made some highly original forms of glass freely formed AT-THE-FIRE.

jelly glass. A small receptacle that usually has a bowl of inverted conical shape, but which may be WAISTED or of DOUBLE-OGEE or PAN-TOPPED form. Many have no HANDLE, but some have one or two single or double loop (B-shaped) handles. The glasses are about 10 to 13 cm. high. They have either a rudimentary STEM or no stem, the bowl resting on the foot (which may be plain, domed or folded), with sometimes an intervening KNOP. After *c.* 1710 the bowl was sometimes bell-shaped. They are without applied decoration, but after *c.* 1745 feature various types of cut decoration. They differ from a DWARF ALE GLASS in that the latter has usually a narrower conical bowl and was never cut. They were made in England in the mid-18th century. They were used, at the end of a meal, for jelly or other similar desserts; sometimes they were arranged on a SALVER. *See* CUSTARD GLASS; SYLLABUB GLASS.

Jersey Glass Co. A glasshouse at Jersey City, New Jersey, established by George Drummer (a glass-cutter from New York City) and associates in 1824. Various partnerships; all including members of the Drummer family, operated it until the 1860s. It made cut, engraved, pressed, and moulded glass of good quality.

Jersey rose. A glass rose found in some paper-weights made by Ralph Barber, *c.* 1905–18, and also by others (*see* PAPER-WEIGHT, *Millville*), elsewhere. The rose is never compact and the tips of its petals are opalescent. It is pink (made from ruby glass) or yellow (rare). It is sometimes accompanied by a bud and three realistic green rose-leaves. Such Millville weights were made on a footed base supporting a cup-like socket in which rested a ball enclosing the rose. A comparable glass rose is found in some rare paper-weights made *c.* 1934–35 by EMIL J. LARSEN at the VINELAND FLINT GLASS WORKS; it is made of deep-red and yellow glass, and has four leaves.

jewel. A small diamond-shaped piece of clear or coloured glass, usually faceted, used to decorate the border of some Irish mirrors (*see* IRISH GLASSWARE). Some, instead of being faceted, are cut with 2 or 3 parallel flutes, sometimes gilded. *See* JEWELLED GLASSWARE.

jewel-and-eye decoration. A type of JEWELLED GLASSWARE with 'eyes' (tiny glass drops) applied to the centre of the large drops. It was made of FRANKISH GLASS, *c*.400–800.

jewel star. A variation of the BRUNSWICK STAR, but cut with the lines that connect the points running to every 6th point rather than every 5th point. It was used *c*.1880–90.

jewelled glassware. Glass objects with decoration of drops of coloured glass applied to the surface by touching the object with a heated glass rod. The drops may be MARVERED or protrude slightly, apparently imitating CABOCHON jewels. Early examples were from Egypt, Cyprus, Gaul, and the Rhineland in the 4th/5th centuries. Some examples in FRANKISH GLASS, *c*.400–800, have small coloured glass 'eyes' (tiny drops) applied to the large drops. Such objects were the forerunners of the NUPPENBECHER of the 14th century. Jewelled glassware was also made at Antwerp in the 16th–17th centuries, and in Germany in the 18th century.

jewellery. Various ornaments, including brooches (FIBULAE), rings, ear-rings, hair-rings, ear-plugs, necklaces, and PENDANTS, worn as personal adornment. These were made with pieces of glass from earliest times in Egypt and Rome, and also in Venice probably before the 11th century. Often they were shaped AT-THE-LAMP. *See* BANGLE; BEAD.

Jewish glassware. Glass objects that are decorated with motifs of religious significance, e.g., the Menorah (7-branched candelabrum). Early examples in the form of hexagonal bottles and small jugs with moulded relief decoration were made in the Eastern Mediterranean region, especially at Sidon *c*.200 AD and in Palestine *c*.200/400 AD. So-called 'sandwich glass', with two layers of clear glass enclosing gold leaf engraved with Jewish symbols, was made in the Roman Empire, *c*.100–200 AD. Amulets with such symbols were made in Palestine, 3rd/4th centuries. Such glassware was not necessarily the work of Jewish glass-workers, although their existence is mentioned in ancient sources, and they are known to have made glassware in Palestine, Syria, and Alexandria; later Jewish glass-makers worked in Venice and in the Rhineland. See *Journal of Glass Studies* (Corning), VIII (1966), p. 48.

Jew's glass. A term formerly used in England to refer to lead glass produced at Birmingham for use in the manufacture of artificial jewellery; the term was in use as late as 1836. *See* PASTE JEWELLERY.

jigger. A small measuring glass for alcoholic drinks, used mainly at bars.

jockey. A small earthenware POT or CRUCIBLE used for small BATCHES and placed on top of a large pot in the FURNACE.

Johnson, Jerom (fl. 1739–61). A London glass-maker and seller who is said to have made many objects of glassware decorated by CUTTING and engraving and who claimed to be the 'inventor' of the styles. Although no single piece of glassware has been identified as made by him, the ware attributed to him is decorated in ROCOCO style, with engraved flowers, hops, and grapes that were symbolic of the intended use, and with rococo cut motifs. His sales announcements, in 1752, for his premises in the Strand, London, and elsewhere, listed a vast variety of glass objects, at wholesale and retail prices. It has been surmised that his successful business was taken over by William Parker, of PARKER'S CUT GLASS MANUFACTORY. He is said to have had an interest in the Cockpit Glassworks, Southwark, London.

jewel-and-eye decoration. Nuppenbecher with blue and yellow decoration, Rhineland, 4th century AD. Ht 8 cm. Römisch-Germanisches Museum, Cologne.

jewelled glassware. Cone beaker decorated with drops of transparent blue glass arranged in similar patterns on both sides, Egypt, *c*.6th century AD. Ht 16·9 cm. Cinzano Glass Collection, London.

joke glass. Flask in the form of a dog, Germany or Netherlands, early 17th century. Ht 13 cm. Kunstgewerbemuseum, Cologne.

joke glass. A type of drinking glass of an unusual form, designed to be difficult to use and to amuse the onlookers. Such glasses may be in the form of a SHOE, BOOT, PISTOL FLASK, BARREL FLASK, POST-HORN, PHALLUS, or of various animals, and include especially the SIPHON GLASS. They are sometimes referred to as TRICK GLASSES. The German term is *Scherzglas*, and the French term for those glasses from which drinking is difficult is *verre à surprise*. *See* PUZZLE JUG; MILL GLASS.

jour, à (or **ajouré**). The French term for open-work or pierced decoration, e.g. VASA DIATRETA.

jug. A wide-mouthed vessel for holding and pouring liquids, usually with a pouring LIP or a SPOUT and with a loop-handle on the opposite side. Generally made of porcelain or pottery, many are of glass. The form is generally globular, baluster-shaped, conical, ovoid, or cylindrical. The lip is found in various forms, some ornamental. Some examples have a hinged LID, but most of those with a thumb-piece are drinking vessels and are preferably designated a TANKARD or a MUG. In the United States a jug is usually called a PITCHER. The French term is *cruche*; the Italian is *brocca* or *boccale*; the German is *Krug. See* EWER; FLAGON; AIGUIÈRE; CÁNTARO; CLARET JUG; DECANTER JUG; CREAM JUG; MILK JUG.

Jugendstil (German). *See* ART NOUVEAU.

juice glass. A small drinking glass in the form of a narrow TUMBLER, the height being usually twice the diameter. They have been made in recent years for drinking fruit and vegetable juices.

K

Kagami Crystal Works. A glasshouse in Tokyo, whose founder and designer was Kozo Kagami, a pupil of WILHELM VON EIFF. It makes pressed glassware and also some cut glass. *See* JAPANESE GLASSWARE.

kalian (Persian). *See* HOOKAH BASE; FLOWER-ENCLOSING FLASK.

kalpis (Greek). A late form of GREEK VASE of the *hydria* type, used for storing or carrying oil or water. It is large and urn-shaped, with a flat bottom and a rounded shoulder with a neck forming a continuous line with the body. It is characterized by two horizontal loop handles on the upper part of the body; sometimes there is also a vertical handle for use when pouring. They were generally made of pottery but some examples exist in glass from the 1st century AD. Some such pieces without the third handle have been used as CINERARY URNS.

Kaltmalerei (German). Decoration in COLD COLOURS.

kantharos (Greek). A drinking-cup with a wide mouth, a stemmed FOOT, and two large vertical loop handles usually extending from the bottom of the bowl upward into a loop above the rim, but sometimes only to the rim. The stem may be short or high. The bowl is sometimes hemispherical but sometimes cylindrical with concave sides. The body of some is in the form of a human head, or with a head front and back. Some have applied decoration; *see* BRÖCKCHEN. They were generally made of Greek pottery but some have been made of ROMAN GLASS in the 1st to 3rd centuries. They have sometimes been erroneously called a KARCHESION which is similar but sometimes has no stem. Both types are associated with Bacchus. *See* HEAD FLASK.

karaba (Persian). A type of large Persian glass FLASK, made at Shiraz, in globular form and jacketed with rushwork.

karchesion (Greek). A drinking cup having a cylindrical bowl, sometimes with concave sides, and having a wide mouth. Originally it had two handles but later examples, 2nd/3rd century, have no handles. It was usually made of Greek pottery but examples in ROMAN GLASS are known.

Kastrup & Holmegaards Glassworks. A Danish glass-factory which since the 1920s has been making modern-style tableware and ornamental glassware in addition to bottles and other ware. It was conceived by Count Christian Danneskiold-Samsøe (d. 1823) to exploit the peat in the Holmegaard's Moor on the Island of Zealand, but his death left it to his widow Henriette to inaugurate the factory in 1825 as Holmegaards Glasvaerk. It at first made green-glass beer bottles, employing glass-blowers, named Wendt, from Norway; but later glass-blowers from Germany and Bohemia were brought in to make tableware. A second factory, Kastrup Glasvaerk, was started by it in 1847 at Kastrup, near Copenhagen, to make bottles while Holmegaards made service glass; both had the same management until 1873 when Kastrup was sold. Another glassworks was begun in 1852 near Aalborg, in Jutland, and it and some small glasshouses in Aarhus and Odense were merged in 1907 with Kastrup, which then controlled the making of all Danish industrial glass except window panes. As a result of a co-operative programme made in 1924 with the Royal Copenhagen Porcelain Co., the artist Orla Juul Mielsen was employed as designer; he was succeeded in 1925 by JACOB E. BANG, who achieved in the 1930s a reputation for designing table-glass and ornamental glass in simple but

kantharos. Bluish glass with ribbing, Roman glass, 1st/2nd centuries AD. Museo Civico, Adria.

karchesion. Greenish Roman glass with white and gold trailing, Rhineland, 2nd/3rd century. Ht 12 cm. Römisch-Germanisches Museum, Cologne.

vigorous blown-glass forms, with deep cutting and etched decoration. Its present designers are PER LÜTKEN who in 1942 succeeded Bang, and Michael Bang (b. 1944), son of Jacob Bang. Jacob Bang returned to the company from 1957 to his death in 1965. In 1954 Holmegaards and Kastrup were merged, to become the present Kastrup og Holmegaards Glasvaerker A/S.

Kaziun, Charles (*c.* 1918–). A contemporary American maker of PAPER-WEIGHTS. Born in Brockton, Mass., he has worked in glassware from an early age, and by 1942 had made his first weight of the type that became his best-known, the 'Kaziun rose'. Thereafter he has made a wide variety of weights, including those in OVERLAY with three colours and a large number with various flower patterns. The pieces are signed 'K'.

Keith, James (fl. 1750–87). An English glass-worker from Newcastle-upon-Tyne who was employed by a glassworks at Nøstetangen, in Norway, founded in the mid-18th century, to introduce there the English glass-making techniques. He remained until he retired in 1787. The ware made was of the types made at Newcastle, the WINE-GLASSES being of three-part construction like the Newcastle BALUSTROIDS.

Kelch. The German term for a CHALICE.

kelp. A type of seaweed, the calcined ashes of which were formerly used as a source of SODA in making glass, especially in Normandy and Scandinavia. It was used by SIR ROBERT MANSELL for ordinary types of glass, the kelp having been gathered, burned, and sold in England and Scotland.

Kempston Beaker. A CONE-BEAKER, so-named from the place in Bedfordshire, England, where it was found in 1863/64. It is in the form of a tall inverted cone of thin light-greenish glass with a slightly everted rim and with a flattened tip. About twenty similar beakers, found in several places in England, are considered to have been made of Teutonic glass in the 5th century AD. Beakers of this type average in height from 22·5 to 26 cm., and the diameter of the mouth is about 9·5 cm. The symmetrical decoration is an unmarvered self-coloured trailing wound horizontally about twenty-two times in the area just below the mouth and then similar trailing making about twelve continuous vertical loops on the lower part toward the rounded or flattened tip. This type differs from other cone-beakers which have thick walls and thick trailing, such as those found in Scandinavia. *See Journal of Glass Studies* (Corning), XIV (1972), p. 48.

Kerzenbecher (German). Literally, candle beaker. *See* WARZEN-BECHER.

kettle-drum bowl. A large bowl of hemispherical shape resting on a stemmed foot, suggestive of the form of a kettle-drum. They occur in ANGLO-IRISH GLASSWARE.

Keulenglas (German). Literally, club glass. A tall thin BEAKER (a type of STANGENGLAS) that is slightly club-shaped and has a pedestal FOOT or a KICK BASE. Some are simply decorated with encircling trailing (but not equidistant circles, as a PASSGLAS, nor a long spiral, as a BANDWURMGLAS), but other very tall examples are decorated on almost the entire surface with numerous small drops of glass, smaller than PRUNTS. They are of brownish or greenish WALDGLAS, up to 42 cm. tall. They were made in Germany and Bohemia in the 15th–16th centuries.

Kew Blas glassware. A type of IRIDESCENT glassware, made in the United States by UNION GLASS CO. from *c.* 1893; it features mainly tan, green, and brown shades in feathery and leafy patterns. The pieces are engraved with the mark 'Kew Blas', an anagram derived from the name

Kempston Beaker. Cone-beaker, blown green glass with trailing, Frankish glass, second half of 5th century. Ht 26·2 cm. British Museum, London.

Kew Blas glassware. Bowl of mother-of-pearl lead glass, Union Glass Works, Somerville, Mass., early 20th century. Ht 10 cm. Corning Museum of Glass, Corning, N.Y.

of W(illiam) S. Blake, the manager of the factory. The ware was imitative of much other American ART GLASS; some forgeries of it were produced, but these can be distinguished by the fact that the mark is etched rather than engraved.

key-fret. A border pattern in the form of lines meeting at right angles in a continuous repetitive design reminiscent of the wards of a key. Sometimes it is termed merely FRET and sometimes 'Greek fret' or 'meander pattern'. It was used on glass decorated by ISAAC JACOBS.

kick. An indentation in the bottom of a glass or in the BASE of any glass object. It may be shallow or deep (conical). When in the bottom of a glass or bottle, it adds stability, but reduces the capacity. *See* KICK-BASE; SPIKED KICK.

kick base. The BASE of any object which is made with a KICK.

Kiessling, Johann Christoph (d. 1744). Court engraver at Dresden, in Saxony, Germany, *c.* 1717–44, who decorated GOBLETS, mainly with hunting scenes.

Killinger, Georg Friedrich (fl. 1694–1726). A glass-engraver in Nuremberg who used WHEEL-ENGRAVING on glass from Bohemia and Nuremberg. His early work, with some DIAMOND-POINT ENGRAVING in combination with wheel-engraving on glass from Nuremberg, was superseded when he used deep cutting on Bohemian glass. Some pieces are signed.

Kimble Glass Co. A glasshouse owned by Evan F. Kimble which in 1931 took over the VINELAND FLINT GLASS WORKS after the death of Victor Durand, Jr. It discontinued the making of ART GLASS except for some pieces, of various colours, made with embedded air bubbles in the style of CLUTHRA GLASS but with fewer bubbles. It also made BUBBLE BALLS in the style of those made by TIFFANY.

King of Prussia glassware. Glassware with a decorative motif featuring Frederick the Great of Prussia (1740–86), allied to England during the Seven Years War (1756–63); he was especially popular in England during the years 1756–57. The King's portrait is engraved on glassware made in England at that time. The close connection which existed at the time between England and Prussia was due to the fact that the English kings were also Electors of Hanover, which territory adjoined Prussia.

King's Lynn. The site, in Norfolk, England, of a glasshouse owned by SIR ROBERT MANSELL in the first half of the 17th century. Another glasshouse there was owned in 1693 by Francis Jackson and John Straw (who also owned in 1693 the FALCON GLASSWORKS in London and who stated in 1695 that they had given up the London factory). An advertisement in 1747 announcing a liquidation sale presumably related to their King's Lynn factory. Another glasshouse in King's Lynn has been said to have been owned in 1753 by Jonas Philips. None of these glasshouses has been identified as the source of the so-called LYNN GLASSWARE. *See* WEDGWOOD GLASSWARE.

Kit-Cat glass. A type of WINE GLASS so-called from its resemblance to the shape of two glasses shown in a painting, *c.* 1721, by Sir Godfrey Kneller portraying two members of the Kit-Cat Club, a club of Whig politicians and men-of-letters existing *c.* 1700–20. (The club was named after Christopher Cat, keeper of the tavern where the club met, and the mutton pies served there were known as 'Kit-Cats'.) The glasses depicted have a rounded funnel BOWL on a true baluster STEM with compressed KNOPS at the top and bottom and a conical folded FOOT; but the term is sometimes erroneously applied to a glass with a TRUMPET

Keulenglas. Club-shaped beaker with milled trailing, Germany, first half of 16th century. Ht 15·2 cm. Kunstgewerbemuseum, Cologne.

key-fret. Blue glass dish with gilt border, signed on underside 'I Jacobs, Bristol', *c.* 1805. W. 19 cm. Courtesy, Sotheby's, London.

Kit-Cat glass. Detail from painting by G. Kneller (the Earl of Lincoln and Duke of Newcastle), *c.* 1721. National Portrait Gallery, London.

Köpping, Karl. Ornamental glass (*Zierglas*) in form of a tulip, made at-the-lamp, designed by Karl Köpping, Berlin, *c.* 1898. Ht 32 cm. Kunstmuseum, Düsseldorf.

BOWL, and a DRAWN SHANK of inverted BALUSTER shape. Such glasses are often incorrectly termed 'Kit-Kat glasses'.

knife. A table knife, sometimes made entirely of glass. Venetian examples, with solid twisted handles, were made in the 16th–17th centuries. *See* KNIFE AND FORK HANDLES; CUTLERY.

knife and fork handles. Glass handles into which were inserted a steel or silver knife or fork. Some were made in OPAQUE WHITE GLASS decorated with polychrome ENAMELLING, and others in blue or brown AVENTURINE GLASS in the 18th century. *See* KNIFE; CUTLERY.

knife-rest. A small, low utensil, shaped somewhat like a dumb-bell, for supporting a knife at the dinner-table. Some were made of porcelain as part of a dinner-service, but some have been made of glass, usually FACETED, or sometimes with twisted millefiori CANES laid lengthwise. They are usually about 9 cm. long. Some were made at Stourbridge.

knop. (1) A component, usually spherical or oblate, of the STEM of a drinking glass, made in many styles, hollow or solid, and used either singly or in groups of the same or different types, and placed contiguously or with intermediate spacing. Some solid knops are made with enclosed TEARS. On early Venetian TAZZE the underside of the body was joined to the STEM or FOOT by a glass ring called a *nodus* (from the Latin, meaning knot); the same word in Dutch, *knoop* or *knop*, was adopted in England. This glass ring became a globular knop or, when it was flattened, a COLLAR or MERESE. The number of knops on a stem grew from two to as many as twelve. *See* WROUGHT BUTTON. (2) A term sometimes used for a FINIAL, but preferably not in connection with glassware.

TYPES OF KNOP

——, *acorn.* A knop in the form of an acorn, with the point down, sometimes placed between two globular ball knops. Sometimes there are two acorns opposed, with the points toward each other.

——, *angular.* A knop of flattened globular shape that bulges at the sides to a rounded point at the circumference and tapers upward and downward to the stem.

——, *annular.* A knop that is in the form of a ring.

——, *annulated.* A knop that is composed of a series of adjacent equal or unequal rings or MERESES placed one above another.

——, *ball.* A knop that is globular.

——, *beaded.* A knop encrusted with tiny glass balls.

——, *bladed.* A knop in the form of a thin ring with a central ridge.

——, *bobbin.* A knop consisting of several small adjacent flattened balls placed vertically in a row (similar to bobbin-turned furniture), sometimes graduated in size.

——, *button.* A flattened version of an annular knop (see above).

——, *cone.* A knop in the form of an upright truncated cone.

——, *cushion(ed).* A knop that is of flattened globular shape, resting on a ring. Variation: with a ring above and below the knop; called a 'double cushion(ed) knop'. *See* CUSHION PAD.

——, *cylinder.* A knop that is cylindrical, drawn inward slightly at the top and bottom.

——, *drop.* A knop that is wider at the bottom than at the top, and is drawn inward at the bottom.

——, *dumb-bell.* A knop shaped like a dumb-bell, placed vertically, being drawn in at the top and bottom.

——, *egg.* A knop that is ovoid, placed vertically.

——, *flattened.* A knop of flattened globular shape.

——, *hollow.* A knop, usually globular, that is hollow (blown) rather than solid. These sometimes contain a coin; *see* COIN GLASS.

——, *melon.* A knop having several vertical raised segments, as on some melons. They are found on some English GOBLETS, and also on some Netherlands goblets of the late 17th century.

——, *mushroom.* A knop that is mushroom-shaped with a domed top.

——, *proper.* A knop shaped like an oblate sphere.

——, *shoulder.* A knop that is cylindrical but bulges at the top and is then drawn inward to meet the stem. Sometimes the knop has moulded decoration or enclosed air bubbles.

——, *swelling.* A knop somewhat cylindrical in form but bulging slightly in the centre.

——, *wide angular.* A knop similar to an angular knop (see above) but wider at the central bulge.

——, *winged.* A knop that has decoration in the form of two or more pincered vertical projections suggesting wings.

Knopfbecher (German). Literally, button beaker. A BEAKER having over all or part of its surface a number of irregularly fairly flat 'button-like' PRUNTS. They were made in Germany or the Netherlands in the 16th century.

Kny, Frederick Engelbert. A Bohemian glass-engraver who migrated to England and worked first at WHITEFRIARS GLASS WORKS and later at Stourbridge from the 1860s at the DENNIS GLASSWORKS of THOMAS WEBB & SONS, where he engraved his best-known work, the ELGIN JUG (completed in 1879). He specialized in ROCK CRYSTAL ENGRAVING and work in INTAGLIO. He signed his engraved work 'FEK'.

kohl tube. A small receptacle for a black substance (kohl) used to darken the edges of the eyelids; some include a small bronze application stick. They are square, cylindrical or flattened pear-shaped, with the interior space being cylindrical (about 7 to 9 cm. in diameter). They were made of coiled glass on a metal rod and are usually decorated with applied threads pulled in a feathery or chevron pattern. They have been attributed to Iranian origin from the 9th to 4th century BC. *See* Dan P. Baraj, 'Rod-formed Kohl Tubes of the Mid-First Millennium BC', *Journal of Glass Studies* (Corning), XVII (1975), p. 23. *See* PALM-COLUMN VASE.

Köhler, Heinrich Gottlieb. A glass-engraver, said to have come from Silesia, who went to Copenhagen in 1746, working for the Saxon Glass Company that made glassware for the Court; he was Court Engraver from 1752. In 1756–57 he went to Nöstetangen Glasverk as designer until 1770. He then established his own glassworks at Christiana, engraving sets of wine glasses until *c.*1780–81, depicting landscapes from personal observation. *See* NORWEGIAN GLASSWARE.

Köpping, Karl (1848–1914). A painter and etcher from Dresden who designed glassware at Berlin in extreme ART NOUVEAU style. His best known design is for at thinly-blown, light, and fragile drinking glass (a form of ZIERGLAS), probably intended as a LIQUEUR GLASS, that has the bowl in the form of an open flower, with a very thin tall stem (the stem of the flower) having applied foliage motifs made AT-THE-LAMP. Such glasses were made by FRIEDRICH ZITZMANN at Wiesbaden, and after 1896 at Ilmenau in Thuringia, and also later by glasshouses in the United States.

Koralli ware. A series of bowls and vases made by RIIHIMÄKI GLASSWORKS from designs of Tamara Aladin. The ware is of opaque glass with coloured abstract patterns.

Kosta Glassworks. One of the leading glassworks of Sweden, founded in 1742 in the glass-making region of Småland, and the oldest still in operation. It was founded by Anders Koskull and Georg Bogislaus Stael von Holstein (the first letters of their names gave the name Ko-Sta). It first made CROWN GLASS for windows, then CHANDELIERS and wine and beer glasses. In 1864 the BODA GLASSWORKS was founded, and it and Kosta, with Åfors and Johansfors, merged in 1970 to form the Åfors Group. It makes many varieties of glass, but is highly regarded for its

knop, acorn. Detail of wine glass, English, *c.* 1710–20. Courtesy, Derek C. Davis, London.

knop, angular. Detail of wine glass, *c.*1715–30. Courtesy, Delomosne & Son, London.

knop, drop. Detail of wine glass, showing knop on drawn stem, *c.*1715–30. Courtesy, Delomosne & Son, London.

knop, dumb-bell. Detail of wine glass, English, *c.*1740–50. Courtesy, Maureen Thompson, London.

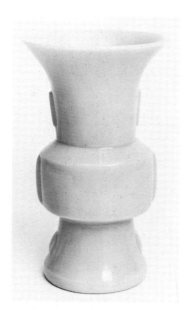

ku. Yellow opaque glass, mark and period of Ch'ien Lung, 1736–95. Ht 25 cm. Victoria and Albert Museum, London.

Ku-yüeh Hsüan decoration (1). Snuff-bottle of semi-opaque white glass with overlay and enamel decoration; mark: Ku-yüeh Hsüan; *c.* 1850. Ht 6 cm. Percival David Foundation of Chinese Art, London.

decorative CUT GLASS and tableware. The Swedish painter GUNNAR G. WENNERBERG designed CASED GLASS vases, 1898–1903. Later designers included EDVIN OLLERS, EWALD DAHLSKOG (from 1926 to 1929), and ELIS BERGH (from 1927). A leading designer in recent years has been VICKE LINDSTRAND (since 1950) who created the so-called ICE BLOCKS, as well as glass sculpture. Among the designers since 1964 are Göran (b. 1933) and Ann (b. 1937) Wärff, who together design and make glassware of much originality.

Kothgasser, Anton (1769–1851). A Viennese miniature painter of the BIEDERMEIER period, who is best known for his enamelled decoration of glass RANFTBECHER. He worked in Vienna, doing porcelain enamelling and gilding from 1784 to 1840 at the Vienna Imperial Porcelain Factory. Kothgasser painted glass from 1812 to 1830, working on STAINED-GLASS WINDOWS with GOTTLOB SAMUEL MOHN. He specialized in flower painting, but also did portraits (one in stained glass of the Emperor Franz I measuring 60 × 52 cm.), scenic subjects, and genre scenes, often adding inscriptions in French. He often used a yellow stain derived from silver, such as was used on stained-glass windows, as well as gilding for borders. His pieces are signed with his name or initials, and sometimes with his Vienna address. He employed apprentices, and his work is often imitated.

Koula Beaker, The. A Bohemian BEAKER of conical shape, wheel-engraved by an unidentified Bohemian engraver known as the 'Master of the Koula Beaker'; the decoration depicts allegorical figures after Martin de Vos. It is so-called from the name of its original owner at Prague. Two other similar beakers are attributed to the same hand, one with allegorical figures and the other, *c.* 1690, with engraved portraits of Leopold I (d. 1705), Holy Roman Emperor, his wife Eleonora Magdalen, and their son Joseph (1678–1711) who became King of Hungary and later Emperor. They are decorated with TIEFSCHNITT of stylized flowers and feathery foliage.

Kraka glass. A type of glass with network design developed at the ORREFORS GLASBRUK by SVEN PALMQVIST, *c.* 1944. The early examples have an opal white glass network between two layers of glass; later pieces have a combination of dark-blue and golden shades. The name was inspired by Kraka, a lady in the Saga of the Vikings, who came to her hero dressed in a fishing net.

Kramer, Baltasar (d. *c.* 1813). A German glass-maker who, with several brothers, emigrated to the United States in 1773. He worked at the Manheim factory of HENRY W. STIEGEL, and later at the NEW BREMEN GLASS MANUFACTORY. When the latter closed, the family went west but was persuaded by Albert Gallatin to return to Pennsylvania and help him start his newly founded NEW GENEVA GLASS WORKS. Kramer (known as Baltzer Kramer) became one of the outstanding early glass-makers of the country.

krater (Greek). A VASE, generally of Greek pottery, for mixing wine and water. It has a wide body and mouth, two side handles, and a small circular base. There were several variations (Corinthian, calyx, bell, volute), but the form later made in glass is usually the calyx type, shaped like a chalice or the calyx of a flower, with two horizontal uplifted loop handles attached at the lower part of the body. Some tiny examples were made of EGYPTIAN GLASS, by the CORE GLASS method, 1500–970 BC.

Krautstrunk (German). Literally, cabbage-stalk. A type of NUP-PENBECHER, having a cup-shaped mouth curving outward from above an encircling THREAD. It was the forerunner of the RÖMER. *See* DOPPELTER KRAUTSTRUNK.

Kreybich, Georg Franz (b. 1662). A seller of glassware from Kamenicky-Senov (Steinschönau) in northern Bohemia who from 1685 to 1719 made 21 trips throughout Europe, first on foot, then by cart and later with a caravan, selling Bohemian glass and who wrote an autobiography of his experiences. He was also a cutter and engraver, adding special decoration and inscriptions to suit the customers.

ku (Chinese). A Chinese vase of beaker shape based on an ancient bronze form, cylindrical, with a bulge midway, and a trumpet mouth. The *ku* was made of Chinese porcelain and also sometimes of opaque coloured glass, including yellow and turquoise.

kuan liao (Chinese). Glass made for the Imperial Court in various techniques, including opaque monochrome (e.g. yellow) glass, CASED and carved glass, and opaque white glass with decoration in translucent enamelled colours.

Kuan-yin. A Buddhist female deity, the goddess of mercy, often depicted in porcelain decoration and figures, but also in figures made of glass in China in the 18th/19th centuries.

Kubiikki ware. A series of glass bowls and vases made by the RIIHIMÄKI GLASSWORKS, square or rectangular in shape, but of varying sizes and heights. They are blown in wooden moulds. The pieces are generally made for household use.

Kufic script. A kind of angular formal Arabic CALLIGRAPHY used as decoration on Islamic pottery and glassware. Sometimes a mock-script was used, suggestive of Arabic writing but consisting of meaningless conventional linear designs. *See* NESKHI; MOSQUE LAMP.

Kugelgravüren (German). Engraving in circular or oval motifs, done with a vertical wheel in Germany by the craftsmen known as *Kugler*, in the 18th and early 19th centuries. The curved motif is termed in German *Kugelschliff*.

Kugler (German). A glass-cutter who specialized in circular and oval cutting (*see* CUT GLASS: *patterns*), combining the technique with other styles of cutting and sometimes also with ENGRAVING. They worked during the 18th and early 19th centuries, making designs with motifs popular in the BIEDERMEIER STYLE.

Kuhnt, Georg Gottlieb (1805–85). A glass-maker of Breslau who made and in some instances signed glass-mosaic panels depicting scenes at the spas of Bohemia, *c.*1850.

Kunckel, Georg Ernst (1692–1750). An engraver who was appointed Court Glass Engraver in 1721 at Gotha, in Thuringia. Of about seventy pieces attributed to him, three are signed and thirteen others are dated. The motifs used by him included pearl festoons and trellised lambrequins, engraved in INTAGLIO.

Kunckel, Johann (1630?–1703). Chief chemist at and Director of the POTSDAM GLASS FACTORY established at Drewitz by Elector Friedrich Wilhelm of Brandenburg (1620–88) from its founding in 1678 until 1694. While there he tested at his private laboratory the formulae collected by ANTONIO NERI, which included one for ruby-coloured glass. Before 1679 he developed the process for making GOLD RUBY GLASS. He tried to preserve his secret but imitators succeeded in making such glass, sometimes substituting copper for gold. Elector Friedrich III, the successor to Friedrich Wilhelm (and later Frederick I), questioned his accounts for the cost of his work and demanded repayment, which (together with the burning down of his factory) obliged Kunckel to sell his home and in 1693 go to Sweden, where he became an authority on mining and was knighted. Kunckel wrote in 1679 his *Ars Vitraria Experimentalis*, an account of contemporary glass-making, in which he forecast the revival of ZWISCHENGOLDGLAS and

Krautstrunk. Beaker with applied prunts and everted lip, Germany, 15th/16th century. Ht 8·5 cm. Victoria and Albert Museum, London.

Kuan-yin. Amber-coloured glass, 18th/19th century. Ht 26·5 cm. British Museum, London.

Kurfürstenhumpen (1). Emperor and Electors, Bohemian, 1611. Ht 23·5 cm. Kunstsammlungen der Veste Coburg.

disseminated knowledge that led to the making of many varieties of glass; in 1689 a second edition was issued. Kunckel made, in addition to gold ruby glass, glass of other colours and OPAQUE WHITE GLASS. His glasses were decorated mainly with FACETING and wheel-engraving.

Kurfürstenhumpen (German). Literally, Electors Beaker. A tall HUMPEN decorated with the enamelled portrayal of the Holy Roman Emperor, together with the seven Electors of the Empire. On early examples the Electors are standing at the side of the Emperor; later ones show them on horseback in two horizontal rows. The design is from old prints, one by Hans Vogel of Augsburg, second half of the 16th century. The earliest recorded glass dates from 1591. The same motif appears on some enamelled GOBLETS with a stemmed FOOT and on some DIAMOND-ENGRAVED *Humpen*. *See* ENAMELLING, *German*.

Kuttrolf (German). A vessel that has a cup-like wide pinched MOUTH or upper container, joined to a globular body by a neck composed of 3 to 5 intertwined thin tubes, so that when liquid is poured from the body it flows only drop by drop and has been said to make a bubbling sound. Most examples have the tubes tilted at an angle to the body, so that the mouth slopes diagonally; but early examples in FRANKISH GLASS are entirely vertical. They were used for slow pouring or dripping, and in some instances for drinking. The name has been said to derive from *Kuttering* (gurgling) or from the Latin *gutta* (drop) or *gutturnium* (a slow-pouring dropper for perfume). Such vessels, in the form of BOTTLES, became popular in the 16th and 17th centuries, and have been made, in differing versions, until the present time. These pieces are derived from the 4th-century Syrian glass SPRINKLERS, the body of which was made into four tubes that joined into a tall neck to provide slow pouring. A variation of the *Kuttrolf* is a tall single-tube flask with a thin curving neck, made in Persia in the 18th and 19th centuries, and called a ROSE-WATER SPRINKLER. The *Kuttrolf* is also called an *Angster*, from Latin *angustus*, narrow. *See* DECANTER, *cluck-cluck*.

Ku-yüeh Hsüan. The Chinese designation of a style of decoration on Chinese glassware (also used on porcelain) of several groups: (1) PEKING IMPERIAL PALACE GLASSWARE; (2) glassware made probably at Peking and bearing the mark '*Ku-yüeh Hsüan*' ('Ancient Moon Pavilion') of the legendary artist who decorated it (*see* HU); and (3) glassware made later at various Chinese factories imitative of such ware. The decoration depicted flowers and figures; it was developed *c*.1720 and was used for groups (2) and (3) later, from *c*.1750 to *c*.1850. The style was used probably contemporaneously on glassware and porcelain, perhaps by the same artists, notwithstanding the view of some modern writers who state that one such ware preceded or copied the other. *See* CHINESE GLASSWARE.

kyathos (Greek). A dipper in the form of a small shallow bowl having one long upright handle, sometimes in the form of a loop. The ancient Greek pottery form (used for transferring wine from a KRATER into a cup) has been adapted in ROMAN GLASS (*see* SIMPULUM). Similar forms, called a PIGGIN, were made *c*.1800 in ANGLO-IRISH GLASSWARE. The German term is *Schöpfgefäss*.

kylix (Greek). A drinking-cup with a calyx-shaped bowl (like a CHALICE or the calyx of a flower), resting on a stemmed FOOT. Usually made of pottery, but some examples were made of ROMAN GLASS.

Kuttrolf. Four intertwined tubes forming neck, Germany, late 16th century. Ht 20·4 cm. Museum für Kunsthandwerk, Frankfurt.

L

label. *See* BOTTLE TICKET.

Labhardt, Christoph (d. 1695). A Swiss cutter and engraver of hardstones who was brought in 1680 to Cassel by the Landgrave Carl of Hesse and did work in baroque style in HOCHSCHNITT and TIEF-SCHNITT on glassware as well as on rock crystal. Upon his death, he was succeeded at Cassel by FRANZ GONDELACH.

Labino, Dominick (1910–). A leading American glass technician and consultant having his own studio at Grand Rapids, Ohio, where since 1963 he has designed and made glassware to his own ideas. He has experimented with the making of replicas of ancient CORE GLASS with TRAILED decoration and applied handles and feet. He has also made glassware in the style of VETRO A RETICELLO, and decorated some glassware by use of colloidal gold. He was a leading figure in establishing the seminars on glass-making sponsored by the Toledo (Ohio) Museum of Glass.

lace glass. A general term applied to VETRO A RETORTI and by some writers to VETRO A RETICELLO. *See* FILIGRANA; FILIGREE; VETRO DI TRINA; PIZZO, VETRO.

lace-de-Bohème glassware. A type of glassware with white enamelled decoration, often on a coloured or satin ground, that was made in Bohemia, *c.* 1880–90, as a cheap imitation of good-quality CAMEO GLASS. Another contemporary Bohemian imitation was coloured glassware with a thin opal FLASHING on which a design was drawn and covered with acid-resisting ink and then placed in HYDROFLUORIC ACID to remove the uncovered areas, leaving a shallow relief.

lace-maker's lamp. A misnomer for an oil lamp, often said to have been used by lace-makers, that has a globular FONT (*c.* 5 to 7·5 cm. in diameter) using a drop burner, and has a round drip catcher, all set upon a stem with a saucer base; it often has a loop handle on one side. Such a lamp was too fragile, expensive, and impracticable for the lace-makers, and the font was too small to serve as a WATER LENS. The lace-makers used any sort of an oil lamp, or even a candle, with the light condensed by a water lens; the lens sometimes was centered to serve several lamps for different workers. Comparable sources of light with a water lens were also called a 'cobbler's lamp'.

lachrymatory. *See* TEAR BOTTLE.

lacy glass. A type of American MOULD-PRESSED GLASS with decoration covering the entire surface of the objects, depicting flowers, foliage, rosettes, fleurs-de-lis, etc., all sightly raised against a diaper background pattern made by small dots in the mould that produced a stippled appearance. The result was an overall sparkling lacy effect. The style was developed *c.* 1830 by DEMING JARVES at the BOSTON & SANDWICH GLASS CO. and was produced by it and the NEW ENGLAND GLASS CO., and possibly also the PHOENIX GLASS WORKS. It has been called 'Sandwich glass'. Similar ware was made in France. The ware included mainly plates and CUP PLATES, but also bowls, trays, creamers, etc., and there was a great variety of patterns.

ladle. A cup-like spoon, usually with a long HANDLE which is sometimes curved, employed for transferring liquids from a vessel.

'lace-maker's lamp', erroneously so-called. English, 18th century. Ht 27·5 cm. Science Museum, London.

lacy glass. Cake tray of colourless pressed glass, probably Boston & Sandwich Glass Co., *c.* 1835. L. 29·8 cm. Toledo Museum of Art, Toledo, Ohio.

lagynos. Probably Near East or Roman glass, 200 BC–AD 100. Ht 21·7 cm. Corning Museum of Glass, Corning, N.Y.

Lalique, René. Mould-blown vase, signed 'R. Lalique', Paris, 1920s. Ht 17 cm. Kunstmuseum, Düsseldorf.

lampada pensile. Bowl of greenish glass with decoration of ground disks, some with points; silver rim with inscription; probably Byzantine, 10th/11th century. D. 36 cm. Treasury of St Mark's, Venice.

Generally made of silver or porcelain, some have been made of glass. Other types are in the form of a Greek KYATHOS and the Roman SIMPULUM; *see* PUNCH-LADLE.

lagynos (Greek). A type of OENOCHOË with a low wide body, angular shoulder, high, thin and vertical neck, circular mouth, and angular handle extending from the top of the neck down to the shoulder. Usually they were made of Greek pottery, but two examples of glass have recently been found. *See Journal of Glass Studies* (Corning), XIV (1972), p. 17.

Lalique, René Jules (1860–1945). A French designer and maker of glassware of designs and styles of his own creation. He started as a designer of exclusive jewellery in ART NOUVEAU style, making also objects of enamel and inlaid glass paste. His commercial success stemmed from the perfume bottles he designed for Coty. He established *c.* 1908 a glass factory at Combs, near Paris, making large quantities of MOULDED, PRESSED and ENGRAVED glass, also some especially thick engraved window panels. He was interested in surface treatment by acid or SAND-BLASTING, resulting in a frosty OPALESCENCE. He was a follower of ÉMILE GALLÉ, making pieces in Art Nouveau style until he developed his own original style. He sometimes used coloured glass, and made some pieces by the CIRE PERDUE process. In 1918 he acquired a glasshouse at Wingen-sur-Moder (Bas-Rhin), which has been continued by his son Marc for all present production, under the name Cristallerie Lalique et Cie. *See* LALIQUE GLASSWARE.

Lalique glassware. A type of glassware made by RENÉ LALIQUE from *c.* 1925 which is colourless but has a pale blue opalescent MAT surface. The ware consisted at first largely of small bottles and ornaments, but later included large MOULDED vases, bowls, and figures. The motifs emphasize plants and animals, and often the female body. The ware is still made today at the Lalique factory. The pieces were signed, but after Lalique's death his initial 'R' was omitted.

laminated glass. (1) Glass that is covered with scales, e.g. the surface of glass that has been buried in damp earth for centuries and that has acquired an IRIDESCENCE. (2) Glass that is shatterproof, as a result of an interlayer between two panes of glass that are sealed together. It was first made by John Wood in England in 1905. Originally the interlayer required that the edges be sealed with pitch, but as this tended to discolour the glass, new materials for the interlayer have been developed which require no sealing. The glass, when struck, cracks radially but the splinters adhere to the interlayer. *See* BULLET-RESISTANT GLASS. The term 'laminated' when sometimes applied to CASED GLASS is incorrectly used.

lamp. A utensil for producing artificial light. Originally this was by means of a light wick floating on oil or protruding from a small spout left for the purpose (as with Roman terracotta lamps). Some lamps are designed to stand on a flat surface or are attached to a base; others are intended for carrying or for suspension by a chain. *See* LANTERN; LANTHORN; OIL LAMP; MOSQUE LAMP; SANCTUARY LAMP; HANGING LAMP; LACE-MAKER'S LAMP; LAMPADA PENSILE; CESENDELLO; SPARKING LAMP; TIFFANY LAMP; WINE-GLASS LAMP.

lampada pensile (Italian) Literally, hanging lamp. A type of HANGING LAMP in the form of a wide shallow glass bowl in which was placed a small oil lamp. The bowl is in the shape of the balance pan of a scale and is suspended by chains attached to the rim. Some were made at MURANO and some probably at Byzantium in the 10th/11th centuries. *See* DISH LIGHT.

lampwork. The technique of manipulating glass AT-THE-LAMP by heating it with a small flame. Examples of pieces so produced range from small figures and objects to large composite three-dimensional scenic groups. *See* GLASS HUNT; GROTTO.

lanceolate. A motif of cut decoration on glassware that is shaped like a spearhead, tapering at each end. It is found in deep CUTTING on some crystal-glass drinking glasses made by the BACCARAT GLASS FACTORY.

Landberg, Nils (1907–   ). A glass-maker at ORREFORS GLASBRUK since 1936, whose early work was in engraving glassware but who later made ware by freehand blowing without use of a mould. He is particularly noted for tall vases and tableware with exceptionally long and very thin stems, sometimes in graduated hues.

Langer, Caspar Gottlieb (fl. 1749). The only engraver of Bohemian-Silesian glassware of the mid-18th century who signed an example of his work. He worked at Warmbrunn.

lantern. A portable or hanging enclosure for a light, with surrounding glass to protect the flame. They are usually in the form of (1) a metal support framing a globular, cylindrical or baluster-shaped glass chimney, with a hook or handle by which it can be suspended, or (2) a square- or polygonal-section metal frame holding four or more vertical panels. Each type has an appropriate means for providing the source of light, e.g. a nozzle for a candle or inflammable fuel and a wick. *See* MOSQUE LAMP; CESENDELLO; LANTHORN.

lanthorn. An early English type of LANTERN. It has a metal framework square or rectangular in section, with a glass panel in each of the four sides and inside a nozzle to hold a candle. It was intended for carrying about, in the house or outdoors, but some have a suspensory hook. Early examples from the mid-17th century had light-green window-glass panels, but later ones had polished PLATE-GLASS panels, some Continental examples being decorated with WHEEL ENGRAVING. The term is probably a combined form of 'lantern' and 'horn' from which the translucent panels of early lanterns were made.

Larsen, Emil J. (b. 1877). A leading American glass-maker, born in Sweden, who emigrated to the United States in 1887. He worked at a number of glasshouses, including DORFLINGER GLASS WORKS until *c.* 1918, PAIRPOINT MANUFACTURING CO., QUEZAL ART GLASS & DECORATING CO. until *c.* 1926, and VINELAND FLINT GLASS WORKS until *c.* 1933. He started his own factory *c.* 1935 at Vineland, New Jersey, making pattern-moulded and free-blown ornamental glassware, and also paper-weights (*see* JERSEY ROSE).

lat(t)esino (Italian). A 16th-century Italian term which is applicable to a type of OPAQUE WHITE GLASS that is similar to LATTIMO but which has a bluish cast and is slightly OPALESCENT. The only enamelled vessel of the 16th century made of *latesino* known in 1974 (according to T. H. Clarke) is a JUG, *c.* 1510–20, in the Musée du Verre, Liège. *See Journal of Glass Studies* (Corning), XVI (1974). Later examples are recorded.

latticino or **latticinio** (Italian, derived from *latte*, milk). Terms that have been often misused by being applied synonymously with LATTIMO (OPAQUE WHITE GLASS), e.g. by W. B. Honey in 1946, and repeated by later English writers without regard to the Venetian usage of the terms. While *lattimo* designates only solid opaque white glass, the two terms *latticino* and *latticinio* have been used in Venice and Murano to apply only to clear glass decorated with embedded threads of glass, usually (*latticino*) white threads. Sometimes spelled *latticino*, sometimes *latticinio*, both spellings were used by early Italian writers of 1564 and 1585, respectively. Today Luigi Zecchin, of Murano, and some other modern glass authorities advocate the abandonment of both terms because of historical confusion, relying solely on the terms FILIGRANA, VETRO A RETORTI, and VETRO A RETICELLO to designate glass decorated

lampwork. Farmhouse in multi-coloured glass, Netherlands (or possibly England), dated 1803. Ht 37.5 cm. Corning Museum of Glass, Corning, N.Y.

lantern. Colourless glass in metal frame; wheel-engraved decoration, and interior candle nozzle, with air vents near top; possibly German, first half of 18th century. Ht 43 cm. Victoria and Albert Museum, London.

with embedded threads. However, the terms *latticino* and *latticinio* have been too frequently used (even with respect to TWISTS in English STEMS) to expect their being wholly discarded now, and moreover they seem serviceable terms if confined to *filigrana* (FILIGREE) glass when the threads are exclusively white, as distinguished from such decoration using some coloured threads (that would be a contradiction of terms), the latter style being included in the broad term *filigrana*. Such usage would include, as *filigrana*, decoration with only a few white threads in the object or in an overall pattern, e.g. vases and bowls, also plates, decorated in white overall *reticello* or *retorti* style that often today are referred to by writers and museums as *latticino* (or *latticinio*). When bands of opaque white glass are applied to the exterior of an object, then COMBED and MARVERED, such decoration is preferably referred to as 'combed', and similarly in the case of 'pulled' decoration. The phrase 'white *latticino*', often seen, is tautological.

lattimo. Slipper, colourless glass, with *lattimo* stripes, Murano, early 19th century. L. 15·5 cm. Museo Poldi-Pezzoli, Milan.

Laub- und Bandelwork. Jug with engraved decoration, Glashütten der Peterhänsel, Adlergebirge, first half of 18th century. Ht 13·3 cm. Kunstgewerbemuseum, Cologne.

lattimo (Italian, derived from *latte*, milk). The same as OPAQUE WHITE GLASS. Usually the entire body of the object is made exclusively of such glass, but sometimes it is made with broad, or even narrow, MARVERED bands or festoons of opaque white glass (*see* COMBED GLASS), as distinguished from glass decorated with embedded threads of opaque white glass, e.g. FILIGRANA glass. *Lattimo* glassware has been made, probably from the mid-14th century and certainly from the mid-15th century, in Venice, and later in Bohemia, Germany, France, Netherlands, Spain, England, and the United States. Decoration with polychrome or monochrome enamelling, in imitation of enamelled porcelain, was done in many countries; from the period 1475–1525 fourteen Venetian examples have been recorded, but from the 18th century and onward examples are numerous (*see* VENETIAN SCENIC PLATES). The German term is *Milchglas*, the French is *verre de lait*. *See* LATESINO. *See* T. H. Clarke, *Journal of Glass Studies* (Corning), XVI (1974), p. 48.

Laub- und Bandelwerk (German). Literally, leaf and strapwork. A late BAROQUE ornament of strapwork and intricate foliage interlacements, sometimes with added CHINOISERIES or other figures. It was used to frame a decorative painting on glass or porcelain, either in gold silhouette or in enamel colours; it was also much used by the wheel-engravers of the baroque period. Its origin is to be found in the designs of Jean Bérain I of France, but in glassware the style was more commonly used in Germany and Bohemia when the books of such designs by Paul Decker and others were exploited as sources of ornament during the first half of the 18th century, particularly by the wheel-engravers and by the HAUSMALER of Nuremberg, Bohemia, and Silesia.

Lauensteiner Hütte. A glassworks founded in 1701 at Lauenstein, near Hamelin in Hanover (Germany) and surviving until 1870. During the 18th century it made glassware with several English features (e.g., drawn stem, thistle-shaped bowl, facet-cut knops, Silesian stem), having employed at the outset an English craftsman of the Tysack family and being thereafter clearly under English influence. The glass was subject to CRISSELLING. The pieces sometimes bear a mark of a lion rampant, with the letter 'C' for 'Calenberg', the principality in which the glasshouse was located.

Lauscha glassware. Glassware made at Lauscha, near Neuhaus, in Thuringia, Germany, from 1595, and especially pieces made AT-THE-LAMP from the 18th century to the present. Many glassworks here still make, in addition to tableware, a large quantity of TOYS, BEADS, and Christmas-tree ornaments of glass.

lava glass. A type of ART GLASS, developed by LOUIS C. TIFFANY, having a rough surface (made by mixing pieces of basalt into the BATCH). The glass has a gold IRIDESCENCE producing a gold pattern in stripes or in abstract designs, to give the effect of molten lava. The pieces are usually of asymmetrical form. It was originally called 'volcanic glass'.

layered glass. A vague term sometimes applied to various types of glassware made of two or more layers of glass. The terms CASED GLASS and FLASHED GLASS are preferably used for those specific types of fused glass, and the term LAMINATED GLASS is preferable for glass rendered shatterproof by an interlayer between two panes of glass. The term has also been applied to (a) glassware decorated by use of a second GATHER over a limited area of the surface, e.g. the LILY-PAD DECORATION, (b) some CHINESE SNUFF BOTTLES with designs in glass of varied colours applied to different areas of the surface, and (c) some Victorian glassware of the first half of the 19th century made of layers of white or coloured glass cut, in the style of CAMEO GLASS, to reveal the different colours.

lazy susan. A popular name for a REVOLVING SERVER, sometimes made of glass.

lead. The source of the LEAD OXIDE used in making LEAD GLASS. Before the use of lead oxide by GEORGE RAVENSCROFT in glass-making, lead was used in the form of LITHARGE, produced by blowing air over the surface of molten lead; when litharge is further oxidized, it becomes red lead. Its use requires special furnace conditions, as its conversion back to metallic lead would discolour the glass and damage the POT (CRUCIBLE).

lead glass. A type of glass containing a high percentage of lead oxide. Lead oxide in a small percentage had been used in Mesopotamia and again, to enhance the brilliance of glass (especially in the manufacture of artificial gems), since before the 11th century, but its importance stemmed from its first use *c.* 1676 by GEORGE RAVENSCROFT, originally as a remedy against CRISSELLING. The METAL is soft. It has a light-dispersing character that gives a brilliance, especially when in the 18th century and later it was decorated with FACETING. It is less fragile than the Venetian SODA GLASS, and so was more suitable for ENGRAVING and CUTTING. It is frequently called in England 'glass-of-lead'. The process was independently discovered in 1781 by the ST-LOUIS GLASS FACTORY. *See* LITHARGE.

lead test. A test for the identification of LEAD GLASS by applying a drop of HYDROFLUORIC ACID, mixed with a drop of ammonium sulphide, to a small area. This results in a black area if lead is present, due to the formation of lead sulphide; SODA GLASS or BARIUM GLASS will show no reaction. *See* ULTRA-VIOLET TEST.

Lebensalterglas. The German term for glassware depicting the AGES OF MAN.

Lechevrel, Alphonse. A 19th-century glass-maker who made CAMEO GLASS at Hodgetts, Richardson & Co., at Stourbridge, and had as a pupil JOSEPH LOCKE.

ledge. The flattened, raised part of a PLATE or DISH that surrounds the WELL. It has been found on Roman pottery and on European pewter dishes which were copied in glass. Sometimes called a 'marli'. *See* RIM.

leech bead. A type of BEAD that is a hollow cylinder and slightly curved, to resemble a leech. Such beads (or a lengthwise section of such a bead, called a 'runner') were sometimes used to cover the bow of a FIBULA. Some were made of Etruscan glass in the 7th century BC, with marvered COMBED DECORATION in a feather pattern.

leech glass. A glass tube to contain a leech (used for application to the human body in bleeding a patient). An 'artificial leech' is a small glass tube with a scarifier, for drawing blood by suction.

lead glass. Covered tankard with gadrooning at bottom and on cover, English, 18th century. Ht 35·8 cm. British Museum, London.

leech bead. Fibula with bronze pin and opaque glass runner with white combed trails, Etruscan, mid-7th century BC. L. (pin) 11·7 cm. British Museum, London.

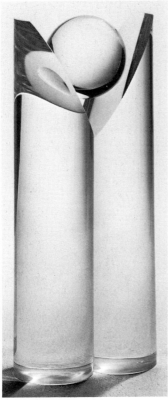

Leerdam glassware. Crystal glass ornament designed by Floris Meydam. Ht 30 cm. Courtesy, Royal Leerdam Glassworks (United Glassworks).

Lehmann, Caspar. Panel with engraved portrait of the Elector of Saxony, Christian II, Caspar Lehmann, *c.*1606. Museum of Decorative Arts, Prague.

leech jar. A receptacle of PHARMACY WARE for keeping leeches. Some are cylindrical; all have some air vents, e.g. a mesh COVER. Examples have been made in glass.

Leerdam glassware. Glassware made at the Royal Dutch Glass Works at Leerdam, near Utrecht, founded in 1765, and since 1937 a unit of the (Dutch) United Glassworks which also has plants at Maastricht and Schiedam. The products include table ware, bottles, ornamental crystal glass, and other types of glassware. Under the direction of P. M. Cochius from *c.*1912, modern glassware was introduced, the designers including K. P. C. de Bazel (from 1915), Cornelis de Lorm (from 1917), C. J. Lanooy (from 1919), Hendrik P. Berlage (from 1923), and, in 1922/23, Chris Lebeau (1878–1945). Berlage established *c.*1925 the Unica Studio to develop modern design, and was succeeded by ANDRIES D. COPIER, who engaged as designers FLORIS MEYDAM from 1935 and WILLEM HEESEN from 1943. Copier was known for his vases of heavy crystal and his blown glass. The Glass Form Centre was opened in 1968 to do experimental and development work. Sybren Valkema (b. 1916) has taught and made glassware at Leerdam.

leg glass. A type of DRINKING GLASS, made in the form of a woman's leg, which had to be drained before setting it down inverted on its rim.

Legras & Cie. A glasshouse founded by August J. F. Legras at Saint-Denis, near Paris. It made CAMEO GLASS embellished with enamelling. It used the marks 'Legras' and 'Mont Joye & Cie.' After World War I it merged with the PANTIN GLASSWORKS, and operated as 'Verreries et Cristalleries de St-Denis et Pantin Réunies', using the marks 'Pantin' and also 'De Vez' (the pseudonym of its Art Director, De Varreux).

Lehman(n), Caspar (1570–1622). Reputedly the first glass-worker in modern times to use, *c.*1590–1605, the technique of WHEEL-ENGRAVING to decorate glass, despite a suggested possibility that such engraving was first resumed at Munich under Duke Wilhelm V during the 1580s. He was born near Lüneburg (Germany) and left in 1586; from then until 1588 he may have been in Munich where, according to a letter, he learned glass-engraving. In 1588 he moved to Prague and worked at the Court of Emperor Rudolf II as a carver of precious stones, being appointed in 1601 as Imperial Gem-Engraver and in 1608 Imperial Gem-Engraver and Glass-Engraver; in 1609, after having been two years in Dresden, he was appointed Royal Lapidary, and was granted the hereditary privilege of glass-engraving. In 1618 he was joined as glass-engraver by GEORG SCHWANHARDT, who four years later succeeded to the privilege. Lehman's work was predominantly wheel-engraving, without polishing and rarely with embellishment by DIAMOND-POINT ENGRAVING. His earlier work included experiments with panels (*see* PERSEUS-ANDROMEDA PLAQUE), some with portraits; later he engraved glass vessels, such as the BEAKER with ALLEGORICAL SUBJECTS which in 1605 he signed 'C Leman'. His work is especially notable for being done on Venetian CRISTALLO glass, although the panels may be POTASH-LIME GLASS which had been well developed in Bohemia before 1600. He founded an engravers' school, but after the suppressed revolt against the Hapsburgs in 1620 his followers migrated, Schwanhardt to Nuremberg, JOHANNES HESS to Frankfurt, and CASPAR SCHINDLER to Dresden.

lehr (sometimes 'leer'). The oven used for ANNEALING glassware. Early lehrs were connected to the FURNACE by flues, but the difficulty in controlling heat and smoke made them impracticable. Later the lehr was a long brick-lined, separately heated steel tunnel through which the glass objects being annealed were pushed slowly; placed in iron pans on a conveyor the ware remained in the lehr for several hours, while being gradually reheated and then uniformly cooled. This type is supposed to have been invented by George Ensell, at Stourbridge, *c.*1780. Later mechanical conveyors were introduced, and today lehrs are entirely electrically heated and operated.

Leighton family. Thomas H. Leighton (1776–1849), an English glass-maker who, after experience in England, Dublin and Edinburgh, emigrated to the United States and in 1826 became Superintendent of the NEW ENGLAND GLASS CO. for almost twenty years. He established the Boston Flint Glass Works, at Cambridge, Massachusetts. Of his seven sons, (1) John H. succeeded him; (2) William, who was proficient in making METALS and colours, developed, c. 1858, a new formula for ruby glass, and in 1855 patented a process for making SILVERED GLASS; in 1864 he left the New England Glass Co. and joined HOBBS, BROCKUNIER & CO. where he developed a new LIME GLASS to supersede the expensive FLINT GLASS for making PRESSED GLASSWARE. The latter's son, William, Jr., at Hobbs, Brockunier & Co., made PEACH BLOW GLASS and developed DEWDROP GLASS and, c. 1883, SPANGLED GLASS. Henry, a grandson of Thomas H., became an expert glass engraver at the New England Glass Co.

lekythos (Greek). A type of small JUG for perfume, oil or ointment. It is tall and slender, with usually a small loop-handle attached to the neck and shoulder, a narrow neck for slow, controlled pouring, a cup-like mouth, and a flat base. Generally made of Greek pottery, some were made of ROMAN GLASS, c. 300–100 BC. Some were used as tomb offerings.

lens. *See* DIMINISHING LENS; MAGNIFYING LENS.

lentil flask. A type of FLASK with a narrowed circular lentil-shaped bowl resting on a flat bottom and with a tall thin neck. They were made in the FAÇON DE VENISE. *See* LENTOID FLASK.

lentoid (or **lenticular**) **flask.** A type of FLASK of Roman CORE GLASS, the characteristic feature of which is a BODY that is of 'lentil' shape, i.e. narrowed globular (or nearly so) shape, somewhat like a PILGRIM FLASK. It has a cylindrical neck and disk mouth, with two small vertical loop handles extending from the shoulder to midway up the neck. Early Egyptian examples have a rounded bottom and no foot. They were made c. 1410–970 BC. Some are undecorated, some have COMBED DECO-RATION in palm-leaf pattern. *See* ARYBALLOS.

letter seal. A glass seal similar to a FOB SEAL except that it has no suspensory ring. *See* WAFER STAMP.

letter-weight. A large type of PAPER-WEIGHT, usually with an oval or rectangular flat base and a HANDLE in the form of a central vertical metal ring or a glass finial. Some have been made in HYALITH glass, and some by APSLEY PELLATT with 'cameo incrustations' (SULPHIDES) depicting prominent persons; others have the base cut with STRAWBERRY DIAMONDS.

Léveillé, Ernest Baptiste. The pupil and successor of EUGÈNE ROUSSEAU who, after Rousseau's retirement in 1885, continued his techniques under the name of Léveillé-Rousseau, producing glassware in the ART NOUVEAU style. After 1900 the factory was bought by Harand & Guignard, and the same production was continued for several years.

Libbey Glass Co. A glassworks at Toledo, Ohio, that resulted from the leasing in 1878 of the NEW ENGLAND GLASS CO. by William L. Libbey (1827–83) and from its transfer by his son, Edward Drummond Libbey (1854–1925), to Toledo in 1888, changing its name to Libbey Glass Co. During its so-called 'Brilliant Period' (1890–1915) it became the largest cut-glass factory in the world. In 1893 it operated the Glass Pavilion at the Chicago World's Fair with 130 craftsmen blowing and cutting glass; much of the glass made there sold as souvenirs, marked 'Libbey Glass Co. World's Fair 1893'. From 1883 to 1940 it made coloured ornamental

Legras glassware. Vase in Art Nouveau style, with overlay decoration, Legras & Cie, Paris, c. 1900–14, signed 'Legras'. Ht 20·2 cm. Kunstmuseum, Düsseldorf.

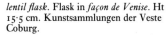

lentil flask. Flask in *façon de Venise.* Ht 15·5 cm. Kunstsammlungen der Veste Coburg.

letter-weight. Hyalith with gilt *chinoiserie* and gilt bronze handle, glasshouse of Count von Buquoy, Bohemia, *c.*1820–30. L. 12·5 cm. Kunstgewerbemuseum, Cologne.

glassware in new techniques developed by JOSEPH LOCKE (e.g. AMBERINA, AGATA, POMONA, MAIZE, PEACH BLOW). After Libbey died in 1925 control passed to his associates, including J. D. Robinson, and on the latter's death in 1929 to his sons Joseph W. Robinson and Jefferson D. Robinson who became respectively President and Vice-President. In 1931 the Robinsons engaged A. DOUGLAS NASH, who introduced in 1933 the high-quality LIBBEY-NASH GLASSWARE. In 1936, after large financial losses, the company was taken over by OWENS-ILLINOIS, INC., operating first as a subsidiary and later as the Libbey Glass Division, making table glass in contemporary styles, much from designs by Freda Diamond (e.g. *Golden Foliage* introduced in 1957) and also ornamental lead crystal glassware from 1940 to 1945 designed by Edwin W. Feurst. Currently the company makes only automatically produced tableware and has made no hand-crafted glassware since its 1940 'Modern American Series'. *See* OWENS, MICHAEL J.

Libbey-Nash glassware. The so-called 'Libbey-Nash Series' of luxury stemware and ornamental glassware created by A. DOUGLAS NASH for the LIBBEY GLASS CO. after 1931. The stemware included 82 different patterns, some made in four choices of colours; the ornamental glassware, of clear and coloured glass, was produced in very limited quantity. The products were of very superior craftsmanship as to cutting and engraving, and the pieces are generally signed with the Libbey name in script, enclosed in a circle. Due to the Depression in the 1930s, the line was discontinued when the company was taken over in 1936 by OWENS-ILLINOIS, INC.

Libisch, Joseph. A Hungarian-born glass-engraver who, after apprenticeships in Vienna and Prague, went to the United States and from 1921 worked at STEUBEN GLASS WORKS, where FREDERICK CARDER considered him to be his best engraver. The GAZELLE BOWL, the first engraved piece designed for Steuben by SIDNEY WAUGH, was first engraved by Libisch (1935); he also first engraved the ZODIAC BOWL (1935) designed by Waugh, the *Acrobats Goblet* (1940) designed by the Russian artist Pavel Tchelitchev, and the *Orchids Bowl* (1954) designed by Sir Jacob Epstein, as well as engraving the VALOR CUP (1940) designed by JOHN MONTEITH GATES and the MERRY-GO-ROUND (1947) designed by Waugh.

lid. A covering for closing the MOUTH of a MUG, TANKARD, TEA-CADDY or other object, usually attached to the BODY by a metal hinge or mounted in a metal frame. *See* COVER.

lily-pad decoration. A style of applied decoration used on glassware in the United States, *c.*1835–60, especially by the CASPAR WISTAR glass-works; it resembles a water-lily pad on a stem and occurs on the lower part of a bowl or a jug. Molten glass was applied to the body of a piece by TRAILING and afterwards tooled upward to form the design, of which there are three varieties: (1) slender vertical stems terminating in a bead-like pad; (2) broader stems with oval or circular pads; (3) curved stems terminating in a flat ovoid pad.

lime. Calcined limestone, used in making certain types of glass. It was in the sea-plant-ash used as an ALKALI in EGYPTIAN GLASS, and it was present as a natural impurity in the soda and potash used for glass-making up to *c.*1660. It was added in various forms, e.g., by adding chalk (carbonate of lime) in making crystal-like glass in Germany, Bohemia, and Silesia in the late 17th century and 18th century. Lime adds stability and produces a lighter and cheaper glass that cools quickly; hence it is used in making bottles, window panes, and electric-light bulbs.

lime glass. A name applied to a type of glass developed in 1864 by William Leighton, at the Wheeling Glass Factory, Wheeling, West Virginia, as a substitute for LEAD GLASS in making inexpensive glass

Léveillé, Ernest Baptiste. Vase in Art Nouveau style, Paris, *c.*1890, signed 'E. Léveillé à Paris'. Ht 19·5 cm. Kunstmuseum, Düsseldorf.

bottles. Its advantages were that it was cheaper, and cooled more quickly, but was lighter and less resonant. *See* LIME.

Lind, Jenny, flask. A type of HISTORICAL FLASK made in the United States in the 1850s, decorated with a relief portrait of Jenny Lind (1820–87), the famous Swedish singer who was brought to the United States by P. T. Barnum in 1850. The flasks, having a flattened globular body and a tall neck, are additionally decorated with a relief wreath and her name. The flasks have a widened lower half that is decorated with a relief lyre. They were made by at least six glasshouses, including the Whitney Glass Works, Glassboro, New Jersey, and in 1925 a FORGERY was made in various colours by the Fislerville Glass Works of south New Jersey. Three varieties are known: a pint flask, a quart flask, and a quart CALABASH.

Lindstrand, Vicke (1904–). A Swedish designer of modern glass who was with ORREFORS GLASBRUK from 1928–41. From 1950 he was with the KOSTA GLASSWORKS, and is noted for his glass sculptures and his ICE BLOCKS.

linen smoother. A glass object to serve as a pressing iron, perhaps for pleating starched ruffles. An example from Spain, 18th century, has a heavy bottom, flat or rounded, from which extends vertically a circular solid HANDLE, REEDED for easier gripping. Similar examples were made in Stourbridge, perhaps as FRIGGERS, in the 18th century; some have crude MILLEFIORI decoration. In England they were sometimes called a 'slicker', a 'slick stone' or a 'smoothing iron'.

liner. An object of glass made to fit closely within a vessel of silver or other metal, so that the contents would not come into direct contact with the metal, e.g. in SALT-CELLARS, BUTTER DISHES, SWEETMEAT GLASSES, WINE COOLERS. *See* SAND-MOULDING.

lion. A decorative motif used on hollow blown stems and on prunts on some glassware. A lion's-mask is found on some Venetian glassware (the lion – the attribute of St Mark – being symbolic of Venice. *See* LION'S-HEAD PRUNT.

lion's-head prunt. A type of PRUNT that was made with a MOULDED lion's-mask in RELIEF, as on some glass made at Venice, and elsewhere in the *façon de Venise*. It is also found on some glassware from the 4th century BC. *See* LION; MASCARON.

lip. The everted pouring projection at the front edge of the MOUTH of a JUG, PITCHER or SAUCE-BOAT. Some are of pinched (TREFOIL) shape. *See* BEAK.

lipped bowl. (1) A cylindrical bowl with a flat bottom and with a pouring lip (sometimes two, opposite each other) on the rim. It has been stated that they were used for four different purposes: (a) as a bowl for water used after rinsing the mouth at a meal; (b) as a FINGER-BOWL; (c) as a WINE-GLASS COOLER; (d) as a bowl for rinsing wine glasses at the table, when a different wine is to be drunk or sediment from port is to be removed from the glass. It is likely that they were used for all such purposes, depending on the level of society of the user and the period when used. They were made in England, some being of blue glass gilded by ISAAC JACOBS. (2) The bowl of a wine-glass having an everted rim; *see* BOWL, *lipped*.

lipper. A glass-maker's tool made of wood in the shape of a truncated cone attached to a handle and used to form a LIP on a JUG, etc.

liqueur bottle. A large bottle with a bulbous BODY which is divided internally into four compartments, each to hold a different liqueur and each having its own neck, mouth, and stopper. Some early-19th-century

lily-pad decoration. Jug (pitcher) of light-green glass, South Jersey style, early 19th century. Ht 19 cm. Toledo Museum of Art, Toledo, Ohio.

linen smoother. Solid glass handle. Ht 14 cm. Museo Arqueologico Nacional, Madrid.

English examples have gilded metal mounts. Similar bottles are made commercially today. *See* MULTIPLE BOTTLE.

liqueur glass. *See* CORDIAL GLASS.

liqueur stand. A stand or rack for holding several DECANTERS, usually made of ORMOLU, GILT BRONZE or SILVER, with partitions to separate the bottles and sometimes with a tall central HANDLE, occasionally embellished as a CANDELABRUM. On some examples, the handle pivots so that its arms prevent the removal of the bottles. The French term is *cave à liqueurs.*

litharge. Lead monoxide, a type of oxide of LEAD that was used in England and on the Continent before 1700 to produce a superior type of glass. It was later used in English glass by GEORGE RAVENSCROFT to avoid the CRISSELLING that had previously resulted from the use of POTASH, instead of BARILLA, as the ALKALI to increase the fusibility of the FLINT employed instead of Venetian pebbles as the source of SILICA. The litharge, although used previously in glass-making in a small proportion, was used by Ravenscroft up to 30 per cent by weight, replacing a portion of the alkali.

Lithyalin glass. A polished opaque glass marbled in red and other strong colours, imitating the appearance of semi-precious stones. It was produced from 1828 until 1840 by FRIEDRICH EGERMANN in Bohemia, by whom it was patented under this name. It was used for a wide variety of objects and in many styles of decoration. One type named CHAMÄLEON-GLAS was introduced in 1835. Imitations of this coloured opaque glass were made at many factories until *c.* 1855.

Littleton, Harvey K. (1922–). A glass technician, designer, and maker, of Verona, Wisconsin, attached since 1951 to the University of Wisconsin as a teacher and studio glass-maker. His work includes free-blown and wrought pieces in his own creative style, in crystal and coloured glass. Among his best-known works are pieces in various forms known as 'Eyes'.

Lobmeyr, J. & L. A Viennese glasshouse, founded in 1823 by Josef Lobmeyr (1792–1855) and carried on after his death by his two sons, Josef (1828–64) and Ludwig (1829–1917). The company was registered in 1860 as J. & L. Lobmeyr. In 1918 Stefan Rath (1876–1960), the son of the sister Mathilde Lobmeyr and August Rath, founded a branch factory at Steinschönau, in Bohemia, called 'J. & L. Lobmeyr Neffe Stefan Rath', where designers included Rath's daughter Marianne (b. 1904), JAROSLAV HOREJC, and Lotte Fink. The Bohemian factory operated until 1945, when it was nationalized by the new régime in Czechoslovakia and is no longer connected with the Vienna company. The Vienna owners, after Stefan Rath, were his son Hans Harald Rath (1904–68) and today the latter's sons, Harald C. (b. 1938), Peter (b. 1939), and Hans Stefan (b. 1943). The factory has done work in HOCHSCHNITT and TIEFSCHNITT, as well as enamelled and iridescent glass, and has made large crystal-glass chandeliers. It has made tableware of fine quality since before 1856, and in the 20th century from designs by ADOLF LOOS, MICHAEL POWOLNY, and Oswald Haerdtl (1899–1959). The factory engaged as artistic director in 1910 JOSEF HOFFMANN who developed the BRONCIT style of glass decoration.

Locke, Joseph (1846–1936). A glass-maker who created many new types and designs of ART GLASS, as well as being an engraver and enameller of glassware. He was born at Worcester, England; at 19 he won a prize for glass design and moved to Stourbridge to work for Guest Brothers and then HODGETTS, RICHARDSON & CO. He studied making CAMEO GLASS, and in 1876 produced a glass copy of the PORTLAND VASE. After working for other English glasshouses, he went in 1882 to the United States and worked for the NEW ENGLAND GLASS CO. at

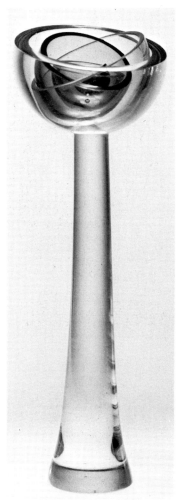

Littleton, Harvey K. 'Eye', made at Val St Lambert, 1972. Ht 35·8 cm. Kunstmuseum, Düsseldorf.

Cambridge, Mass., and then for LIBBEY GLASS CO., at Toledo, Ohio, where he created AMBERINA, then Plated Amberina, and later POMONA, WILD ROSE (PEACH BLOW), AGATA, and MAIZE GLASSWARE. He made a few pieces of cameo glass. In 1891 he moved to Pittsburgh, became head designer for the United States Glass Co., and also opened his own workshop as Locke Art Glass Co. where he continued to make glassware until his death; some pieces are signed 'Locke' or 'Locke Art'.

log glass. A type of SAND-GLASS, used on ships, that ran for 14 seconds or 28 seconds, and was used to time the sailing speed by the length of line paid out while the sand was running through.

Lomonosov, M. V. (d. 1765). A Russian glass technician who was commissioned by the Government in 1751 to teach the making of coloured glass. His pupil, Peter Druzhinin, continued his experiments for about twenty years. Lomonosov established his own factory in 1754 in St Petersburg where, until the factory was closed soon after his death, SMALTO in many different colours was produced. He also established a studio that operated from c. 1754 to 1769 to make MOSAICS, producing about forty important, and some huge, pieces, using brilliant colours rather than imitating oil painting. *See* RUSSIAN GLASSWARE.

looped trailing. A term used to describe decoration in the form of a series of vertical looped THREADS extending above the rim of a vessel, e.g., a SWEETMEAT GLASS. As the loops are not applied to the surface of the piece, the term THREADING is to be preferred.

Loos, Adolf (1870–1933). A Viennese architect who designed industrial products, and in 1921 designed a set of drinking glasses later produced by J. & L. LOBMEYR.

Lorraine method. *See* BROAD GLASS.

lost-wax process. *See* CIRE PERDUE.

Lotharingian glassware. Glassware made in the Rhineland, and especially at Cologne, including early ROMAN GLASS and later WALDGLAS. (Lotharingia is the former name of Lorraine, and it originally included modern Alsace-Lorraine, the Low Countries, and part of north-west Germany.)

lottery glass. A drinking glass engraved with a scene relating to a lottery and an inscription of good wishes. Lotteries were used in the mid-18th century as a means of selling the wares of the smaller German porcelain factories. Such glasses, made at Newcastle-upon-Tyne, England, were engraved in the Netherlands.

lotus-bud cup. A type of CUP that is shaped like a lotus-bud or an inverted bell. They were made during the latter part of the XVIIIth Dynasty (c. 1400–1320 BC) and were among the first artistic objects made of EGYPTIAN GLASS.

Lötz glassware. Glassware made at the factory founded in 1836 at Klostermühle, in Bohemia, by Johann B. Eisenstein, and acquired c. 1840 by Johann Lötz (1778–1848). After Lötz's death the factory was continued by his widow Suzanne (under the new name 'Johann Lötz Witwe'; German, *Witwe* = widow). His nephew, Max Ritter von Spaun, managed the factory from c. 1870 and was responsible for a great increase in the quantity and quality of its production. From the 1890s until c. 1900, it made ART GLASS imitating semi-precious stones, e.g., agate and aventurine. It obtained a patent for making IRIDESCENT GLASS (similar to the later FAVRILE GLASS of LOUIS COMFORT TIFFANY) and exported much of such glassware in ART NOUVEAU style to the United States. After c. 1900 it made black-and-white glassware until it suspended operation c. 1914. It was revived after World War I but

lottery glass. Bowl of wine glass, depicting lottery wheels, Newcastle glass, Dutch engraved. Ht 20·3 cm. Cinzano Glass Collection, London.

Lötz glassware. Bowl with iridescent decoration, Lötz Witwe, Klostermühle, Bohemia, c. 1900. Ht 9 cm. Kunstmuseum, Düsseldorf.

loving cup. Stem with hollow knop enclosing 1816 coin, probably New England Glass Co., *c*. 1825–50. Ht 24·7 cm. New Orleans Museum of Art (Billups Collection).

destroyed by fire in 1932. Some pieces made for export after 1891 are signed 'Lötz-Austria', and some for domestic sale 'Lötz Klostermühle' with crossed arrows in a circle.

loving cup. A large (8-oz. capacity or more) ornamental drinking vessel with two or more HANDLES and, usually, a stemmed FOOT, to be passed around at a banquet or similar gathering and from which a number of persons drink in turn.

low relief. RELIEF decoration where the projecting portion is only slightly raised, as on a coin. The French term is *bas relief*; the Italian is *basso rilievo*. The German term *Hochschnitt* embraces all degrees of decorative relief work, when executed on the wheel, above the surface of the object. *See* HIGH RELIEF; MEDIUM RELIEF.

Lubbers process. A process of making SHEET GLASS by drawing a cylinder of molten glass, up to 50 feet (over 15 m.) long and 3 feet (90 cm.) wide, from a tank of molten glass by an iron ring to which the glass adheres. It was developed in the United States, and superseded the initial stage of making BROAD GLASS by blowing a cylinder of glass.

Lucas, John Robert (1754–1828). The founder in 1788 of a glasshouse at Nailsea, near Bristol, England, who was a bottle-maker reputed to have sought to escape some of the Excise Duty by making common domestic vessels in bottle-glass which was taxed at a lower rate. The factory continued until 1878, having been under the management of ROBERT LUCAS CHANCE from 1810 to 1815, before he took over in 1824 the SPON LANE GLASSWORKS. The factory has been found by documentary and archaeological evidence to have produced only window glass, but none of the glassware commonly attributed to Nailsea. *See* NAILSEA GLASS HOUSE; NAILSEA GLASSWARE.

luck. An object on which the prosperity of a family, etc., is supposed to depend. The best-known glass example is the LUCK OF EDENHALL.

Luck of Edenhall, The. A famed flared BEAKER of ISLAMIC GLASS made and decorated in Syria (possibly at Aleppo) in the 13th century. It is of clear glass, with coloured enamelled decoration of finely drawn and stylized foliate and geometric forms in Islamic style, in white, blue, and red. It is accompanied by an English cut-leather case dating from the 14th century. It has been stated that it was probably brought to England from Palestine by a Crusader toward the end of the 14th century. It was long in the possession of the Musgrave family, of Edenhall, near Penrith, Cumberland. The legend is that it was left by fairies on St Cuthbert's Well, a spring at Edenhall, the luck of the family to be dependent upon its safe-keeping. When one of the family seized it from the fairy king, he warned, 'When this cup shall break or fall/Farewell to the luck of Edenhall'. The name was used at least as early as 1785 and published in 1791. It was also used in a ballad, *Glück von Edenhall*, by the German poet Johann Ludwig Uhland (1787–1862) and again as the title of a ballad written in 1826, as well as a ballad so-named and written by Henry Wadsworth Longfellow in 1842. The beaker has been owned by the Victoria and Albert Museum since 1958. A similarly shaped and enamelled beaker (also with a leather case) is in the Kunstgewerbe-museum, Cologne. *See* LUCK; ENAMELLING, ISLAMIC; ALEPPO GLASSWARE.

Ludlow bottle. *See* CHESTNUT FLASK.

lug. An ear, knob, solid scroll or other attachment for lifting or holding an object, but not including a HANDLE (which has space for the fingers or hand to be passed through). Some lugs have a small opening for a suspensory strap, e.g. on a PILGRIM FLASK.

Luck of Edenhall. Enamelled beaker, Syrian (probably Aleppo), *c*. 1250. Ht 15·7 cm. Victoria and Albert Museum, London.

Luise glass. A drinking glass painted *c*. 1810 by SAMUEL MOHN which shows an acrostic by depicting five flowers, the initial letters of the German names spelling 'Luise'.

luminor. A type of night light made of clear glass in the form of an ornament to be placed on a stand concealing an electric light that provided transmitted light from below. Examples were designed by FREDERICK CARDER at the STEUBEN GLASS WORKS in the 1920s; a frequent Steuben form was a bird, made of moulded glass, with the feathers and details added by cutting. *See* BUBBLE BALL.

Lundgren, Tyra (1911–   ). A Swedish potter who from 1934 to 1940 designed glassware for VENINI, notably a glass dish in the form of a leaf with enamelled veining.

Lundin, Ingeborg (1921–   ). A glass-designer at ORREFORS GLAS-BRUK since 1947. She favours simple lines, using at times shallow engraving and a few colours to create original designs.

lustre. (1) A glass GIRANDOLE, CANDELABRUM or CANDLESTICK decorated with prismatic pendent DROPS of glass or crystal. *See* VASE LUSTRE. (2) A pendent drop of glass or crystal, smooth or FACETED, and in various forms, e.g., icicle (*see* ICICLE GLASSWARE), pear-shaped, etc.

lustre painting. Painting on the surface of glass with metallic oxide pigments, comparable to lustre painting on pottery and producing the same metallic iridescent effect. It has been used as a ground completely covering the piece or as a design or pattern. Oxides of gold, copper or silver, and later of platinum or bismuth, were dissolved in acid and, after being mixed with an oily medium, were painted on the glass; firing in a REDUCING ATMOSPHERE, smoky and rich with carbon monoxide, caused the metal to fuse into a thin film, producing a non-palpable evenly distributed metallic flashing. Gold yielded a ruby colour, platinum a silver effect, and silver a straw-yellow. This style of decoration appears on ISLAMIC GLASS from the 9th to the 11th centuries; most specimens have been found in and are attributed to Egypt. In modern times, lustre painting has been used on the FAVRILE glassware of LOUIS COMFORT TIFFANY, as well as LÖTZ GLASSWARE and glassware of FREDERICK CARDER at STEUBEN GLASS WORKS. Such lustre painting must be distinguished from natural IRIDESCENCE. *See* STAINING.

Lütken, Per (1916–   ). A Danish glass-designer who succeeded JACOB E. BANG as chief designer at Holmegaards Glasvaerk in 1942 (*see* KASTRUP & HOLMEGAARDS GLASSWORKS). He has made objects with simple lines and occasional CASING in another colour.

Lutz, Nicholas. A skilled glass-blower, born at St-Louis, in Lorraine, France, where at 10 he worked as an apprentice at the ST-LOUIS FACTORY. Soon after arriving in the United States, he worked at the BOSTON & SANDWICH GLASS CO., and when it closed in 1888, he went to the MT. WASHINGTON GLASS CO., and later to the UNION GLASS CO. He made glassware in the Venetian FILIGRANA style, some pieces with added ornamentation of PRUNTS moulded with a cherub's head; he also made PAPER-WEIGHTS, particularly types with fruit and flowers. *See* STRIPED GLASSWARE.

Lycurgus Cup. A glass CUP tapering slightly inward toward the base and depicting, in VASA DIATRETA technique and in HIGH RELIEF, in an encircling frieze, five figures who have been identified as representing the myth of the punishment of Lycurgus by Bacchus (Dionysus). The glass (soda-lime type) is normally dull pea-green with an area of yellow-green, but it appears wine-red when viewed by transmitted light (also faintly so in direct light), the change being attributed to the presence of a tiny percentage (0·003/0·005%) of colloidal gold. Recent analysis of the

luminor. Colourless cut glass eagle designed by Frederick Carder, late 1920s, for Steuben Glass Works. Ht 20 cm. Corning Museum of Glass, Corning, N.Y.

Lycurgus Cup. Roman glass with
diatreta decoration; colour opaque
green, but wine-red when viewed in
transmitted light; 4th century AD (with
later silver rim. Ht 16·5 cm. British
Museum, London.

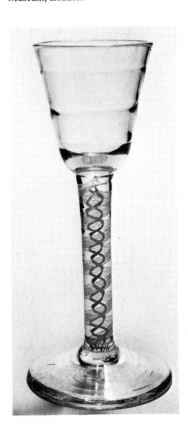

Lynn glassware. Wine glass with
ribbed rounded funnel bowl,
*c.*1745–80. Ht 14 cm. Norfolk
Museums Service (King's Lynn
Museum).

glass shows traces of silver and other minerals, which may contribute to the dichroic effect. The cup is of ROMAN GLASS of the 4th century. It has a metal rim, affixed at a later date, which is believed to conceal some chips; it is surmised that originally the cup was higher. Before the cup was purchased in 1958 by the British Museum, it had an attached modern metal base which has since been removed to reveal the remnant of the original foot in open-work, forming a small base-ring. *See Journal of Glass Studies* (Corning), V (1963), pp. 9, 18. *See* CAGE CUP; DICHROMATIC GLASS; TRANSMUTATION COLOUR. [*Plate VI*]

Lynn glassware. Drinking glasses (usually WINE-GLASSES or TUMBLERS) and DECANTERS that are characterized by several (two to eight) contiguous bands of encircling very shallow horizontal RIBBING. Such rings on decanters are wider and fewer than on glasses. The rings were probably produced by the PUCELLAS in the making. Such ware was attributed in 1897, without any stated evidence, by Albert Hartshorne (*Old English Glass*, pp. 251, 278) to KING'S LYNN, and the pieces have since been generally referred to as 'Lynn glassware'. They have sometimes been also attributed to nearby Norwich (hence sometimes called 'Norwich glassware') but there is no record of any glasshouse ever being there; a possible explanation is a warehouse there that was owned by a King's Lynn glasshouse. It has been suggested by Francis Buckley, *Old English Glass* (1925), that such ware may have been made at Yarmouth, near King's Lynn, *c.*1728–58.

Lysle, Anthony de. A French independent pewter- and glass-engraver, said to have come from France to London, *c.*1582, and to have possibly done some of the DIAMOND-POINT ENGRAVING on certain of the GOBLETS attributed to the factory of JACOPO VERZELINI, especially two that bear the engraved motto of the Pewterers' Company. A goblet dated 1578, found at Poitiers, France (now in the Cluny Museum, Paris), and bearing the engraved fleur-de-lis of France has also been thought to have been decorated by him. *See* CLUNY GOBLET; VERZELINI GLASSWARE.

M

magnifying lens. A polished glass convex lens that diffuses the light and enlarges the image; the opposite effect is produced by a DIMINISHING LENS. They are used in VISION AIDS (e.g. spectacles), magnifying glasses, microscopes, binoculars, telescopes, etc.

Maigelein (German). An early type of small hemispherical CUP, resting on a base with a high pointed KICK (this is lower on later examples). A few are undecorated, but usually they have vertical or spiralling ribs, and most have spiralling ribs running in opposite directions and forming a network pattern (*Kreuzrippenmusterung*) of thick or fine ribbing. They were MOULD-BLOWN, the process having been to blow the molten glass into a ribbed mould and then twist the glass; and for network patterns, to blow into a second mould and then twist again, this time in the opposite direction. They were made of WALDGLAS in the 15th and 16th centuries.

maize glassware. A type of ART GLASS, first developed by JOSEPH LOCKE for the NEW ENGLAND GLASS CO., which continued to be made at the LIBBEY GLASS CO. It was patented in 1889. The body of the object is of opaque white or cream colour, with an overall moulded pattern of vertical rows of maize (sweetcorn), with some wavy tips of husks in green or brown (sometimes red or blue) rising from the bottom of the object.

maize glassware. Jug (pitcher) of opaque white glass with blue and gilded leaves, New England Glass Co., *c.* 1888/89. Ht 22 cm. Toledo Museum of Art, Toledo, Ohio.

mallet flask. A type of FLASK in the shape of a mallet, with a hexagonal body and a long thin vertical neck. An example has been found at Pompeii.

Maltese Cross bottle. A type of SQUARE BOTTLE or HEXAGONAL BOTTLE whose sides are decorated in relief, one side having a Maltese Cross which sometimes is placed above a relief column or altar. They, together with comparable glass vessels with Christian or Jewish symbols, have been attributed to Syria, 4th/5th centuries, but more recently to Jerusalem in the 6th/7th centuries. *See* Dan Barag, 'Glass Pilgrim Vessels from Jerusalem', *Journal of Glass Studies* (Corning), XII (1970), p. 35. *See* CHRISTIAN GLASSWARE.

mallet flask. Roman glass, found at Pompeii, *c.* 1st century AD.

Mal'tsev Glass Factory. A former Russian privately owned glasshouse near Moscow, founded in 1724 by Vasilii Mal'tsev. When it was ordered to be closed in 1747 to prevent fires in the Moscow vicinity, his heirs opened two glasshouses near Gusev and in 1792 they founded a third one at Diat'kov where management of all three operations was moved in the mid-19th century. It is believed that it produced mainly utility glassware, but also special quality pieces. Later in the 19th century the factories were separated, that at Diat'kov going to S. Ivan Mal'tsev, son of Akim Mal'tsev, and those at Gusev to I. S. Mal'tsev. Both factories changed from making enamelled glass to cut crystal glass, but towards the end of the 19th century they again made enamelled glass in Russian style, some from designs by the artist Elisabeth Boehm. *See* RUSSIAN GLASSWARE.

Mandarin Yellow. A type of ART GLASS developed *c.* 1916 by FREDERICK CARDER at STEUBEN GLASS WORKS; it is translucent and coloured a monochrome yellow – inspired by the yellow Chinese porcelain of the Ming dynasty (1368–1644). It was very fragile, due to internal stresses, so that only a small quantity was made. Other hues of yellow used by Carder for transparent glass included amber and topaz. *See* CHINESE GLASSWARE; YELLOW GLASS.

Mano sul Seno, La. Designed and made by Alfredo Barbini, 1948. W. 25 cm. Cenedese glassworks, Murano.

manganese. A metallic agent, used as manganese oxide (pyrolusite) in colouring glass, producing amethyst or purple hues. It was used in medieval times to produce pale rose-red to pink colours. When combined with COBALT, it produced a black glass (*see* HYALITH GLASS). It was also used as a DECOLOURIZING AGENT, to make colourless transparent glass by counteracting the greenish or brownish colour imparted to untreated glass by traces of iron (present as an impurity in the SILICA in the BATCH), functioning as a chromatic neutralizer. It was used by the Romans in the 1st century AD; manganese and ANTIMONY were then apparently used contemporaneously, the former superseding the other in the 5th century AD. The Venetians perhaps rediscovered its use in the 15th century when they made CRISTALLO that approached the colourlessness of natural crystal. When used as a decolourizer it is called 'glass-maker's soap'.

Mano sul Seno, La (Italian). Literally, hand on breast. A glass ornament in such form designed and executed by ALFREDO BARBINI in 1948 for the CENEDESE glassworks, Murano.

Mansell, Sir Robert (1573–1656). An English admiral who, after he retired, acquired in 1617 an interest in the glasshouse of Sir Edward Zouch. In 1615 the use of wood as fuel for glasshouses was prohibited by Royal Proclamation, and Mansell foresaw the potential of coal as a fuel. He built, *c*. 1617, a new glasshouse (*see* BROAD STREET GLASSHOUSE). He bought out his partners by 1618 and obtained in 1623 a patent granting him a monopoly for all sorts of glassware made by the use of coal, which he maintained, against many attacks, until 1642. He employed many Italian workmen, including ANTONIO MIOTTO, PAOLO MAZZOLA, and from 1630 some craftsmen from Mantua, including the DAGNIA FAMILY. He made flat glass for mirrors and spectacles, bottles, lenses, etc. He was primarily an industrialist and financier, rather than a glass-maker, and he acquired glasshouses throughout England and Wales, and eventually also in Scotland, and attempted to start one in Ireland.

marble, glass. A small ball of glass. They are made industrially of (a) colourless glass for making GLASS FIBRE, and (b) decorated glass (with coloured TWISTS) or coloured glass for playing the game of marbles. The process for making marbles, used from the early 16th century, is to break a glass rod into small short pieces, then place them in a mixture of sand and charcoal in an iron vessel that is heated and rotated, so that the softened edges are rubbed off the pieces, which are then smoothed and polished. *See* BALL-STOPPER.

marbled glass. (1) Glass decorated with haphazard streaking of two or more colours to resemble marble. It was especially made in Venice in the 15th–17th centuries. *See* AGATE GLASS; ONYX GLASS; LITHYALIN GLASS; CALCEDONIO. (2) A name used sometimes in the United States for SLAG GLASS.

margarite (Italian). An ancient (and now obsolete) term for CONTERIE.

Marinot glassware. Flask, colourless glass with embedded air bubbles, by Maurice Marinot, 1931. Ht 12·6 cm. Victoria and Albert Museum, London.

Marinot, Maurice (1882–1960). A French artist, of Troyes, who began glass-making in 1911 and became an outstanding studio glass-maker, designing, making, decorating and signing all his pieces. He first made bottles, bowls, flasks, etc., in glass coloured in the style of the Fauve painters, but from 1923 he began making pieces in massive and abstract form, with engraved, cut and acid-etched abstract patterns and with interior decoration of streaks of opaque glass of different colours and of gold specks, and especially with innumerable tiny air bubbles embedded in the glass. From 1925 until, for reasons of health, he retired in 1937 from glass-making, he was highly acclaimed. His studio, where he had resumed painting, was destroyed, with much of his glassware, by bombing in 1944.

marks. Identifying marks by glasshouses, glass-makers, and decorators, used much less often on glassware than on ceramic ware. Identification is more often by the METAL and the style of the piece and the decoration. But marks have been used from ancient to modern times. *See* FACTORY MARK; GLASS-MAKER'S MARK; CHINESE REIGN-MARK.

marqueterie de verre (French). A type of decorative glass developed by EMILE GALLÉ, *c.* 1897, that was inspired by marquetry decoration in wood. Shaped pieces of hot glass were pressed into the body of a glass object of contrasting colour, so that the resulting surface is flat. The inlaid pieces may be further decorated by engraving or carving, and sometimes there is added decoration of applied glass.

marriage glass. A drinking glass made and decorated to commemorate a marriage, with appropriate decoration and inscriptions. English examples, made from the 16th century, are often in the form of a GOBLET (*see* DIER GLASS; ROYAL OAK GOBLET) or a FLUTE. Bohemian and German examples are generally in the form of a HUMPEN or a STANGENGLAS (sometimes in pairs). Some French *verres de mariage* are extensively decorated with engraved inscriptions and the names and coats-of-arms of the couple. Some rare Venetian examples, in the form of a goblet, have enamelled decoration depicting Biblical subjects and portrait heads of the betrothed or of their coats-of-arms. The Italian term is *coppa nuziale*. *See* WEDDING CUP.

Martinuzzi, Napoleone. An Italian sculptor who designed glassware for VENINI from 1928, and also for Vetraria Vistosi of Murano.

marver. A polished iron or marble table upon which MOLTEN GLASS, gathered on the end of an iron tube (BLOW-PIPE) or rod (PONTIL), is rolled into a globular or cylindrical symmetrical mass. The process is known as 'marvering'. As the PARAISON is very much cooler than when it was attached to the blow-pipe or the pontil, it does not adhere to the marver.

mascaron. A decorative motif in the form of a mask, e.g., a lion's-head mask, found on glassware made in Venice or elsewhere in the FAÇON DE VENISE, including that made in England by JACOPO VERZELINI.

mascot. A decorative object to be placed at the front of the bonnet (hood) of an automobile. Some were made of glass, *c.* 1908, to indicate the make of the automobile or to be merely decorative, especially the individually-produced pieces made by LALIQUE in the mid-1920s with birds or figures mounted on a base or raised above an electric light (some with coloured filters that occasionaly revolved).

mask. A representation, usually in RELIEF, of a face as decorative ornament. Masks were originally derived from ancient Greek and Roman sources; often grotesque, they depict human beings, satyrs, and animals. Masks occur in two forms: three-dimensional as ornamental ware, and in low relief as decorative ornament. In glassware, early examples in PÂTE-DE-VERRE were made in Egypt. In Roman times they occur on the front of VASES or at the lower terminal of handles. Later Venetian glassware sometimes has a mask on a PRUNT. *See* MASCARON.

Masonic glassware. Various glass objects decorated with Masonic emblems, insignia, and inscriptions, such as FIRING GLASSES and German BEAKERS. In the United States many glass FLASKS were decorated with various Masonic devices and emblems.

mastos (Greek). Literally, breast. A hemispherical drinking cup in the form of a woman's breast; it could be rested only on its rim. The Greek pottery examples were copied in glass, frequently with interior cut lines.

marqueterie de verre. Vase by Emile Gallé, Nancy, *c.* 1898, signed 'Gallé'. Ht 44 cm. Kunstmuseum, Düsseldorf.

mastos. Drinking cup of yellowish-green glass with blue tip, Roman glass, 4th/5th century. Ht 12·5 cm. Wheaton College, Norton, Mass.

measuring glass. Probably made in Germany, 19th century. Ht 16·6 cm. Kunstgewerbemuseum, Cologne.

measuring spoon. Pressed glass, *c.* 1910, L. 10·2 cm. Collection of Dr John W. Scott, Toronto.

mat (or **matt**). Having a dull finish; not glossy. To produce this effect on glassware several methods are used, e.g. the use of HYDROFLUORIC ACID or unpolished shallow engraving. *See* MATTSCHNITT.

Matagama bead. A type of BEAD found only in Japan and so presumably made there during the 3rd/6th centuries. It is shaped like a comma. Although such beads were mainly made of stone, examples in glass are known. They were probably used for necklaces.

Matsu-no-ke. A style of decoration developed by FREDERICK CARDER at STEVENS & WILLIAMS LTD in the 1880s and used by him from 1922 at STEUBEN GLASS WORKS. It depicts gnarled branches of an ornamental pine tree (*matzu* in Japanese), either in clear crystal applied to the exterior of some ornamental objects with handles of various transparent colours or in iridescent surface decoration.

Mattoni, Andreas Vincenz Peter (1779–1864). A glass-engraver who founded a school at Karlsbad, Bohemia. Among his pupils were Anton Heinrich Pfeiffer (1801–66), Rudolf Hiller (1827–1915), Anton Rudolf Dewitte (1824–1900), and LUDWIG MOSER (1833–1916). In turn, a pupil of Pfeiffer was Anton Urban (1845–1909).

Mattschnitt. Literally, mat engraving. The German term for WHEEL-ENGRAVED decoration in intaglio (TIEFSCHNITT) that is not polished and has a MAT finish, in whole or in part.

Mauder, Bruno (1877–1948). A glass-maker, born in Munich, who taught glass-making at Zwiesel, in Bavaria. He departed from the ART NOUVEAU style to make glassware in naturalistic shapes, and exerted great influence on modern German glass-making. He also experimented with producing CRYSTALLINE textures and frosting on the surface of glass by the use of HYDROFLUORIC ACID.

Mäuerl, Anton Wilhelm (1672–1737). An artist of Nuremberg who is notable for ENGRAVING glassware, especially in the use of LAUB- UND BANDELWERK, *c.* 1719. He was in England from 1699 to 1710, perhaps doing engraving.

Mazzola, Paolo (fl. 1640–55). An Italian glass-worker who was brought from Murano to England by SIR ROBERT MANSELL and worked for him from *c.* 1640 to *c.* 1655, by which date he had moved to Liège. He is said to have been a maker of WINGED GLASSES.

mead glass. A type of English drinking glass for drinking mead (a fermented beverage, popular in England and elsewhere in the 17th and 18th centuries and later, made of honey and water, with yeast and sometimes added sherry or various flavouring ingredients). The form of the glass is controversial, but the type traditionally but dubiously associated with mead drinking has a shallow globular BOWL with a slightly inverted RIM, set on a stemmed FOOT; early examples have a baluster stem, later ones an air-twist stem (*see* STEM). Another type was like a TUMBLER with a KICK BASE.

meander. *See* KEY-FRET.

Measey, Michael. A member of the GLASS SELLERS' COMPANY. *See* GREENE, JOHN.

measuring bottle. A type of bottle whose capacity is a specified quantity of liquid. Some, of circular or polygonal section with a high neck and a funnel-shaped or a PAN-TOPPED mouth, were made, in Normandy and southern France in the 18th century, in several sizes for measuring oil and wine.

measuring glass. A graduated glass for measuring liquids. They have been made in many shapes and sizes, generally in the form of a TUMBLER or a conical receptacle, and some with a stemmed FOOT and pouring lip. They have been made probably at least from before the 19th century. The German term is *Messgefäss*.

measuring spoon. A type of SPOON marked with gradations for use in measuring medicines, etc. They have a short curved handle on which they rest. *See* MEDICAL SPOON.

Medaillonbecher (German). Literally, medallion beaker. A type of Bohemian cylindrical DRINKING GLASS decorated in a style related to ZWISCHENGOLDGLAS and ancient Roman FONDI D'ORO. The technique was developed in Bohemia and perfected by JOHANN JOSEF MILDNER in the 1780s; signed examples are known from 1787 to 1808. He made a double-wall TUMBLER or BEAKER of colourless glass or OPAQUE WHITE GLASS, cut a circular recess into the outer wall, and inserted flush with the surface a MEDALLION (decorated on the reverse in engraved gold or silver leaf, and also red lacquer) depicting miniature scenic views, still life, figures of saints, monograms or portraits. Sometimes he decorated parchments similarly and enclosed them in the glass recess. He signed the reverse of the medallions, adding the name of the sitter for a portrait and the date. He sometimes added a motto, verse, or dedication (or another medallion) on the other side of the glass, and also another medallion in the base. *See* ÉGLOMISÉ; GOLD ENGRAVING.

medallion. (1) A flat thin tablet, of glass or ceramic ware, usually circular or oval, bearing a portrait, design, or inscription, which may be in gold leaf, wheel-engraved, painted or in RELIEF. They were used principally for personal wear or cabinet display, as distinct from a PLAQUE used mainly for wall or furniture decoration. *See* TASSIE MEDALLION; SULPHIDE; PHALERA; MEDAILLONBECHER. (2) A similar glass object, used from Roman times and stamped as a weight for measuring money or goods, or applied to a vessel as a guarantee of a certain capacity. In England and elsewhere in the 17th–19th centuries comparable medallions were used as seals on bottles. *See* SEAL, BOTTLE.

medical glassware. Various objects of glass used by infants or sick persons, or for preparing, containing or dispensing medicines in pharmacies or hospitals. See PHARMACY GLASSWARE; ALBARELLO; FEEDING BOTTLE; FEEDING CUP; DRUG JAR; MEASURING GLASS; MEASURING SPOON; MEDICAL SPOON; MEDICINE BOTTLE.

medical spoon. A SPOON having, instead of the usual long handle, a loop-handle whose terminal curls under the bowl of the spoon and serves as a foot so that when filled the spoon can rest horizontally without tipping sideways. *See* MEASURING SPOON.

Medici Papal glassware. Various glass objects, e.g., dishes, TAZZE, and EWERS, decorated in enamelling and gilding with the coat-of-arms of one of the Medici Popes, probably either Leo X, 1513–21, or Clement VII, 1523–34. They were made of Venetian glass, *c.* 1513–34, and it has proved impossible to determine for which Pope they were made.

medicine bottle. A small bottle, made for use by an apothecary. They were made in Germany in various sizes, shapes, and styles from the 16th century. The German term is *Arzneiflasche*. *See* POISON BOTTLE.

medium relief. RELIEF decoration of a height from the surface intermediate between HIGH RELIEF and LOW RELIEF. The French term is *demi-relief*; the Italian is *mezzo rilievo*. The German term *Hochschnitt* embraces all degrees of wheel-engraved relief work.

Meges. A glass-maker who made MOULD-BLOWN glassware in Syria, 1st century AD. Two BEAKERS are known with his name moulded on them, similar to beakers similarly shaped and marked NEIKAI(O)S or JASON.

Medaillonbecher. Beaker with portrait of J. von Fürnberg, once owner of Gutenbrunn, painted in colours on parchment, Johann Mildner, Gutenbrunn, 1792. Ht 11·5 cm. Museum für Kunst und Gewerbe, Hamburg.

medallion. Engraved portrait of Count Karl-Friedrich von Sachsen-Weimar, Dominik Biemann, Bohemia, *c.* 1835. Ht 13·3 cm. Kunstmuseum, Düsseldorf.

Meisterstück. Colourless glass with coloured and gilt trailing and with cresting, Cologne, 3rd century AD. Ht 27 cm. Römisch-Germanisches Museum, Cologne.

merese. Detail of goblet with bowl wheel-engraved in *intaglio* (*Tieftschnitt*) and with groups of mereses on stem, Georg Friedrich Killinger, *c.* 1694–1726. Museum für Kunst und Gewerbe, Hamburg.

Meisenthal, Verrerie de. A glasshouse at Meisenthal, in Lorraine (now in France), which made glass in ART GLASS style. It was also known as Burgun, Schverer & Cie. It sold blanks to Désiré Christian & Sohn, of Meisenthal, which decorated in Art Glass style, and which used its own name as its mark. These factories are sometimes further confused by references to the 'Vereinigten Glashütten von Vallerysthal', in a nearby town in Lorraine, and which used the mark 'Vallerysthal'.

Meisterstück (German). Literally, masterpiece. A footed PILGRIM FLASK with SNAKE TRAILING, made at Cologne in the 3rd/4th century AD, so-called by the expert Fritz Fremersdorf because of the outstanding quality of its workmanship. The flat sides are decorated with trailed leaf motifs in gilt and garlands in multi-coloured glass. The edges and handles are decorated with pincered CRESTING. Other pilgrim flasks of like form but slightly different decoration are sometimes so-called as a group, including particularly one dating from the 3rd/4th century found in the Netherlands.

Melort, Andries (1779–1849). A Dutch engraver, of Dordrecht and The Hague, who engraved flat sheets of glass with STIPPLING, copying works by Dutch painters, *c.* 1838. His son, S. J. Melort, was also a glass-engraver.

membrano, vetro (Italian). A type of ornamental glassware developed by VENINI *c.* 1970, of which the characteristic feature is that the object (e.g. a large bowl or vase) is divided part-way internally into two (sometimes four) sections by means of a thin glass partition. Imitations have been made by pushing up from the bottom a partitioning narrow cone into the body of the piece.

memento mori (Latin). Literally, remember you must die. A motif used as decoration on various objects, occasionally made of glass, in the form of a reminder of mortality, e.g. a death's-head, or as a shape (e.g. a coffin) for some small object such as a SNUFF-BOX.

menu-card holder. A small decorative object with a slot for holding a menu-card or place-card. Some were made of glass by LALIQUE.

Menzel, Johann Sigismund (1744–1810). A glass-worker at WARM-BRUNN in Silesia who, in the late 18th and early 19th centuries, made objects in ZWISCHENGOLDGLAS. He also decorated goblets with inserted medallions with silhouette portraits in black; *see* MEDAILLONBECHER.

mercury band. A thin air thread, of silvery appearance, encircling some PAPER-WEIGHTS near the rim and usually on both sides of a TORSADE, especially found on a mushroom weight made by BACCARAT or ST-LOUIS.

Mercury bottle. A MOULD-BLOWN coloured-glass BOTTLE of square section and with a high thin neck and a flat disk-shaped mouth. The bottom of some examples has relief decoration depicting the god Mercury, and sometimes also bears the initials of the glass-maker or symbols or initials indicating the contents. They were made in the 1st–4th centuries AD. *See* SIDONIAN GLASSWARE.

mercury glass. The same as SILVERED GLASS.

merese. A type of COLLAR, usually with a sharp edge and placed between the BOWL and the STEM, or the stem and the FOOT, of a WINE GLASS, although they sometimes are found along the stem in several pairs or sometimes in groups of up to ten.

Merovingian glassware. *See* FRANKISH GLASSWARE.

Merry-Go-Round, The. A large covered BOWL resting on scrolls on a heavy BASE, and having an ENGRAVED encircling decoration depicting a merry-go-round. It was designed in 1947 by SIDNEY WAUGH and made by STEUBEN GLASS WORKS, to be presented by President and Mrs Harry Truman to Queen Elizabeth II (then Princess Elizabeth) on the occasion of her marriage to Prince Philip in November 1947. The engraving was by JOSEPH LIBISCH.

merrythought decoration. A style of decoration of trailed glass in the form of a merrythought or wishbone, as sometimes used on FRANKISH GLASSWARE.

Mesopotamian glass. Glass made in the region north-east of Syria and west of Persia, between the Tigris and Euphrates rivers. Evidence has been found indicating the making of glass here by 2000 BC, and fragments of glass vessels found at sites near Nineveh have been attributed as probably Mesopotamian, *c.* 1525–1475 BC, i.e. before the earliest glass vessels attributed to Egypt, *c.* 1470. It is probable that glass-making continued here until *c.* 1150 BC and then the industry revived in the 9th century under the influence of the Phoenician traders, but lapsed again in the 4th century until several centuries later. According to recent finds, MOSAIC glass was made in Mesopotamia from the 15th century BC. *See* GLASS: *pre-Roman history.*

Messgefäss (German). The German term for a MEASURING GLASS.

metal. The fused material, in molten or hard state, made up of various essential ingredients (e.g. SILICA and ALKALI), from which glass is made. The metal produces glass of varying characteristics depending on the quality and proportions of the ingredients.

metal-cased glassware. A glass object encased in a metal covering, usually of gold or silver. Some such drinking vessels are known from the 6th century BC. A glass ALABASTRON, from the 1st century BC, was encased in silver with *repoussé* decoration; today the worn-off metal reveals the glass on the lower quarter of the piece. *See* SILVER-CASED GLASSWARE.

metallic oxide. The oxide of a metal employed as a pigment in imparting a colour to glass and in making ENAMEL COLOURS to decorate the surface of glass. Oxides for surface decoration were suspended in an oily medium to make them more suitable for application with a brush, and the actual colours were developed during firing. The oxides most commonly used are those of ANTIMONY, COBALT, COPPER, IRON, and MANGANESE; also used are CHROMIUM, GOLD, NICKEL, SELENIUM, SILVER, TITANIUM, and URANIUM, and TIN OXIDE is used as an opacifying agent. The resultant colour from any metal oxide depends on the nature of the glass itself and the purity of its ingredients and also on the furnace conditions, as to the degree of heat or the existence of a REDUCING ATMOSPHERE.

methylated spirit lighter. A bottle having an iron plunger attached to the STOPPER and forked at the other end to hold a wad of cotton wool, the wool to soak in methylated spirit in the bottle and to be touched to a flame by the plunger and used as a taper to light a cigar or a pipe. They were made in England, *c.* 1870–90.

Meydam, Floris (1919–). A glass-designer at the Leerdam glassworks (*see* LEERDAM GLASSWARE). He is known for his original dishes of free form and varied colours.

Meyer, Johann Friedrich (1680–1752). An artist in Dresden and the Court enameller, who decorated glass with OPAQUE ENAMEL that was noticeably palpable.

Mildner, Johann Josef (1763–1808). An Austrian glass-cutter from Gutenbrunn, in Lower Austria, who from 1787 to 1808 revived and

Merry-Go-Round, The. Presentation bowl designed by Sidney Waugh for Steuben, 1947. D. 25 cm. Courtesy, Steuben Glass Works.

Mildner, Johann Joseph. Beaker with medallion, J. J. Mildner, Gutenbrunn, *c.* 1800. Kunstmuseum, Düsseldorf.

millefiori. Bowl of green glass with rope edge, Roman glass, 2nd/3rd centuries, Ht 7·5 cm. Victoria and Albert Museum, London.

millefiori mosaic glass. Bowl with rope edge, having polychrome stripes and coloured twists, Alexandria, 1st century AD. Museo Civico, Adria.

elaborated the technique of making ZWISCHENGOLDGLAS by developing the MEDAILLONBECHER. His glasses have medallion panels and borders decorated on the inner side in gold-leaf and red lacquer, and inserted into spaces cut out to receive them. He did portraits, and also depicted figures of saints, landscapes, and still life, and sometimes did miniature portraits painted in colours on parchment that were inserted in the medallions. Medallions were also inserted sometimes in the bottom of tumblers. He also occasionally used DIAMOND-POINT ENGRAVING.

milk-and-water glass. A type of glass somewhat similar to OPAQUE WHITE GLASS, but semi-opaque and white in reflected light. It was normally opacified by the inclusion of ASHES from calcined bones. It must be distinguished from certain opaque white glass that can look milk-and-white if made with not enough opacifier or if the process was not continued long enough. The milk-and-water glass was made in Bohemia and Germany in the late 17th to early 19th centuries, and was decorated with enamelling. It was also made in England in the early 19th century, and was popular for making COUNTRY-MARKET GLASSWARE. The German term is *Beinglas* (literally, bone glass).

milk glass. The same as OPAQUE WHITE GLASS. The German terms are *Milchglas* and *Porcellein-Glas*; the Italian is *lattimo*; the French is *blanc de lait.* [*Plate IX*]

milk-jug. A JUG or PITCHER of medium size (about 18 cm. high) for serving milk. They were made of glass in many shapes and styles, larger than a CREAM-JUG and smaller than a water-jug.

mill glass. A drinking glass that rests on the rim of its bowl when unfilled and upturned and has attached to the end away from the bowl a small silver windmill. Such glasses included tall FLUTES and short glasses with a rounded funnel bowl. They were made in the Low Countries in the 16th and 17th centuries, and called there *molenbekers*. One unusual example, made as a JOKE GLASS, has attached on the silver mill a dial, which registers the number of times that the glass is emptied by the drinker, and sails that are set in motion by blowing through a glass projecting tube.

milled. Decorated with closely spaced parallel grooves, as around the edge of a coin; such decoration is found around the foot of some glasses and, as added decoration, on some TRAILING.

millefiori (Italian). Literally, a thousand flowers. A style of decorating glass with slices of coloured CANES embedded in clear molten glass, usually in flower-like designs. The style was known at Alexandria in the 1st century BC and at Rome in the 1st century AD, but it was so-called only after it was revived and modified at Venice in the 16th century. (The name has an affinity with *millefleurs*, the French 18th-century term for porcelain from China and Meissen whose decoration is a multitude of tiny painted or moulded flowers.) The Roman technique, fusing in a mould the slices in translucent or opaque glass, was similar to that used earlier in making glass MOSAIC. The Venetian process involved placing the slices of the canes in a mould, in a closely arranged pattern, and then embedding them in a clear molten glass matrix. Venice (or possibly Bohemia) was the first to use the process in making PAPER-WEIGHTS, and the same process was used in mid-19th-century France, as well as in England and the United States, to make *millefiori* paper-weights. *Millefiori* glass is found principally in paper-weights, but also (especially in ware made in England at Stourbridge) in the heavy glass base of some DRINKING GLASSES, VASES, INK-BOTTLES, SCENT-BOTTLES, JUGS, and SHOT-GLASSES, as well as in HAND-COOLERS, the base and bowl of some TAZZE, STOPPERS, and the shank of SEALS. *See* PAPER-WEIGHT, *millefiori* and *mushroom*; MILLEFIORI MOSAIC.

The millefiori technique was used by FREDERICK CARDER, from *c.* 1909 until the mid-1920s, at STEUBEN GLASS WORKS, but not usually to produce the traditional multi-coloured scattered 'thousand-flower' patterns; rather, he used slices of CANES to make a single-flower pattern on some AURENE GLASS vases, but mainly to make geometric designs on plates with regularly placed coloured canes and to make random patterns on bowls with two or more basic colours.

millefiori mosaic. A type of MOSAIC glass made with CANES such as are used for making MILLEFIORI glassware, but having the canes cut lengthwise or diagonally instead of crosswise, so that the appearance of the slices is not of a stylized flower but of mingled stripes of various colours. Sometimes such ware has been included under the term 'millefiori', but as it lacks the characteristic floral appearance of the pieces made with canes cut crosswise, it is deemed preferable to designate it as a special type of mosaic glass. It was made at Alexandria in the 1st century, and later at MURANO.

milleocchi, vetro (Italian). Literally, thousand eyes glass. A style of glass developed in 1954 at the Fabbrica Salir, of MURANO the characteristic feature of which is an incised decoration of numerous adjacent small circles in haphazard patterns; it is used for some tall conical vases. The designer was Vinicio Vianello.

Millville glassware. Glassware made at a glasshouse founded at Port Elizabeth, New Jersey, by James Lee and others, *c.* 1800, and later, *c.* 1806, at a new factory founded by them at nearby Millville. After several changes of management, it was acquired, *c.* 1844, by Whitall brothers and known from 1849 as Whitall Brothers & Co, and then in 1857 as Whitall, Tatum & Co. It was later acquired by the Armstrong Cork Co. Its main production was window glass and glass insulators, but from *c.* 1863 it also made PAPER-WEIGHTS. *See* JERSEY ROSE.

Milton Vase, The. A CAMEO GLASS vase of blue glass overlaid with opaque white glass, of truncated conical shape and with two scroll handles. It is decorated with a carved scene of the avenging angel, Raphael, on one side, and Adam and Eve on the other, inspired by Milton's *Paradise Lost*. The shape was designed by PHILIP PARGETER, and the decoration by Pargeter and JOHN NORTHWOOD, who engraved it and signed and dated it 1878. It was on loan to the British Museum from 1959 until 1975, when it was sold in London at Sotheby's (Belgravia) for £26,000.

miniature. A tiny replica of an object. Glass examples include JUGS, BASKETS, DECANTERS, FLASKS, TEA SERVICES, animal figures, plates of fruit, and many small objects made AT-THE-LAMP. Some were made possibly as salesmen's samples, e.g. some tiny decanters 9 cm. high. *See* TOY; FRIGGER; PAPER-WEIGHT, *miniature*.

Minter Glasshouse. A Russian privately owned glasshouse near Moscow, founded in 1719 by A. Minter. It produced a huge quantity of glass dishes and probably CROWN GLASS window panes. Its closing date is unknown. *See* RUSSIAN GLASSWARE.

Miotti family. A family of glass-makers at MURANO, important principally in the 17th and 18th centuries. The founder was Antonio, recorded in 1542. He was followed by his three sons: Peregrin, Bastian (1543–94), and Alvise (1553–99). Alvise's grandson, Daniele (1618–73) took over the AL GESÙ GLASSWORKS *c.* 1663, and the factory continued under descendants until it closed in the latter part of the 18th century. *See* MIOTTO, ANTONIO. *See* Luigi Zecchin, 'I Muranesi Miotti', *Journal of Glass Studies* (Corning), XIII (1971), p. 77.

Miotto (or Miotti), Antonio (fl. 1616–23). An Italian glass-maker from MURANO (possibly connected with the MIOTTI FAMILY that operated the AL GESÙ GLASSWORKS) who worked in Middelburg, Netherlands, from 1605 until *c.* 1621 when he was brought to England by SIR ROBERT

Milton Vase. Blue and white cameo glass, Philip Pargeter and John Northwood, 1878. Ht 34 cm. Courtesy, Sotheby's Belgravia, London.

MANSELL to manage his Broad Street factory in London. After several years he was led by personal disagreements to leave London and return to Middelburg, and subsequently he worked in glasshouses in Brussels and Namur.

mirror. A looking-glass, used to reflect an image. Mirrors were made for decorative as well as functional purposes, in many forms, including some to be attached to or hung framed on a wall, some for coaches, and some as hand-mirrors (see below). *See* CHIMNEY GLASS; PIER-GLASS; CHEVAL-GLASS; SCONCE; IRISH GLASSWARE. Some were decorated on the surface or on the reverse of the glass. *See* MIRROR DECORATION; MIRROR PAINTING.

Ever since Narcissus looked too long at his reflection in the pool it can be assumed that some sort of mirror was sought. It has been said that mirrors were made in ancient times, and that in Egypt in the 1st century BC mirrors were made of silvered glass. The Romans made some of polished metal, but also some of lead-backed glass. In Germany and Lorraine as early as the 12th century there were mirrors of glass backed with lead, tin or other metal. In Nuremberg convex mirrors were made. In Venice, *c.* 1507, leave was granted to the Dal Gallo brothers of Murano to make glass mirrors 'by a secret process', and soon thereafter mirrors of BROAD GLASS were made by Murano mirror-makers (*specchiai*) who by 1569 formed a guild. From the 17th century, mirrors were in great demand throughout Europe for decorating cabinets and coaches and, especially, large mirrors for wall decoration (as the French mirrors, from *c.* 1678, at Versailles), and competition for Venetian mirrors was great. In France various companies, were formed which were later merged into the present ST-GOBAIN FACTORY. In England mirrors were made from 1621 by SIR ROBERT MANSELL and by 1663 by the DUKE OF BUCKINGHAM. In France in 1691 the method of making CAST GLASS was perfected, and from 1773 mirrors of cast glass were produced by the British Cast Plate Glass Co. Mirrors made in Venice in the 16th century were backed with an amalgam of tin and mercury, but after 1840 the Venetians used a deposit of silver or platinum. In Spain the Royal Factory at LA GRANJA DE SAN ILDEFONSO made large and good-quality mirrors, in the last quarter of the 18th century, for palaces and wealthy homes. The Italian term is *specchio*; the German is *Spiegel*; the French is *miroir*.

mirror. Venetian mirror with glass frame, 18th century. W. 1·65 m. Museo Vetrario, Murano.

mirror. Engraved mirror, Venetian, 18th century. W. 78 cm. Museo Vetrario, Murano.

TYPES OF MIRROR

———, *convex.* A mirror with a convex surface that reflects a reduced image. In Nuremberg, in the 15th century, they were made by blowing into a glass globe, while hot, a metallic mixture with resin or salt of tartar; when cooled, the globe was cut into circular convex mirrors. It has been said that a similar process was used in Lorraine in the 14th century. In England, during the late 18th century, they were made in REGENCY STYLE with carved and gilded wooden frames having a floral or figure cresting; these were hung on the end walls of a living-room to reflect its entire contents. Sometimes called a 'bull's-eye mirror'.

———, *framed.* A mirror in a frame to be hung on a wall, for decorative as well as utilitarian purposes. Those made in England at the end of the 17th century had decorative cut borders made by DIAMONDING, but soon thereafter separate glass frames were used, mitred at the corners and surmounted by a glass cresting. In the early 18th century glass frames were superseded by wood frames richly carved or decorated with gesso work, and after *c.* 1730 these were made in various styles, e.g. ROCOCO, Chinese, and Gothic, and later in ADAM STYLE. In the 19th century, the styles were REGENCY and Empire. Prized today are mirrors with frames in the style of Chippendale or Sheraton. Framed mirrors were also made in Venice from the 17th century, often square or rectangular, with an engraved glass surround and a gilded wood crest. The Italian term is *specchio a cornice*. *See* PIER-GLASS.

———, *hand.* A small mirror set in a frame with a projecting handle, used mainly at a dressing table. The frames have been made in many shapes and styles, and of various materials, some of decorated glass (especially at Venice) but also of silver or other metal, of painted or lacquered wood, etc.

mirror decoration. Decoration, usually on the surface of a mirror, in various styles and methods; some have painted, wheel-engraved, or gilt decoration, and rare examples have TRANSFER-PRINTING or an attached MOULDED portrait. *See* MIRROR PAINTING.

mirror monogram. A monogram executed in reverse so that its reflection in a mirror appears as normal. Such monograms are found on wine glasses of the 18th century. The German term is *Spiegelmono-gramm.*

mirror painting. Painted decoration on the reverse of the glass of a MIRROR. It probably originated in England near the end of the 17th century. The silver was scraped off the mirror where the decoration was intended, and a painting in oils was substituted. Being painted on the reverse, the technique required painting in reverse, doing the details first and the background last (sometimes called 'back painting'). *See* PANEL.

——, *Chinese.* Mirror painting (sometimes called 'painting on glass' or 'back painting') as done in China from *c.* 1766 until the early 19th century. As in England, the design was first traced on the back of the mirror, then the amalgam was removed with a tool so as to leave the design area clear, and the picture was painted in with oil paints. Sometimes the painting was on the back of thin clear glass, as the Chinese thought that the thickness of mirror glass modified their colours. Canton was the centre for such work, and the mirrors and glass were imported from England, as were the prints sent to be copied. The painters were usually not professional artists and their names are not known; they were not highly regarded, and the glass paintings were displayed mainly in places of public amusement.

mitre-cut. A style of cutting on glass with a sharp groove, made with a V-edged wheel (*see* CUT GLASS).

mixer. The workman in a glassworks who mixes the BATCH. Also called the 'founder'.

mixing glass. A type of glass, e.g. a TUMBLER or a GOBLET with stemmed FOOT, made in various shapes as a receptacle for mixing ingredients; the characteristic feature is the presence of a pouring lip.

mixture. The same as BATCH.

modiolus (Latin). A drinking-cup in cylindrical form, or with the side sloping outward slightly toward the rim, and having one small loop-handle (or a pair together) just below the rim. Some rest on a FOOT-RING, some on small ball feet. Usually made of pottery, some were made of ROMAN GLASS in the 1st century AD.

Mogul (Mughal) glassware. Glassware made in India from the 16th century, largely under Persian influence as to style. *See* INDIAN GLASSWARE.

Mohn, Gottlob Samuel (1789–1825). The son of SAMUEL MOHN, born at Weissenfels, Germany. He began as a HAUSMALER and worked on STAINED-GLASS WINDOWS. He worked with his father in Dresden until 1811 when he moved to Vienna, where later ANTON KOTHGASSER worked with him. He decorated glass TUMBLERS with TRANSPARENT ENAMEL painting, including SILHOUETTES and ALLEGORICAL SUBJECTS, but principally scenic views, usually framed in an ornamental gilt border. He secured the patronage of the Emperor, and became Court Painter at the Imperial Palace at Laxenburg, working on stained glass

modiolus. Blue glass, with white rim and handle, 1st century AD. Ht 9·1 cm. Römisch-Germanisches Museum, Cologne.

Mohn, Gottlob Samuel. Beaker with transparent-enamel scene of Dresden; Dresden, before 1811. Ht 10·5 cm. Kunstgewerbemuseum, Cologne.

Mohn, Samuel. Beaker with transparent-enamel scene, signed '*Mohn px 1808*', Dresden. Ht 9·5 cm. Kunstgewerbemuseum, Cologne.

'*molar*' *flask.* Scent bottle, Persian, 9th/10th centuries. Ht 9 cm. Victoria and Albert Museum, London.

windows. After his father's death, he continued (from Vienna) to operate the family factory at Dresden.

Mohn, Samuel (1762–1815). A glass-decorator known mainly for his development of the use of TRANSPARENT ENAMEL painting to replace the former opaque enamels. He was born near Merseburg, Germany, and started as a HAUSMALER of SILHOUETTES on porcelain. He worked on glass in Dresden from 1805 until his death; from 1812 he also had a glass workshop in Leipzig. His son GOTTLOB SAMUEL MOHN worked with him until moving in 1811 to Vienna. He decorated TUMBLERS, his principal motifs being city scenes, palaces, and tourist sights, with usually an encircling coloured floral garland below the rim and the name of the scene at the base. He also did colourful butterflies and flowers. His factories produced much glassware with the help of apprentices and the use (after 1809) of transfer prints for the outlines of the paintings; pieces made to his requirements were signed by him and after 1912 also by the artist who painted them. *See* LUISE GLASS.

Mohs' scale. A scale for measuring the hardness (resistance to abrasion) of a specimen of mineral or other substance, prepared in 1812 by the mineralogist, Friedrich Mohs (1773–1839), being a series of ten specified minerals ranging from talc (1) to diamond (10) – the intervals between the listed minerals not being equal – against which the specimen is rated. On this scale glass is rated 5·5, slightly harder than ordinary steel but softer than rubies.

moil. The unused waste molten glass left on a BLOW-PIPE or PONTIL after the removal of the PARAISON or the object being made. Also called 'overblow'. It is mixed with broken glass in making CULLET.

'molar' flask. A glass flask, of square section and angled shoulder, from which extends a wide vertical neck to a slightly everted mouth; the body is extended to form four pointed feet, so that the flask somewhat resembles an extracted molar tooth. The body was decorated by cutting with a variety of geometrical patterns. They were made in Persia, and probably also Mesopotamia and Egypt in the 9th/10th centuries.

molten glass. Glass in a melted state after FUSING the ingredients at a high temperature until the BATCH has become liquid (which allows gases to become driven off and impurities removed), then allowing it to cool until it reaches a stage when it is PLASTIC and DUCTILE. It is sticky and tenacious, and adheres readily to an iron BLOW-PIPE plunged into it and then withdrawn, as well as to other glass objects previously heat-softened in readiness.

Monart Glass. A type of glassware, in ART GLASS style, developed in the 1920s by Salvador Ysart (1887–1956) at the MONCRIEFF GLASSWORKS, in which objects of heavy glass are decorated with streaks and splashes of various colours. Some examples were designed by the wife of John Moncrieff. The ware was made only by Salvador Ysart and his son PAUL YSART.

Moncrieff Glassworks. A glasshouse (originally the North British Glass Works) founded at Perth, Scotland, by John Moncrieff, *c.* 1864. It engaged in 1922 Salvador Ysart (1887–1956), a glass-maker who in 1904 had migrated from Barcelona with his three sons. The factory made decorative ART GLASS called 'MONART GLASS'. He was succeeded by his son PAUL YSART in 1948 when he and his other sons left to found VASART GLASS, LTD.

monkey flask. A type of FLASK in the form of a monkey seated on a wicker chair and playing a syrinx (Pan-pipe); above the head extends a vertical neck with a trumpet mouth. They were made of ROMAN GLASS in the Rhineland, 3rd/4th century.

monochrome. Decoration employing a single colour. *See* POLY-CHROME; GRISAILLE; CAMAÏEU, EN.

monogram. A character composed of two or more initial letters interwoven, usually representing a name, and employed as an identifying ornament; sometimes called a 'cipher'.

Monogrammist HI, The. *See* JÄGER, HEINRICH.

Monopolists, The. English glass-makers and entrepreneurs who successively obtained patents and licences granting them a monopoly in connection with the making of crystal glass and the use of certain processes for coal-burning glass furnaces. They included, after JEAN CARRÉ and JACOPO VERZELINI, SIR JEROME BOWES, WILLIAM ROBSON, Sir Percival Hart, Edward Forcett, Sir William Slingsby, Sir Edward Zouch, THOMAS PERCIVAL, SIR ROBERT MANSELL, and the DUKE OF BUCKINGHAM. *See* HAY, SIR GEORGE. The monopoly system was nominally ended by 1660. *See* Eleanor S. Godfrey, *The Development of English Glassmaking, 1560–1640* (1975).

monteith. A large circular or oval bowl, to contain iced water for the purpose of cooling wine glasses. The rim has a series of scallops, sometimes alternately bent outward, for the purpose of suspending wine-glasses (usually at least six) by the foot so that the bowl of each could be cooled by immersion before use. They were originally of silver, called a 'chilling bowl', the earliest known dated silver bowl being from 1684, and the earliest mention of the name being 1683, said to be derived from a Scotsman named Monteith (or Monteigh) who, at Oxford in the reign (1660–85) of Charles II, wore a cloak scalloped at the bottom. Such bowls have been made of silver, porcelain, and faience in England and on the Continent; two examples in glass are recorded, both identically shaped but only one engraved. The term was in 1941 unjustifiably applied by E. Barrington Haynes (and later by dealers) to a BONNET GLASS; it has also been incorrectly applied to an Irish scalloped-rim canoe-shaped fruit (or salad) bowl (*see* IRISH GLASSWARE). The French terms are *seau à verres, seau crénelé, verrière,* and *rafraîchissoir. See* WINE-GLASS COOLER. *See* Jessie McNab, 'Monteiths', *Antiques Magazine*, New York, August 1962, p. 156.

monteith. Colourless glass engraved in mat intaglio, dated 1700. Ht 17·5 cm. The Metropolitan Museum of Art, New York. Bequest of Florence Ellsworth Wilson, 1943.

Mooleyser, Willem (fl. 1685–97). A ROTTERDAM engraver who decorated bowls, flasks, GOBLETS and RÖMERS in DIAMOND-POINT ENGRAVING. Some pieces are signed 'WM' and are dated.

Morelli, Alessio (Aloisio; fl. 1667–72). A glass-maker at MURANO. *See* GREENE, JOHN.

Moresques. A term sometimes applied in England in the 19th century to decorative motifs similar to ARABESQUES, especially those derived from Spain and Sicily, both of which were, at one time, under Saracen domination. Similar motifs based on Roman decorative motifs are more usually termed GROTESQUES.

Morgan Cup. Blue and opaque white glass, cased and cut, Roman glass, 1st century BC. Ht 6·2 cm. Corning Museum of Glass, Corning, N.Y.

Morgan Cup. A semi-globular DRINKING BOWL of blue glass that is considered to be the earliest example of CAMEO GLASS to have survived intact. It is decorated with an encircling white relief frieze depicting figures in a Dionysiac rite, and on the bottom there is an 8-pointed rosette. It is of ROMAN GLASS, attributed to the 1st century BC.

Morgan Vase. A Chinese porcelain vase with peach-bloom glaze, made during the reign of K'ang Hsi (1662–1722) in the Ch'ing dynasty, which was sold in 1886 from the collection of Mrs Mary Morgan. A glass vase in imitation of the porcelain original was made of PEACH BLOW GLASS by Hobbs, Brockunier & Co. in the late 19th century; other factories also made imitative versions of the colour and of the particular vase.

mortar beaker. South Tyrol, *c.* 1500. Ht
10 cm. Kunstsammlungen der Veste
Coburg.

mosaic. Bowl of multi-coloured
translucent glass, Alexandria, 1st
century AD. Ht 10·7 cm. Victoria and
Albert Museum, London.

mosque-lamp globe. Islamic glass with
names and titles of the Sultan and his
'napkin' blazon, 14th century. Ht
14·5 cm. Victoria and Albert Museum,
London.

mortar. A circular beaker-like vessel in which various materials can be
triturated with a PESTLE. Although usually of a heavy resistant ware,
glass examples were made in Germany in the 16th century and in
England in the 18th century and later. The German term is *Mörser.*

mortar beaker. A type of BEAKER with waisted side, tapering toward
the bottom. They were made in the South Tyrol, *c.* 1500. The German
term is *Mörserbecher.*

mosaic. (As applied to portable glass objects rather than architectural
mosaic embedded in cement in floors, walls, and ceilings of churches and
rooms.) A PLAQUE, BOWL, JUG or other object formed of or decorated
with many small adjacent pieces of vari-coloured glass. The process of making it involved placing together blobs of molten glass of
different colours and reheating the mass until they fused but did not
mix; then the hot plastic mass was drawn into a long thin CANE which
was cut, horizontally or diagonally, into thin slices. The slices, of various
colours, shapes, and sizes, were then placed contiguously on a core in the
shape of the desired vessel, covered by an outer mould to retain the
shape, and reheated until the edges of the slices were fused together;
when the mould was removed, the surface was ground smooth. Some
pieces were made as plaques. Some mosaic was made flat on a clay
surface and, when made DUCTILE by reheating, could be blown or
shaped into various objects, including parts of jewellery. The resultant
pieces showed a mottled multi-coloured (not necessarily flower-like)
pattern that is sometimes called *millefiori* but which differs from the
effect created by the Venetian process of tightly arranged flower-like
canes of true MILLEFIORI glass, embedded in clear molten glass rather
than in translucent or opaque glass and fused together. *See* MILLEFIORI
MOSAIC.   *[Plate VII]*
    The architectural process of embedding TESSERAE in cement was
developed by the Greeks and was extensively used in the Roman and
Byzantine period; but portable mosaic objects, formerly said to have
been introduced at Alexandria in the 1st century BC (*see* MURRHINE
GLASSWARE), have recently been found in Mesopotamia and Iran, from
the 15th century BC. Early examples of mosaic glass are the rare
Byzantine icons of the 1st century AD. Some objects in mosaic glass were
produced by the Romans from the 1st century BC, and Venice from the
mid-18th century made mosaic jewellery, including brooches and insets
for rings, as well as such objects as handles for knives. Examples of
mosaic work made by the Aztecs (or Miztecs) in the 14th century are
known. *See* OPUS SECTILE; TESSERAE; LOMONOSOV, M. V.

Moser, L. Kolo(man). A glass-worker and designer who founded a
studio at Karlsbad (Karlovy Vary) in Bohemia in the early 1900s and
made glassware in ART GLASS style. His work included CAMEO GLASS
with enamelled decoration and also some decorated with many crowded
pieces of applied coloured glass. He designed iridescent glass, *c.* 1902,
for J. Lötz Witwe (*see* LÖTZ GLASSWARE). He also made a type of one-
colour glass called 'Alexandrit', wholly unlike the ALEXANDRITE made
by STEVENS & WILLIAMS, LTD., and THOMAS WEBB & SONS.

Moser, Ludwig (1833–1916). A glass-engraver from Karlsbad (now
Karlovy Vary), Bohemia, who was a pupil of ANDREAS MATTONI. He
became a glass-merchant and supplied the Imperial Court, and in 1857
founded a glass-factory at Meierhöfen (Nové Dvory), near Karlsbad.

mosque lamp. A type of LANTERN of ISLAMIC GLASS of inverted bell-
shape, with an oil receptacle inside and having side loop-handles (three
on early examples, six on those from the 14th century) from which they
were suspended by chains from the ceiling of Egyptian mosques.
Sometimes the chain passed through a MOSQUE-LAMP GLOBE. The
lamps were usually ornamented with gilding and vivid ENAMEL colours,
with the name of the Sultan, armorial blazons, dedicatory inscriptions,
floral designs of Chinese derivation, and NESKHI script with quotations

from the Koran; but some smaller examples are entirely undecorated. The undecorated lamps are said to have come originally from Syria in the 11th century, and these decorated to date from after *c*. 1280. Some were made of Venetian glass for export to the Near East in the 15th and 16th centuries; only rarely is the artist's identity shown on these. Those from the late 13th to the mid-14th centuries have the finest workmanship. Copies have been made in the 19th and 20th centuries, and are sometimes called an 'Islamic Vase'. The Italian term is *lampada da moschea*; the German is *Moscheelampe*. See BROCARD, JOSEPH.

mosque-lamp globe. A large ISLAMIC GLASS globe through which the chains suspending a MOSQUE LAMP were gathered and united into a single chain for suspension from a high ceiling. The globes were made of gilded and enamelled glass, and were sometimes inscribed with the names and titles of the Sultan.

Moss Agate glass. A type of ART GLASS developed in the late 1880s by JOHN NORTHWOOD in collaboration with FREDERICK CARDER at STEVENS & WILLIAMS, LTD., and later made by Carder at STEUBEN GLASS WORKS, *c*. 1910 to 1920. It was made by casing a PARAISON of clear SODA GLASS with a layer of clear LEAD GLASS, then covering the piece with powdered glass of variegated colours, and adding another casing of clear lead glass; the object was then shaped and heated, and cold water was injected into it, causing the soft soda glass to crackle, and then it was quickly emptied and lightly reheated so that the interior cracks were united but left a crackled network. Steuben pieces so made are usually red, brown or yellow, but rare examples are blue, all with variegated colour shading as a result of the powdered glass, so that no two pieces are identical. Similar ware was produced in France by EUGÈNE ROUSSEAU and E. LÉVEILLÉ, and was copied elsewhere.

Mother-of-Pearl glassware. A type of ART GLASS that has a double FLASHING, the inner opaque layer being blown into a patterned MOULD and then covered with a flashing of clear or coloured glass (trapping air bubbles in the moulded spaces) and then covered again with a clear glass flashing that is later acid-dipped to provide a satin surface. Among the patterns for the inner layer (white or coloured) are the 'Herringbone', the 'Swirl', the 'Diamond-quilted', the 'Feather', the 'Moiré', the 'Zipper' (close vertical notched ribbing), and the 'Peacock Eye'; these patterns show through the outer flashing. Some examples have added enamelled or applied decoration. 'Rainbow' type has several blended colours. Such ware was made by the PHOENIX GLASS CO., Pittsburgh, the MT. WASHINGTON GLASS CO., and others.

mould. A form used in shaping glassware, by the processes of making MOULD-BLOWN and later MOULD-PRESSED ware. Some moulds were used to give the glass its final form, others to give a patterning which was subsequently worked over and sometimes occasionally almost eliminated, e.g. by ribbing. Many open-shaped glasses made before the use of BLOWING are thought to have been moulded, some perhaps by the CIRE PERDUE process. Moulds were originally made possibly of FIRE-CLAY, then of wood or carved stone, and later of metal. Moulds were used from earliest times, sometimes made in one part, much later in two parts that were hinged; they were then used for making mould-blown glassware. Still later there were three-part and four-part moulds, used for complicated forms. The pieces so made were of a great variety of forms, e.g., bottles with human faces, bunches of grapes, etc. Some parts of objects (e.g. handles) were made in a separate mould and joined to the body of the piece. After *c*. 1825 metal moulds were used for mould-pressed glassware. See SAND-MOULDING; DIP-MOULD; PIECE-MOULD; BLOWN-THREE-MOLD GLASS.

mould-blown glass. Glassware made by the process of blowing the molten glass PARAISON (bubble) into a MOULD. The interior wall of mould-blown glass is in the shape of the outer form, so that the blown

mosque lamp. Lamp with gilt and enamel decoration, Syria, mid-14th century. Ht 38·5 cm. Kunstgewerbemuseum, Cologne.

Moss Agate glassware. Vase designed by Frederick Carder for Steuben Glass Works, 1920s; blue matrix with variegated shadings. Ht 38 cm. David Williams Collection. Courtesy, Corning Museum of Glass, Corning, N.Y.

mould-blown glass. Dark-green hexagonal bottle, Aleppo, Syria, 4th/5th century. Ht 22·5 cm. British Museum, London.

object could be reheated and further blown to enlarge its size. Mould-blown glass was made in Egypt in the 2nd to 4th centuries, and later universally. In the United States mould-blown glass was made near the end of the 19th century in preference to FREE-BLOWN GLASS, and was adapted to machines for making bottles and electric-light bulbs; it was extensively made in the Middle Western states to make a large variety of glassware, with LOW RELIEF decoration in imitation of English CUT GLASS of the early 19th century. The method was largely superseded, *c*.1827, by the process of PRESSING which led to mass-produced MOULD-PRESSED GLASS.

mould-pressed glass. Glassware made by the process of PRESSING, as distinguished from MOULD-BLOWN GLASS. It was originally made, during the Revived Rococo period, *c*.1827 and onwards, in the United States to copy elaborate CUT GLASS, and was mass-produced by the BOSTON & SANDWICH GLASS CO. and by rival companies. Such glassware was also made in England from the 1830s onward. Early American mould-pressed glass sometimes shows the seam marks from the moulds, as it was not normally hand-finished. *See* BARIUM.

moulded glass. *See* MOULD-BLOWN GLASS; MOULD-PRESSED GLASS; BLOWN-THREE-MOLD GLASS.

mount. A decorative metal attachment, functional (e.g., as a base) or ornamental, on a VASE or other object of porcelain or glass of superior quality. In the 18th century such mounts were made of GILT BRONZE, ORMOLU, silver or pewter. MUGS were commonly given silver or pewter LIDS, and lids of some boxes were attached by a hinged metal mount.

Mount Washington Glass Co. A glasshouse established at South Boston, Mass., in 1837 by DEMING JARVES for his son George D. Jarves. It became Jarves & Cormerais in 1850. It engaged William L. Libbey who, with Timothy Howe, became owner in 1860, and sole owner in 1866. Libbey moved the business to New Bedford, Massachusetts, in 1869, and in 1870 sold the Company. It resumed its original name in 1871, and in 1894 was acquired by PAIRPOINT MANUFACTURING CO. The company made mould-blown, mould-pressed, and cut glass. *See* BURMESE GLASS; PEACH BLOW GLASS.

mouth. (1) The orifice of a bottle, jug, jar or pitcher from which the contents are poured. The mouth may or may not have a LIP. Some are of special shape, e.g., flared mouth. (2) The side opening of a CROWN POT from which the MOLTEN GLASS is removed for working.

muff. The cylinder of glass made in the process of producing BROAD GLASS and before it is split. Such glass is sometimes called 'muff glass'.

muffineer. A CASTOR (with a perforated cover or cap which is sometimes of silver), for sprinkling sugar on muffins, etc.

muffle. A fire-clay box in which glass or porcelain objects were enclosed, when placed in the muffle kiln, to protect them from the flames and smoke while being subjected to low-temperature firing, especially in the process of applying ENAMEL COLOURS and fired GILDING at the temperature of 700–900° C.

mug. A drinking vessel resting on a flat BASE without a STEM and having a HANDLE and a circular RIM. Some have a LID with a thumb-rest, usually of metal, such as silver or pewter. Glass mugs were made in a variety of forms, usually cylindrical but sometimes WAISTED, barrel-shaped, inverted bell-shaped, DOUBLE-OGEE shaped, and occasionally 'square' (i.e. cylindrical but having equal diameter and height, so as to produce a square silhouette). Examples in OPAQUE WHITE GLASS, with enamelled decoration, were made in Spain at LA GRANJA DE SAN ILDEFONSO, *c*.1775–85, and others in England in the third quarter of the

18th century, and (sometimes in miniature form) in Bohemia. A mug is sometimes erroneously referred to as a TANKARD. *See* BEAKER; STEIN.

Müller Frères. A glasshouse founded *c.* 1910 near Lunéville (Lorraine) by Henri Müller and his brother Désiré. It combined the glassworks previously founded *c.* 1900 by Henri Müller at Croismare, near Lunéville. Both factories specialized in ART GLASS with an OVERLAY of two or three colours, and with etched decoration made by the use of hydrofluoric acid.

multi-layered cased glass. A type of CASED GLASS composed of three or more layers of different colours and carved in the manner of CAMEO GLASS to reveal the various colours in different areas of the decoration. Some VASES and CHINESE SNUFF BOTTLES are so made and decorated.

multiple bottle. A BOTTLE that is divided internally into separate compartments, each with its own MOUTH or SPOUT. Examples exist from early Roman times with three or four such compartments. Relatively modern bottles, some from the early 19th century, have four compartments, each with a separate small mouth and STOPPER. *See* DOUBLE BOTTLE; TRIPLE BOTTLE; DOUBLE CRUET; LIQUEUR BOTTLE.

Mulvaney, Charles, & Co. The principal glasshouse of Dublin, 1785–1835. Some of its ware is marked 'CM & CO'. It was probably closely associated with Jeudwin, Lunn & Co. of Dublin. *See* DUBLIN GLASSWARE.

Münzegewicht (German). Literally, coin weight. A disk-shaped weight made of coloured glass with moulded decoration or inscription. They were made of late ROMAN GLASS, of Egyptian glass in the 10th/11th centuries, and of later ISLAMIC GLASS.

Murano. A compact group of small islands about five kilometres from Venice where glass-makers had settled before the 11th century and where, since *c.* 1292 (when glass-making was banned in Venice) practically all the glassware that has come to be known as VENETIAN GLASSWARE has been produced. The reason for forcing the glasshouses out of Venice was the fear of fire spreading from the furnaces, but also an attempt to preserve the secrets of the Venetian processes by the stringent restrictive laws that were severely enforced (although not always successfully, as many glass-makers were able to leave and go to L'ALTARE, Florence, etc. as well as to the Netherlands, Germany, and elsewhere). There are today numerous glasshouses on Murano, of which well-known ones are BAROVIER & TOSO (with predecessors dating from the 15th century), SALVIATI (from 1859), VENINI (from 1921), CENEDESE (from 1945), and ALFREDO BARBINI (from 1950).

Murano glassware. Glassware made in MURANO but generally called VENETIAN GLASSWARE. Early ware was mainly BEADS, CANES, and GLASS CAKES, then PANES, MIRRORS, and utilitarian glassware, but from the 15th century artistic glassware was introduced, with new varieties of glass (e.g., AVENTURINE, CALCEDONIO, AGATE GLASS, ICE GLASS); CRISTALLO and MILLEFIORI glass were developed, and emphasis was placed on coloured glass and manipulated decoration rather than engraving or enamelling. In the 20th century new techniques were invented, especially for surface treatment of glassware, and new styles were introduced, such as massive OVERLAY glass and hand modelling of molten glass blocks, along with the production of much gaudy tableware and ornamental figures for the tourist trade. CHANDELIERS and other lighting fixtures have long been a speciality. For a discussion of Murano glassware, with many illustrations, *see* Astone Gasparetto, *Vetro di Murano*, 1958.

Murray, Keith (1893–   ). A New Zealand architect who from *c.* 1925 became interested in glass techniques and decoration. In 1932 he

muffineer. Irish, probably Dublin, 1797, with silver cover. Ht 19·5 cm. Collection of Phelps Warren, New York City.

mug. Cased glass with decoration of windows and polychrome enamelled sprays, St Petersburg Imperial Glass Factory, second quarter of 19th century. Ht 9·5 cm. Hermitage Museum, Leningrad.

Münzegewicht. Coloured glass weight. D. 2·2 cm. Kunstgewerbemuseum, Cologne.

designed some experimental table-glass for WHITEFRIARS GLASS WORKS and was engaged part-time in the 1930s at STEVENS & WILLIAMS, LTD, where he designed table-services, ornamental glassware, and large bowls, vases, and dishes of heavy METAL with cut decoration.

murrhine glassware. Glass objects (usually a bowl or sometimes a vase) made of MOSAIC glass at Alexandria, *c.* 200 BC–AD 100, and highly treasured in ancient times. (They were inspired by and imitative of the ancient murrhine bowls once thought to have been made of some kind of pottery, even Chinese porcelain, and later believed to have been made of hardstone but, recently, of the mineral fluorspar; such ancient ware emitted an aroma that is now said to have come from the gum-resin needed to bond the fluorspar in the cutting process.) They were made in the manner of mosaic, from horizontal slices of rods or CANES of coloured glass that had been fused together over a mould. The technique thus differed from that used later by the Venetians in making what they called MILLEFIORI glass by embedding slices of canes in a glass matrix.

murrina (Italian). A type of modern MOSAIC glassware. The pieces have many inserts of vari-coloured glass, embedded as large pieces or as long and wide streaks, as distinguished from the embedded horizontal slices of CANES in usual MILLEFIORI glassware. Such glass was made at MURANO in the 19th century and thereafter, in imitation of the so-called MURRHINE GLASSWARE made at Alexandria. The name is said to be derived from the ancient murrhine bowls or the Italian word *murra*, the material of which they were once supposed to have been made. *See* MILLEFIORI MOSAIC.

musical glasses. Glasses that, due to their resonant quality, were used as musical instruments. In Germany in the 18th century TUMBLERS and WINE-GLASSES (up to 37, and later 120) of different sizes were filled with water, and sometimes mounted on a spindle, then the rims were rubbed by a player using his finger tips (or later pads or leather hammers) to produce varying sounds. Later there were glasses ground to produce the right pitch when rubbed with wetted fingers. Benjamin Franklin (1706–90) adapted this into what he called an 'Armonica', the forerunner of the harmonica. Such musical glasses are still played today in theatres.

musical instruments. Various objects in glass, used to produce musical sounds, e.g., MUSICAL GLASSES, BELLS, HORNS, dulcimers. Some musical instruments were copied in form only to serve as ornaments, e.g. a VIOLIN, BUGLE, TROMBONE, COACH-HORN, and others as DRINKING GLASSES, e.g. POST-HORN.

mustard-jar. A small jar for serving prepared mustard, having a cover with an indentation for the spoon; some jars intended for dry mustard have a cover without indentation. Often made of porcelain or faïence, some were made of glass. The French term is *moutardier.*

Mutzer glassware. BLOWN-THREE-MOLD glass vessels once thought to have been made by Frederich Mutzer or his son Gottlieb, the former being an emigrant glass-blower from Germany in the early 19th century who worked in the Philadelphia area. It has been suggested that the ware was made in the 20th century by the Clevenger Factory at Clayton, New Jersey. Recently it has stated that the pieces are modern FAKES.

Mycenaean glass-paste beads. BEADS of glass paste made at Mycenae, *c.* 1300 BC, that are in the form of small thin tablets of which the ends are ribbed and perforated for threading. They were circular, rectangular or triangular, and usually of blue or pale-yellow glass. They are decorated with relief Mycenaean motifs, such as rosettes, ivy, and spirals. They were formerly believed to have been used for NECKLACES and as decoration on garments, but a recent view is that they were also

Mycenaean glass-paste beads. 24 blue beads, Mycenae, *c.* 1300 BC. L. (each) 3·2 cm. Römisch-Germanisches Museum, Cologne.

used to adorn diadems and the skulls of skeletons. See *Journal of Glass Studies* (Corning), X (1968), p. 1.

Myers, Joel P. (1934–   ). A former designer (1963–70) at BLENKO GLASS CO. who became a teacher and studio glass-maker at Illinois State University. He developed a style of bottle with chrome oxide MARVERED into the PARAISON which, after being blown, is covered with crystal glass and is then additionally blown to produce a cased bottle. He has also produced a series of thin tall drinking glasses, called 'Tallest Form', which have bowls of various forms and textures and usually a domed foot.

Myra-Kristall. A type of ART GLASS developed in 1925 by KARL WIEDMANN (b. 1905) at the glassworks of the WÜRTTEMBERGISCHE METALLWARENFABRIK (W.M.F.). It is in the style of the glassware of LOUIS COMFORT TIFFANY.

Myth of Adonis, The. A unique oval casket made by STEUBEN GLASS, INC., in 1966. It consists of engraved crystal glass panels mounted in an ornamental frame of 18-carat white gold. The panels depict the myth of Adonis, the eight principal panels showing his life, death, and rebirth, and the eight small panels symbolizing the seasons; the upper panels relate to the earth and the heavens, the lower ones to the underworld. The cover, ornamented with green-gold leaves, has a crystal finial, emerald cut. The piece was designed by DONALD POLLARD, the engraving design was by Jerry Pfohl, and the engraving was done by Roland Erlacher. The goldwork on the frame was made by Cartier, Inc.

mythological subjects. Decoration on glassware, usually WHEEL-ENGRAVED, depicting persons or events of Greek or Roman mythology or legend, e.g. Perseus and Andromeda (*see* PERSEUS-ANDROMEDA PLAQUE), Thetis and Peleus (*see* PORTLAND VASE), Lycurgus (*see* LYCURGUS CUP), Pyramus and Thisbe (*see* FAIRFAX CUP), as well as Apollo and Athena, Bellerophon and Pegasus, Actaeon and Artemis, etc. *See* ALLEGORICAL SUBJECTS.

Myth of Adonis. Oval casket by Steuben Glass Works, 1966. Ht 17 cm. Collection of Brady Hill Company, Birmingham, Michigan. Courtesy, Steuben Glass, New York.

N

Nailsea Glass House. An English glassworks at Nailsea (Somerset), about seven miles from Bristol, founded in 1788 as Nailsea Crown Glass & Bottle Manufacturers, by JOHN ROBERT LUCAS with William Chance who was later joined in partnership by Edward Homer. In 1790 a second glasshouse was founded nearby by William Chance and others. In 1810 they came under the management of ROBERT LUCAS CHANCE (son of William Chance and nephew of John Robert Lucas) who later bought British Crown Glass Co. at Smethwick, near Birmingham, and founded Chance Brothers, which acquired it and the Nailsea factories in 1870. Originally the Nailsea factory made mainly CROWN GLASS for windows, and then SHEET GLASS, also bottles; later it made household ware, flasks, and FRIGGERS, and a subsidiary at nearby Stanton Drew made bottles. A great amount of glassware (preferably termed 'Nailsea-type') has been attributed to Nailsea, but it – especially colourless FLINT GLASS – is now considered to have been mostly made at Bristol, Stourbridge, and elsewhere in England. Some glassware, however, is accepted as produced at Nailsea, having recognizable characteristics. *See* NAILSEA GLASSWARE.

Nailsea glassware. Glassware of the so-called Nailsea type, but now mostly attributed to unidentified glassworks other than the NAILSEA GLASS HOUSE; some Nailsea-type glassware was made at nearby Rockwardine. It embraces ware of three types: (1) FLECKED GLASS-WARE; (2) FESTOONED GLASSWARE; and (3) coloured (pale green) or clear glassware with simple white FILIGRANA decoration. The last group included FLASKS (some being GEMEL FLASKS and BELLOWS FLASKS) and other household and decorative ware. Also attributed to Nailsea are various objects such as WITCH BALLS, WALKING STICKS, ROLLING PINS, POLE-HEADS, etc. Some BELLS of multi-coloured glass, sometimes attributed to Nailsea, are now said to be certainly made elsewhere. *See* Keith Vincent, *Nailsea Glass* (1975).

Napf. The German term for a bowl.

Napoli glass. A type of ART GLASS that is clear, with different patterns of decoration on the inside and the outside. It was made by the MOUNT WASHINGTON GLASS CO.

nappie. A circular dish with a flat bottom and slightly curving sides. It is a large form of a SAUCE DISH. They were made in the United States, especially of LACY GLASS by the BOSTON & SANDWICH GLASS CO.

Nailsea-type glass. Flask of blobbed glassware, *c.* 1800–20. Courtesy, Maureen Thompson, London.

Nash, A. Douglas (d. 1940). A son of ARTHUR J. NASH who worked with his father at the LOUIS C. TIFFANY factories at Corona, Long Island, New York, until Tiffany withdrew from active participation in 1919. He purchased the assets of the company and operated as A. Douglas Nash Associates until 1920 when Louis C. Tiffany Furnaces, Inc. was formed. When Tiffany withdrew his financial support in 1928, Nash, with his father and others, formed A. Douglas Nash Corporation, which continued to make ART GLASS in the style of Tiffany. It also made new types developed by Nash, including varieties of tinted glass and lustred glass, and also ware with embedded air bubbles in the style of CLUTHA GLASS. Nash also developed CHINTZ GLASSWARE and SILHOUETTE GLASSWARE. He used the marks 'Corona', 'ADNA', and 'Nash'. After his company failed in 1931, he went to Toledo, Ohio, and joined LIBBEY GLASS CO., as designer and technician, to assist it to regain its position in the luxury glass field. There he designed the LIBBEY-NASH GLASSWARE. He left Libbey in 1935 and worked for various firms until his death.

Nash, Arthur J. (1849–1934). An English glass-maker and technician who was at one time manager of the WHITE HOUSE GLASS WORKS, at Stourbridge, England. He was brought to the United States, *c*.1895, by LOUIS COMFORT TIFFANY, and first did experimental work in glass in Boston. Later he joined Tiffany at the Corona glasshouse, together with his sons, A. DOUGLAS NASH and LESLIE NASH, and there he developed CYPRIOTE GLASS and many other forms of ART GLASS. His technical skill accounted for much of Tiffany's success.

Nash, Leslie. A son of ARTHUR J. NASH and brother of A. DOUGLAS NASH. He worked for the companies of LOUIS COMFORT TIFFANY. After leaving Tiffany he started his own glass factory at Woodside, Long Island, New York, and bought BLANKS for decoration. He was not connected with A. Douglas Nash Co., his brother's factory. *See* TIFFANY GLASSWARE: *Favrile fabrique*.

natron. Sodium carbonate. Pliny, in the 1st century AD, related a tale, now considered apocryphal, that some Syrian merchants placed cakes of natron on the sands by the River Belus as a support for their cooking pots and that overnight the sand and the soda were fused by the heat to produce the first known glass. Natron was used as the soda constituent in making EGYPTIAN GLASS.

naval subjects. Decoration on glassware, engraved or enamelled, depicting naval engagements, e.g. the Dutch victory over the Swedish fleet in 1659, WHEEL-ENGRAVED on a Dutch WINGED GLASS; the battle of the Medway in 1667, enamelled on a green glass RÖMER; and the Battle of the Dogger Bank in 1781 engraved on a Dutch glass. *See* SHIP SUBJECTS; PRIVATEER GLASSWARE.

Navarre, Henri (b. 1885). A French glass technician and studio glass-maker who, *c*.1925–35, made pieces AT-THE-FIRE and some tinted transparent glassware. He was influenced by the ART DECO style of MAURICE MARINOT.

nécessaire (French). A small travelling case containing a drinking glass and a glass knife, fork, and spoon. Some were made (including the case made of glass) in Germany and at Murano in the early 18th century, with enamelled decoration.

neck. That part of a BOTTLE, VASE or other vessel that is between the MOUTH and the SHOULDER or BODY. It may be either long or short, and with sides parallel or sloping. Some have interior threading for a screw STOPPER. The French term is *col*. *See* TRUMPET NECK; CONTRACTED NECK.

necklace. A string of BEADS, round or oblate, or of flat pierced small plaques or disks of multi-coloured glass or glass paste (*see* PÂTE-DE-VERRE). Beads, and therefore necklaces, were made in many places almost throughout glass history. Early examples in glass paste were made at Mycenae *c*.1300 BC (*see* MYCENAEAN GLASS-PASTE BEADS) and of glass by the Egyptians in the 4th/3rd centuries BC. They were made of ROMAN GLASS in the 1st century AD. In China some Buddhist religious necklaces of green glass imitating jade were made in the 18th century.

needle. A small pointed instrument, some, used for sewing, having an eye at the end opposite the point. Some without an eye were made of coloured glass in various parts of the Roman Empire, occasionally decorated with opaque threads wound around one end. These are said to have been probably used as hair ornaments.

needle-case. A small receptacle for holding sewing NEEDLES. Some are cylindrical, divided midway into two sections, the half that is the cover fitting over a thinner projecting extension of the lower half. Some were made of enamelled OPAQUE WHITE GLASS in Bohemia. *See* GALANTERIE.

Nailsea-type glass. Hand-bells, mid-18th century. Courtesy, Maureen Thompson, London.

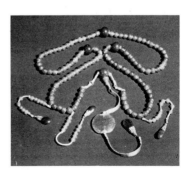

necklace. Buddhist religious necklace of green glass imitating jade, Chinese, probably 18th century. Courtesy, Hugh Moss, Ltd, London.

nef. A table ornament in the form of a fanciful rigged sailing ship. One well known example, entirely of glass, was made in 1549 at Beauwelz, Belgium, and presented to Emperor Charles V (1500–58) and his son Philip on their visit there; it was 125 cm. long and embellished with glass figures of animals. Others were made in the Netherlands in the 17th century and in England in the early 18th century. Occasionally spun glass (GLASS FIBRE) is incorporated, especially to make the surrounding sea, and some elements are sometimes of coloured glass. *See* NEF EWER; BOAT; SHIP.

nef ewer. A type of EWER made entirely of glass in the form of a NEF, with the spout of the blown bowl (the hull) forming the prow and pincered lattice-work the rigging, and usually having sailors at the prow or in the rigging, a lantern at the stern, and a dolphin or a cornucopia at the masthead (such rigging, etc., being made AT-THE-LAMP). The bowl rests on a stemmed foot. Examples are usually made of clear glass with some elements in coloured glass. The exclusive right to make such objects at MURANO was granted in 1521 to Arminia (Ermonia) Vivarini, the daughter of the painter Alvise Vivarini; they were also made in the 17th century. The Italian term is *navicello. See* COLINET FAMILY.

nef ewer. Venetian, mid-16th century. Ht 29 cm. Museo Vetrario, Murano.

Nelson glassware. Goblet with engraving of H.M.S. *Victory, c.* 1805. Ht 15 cm. Courtesy, Delomosne & Son, London.

Nehou, Louis-Lucas de. The nephew of RICHARD-LUCAS DE NEHOU, whose glass factory, founded in 1653, was at Tourlaville, near Cherbourg, in Normandy; he left to join the factory of Pierre de Bagneux (founded in 1684). He thereafter became the Director from 1688 to 1702 of another glassworks at Tourlaville established in 1688, under the patronage of Louis XIV, by a speculator, Abraham Thévart, to compete with Bagneux and to develop the process of making CAST GLASS invented by BERNARD PERROT. In 1693 Nehou set up another glassworks, the Manufacture Royale des Grandes Glaces, at the château of St-Gobain in Picardy, and these two factories, and another that had been founded in 1665 by JEAN-BAPTISTE COLBERT in the Faubourg St-Antoine in Paris, were combined in 1695 to form the Manufacture Royal des Glaces de France, of which Nehou was Director from 1710 to 1728; it became known in 1830 as the ST-GOBAIN GLASS COMPANY. The Tourlaville operations were terminated in 1787.

Nehou, Richard-Lucas de (d. 1675). The founder of a glassworks established in 1653 at Tourlaville, near Cherbourg, in Normandy, where at one time his nephew, LOUIS-LUCAS DE NEHOU, worked. It was united in 1667 with a glassworks, the Manufacture Royale des Glaces de Miroirs, set up in 1665 by JEAN-BAPTISTE COLBERT in Paris.

Neikai(o)s. A glass-maker who made MOULD-BLOWN glass in Syria in the 1st century AD. His name is moulded on some pieces. Sometimes a BEAKER so marked is similarly shaped to two marked MEGES.

Nelson, Lord, glassware. Glassware, mainly MUGS, RUMMERS, and GOBLETS, commemorating Horatio, Lord Nelson (1758–1805) with decoration of a portrait, a depiction of his ship *Victory*, or other memorabilia.

Neo-classical style. The revival, in the second half of the 18th century, of classical decoration, based on that of Greece and Rome, which followed the ROCOCO style. Neo-classicism was to some extent a product of the excavation of Pompeii in the 1750s, and the term, in its early phases, is more or less interchangeable with the term ADAM STYLE in England. In France, it is often called *Louis Seize* style. It is notable for the employment of SWAGS, rams'-heads, and similar classical motifs. After the Revolution of 1789 in France, the early Neo-classical style developed first into the *Directoire* style (which is principally a furniture style) and then into the EMPIRE STYLE; the latter was contemporary with the English REGENCY STYLE, which owes much to it.

In glassware the BAROQUE and ROCOCO styles of glass engraving were superseded by a taste for glass with FACETING, and also with

engraving in neo-classical style. Cylindrical TUMBLERS and straight-
sided GOBLETS of simple form replaced in Germany and elsewhere the
earlier graceful but richly decorated forms. In England, slender forms
were favoured; the stems of wine glasses became shorter and cutting
slighter.

Neri, Antonio (1576–1614). A Florentine chemist and priest who was
the author of *L'Arte Vetraria*, published in Florence in 1612. It was the
first printed book on the art of the glass-maker, giving many formulae
for coloured glass and other glass recipes based largely on his experience
in glass-making in Florence and in Antwerp glassworks founded by
Venetians. The book was translated into English by Christopher
Merret(t) in 1662, and into Latin (published in Holland in 1669),
French (1752), and German (1679), where it formed the greater part of
JOHANN KUNCKEL's book. Neri mentioned the beauty of glass
containing lead, later perfected in England in 1676 by GEORGE
RAVENSCROFT. Neri originally worked in Florence under the patronage
of Don Antonio de' Medici, *c.* 1601, then in Pisa, *c.* 1603, and moved to
Antwerp in 1604, but returned to Florence by 1611.

Nes, G. V. (fl. 1687). A Dutch glass-engraver. One RÖMER decorated in
DIAMOND-POINT ENGRAVING and signed 'G. V. Nes' is known.

Neskhi. A type of Arabic cursive script, used as decoration on ISLAMIC
GLASS. *See* CALLIGRAPHY; KUFIC SCRIPT; MOSQUE LAMP.

nestoris. A globular JAR with two vertical handles attached at the
shoulder and at the rim of the mouth; the handles extend upward above
the rim and are characterized by two knobs, one at the junction with the
body and the other at the apex. They were made of Greek pottery
(sometimes called a *torzella*) but very similar jars have been made in
glass, one having a third knob midway between the shoulder and the
apex.

Netherlands glassware. Glassware is recorded from glasshouses at
Middelburg from 1581 (after the Netherlands became separated from
the southern part of the Low Countries that is now Belgium – *see* SOUTH
NETHERLANDS GLASSWARE) until 1609 (when Dutch independence
from Spain was established); but no examples (except beakers
attributed to ANTONIO MIOTTO) have been identified from this early
period. From 1609, many glassworks were started in North Netherlands, at Leiden, The Hague, Dordrecht, Rotterdam, Amsterdam,
Delft, and elsewhere. At first, when dependent on Venetian and Altarist
workmen, the ware was influenced by German and English styles and
techniques; but during the 17th and 18th centuries the Dutch excelled
in developing three styles of decoration, mainly by the work of talented
amateur engravers: (1) DIAMOND-POINT ENGRAVING, from *c.* 1575 to
c. 1690, on thin Dutch glasses before English glasses were imported (*see*
VISSCHER, ANNA ROEMERS; VISSCHER, MARIA TESSELSCHADE ROEM-
ERS; SCHURMAN, ANNA MARIA VAN; MOOLEYSER, WILLEM;
HEEMSKERK, WILLEM JACOBZ VAN); (2) from *c.* 1690 to *c.* 1750, WHEEL
ENGRAVING in German and Bohemian styles, featuring COM-
MEMORATIVE GLASSWARE and designs of ships and heraldic devices (*see*
SANG, JACOB; SCHRÖDER, C. C.); and (3) from *c.* 1750 to *c.* 1800,
STIPPLING (*see* GREENWOOD, FRANS; WOLFF, DAVID; SCHOUMAN,
AERT; HOOLART, G. H.; BLIJK, JACOBUS VAN DEN). Some ÉGLOMISÉ
work was done by ZEUNER. Modern glassware is being made at the Royal
Dutch Glass Works (a unit of the United Glass Works) at LEERDAM.

netted glassware. BEAKERS, TUMBLERS, etc., that are wrapped in a
network of thin wire. Several REICHSADLERHUMPEN bearing an
enamelled date '1616' and having such a wire mesh covering were made
in the 19th century, possibly at the factory of FRITZ HECKERT or
elsewhere in Bohemia. They are to be distinguished from the netted
glass vessels that were blown into a wire mesh, made by Heckert at

nestoris. Jasper glass, with black knobs,
German, 18th century. Ht 15·7 cm.
Musée de Verre, Liège (photo A.C.L.,
Brussels).

netted glassware. Armorial beaker and
Humpen. Germany, late 19th century.
Ht 27·4; 20 cm. Corning Museum of
Glass, Corning, N.Y.

Petersdorf, in Bohemia, *c.* 1869. The German term for such netted glass is *Drahtglas*. *See* RETICULATED GLASSWARE.

network beaker. *See* VASA DIATRETA.

network glass. A suggested English term for VETRO A RETICELLO or *Netzglas*.

Netzglas (German). Literally, net glass. The same as VETRO A RETICELLO.

Neumann, Johann (fl. 1680–97). A glass-engraver from Zittau, who was Court Engraver to Duke Johann Adolf I of Saxe-Weissenfels, in Saxony.

Nevers figure. A type of small glass FIGURE made in the late 16th and 17th centuries (first mentioned in 1605) at Nevers, France (but also elsewhere in France); it is of an opaque fusible glass material (sometimes misleadingly called ENAMEL GLASS). Examples vary from 2·5 to 15 cm. high, and are exceedingly detailed as to features and fabrics. Single figures often have stands of trailed glass threads; a figure without such a stand has usually been broken from a GROTTO. Some animals are MOULD-BLOWN, with applied thin glass threads (*verre frisé*). Occasionally large groups were made, such as a Crucifixion scene or several figures with animals. They were made of portions of glass RODS softened AT-THE-LAMP and then manipulated with pincers or other instruments, and often fastened on an armature of copper wire. The names of some of the artists are known, including Jean Prestereau (1595) and his son Léon. Such ware, being made of white opaque material, might be casually mistaken for porcelain or faience. Most examples display coloured details. Similar figures were made in Venice, Germany, England, and Spain in the 17th century and later; they are not readily distinguishable or datable. They are sometimes called '*verre filé de Nevers*'. *See* NEVERS GLASSWARE; FOUR SEASONS, THE.

Nevers figures. Woodman and wife, opaque white glass with blue and black enamel decoration, made at-the-lamp, Nevers, 18th century. Ht 9 cm. Kunstmuseum, Düsseldorf.

Nevers glassware. Glassware made at Nevers, France, from the 16th century, when Ludovico Gonzaga by his marriage to Henrietta of Cleves became Duke of Nevers, and brought to the town Italian glass-makers of the Saroldi (Šarode) family from L'ALTARE. They and other Italians and their descendants continued to make glass of various kinds in Italian style until the 18th century. The local speciality was the NEVERS FIGURE.

New Bremen Glass Manufactory. A glassworks at New Bremen, Maryland, founded in 1784 by John Frederick Amelung when he came from Bremen, Germany, and which made superior glass with fine wheel-engraved decoration. Due to small demand, the factory was sold in 1795. It is sometimes referred to as the 'Amelung Glassworks'. It is the only American factory of its period of which inscribed and dated pieces are known, often presentation glasses; but a number of unsigned pieces have also been definitely attributed to this factory, based on the quality of the glass and of the engraving. Its best known piece is the so-called 'Bremen Pokal', made and engraved in 1788 and sent to Germany to demonstrate its maker's skill.

New England Glass Co. An important company that acquired in 1817 the property at East Cambridge, Massachusetts, of the Boston Porcelain & Glass Co. (founded in 1814) and was incorporated by DEMING JARVES and associates in 1818. In 1825 Jarves left to found the BOSTON & SANDWICH GLASS CO., being succeeded by Henry Whitney as general manager. It made MOULD-PRESSED GLASS, making plates, etc. in LACY GLASS, and also MOULD-BLOWN GLASS, PAPER-WEIGHTS, and a great variety of cut and engraved LEAD GLASS in Venetian and English styles, as well as new coloured glass developed by JOSEPH LOCKE. In 1878, after suffering from the competition of LIME GLASS, it was leased to William L. Libbey (1827–83) who had been its agent from 1872. When it was closed in 1888 by strikes, his son Edward Drummond Libbey

(1854–1925) moved the business to Toledo, Ohio, the old company was dissolved in 1890, and the new business became the LIBBEY GLASS CO. Until it moved to Toledo LOUIS F. VAUPEL had for 28 years been its master engraver and cutter. *See* AMBERINA; PEACH BLOW; POMONA; AGATA; MAIZE GLASSWARE. *See* Lura Woodside Watkins, *Cambridge Glass* (1930).

New Geneva Glass Works. A glasshouse founded in 1797 in Fayette County, Pennsylvania, by Albert Gallatin (1761–1840), not a glassmaker but a statesman (U.S. Secretary of the Treasury, 1801–13). He, with James Nicholson, financed this first glasshouse west of the Allegheny Mountains, employing the Kramer family of German glassworkers, including BALTASAR KRAMER, all of whom had left the then closed NEW BREMEN GLASS MANUFACTORY. The ware initially included bottles and window glass, but later moulded and blown household ware. The finer products have been called 'Gallatin-Kramer glass' or 'Kramer Family glass'. Gallatin and Nicholson sold their interest in 1813 to the Kramers who, in 1807, moved the factory across the river to Greensboro, so as to use coal fuel. They sold it before 1830, and it burned down in 1847. No specimens have been identified.

Newcastle glass. Glassware of many varieties and styles made in numerous factories at Newcastle-upon-Tyne from the 17th century to the present day. Early glass-makers included the DAGNIA FAMILY, and other glass-workers who were brought by SIR ROBERT MANSELL to his large factory there. The glass has been long noted for its softness, weight, and colour, which made it very suitable for ENGRAVING; much was exported in the 18th century to Holland for decoration with WHEEL ENGRAVING and STIPPLING. Newcastle was an important centre for window glass, but it also produced glassware used for ornamental engraved decoration, such as stemmed wine glasses, JACOBITE GLASSWARE and WILLIAMITE GLASSWARE and in the 19th century CASED and PRESSED GLASS, as well as much COUNTRY-MARKET GLASS. Its glass was used for ENAMELLING by the BEILBYS. The term 'Newcastle glass' is often used to refer to LIGHT BALUSTER glasses of the type decorated by the Dutch.

newel ornament. A solid glass ball (7.5 to 10 cm. in diameter), used to decorate the top of a newel post. Some were made of MILLEFIORI glass. It was usually made with an integral glass peg by which it was dowelled into the newel post, but such pegs have often been broken off. The French term is *boule-de-rampe*.

nickel. A metallic agent used in colouring glass, producing in POTASH glass a deep violet or purple, but in SODA-LIME GLASS it produces yellow. It was first used as a DECOLOURIZING AGENT in the second half of the 18th century.

nipple shield. A small conical object with a central hole that is intended to be placed over a woman's teat to protect her during feeding. Some have been made of glass. *See* BREAST GLASS.

nipt-diamond-waies. A style of network decoration on English glassware, somewhat in the form of diamonds or lozenges made from applied or mould-blown glass threads pincered together at regular alternate intervals (hence 'nipped diamond-wise'). The term was used by GEORGE RAVENSCROFT in his price list of 1677, and pieces so decorated were made in London and elsewhere until well after 1700. The style was used at Newcastle-upon-Tyne in the late 18th century. A similar type of decoration is found on some ROMAN GLASS of the 2nd to 4th centuries and on many types of later ware. The term is often abbreviated 'NDW'.

Non-Such Flint Glass Works. A glasshouse at Bristol, England, operated from 1805 until *c.* 1835 by ISAAC JACOBS. Some gilding may have been done there by MICHAEL EDKINS.

nipt-diamond-waies. Jug of soda glass, heavily crisselled, Savoy Glasshouse, London, before 1675. Ht 21·9 cm. New Orleans Museum of Art (Billups Collection).

Normandy method. *See* CROWN GLASS.

Norsk Glasverk. *See* NORWEGIAN GLASSWARE.

Northwood, John (1836–1902). An English glass-maker who specialized in making CAMEO GLASS. At the age of twelve, he began working for W. H., B. & J. RICHARDSON at Stourbridge, and later for Benjamin Richardson I. In 1860, at nearby Wordsley, he founded (with his brother Joseph as business manager) J. & J. Northwood, and later, in 1881/2, he became technical adviser to STEVENS & WILLIAMS, LTD. In 1873, after nine years' work, he completed the ELGIN VASE, and in 1876 he won a prize for making (after over twenty years of experimentation) a glass copy of the PORTLAND VASE. In 1878 he completed the MILTON VASE and in 1882 the PEGASUS VASE (for THOMAS WEBB & SONS). He invented, *c*. 1880, a machine to make herring-bone decoration in glassware by pulling glass threads embedded in a molten glass object. His work was continued by his son JOHN NORTHWOOD II and by his followers (including GEORGE and THOMAS WOODALL), who founded the Northwood School at Stourbridge; designing was undertaken there for several firms until *c*. 1890, when methods of decoration less costly than cameo work were introduced. *See* John Northwood II, *John Northwood* (1958).

Northwood, John, II (1870–1960). The son of JOHN NORTHWOOD, who made CAMEO GLASS inspired by the work of his father. In 1906 he completed a cameo PLAQUE 38 cm. in diameter depicting the 'Birth of Aphrodite'. *See* SILVERIA GLASSWARE.

Norwegian glassware. Glassware has been made since the mid-18th century in Norway, first by Nöstetangen Glasverk (tableware and chandeliers, from 1741 to 1777), then by Hurdals Verk (CROWN GLASS from 1755, and later chandeliers and tableware) and HADELANDS GLASVERK (bottles from 1765). In 1753 the glass industry was reorganized by Casper von Storm, who received in 1760 a state monopoly for Norway and Denmark that continued until 1803, after which several privately-owned glassworks were founded. Production was greatly influenced by JAMES KEITH, who was brought from England in 1775, and by HEINRICH GOTTLIEB KÖHLER who joined Nöstetangen in 1756/57. From 1777 Hurdals Verk took over the making of fine glass from Nöstetangen Glasverk, but transferred it in 1808 to Gjövik Glasverk, which closed in 1843. The factories were all taken over in 1852 by new owners who introduced industrial methods, and their descendants, the Berg family, control them today. The Hadelands factory was enlarged and became the only factory until recent times that made tableware and ornamental glassware; it engaged SVERRE PETTERSEN as designer. A recent factory is Norsk Glasverk, at Magnor, making decorative glassware and tableware since 1950. For a discussion of Norwegian glassware, *see* Ada Buch Polak, *Gammelt Norsk Glass* (1953), with an English summary.

Norwich. *See* LYNN GLASSWARE.

Nöstetangen Glasverk. *See* NORWEGIAN GLASSWARE.

Notsjö Glassworks. A Finnish glassworks, located at the village of Nuutajärvi; founded in 1793, it became the leading Finnish glasshouse in the second half of the 19th century, and since 1950 it has belonged to the Wärtsilä Group, a large industrial concern. Since 1971 its glass products have been marketed under the trade name 'Arabia'. The products are mainly tableware in many modern patterns, but some ART GLASS is produced, including plates with concentric circles of various colours designed by KAJ FRANCK. Other designers include Oiva Toikka (b. 1931) and Heikki Orvola (b. 1943). *See* FINNISH GLASSWARE.

Novogrudok glassware. Glassware excavated in 1958/62 at Novogrudok, in southern Russia, and attributed to a glassworks there from the

Notsjö glassware. 'Lollipop Trees', glass sculpture by Oiva Toikka, early 1970s. W. 1·20 m. Notsjö Glassworks, Nuutajärvi, Finland.

12th century. The ware, mostly fragments, consists of (1) BEAKERS (called STOPKA); (2) thin elongated blue BOTTLES with a narrow neck, a small mouth, and a rounded bottom, some decorated with a painted dove; and (3) pieces of a BEAKER of the type known as a HEDWIG GLASS. It has been stated by B. A. Shelkovnikov that the fragments are so similar as to indicate that they all came from a single glassworks. The glass is varied-coloured, colourless, and opaque white – and some pieces are double-layered glass; some examples are painted with gilt or with white and red enamel.

Nuppenbecher (German). Literally, drop beaker. A type of BEAKER made of WALDGLAS, cylindrical or slightly conical, tall or short, that is characteristically decorated with an overall pattern of numerous large drops of glass on the lower part. The drops were melted on by touching the vessel with heated glass at the end of an iron rod, after which the drops were sometimes drawn upward or outward to a point; on some examples of colourless glass the drops are decorated by the addition of a small opaque coloured drop (PEARL) as an 'eye'. Such beakers were made in Germany and Bohemia, and probably also in the Netherlands, from the 14th century, with the early examples having the drops arranged irregularly, but by the 15th and 16th centuries in diagonal or vertical rows. The smaller glasses vary in height from 6 to 10 cm., these having one or two horizontal rows of drops, but some examples may be up to c.26 cm. in height (Bohemian examples up to c.46 cm.), these having 5 to 9 rows of drops. Many examples have an applied crinkled or toothed FOOT-RIM. A few *Nuppenbecher* have a footed stem, occasionally decorated with threading or engraving, some have a KICK-BASE, and some a COVER. Some related beakers, instead of having straight sides, have a cup-shaped mouth curving outward from above an encircling thread; these have been termed a KRAUTSTRUNK. *See* WARZENBECHER.

Nuremberg glassware. Glassware decorated at Nuremberg since the 16th century, mainly with shallow WHEEL ENGRAVING and with ENAMELLING in SCHWARZLOT. Glass engraving at Nuremberg dates from 1622 when GEORG SCHWANHARDT returned from Prague. Other leading glass engravers at Nuremberg were his sons GEORG and HEINRICH and GEORG FRIEDRICH KILLINGER, HERMANN SCHWINGER, JOHANN HEEL, ANTON WILHELM MÄUERL, HANS WOLFGANG SCHMIDT, PAULUS EDER, and CHRISTOPH DORSCH. Such engraving continued until c.1725, but from c.1700 was progressively superseded by BOHEMIAN GLASSWARE which, being POTASH GLASS rather than SODA GLASS, was more suitable for deep cutting (*see* CUT GLASS). The engraved decoration at Nuremberg, due to the quality and thinness of its glass, was shallow, sometimes with some supplemental DIAMOND-POINT ENGRAVING, and in the style known as BLANKSCHNITT; the glasses were often made with a hollow-knopped STEM and with MERESES. Enamelling was also done at Nuremberg, especially in *Schwarzlot* after its introduction by JOHANN SCHAPER, c. 1655; others who did like work were JOHANN LUDWIG FABER and ABRAHAM HELMHACK; translucent colours were also used there.

Nyman, Gunnel (1909–48). A glass-designer (originally a designer of furniture) who worked for RIIHIMÄKI GLASSWORKS after the death of HENRY ERICSSON; after achieving recognition at the 1937 Paris Exhibition, she designed also for other factories, such as IITTALA GLASSWORKS and NOTSJÖ GLASSWORKS. She used a variety of techniques and styles, often featuring vases with part 'folded' over the other side of the body, and also pieces decorated by cutting and SAND-BLASTING.

Nuppenbecher. Clear glass with coloured 'eyes' (pearls), German, c.1650. H. 21·5 cm. Kunstsammlungen der Veste Coburg.

occhi, vetro. Vases of coloured opaque glass, with mosaic pattern in clear glass, Venini, Murano.

œnochoë. Roman core glass with combed decoration, *c.* 100 BC. Ht 12·6 cm. Römisch-Germanisches Museum, Cologne.

obsidian. A volcanic mineral that is the earliest from of natural glass. It is usually black or very dark, the splinters of which are transparent or translucent and have a bright lustre. It has been found in Mexico and was much used in South America for primitive tools. It was used by the Egyptians in pre-dynastic times. Some BLACK GLASS, due to the similarity, is called 'obsidian glass'.

Ochsenkopf (German). Literally, ox-head. A type of BEAKER (sometimes a GOBLET or a miniature TUMBLER) decorated with an enamelled depiction (in several versions) of an ox-head above a symbolic fir-covered mountain from which flow the four rivers, the Main, Naab, Saale, and Eger. The mountain, called the *Ochsenkopf*, is in the Fichtelgebirge region of Upper Franconia in northeast Bavaria. The mountain, on top of which is often depicted a church or house, is usually encircled by a painted chain fastened with a padlock, which are said to protect its riches, all as explained by a lengthy inscription on the reverse. Added decoration on some examples includes a coat-of-arms, symbolic motifs, flowers, and figures. The pieces were made mainly at Bischofsgrün, in Upper Franconia, from 1656 until the late 18th century, and are sometimes called a '*Fichtelgebirgsglas*'. *See* ENAMEL-LING, *German*. [*Plate X*]

occhi, vetro (Italian). Literally, glass with eyes. A type of ornamental glass developed by VENINI, the characteristic feature of which is the coloured or white opaque body decorated with a mosaic pattern of pieces of clear glass of various irregular shapes and sizes, arranged more or less in rows.

œil-de-perdrix (French). Literally, partridge eye. A diaper pattern developed and principally used at Vincennes-Sèvres for decorating porcelain after *c.* 1752; it consists of dotted circles, hence the name. It was used as a ground decoration on some OPAQUE blue glass made at Bristol, *c.* 1840, to simulate Sèvres porcelain. *See* SÈVRES-TYPE VASE.

œnochoë (Greek). A JUG with an ovoid BODY, a vertical loop handle, a flat base, and usually a TREFOIL (pinched) LIP. Examples occur in many shapes and sizes, some having a circular mouth, a beaked lip or a spout. There are two types: (1) with the neck set off from the shoulder; and (2) with a continuous curve from the neck to the body (the OLPE is a form of this type). Those of Greek pottery were made to transfer wine from a KRATER to drinking cups. There are glass *œnochoë* of small size (4 to 9 cm.) used as receptacles for toilet preparations; these, made in the eastern Mediterranean region during the 15th–14th centuries BC and later, *c.* 600–400 BC, and of ROMAN GLASS, *c.* 100 BC–AD 100, are CORE GLASS vessels having a pinched lip and usually COMBED DECORATION.

ogee. A shape or ornamentation in the form of a double curve, as in the letter S. It is sometimes continued to form a DOUBLE OGEE.

Öhrström, Edvin (b. 1906). A Swedish sculptor who was a designer at ORREFORS GLASBRUK from 1936; he worked with GRAAL GLASS, developed ARIEL GLASS, and designed for abstract ENGRAVING.

oil gilding. *See* GILDING, *unfired*.

oil lamp. Any type of LAMP using oil and a wick. Glass oil lamps were made at MURANO in the 16th–17th centuries in fantastic forms, e.g., shaped like a running horse or other animal. Glass oil lamps made in the early 18th century were in the form of a flattened glass globe (FONT) with a hole in its top for a wick, the globe resting on a stand having a high stemmed FOOT, and having a loop-handle attached to the stem near its top and bottom; some had a drip-pan to catch any spilled oil. Later examples were in the form of a globe, with multiple (2 or 3) projecting spouts for wicks, and usually having one side handle. Examples from Spain, 18th century, are of varied shapes, some with multiple spouts for wicks. In the United States they were made in a great variety of shapes and sizes. *See* WINE-GLASS LAMP; SPARKING LAMP; WHALE-OIL LAMP; PEG LAMP; SINUMBRA LAMP.

ointment jar. A small shallow BOWL with a COVER used principally to store ointments and unguents (some were for cosmetics and dentifrices). There are circular examples in ROMAN GLASS from the 1st century AD, about 4 cm. in diameter. *See* PYXIS.

Oliva, Ladislav (1933–   ). A Czechoslovakian glass-designer and technician who has specialized in designing LEAD-GLASS vases and plates for decorating by SAND-BLASTING, creating unique and original patterns. His pieces have been executed at Nový Bor since 1959.

Olive Glassworks. A glasshouse founded in 1781 at Glassboro, New Jersey, by five Stanger brothers, former employees of the WISTAR glassworks. It made SOUTH JERSEY TYPE GLASSWARE. Some of its workmen founded in 1813 the Harmony Glassworks, at Glassboro, in which an interest was acquired *c.*1835 by Thomas H. Whitney who, with his brother, Samuel A. Whitney, changed the name to Whitney Glassworks. It later became a unit of the Owens Bottle Co., and then of Libbey-Owens-Ford Co.

Ollers, Edvin (1888–1960). A Swedish painter who was a glass-designer from 1917 at KOSTA GLASSWORKS, where he departed from the tradition of CUT GLASS to make BLOWN GLASS with embedded air bubbles. He later designed PRESSED GLASS for the Elme Glassworks.

olpe (Greek). A wine JUG resembling the OENOCHOË, having a pear-shaped body and a vertical loop-handle extending from the shoulder to the rim. It has a large circular mouth, sometimes with a TREFOIL (pinched) LIP. Generally they were made of Greek pottery, but examples in eastern Mediterranean glass, *c.*400 BC, are known.

omom. A PERFUME SPRINKLER, made of Syrian glass in the 13th century, having an oblate spherical body with a KICK BASE and a tall, slender, attenuated vertical NECK. It is a generally used Islamic shape, which also occurs in Syrian 13th-century enamelled and gilt glassware. At the base of the neck there is sometimes an attached ornament. Known examples have COMBED DECORATION or enamelled decoration with mixed Islamic-Chinese motifs. They are usually between 12 cm. and 20 cm. in height. The name is said to be a faulty transliteration of the Syrian word *qumqum*.

omphalos bowl. A shallow BOWL with a low rim that has a flat bottom with a central concavity and corresponding bulge on the interior. (The word *omphalos* – = navel – is also used to refer to the boss of a shield.) Those of glass are similar to the Greek *phiale mesomphalos* and the Roman *phalera* made of pottery. They were probably made from the 8th century BC.

onyx glass. (1) A type of glass that is dark coloured, usually brown, with streaking of white and other colours, made by mixing molten glass of various colours. *See* AGATE GLASS. (2) A type of decorative glass patented by George W. Leighton in 1889 and made by Dalzell, Gilmore, Leighton & Co., of Findley, Ohio. It was made by mixing in the BATCH metallic elements that produced a silver or other lustre when heated.

Ochsenkopf. Goblet dated 1737, Bischofsgrün, Franconia. Ht 20·8 cm. Kunstsammlungen der Veste Coburg.

omom. Perfume sprinkler of greenish glass with red and blue enamelling, Syrian, probably from Aleppo, 13th/14th century. Ht 12·8 cm. Wheaton College, Norton, Mass.

omphalos bowl. Opaque red glass with flaky weathering, Roman glass, 1st century AD. D. 14·6 cm. Wheaton College, Norton, Mass.

The rims of such objects were not FIRE POLISHED (due to the risk from reheating) but were ground and so remained somewhat rough.

opal glass. A TRANSLUCENT white glass, partially opacified with oxide of tin. Seen by transmitted light it shows brownish or reddish tones ('sunset glow'). It was made in Venice in the 17th/18th centuries, and in Germany and Bohemia (opacified with ASHES of calcined bones) in the late-17th/19th centuries. It was also made in England in the 19th century for COUNTRY-MARKET GLASS; although often attributed to Bristol, it was made in many glassworks throughout the country. Some examples are of milk-and-water hue rather than pure white. Many examples of opal glass found in Spain were probably imported. *See* CRANBERRY GLASS; OPALINE; OPAQUE WHITE GLASS.

opalescent glass. (1) The same as IRIDESCENT GLASS. (2) A type of iridescent glass developed by FREDERICK CARDER at STEUBEN GLASS WORKS having the appearance of an opal gemstone. The effect was produced by suddenly cooling the glass object with a direct jet of compressed air, extending over the entire piece or only the area of the design, and then reheating it. It was made in four basic colours (green, pink, yellow, and blue); a variation shaded from opal to yellow was called 'Star Opal'. (3) A type of ART GLASS made in the late 19th century having a raised OPALESCENT white design. The process involved covering a PARAISON of coloured glass with a layer of clear glass containing bone ash and arsenic; this was blown into a patterned mould to produce the raised design and then was reheated which caused the raised portion to become opalescent against the original coloured ground. Such ware was made inexpensively by several glasshouses in the United States, and a version with raised white or coloured design was patented in England in 1889 by Thomas Davidson of Gateshead-on-Tyne.

opaline. A slightly TRANSLUCENT type of glass, opacified with ASHES of calcined bones and coloured with METALLIC OXIDES, usually pastel hues but sometimes whitish like (but less dense than) OPAQUE WHITE GLASS and sometimes of strong colours (e.g. dark blue) or black. The name was derived from *'opalin'* applied at the BACCARAT FACTORY, *c.* 1823. It is said to have been made first at MURANO in the 17th century, inspired by Bohemian coloured glass, and later made in France and since *c.* 1932 again in Murano. The early French examples resemble opaque white glass but later pieces are closer to MILK-AND-WATER GLASS (*Beinglas*). The best French pieces were made *c.* 1840 to *c.* 1870, at the factories of Baccarat, ST-LOUIS, and CHOISY-LE-ROI. It is made in many hues, e.g. pale-green, turquoise, deep-blue, pink, coral, ruby-red, yellow (rare), *bulle de savon* (savonneuse – soap-bubble rainbow hues), *gorge-de-pigeon* (pigeon's neck). In 1612 ANTONIO NERI suggested in his book a formula for making glass in a peach-bloom colour, using MANGANESE. Opaline has been used for making CARAFES, VASES, CANDLESTICKS, but mainly BOXES; some have gilt-bronze MOUNTS. Opaline has been MOULD-BLOWN or MOULD-PRESSED, and sometimes (before *c.* 1835) decorated with COLD COLOURS and (later) ENAMELLING. It was made in the United States after *c.* 1830 by the BOSTON & SANDWICH GLASS CO. *See* OPAL GLASS; PÂTE-DE-RIZ. For a full dicussion of opaline, its makers, forms, and colours, *see* Yolande Amic, *Opalines Françaises du XIX Siècle* (1950). [*Plate VIII*]

opaque. Not transmitting light; some glass so designated is in fact slightly TRANSLUCENT, especially objects having a thin body. *See* TRANSPARENT.

opaque enamel. An ENAMEL that is not TRANSPARENT. It was usually produced by adding tin-oxide (as in the white glaze on faience, maiolica, and delftware). When used to decorate the surface of glassware, it was applied thickly and hence palpable. It was the principal medium for

opaline. Ewer and basin, white opaline with gilt-bronze mounts, French, *c.* 1830. Ht 30·5 cm. Courtesy, Delomosne & Son, London.

decorating glass by ENAMELLING until the development of TRANS-
PARENT ENAMEL, *c.* 1810.

opaque white glass. A type of opacified glass that has the appearance
of white porcelain. It is opacified usually with TIN OXIDE (as was the
white glaze on faience). Experiments to produce it were carried out at
Venice, and succeeded in the latter part of the 15th century. Some rare
16th-century specimens survive. It became popular with the demand in
the 17th and 18th centuries for porcelain from China and later from
European factories, especially objects made with ENAMELLED decoration as done on porcelain. Such glassware was made and decorated
in Venice, France, Germany, Bohemia, and China in the late 17th and
early 18th centuries. The Venetian glass was sometimes decorated with
iron-red enamelling EN CAMAÏEU or with gilt. The motifs were ROCOCO,
often CHINOISERIES, WATTEAU SUBJECTS, and MYTHOLOGICAL
SUBJECTS. The objects included PLATES, TUMBLERS, CANDLESTICKS,
VASES, etc. Such white glass was among the ware ordered in 1667 from
Venice by JOHN GREENE; later it was made in England from the late 17th
century at various places (much of it once attributed erroneously to
BRISTOL), particularly small objects, e.g. SCENT BOTTLES, and for some
of these the enamel decoration has been attributed to MICHAEL EDKINS.
Opaque white glass was also made in Spain, and in Bohemia or Germany
for the Spanish and Portuguese market, and in Venice for use in white
FILIGRANA decoration. It is often referred to as 'milk glass'. The Italian
term is *lattimo*; the German *Milchglas* or *Porcellein-Glas*; the French
blanc-de-lait. *See* OPAL GLASS; MILK-AND-WATER GLASS; OPALINE;
BASDORF GLASS; LATESINO.   [*Plate IX*]

opaque white glass. Vase with *chinoiserie*
decoration, Staffordshire, *c.* 1755–60.
Ht 14 cm. Victoria and Albert
Museum, London.

optic blown. *See* OPTICAL GLASS (2).

optical glass. (1) High-quality LEAD GLASS used for making lenses,
prisms, etc. (2) Glass that has been MOULD-BLOWN and then
additionally blown to increase the size of the object or to soften or
modify the lines of the pattern by blowing into a plain mould. It is
sometimes referred to as 'optic blown'.

opus interrasile (Latin). Literally, work in low relief or embossed. *See*
SILVER-CASED GLASSWARE.

opus sectile (Latin). A MOSAIC panel made with pieces of glass which
were not fused but were embedded in cement.

opus sectile. Mosaic panel of Roman
glass, 100 BC–AD 100. W. 16 cm.
Victoria and Albert Museum, London.

Orlov Glasshouse. A glasshouse at Miliatino, Russia, purchased
before 1828 by M. F. Orlov, which produced much artistic glass and
tableware until after 1839. It was later acquired by S. I. Mal'tsev, but
from 1884 to 1894 it was taken over by the State owing to debts. *See*
RUSSIAN GLASSWARE.

ormolu (from French *or moulu*). Literally, ground (i.e. pulverized)
gold. An alloy of copper, zinc, and tin having the colour throughout of

gold. It is similar in appearance to gilded brass (an alloy of copper and zinc) and gilded bronze (an alloy of copper and tin), and originally the term referred to the latter. Today the term is also loosely applied to brass or bronze gilded by the process of fire gilding or mercuric gilding or even to copper so gilded (a cheap soft imitation). Ormolu has been used in England as a MOUNT or an embellishment to objects of porcelain or glass, but not to the extent that gilded bronze has been used in France. Many ormolu mounts, used for a wide variety of objects, were made by Matthew Boulton (1728–1809) of Birmingham, in the style of the French work. Mounts of this type in France (*bronze doré d'or moulu*) were usually made of bronze gilded by the mercuric process (*dorure d'or moulu*, gilding with gold paste). *See* Nicholas Goodison, *Ormolu* (1974).

Orpheus Beaker. An engraved covered BEAKER that is the most celebrated work of GOTTFRIED SPILLER. It depicts Orpheus wearing a laurel-wreath at the foot of a tree-trunk and playing his lyre to a group of beasts. Although unsigned, it has been attributed since 1786 to Spiller, and serves to identify his other work. A vase with the same motif was designed by VICTOR PROUVÉ for ÉMILE GALLÉ.

Orrefors glassware. Glassware of several types made by Orrefors Glasbruk A–B, a leading Swedish glasshouse at Orrefors, in the main glass-producing Småland region. The factory started in 1726 as an ironworks which was later abandoned and in 1898 converted into a glassworks, producing medicine and ink bottles, window panes, and cheap domestic glass. It was acquired in 1913 by Johan Ekman who, with his manager Albert Ahlin, decided to change to decorative glass. They brought in as designers SIMON GATE (in 1916) and EDWARD HALD (in 1917) who learned glass techniques from the factory's master glass-blower, Knut Bergqvist (from 1914) and who extended the techniques of EMILE GALLÉ and created new types of ART GLASS. (*See* GRAAL GLASS; ARIEL GLASS; KRAKA GLASS; RAVENNA GLASS.) Later designers included VICKE LINDSTRAND (at Orrefors 1928–41), EDVIN ÖHRSTRÖM (from 1936), SVEN PALMQVIST (from 1936), NILS LANDBERG (from 1925), INGEBORG LUNDIN (from 1947), JOHN SELBING, GUNNAR CYRÉN (from 1959), Carl Fagerlund (b. 1915), and Eva Englund (b. 1937). In 1946 the company was purchased by Henning Beyer. It has an affiliated glassworks at Sandvik that produces utilitarian glassware. The company has made much engraved glassware but is best known for its tableware and ornamental CASED glass. It is also a producer and exporter of large CHANDELIERS and stemware.

osculatory. A pax; an early Catholic ecclesiastical tablet bearing a religious representation, which was kissed first by the priest and then by the congregation. Some were made of Italian glass, decorated with DIAMOND-POINT ENGRAVING, especially in the 15th and 16th centuries.

osier glassware. Glassware PRESS-MOULDED with an overall pattern in the form of interlaced willow-twigs, to resemble basketwork. It was made of OPAQUE WHITE GLASS in the late 19th century, especially by SOWERBY'S ELLISON GLASSWORKS.

Osler, F. & C., Glasshouse. A large glassworks at Birmingham, England, founded in 1807, which specialized in large objects in which glass was used in conjunction with metal framework. It made tableware and ornamental glass, much being exhibited at the CRYSTAL PALACE in 1851, but was best known for its large constructions, such as CHANDELIERS 20 ft (6·15 m.) high and the fountain 27 ft (8·30 m.) high, made for the Great Exhibition.

overblow. The excess portion of the PARAISON attached to the BLOW-PIPE or PONTIL, which is removed when the object itself has been CRACKED-OFF. It is also called 'moil'.

overlay. (1) The outer layer of glass on CASED GLASS. Some paper-weights (*see* PAPER-WEIGHTS, *overlay*) and other overlay objects have

overlay. Covered goblet, blue overlay engraved by Karel Pfohl, Bohemia, *c.*1850. Museum of Decorative Arts, Prague.

two outer layers of different colours ('double overlay') and sometimes a thin FLASHING of clear glass to protect the coloured overlays (called 'encased overlay'). Some CHINESE SNUFF BOTTLES are made of overlay glass with two to four layers of different coloured glass and with CAMEO cutting to produce a design against the ground colour; others have a single overlay of several colours applied to different areas of the same ground. (2) A type of decoration, sometimes called 'overlay type', made by attaching and fusing to the body of an object ornamentation in the form of glass flowers, fruit, leaves, etc.; this was made in the 19th century by several English and American glasshouses.

oviform. Ovoid or egg-shaped.

Owens, Michael J. A glass-technician of Toledo, Ohio, who in 1901, while superintendent of production at the LIBBEY GLASS CO., invented the automatic bottle-blowing machine which gathered the molten glass by suction and formed the bottle in a single operation.

Owens-Illinois, Inc. A leading glass-producer of the United States. It grew from the Owens Bottle Machine Co., founded in 1903 at Toledo, Ohio, by MICHAEL J. OWENS with financing provided by his associate, Edward Drummond Libbey, who controlled the LIBBEY GLASS CO., also of Toledo; Owens, the superintendent of production at the Libbey Glass Co., had invented in 1901 the automatic bottle-blowing machine, and the company developed inventions for automatic manufacture of tumblers and bottles. In 1929 it merged with Illinois Glass Co. to become the Owens-Illinois Glass Co. (later Owens-Illinois, Inc.). In 1936 it acquired the Libbey Glass Co. which is now its division making tableware.

ox-head glassware. Drinking glasses, usually HUMPEN, with enamelled decoration of an ox-head among fir trees; see OCHSENKOPF.

[Plate X]

overlay. Chinese snuff bottle of snowflake glass with red overlay, 18th century. Ht 7 cm. Courtesy, Hugh Moss, Ltd, London.

P

PF (or **PP**) **glassware.** Opaque white glassware enamelled by an unidentified hand (one piece is initialled 'PF' or 'PP'), featuring floral bouquets, and also peacocks and CHINOISERIES on a 'floating raft'. It has been said that the decorator was possibly from Newcastle, and that his work has been erroneously attributed to MICHAEL EDKINS. *See* R. J. Charleston, 'A painter of Opaque-white glass,' *Glass Notes, Arthur Churchill, Ltd*, 13 (1953), p. 13.

pad-base. A base of a vessel made by applying to the bottom an additional small gathering of molten glass and spreading it out to form a ring. *See* BASE-RING.

pad pontil mark. A type of PONTIL MARK resulting from applying an extra GATHER of glass to the bottom of an object before it was removed from the BLOW-PIPE to the PONTIL. It was intended to reinforce a thin base.

painted decoration. Decoration on glassware by means of (1) COLD COLOURS, or (2) ENAMELLING.

painting. (1) The decoration of glass mirrors (*see* MIRROR PAINTING) by painting on the surface, as was done in England in the 18th century. Dutch artists specialized in the work. (2) Painting on the front or the reverse of glass. *See* PANEL; SCENIC DECORATION; VENETIAN SCENIC PLATES.

pair. Two objects of the same kind and form, e.g. figures, suited to each other and intended to be used as a unit or to stand together. Pairs are not necessarily identical but are related.

Pairpoint Manufacturing Co. A glasshouse at New Bedford, Mass., founded in 1865, which took over in 1894 the MOUNT WASHINGTON GLASS CO. From 1952 it was known as Gunderson-Pairpoint Glassworks; in 1957 it moved to East Wareham, Massachusetts, as the Pairpoint Glassworks, and ceased operating in 1958. *See* PAPERWEIGHT, *Pairpoint*.

pall. A ceremonial covering to be spread over a coffin. Some were made in China in the late Chou dynasty (*c*.400–240 BC), perhaps at Ch'ints'un, consisting of many small rectangles of glass (originally with inlaid gilt ornamentation) stitched together vertically and horizontally by small loops of wire (in the same manner as the jade funeral suits of Prince Liu and his wife Lady Tou, made in the early Han dynasty, 2nd century BC).

pallet. A glass-maker's tool, similar to a wide spatula, having a square piece of metal or wood attached to a handle. It was used to flatten the BASE of a JUG or other such vessel that was being made. *See* TOOLS, GLASS-MAKERS'.

palm-column vase. A thin vertical cylindrical vase decorated at the top with spreading palm fronds. They were made of CORE GLASS with COMBED DECORATION, and sometimes have a thin accompanying wooden stick for the application of eye paint (kohl; *see* KOHL TUBE). They were made of EGYPTIAN GLASS in the late XVIIIth and XIXth Dynasties (*c*.1400–1250 BC).

palm cup. A small CUP of hemispherical shape that can be held in the user's palm. It has no HANDLE and sometimes has a KICK in the BASE.

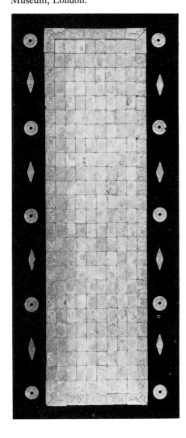

pall. Glass plaques with incised designs, Chinese, Chou dynasty, *c*.400–240 BC. L. 1·53 m. British Museum, London.

IX *Opaque white glass* (milk glass). Group of 18th-century objects with enamelled decoration, made in Bohemia, France and Germany. Courtesy, Christie's, London.

X *Ox-head glass* (*Ochsenkopfhumpen*). Beaker of clear glass with enamelled decoration, Bischofsgrün, Upper Franconia, 1716; ht 24 cm. (9½ in.). Courtesy, Cinzano Collection, London.

Undecorated examples have been made of FRANKISH GLASS. Similar shaped pieces were made later of ISLAMIC GLASS and, in the 16th century in the Rhineland, of WALDGLAS; these are sometimes decorated with MOULDED rosettes in raised dots. They were also made in England in the mid-19th century, marked 'Elfin' with gilt inside.

palm-leaf decoration. *See* COMBED DECORATION.

Palme, Franz Joseph. A Bohemian engraver who worked at the DENNIS GLASSWORKS of THOMAS WEBB & SONS from 1882. He specialized in engraving animals, and signed much of his work.

Palmqvist, Sven (1956–). A Swedish glass-designer, working at ORREFORS GLASBRUK from 1936, who developed the techniques for making KRAKA GLASS and RAVENNA GLASS. He also invented a process of making glass bowls by centrifugal force without any hand finishing. Among his outstanding works are a glass wall 'Light and Darkness' erected at Geneva, made from about 200 blocks of Ravenna Glass, and a similar project in the form of a column exhibited in 1967 at Montreal.

palm cup. Bluish glass with fluting, Anglo-Saxon, 6th/7th century. Ht 7·5 cm. British Museum, London.

Palmqvist, Sven. Kraka glass vase. Courtesy, Orrefors Glasbruk.

pane glass. Glass in flat sheet form, e.g. most WINDOW GLASS. It was first made by the process of CASTING. Later different techniques were employed to make CROWN GLASS, BROAD GLASS, and PLATE GLASS. A recently developed method is that used to make FLOAT GLASS. Flat sections cut from all such types are termed 'panes'. *See* RULLI.

panel. A flat tablet, of ceramic ware or glass, generally used as a fixed decoration set into or suspended on a wall. Glass examples were decorated in RELIEF, with ENAMELLED scenes or with MOSAIC designs. Some decorated with relief may have been used as PLAQUES. Mosaic examples were made at Alexandria, 100 BC–AD 100; *see* OPUS SECTILE. Some made at MURANO of OPAQUE WHITE GLASS are enamelled with BIBLICAL SUBJECTS or with scenic views in the style of Canaletto and other painters. *See* SCENIC DECORATION.

panelling. A decorative pattern of a series of adjacent or close parallel vertical depressions in the form of arches, with a rounded or oval top. The panels were made by blowing the PARAISON into a DIP-MOULD made with the design. The panels were deeper when made in a piece formed with two GATHERS of molten glass. The pattern was used in 18th-century American glassware, especially vases, some of which may have been made at the STIEGEL glassworks. It was also used in England in the early 19th century. Sometimes called 'sunken panels'.

Pantin Glassworks. A glasshouse founded in 1851 by E. S. Monot at La Villette, near (now in) Paris and moved in 1855 to Pantin (Seine), near Paris. After several changes of name (including Monot & Stumpf; Monot Père et Fils & Stumpf; Stumpf, Touvier, et Viole(t)e & Cie.), it became, c. 1900, the Cristallerie de Pantin, operated by Saint-Hilaire, Touvier, de Varreux et Cie. It is known to have made PAPER-WEIGHTS; some unidentified weights, with flowers, lizards, etc., may have been made by it. After World War I it merged with LEGRAS & CIE, and operated as 'Verreries et Cristalleries de St-Denis et Pantin Réunies', using the marks 'Pantin' and 'De Vez' (the pseudonym of its Art Director De Varreux). It also used the mark 'Degué', the name of one of its master glass-makers.

pan-topped. The shape of the BOWL on some glass pieces, the curved upper part of the bowl being wider than the curved lower part and being somewhat pan-shaped, e.g., the bowl of a pan-topped drinking glass or SWEETMEAT GLASS or the rim of a DECANTER that is so-shaped. A bowl that is 'saucer-topped' is somewhat shallower, and one that is 'cup-topped' or 'bucket-topped' is progressively deeper.

paper-weight. A small decorative object, usually of solid glass and therefore heavy for its size, made originally for the purpose of holding down papers on a desk, but in recent years they are much sought as collectors' items. They are usually circular (about 8 cm. in diameter) but some are pentagonal, scalloped, etc. They are generally high-domed and magnify the interior motif, but some rare ones are flat-domed and some modern ones are fancifully shaped. Prized examples are of OVERLAY glass with WINDOWS that reveal (and reduce the apparent size of) the interior motif. Some are decorated with slices (set-ups) of millefiori CANES or subject motifs, often superimposed on an intricate GROUND or CUSHION or with a diamond- or star-cut bottom. Values depend more on rarity of decoration than the infrequently marked dates. *See* STAVES; BRUISE.   [*Plate XI*]

The earliest were made in MURANO and Bohemia, *c.* 1843/45. The earliest recorded dated example was made with FILIGRANA decoration at Murano by PIETRO BIGAGLIA, who also made millefiori weights. The finest were made in France from 1843 onwards by BACCARAT, ST-LOUIS, and CLICHY. English paper-weights were made from *c.* 1848, and American paper-weights from *c.* 1851. There was a revival of interest *c.* 1878, and weights are still being made. The designs of paper-weights are so intricate and multi-coloured that no two are identical.

Paper-weights are still produced in several countries; the modern weights are often in the style of French examples made *c.* 1845–55, but some are of original design. Weights are still being made: (1) in France at BACCARAT and ST-LOUIS; (2) in England and Scotland (*see* paper-weights, *British* (*modern*) below); (3) in Czechoslovakia, some with large floral motifs in bright colours; (4) among notable pieces being made in the United States are those of CHARLES KAZIUN and PAUL J. STANKARD; (5) in Sweden weights are made at Orrefors (*see* ORREFORS GLASSWARE; some of these are square or circular, with a heart-shaped or circular motif designed by GUNNAR CYRÉN or INGEBORG LUNDIN), and by KOSTA GLASSWORKS (these being cylindrical in form). Also, since *c.* 1930, some have been made in China and Japan; these are readily identifiable, being made of light SODA GLASS high in alkali content and so not as clear as crystal glass.

The French term is *presse-papier*, the Italian *fermacarte*, and the German *Briefbeschwerer*. *See* LETTER-WEIGHT; DOOR-STOP; MERCURY BAND; TORSADE. *See* Paul Hollister, Jr., *Encyclopedia of Paper-weights* (1964), for the history, methods of production, sources, and varieties, with many illustrations; also, Patricia McCawley, *Glass Paper-weights* (1975).

TYPES OF PAPER-WEIGHT

——, *American*. Paper-weights have been made from 1851 by several factories in the United States, including the BOSTON & SANDWICH GLASS COMPANY, NEW ENGLAND GLASS COMPANY, GILLINDER GLASS COMPANY, and WHITALL, TATUM & CO. The decoration is usually in the style of French paper-weights. Modern examples are made by CHARLES KAZIUN and PAUL J. STANKARD.

——, *animal*. A paper-weight whose main decoration is a glass representation of one of a variety of animals, reptiles, birds, etc., made in a mould or hand-made AT-THE-LAMP and embedded in clear glass. Among the motifs are a salamander, a squirrel, a caterpillar, a duck, a butterfly, a parrot, etc.; *see* snake paper-weight, below. These weights are to be distinguished from MILLEFIORI examples where some individual CANES have a motif of an animal. Animal weights, highly prized, were made at BACCARAT, ST-LOUIS, and CLICHY, and they are sometimes included in the general category 'subject paper-weights'.

——, *Baccarat*. A paper-weight made by the BACCARAT GLASS FACTORY. Some are dated from 1846 to 1853, the most usual date being 1848 and the rarest 1849. Most dated examples are also marked 'B'. The majority are of the millefiori type, and many are of the flower and bouquet types (the latter characterized by crossed stalks); others (the most valuable) are subject paper-weights and some have INCRUSTATIONS. Some are OVERLAY, and some have a rare GROUND (e.g., 'carpet', 'star-dust', 'gauze'). Many of the CANES have a SILHOUETTE of persons or animals, and others are of ARROWHEAD type. There are a few magnum and miniature weights. Most have a star-cut base.

paper-weight, barber's pole. Panelled weight with blue-and-white ribbon twists, George Bacchus & Co. D. 6·8 cm. Courtesy, Christie's, London.

———, *barber's-pole*. A paper-weight that is in the general form of the chequer or panelled type, but having the FLORETS separated by sections of white FILIGRANA (LATTICINO) glass encircled by a twist of opaque coloured glass, not necessarily red but sometimes blue or other colour. They were made by CLICHY.

———, *basket*. A paper-weight whose motif is an enclosed basket formed by a circle of STAVES enclosing MILLEFIORI canes. They were made in France by CLICHY and ST-LOUIS. In some Bohemian examples, the staves are topped by twisted ribbons and the basket has an arched handle of twisted ribbons. Such weights are to be distinguished from a rare type of Clichy weight made in the form of a basket with vertical staves enclosing millefiori canes.

———, *Bohemian*. A paper-weight made in Bohemia or Silesia, from *c*.1845, the earliest dated piece being 1848, by which date the examples compare favourably with the weights of the three great French factories.

———, *bottle-green*. *See* DOOR-STOP.

———, *bouquet*. A paper-weight the central motif of which is not an individual glass flower (*see* flower paper-weight, below) but two or more flowers (natural or stylized) with leaves, arranged in a bouquet or graland. Some have a FLAT BOUQUET, others a STANDING BOUQUET (often within a TORSADE). Some are of the overlay type with windows; a few have faceting on the sides. Such weights by BACCARAT and sometimes CLICHY have crossed flower stalks; examples by Clichy often have the flowers tied at the base by a ribbon. Also termed a 'spray paper-weight'. *See* posy paper-weight and tricolore paper-weight, below.

———, *British (antique)*. (1) Stourbridge: millefiori weights made at Stourbridge from *c*.1849 in the style of CLICHY weights but with noticeable differences. (2) GEORGE BACCHUS & SON: millefiori weights and some with white FILIGRANA decoration, inspired by Bohemian glass, made since *c*.1848. (3) WHITEFRIARS GLASS WORKS: weights with coloured FLORETS and sometimes faceted paper-weights, said to have been made *c*.1848. (4) ISLINGTON GLASSWORKS: weights somewhat similar to those made at Stourbridge; examples have a CANE with a silhouette of a horse, and are marked 'I G W'.

———, *British (modern)*. A contemporary paper-weight made in the United Kingdom. Makers include: (1) WHITEFRIARS GLASS WORKS; (2) PAUL YSART; (3) MONCRIEFF GLASSWORKS; (4) STRATHEARN GLASS LTD. (5) PERTHSHIRE PAPER-WEIGHTS, LTD: (6) CAITHNESS GLASS, LTD. (7) Wedgwood (*see* WEDGWOOD GLASSWARE).

———, *British (Victorian)*. Paper-weights were popular in the Victorian period; the usual form, with a domed clear magnifying glass, is decorated by having on the bottom a coloured print (or later, *c*.1890, a photograph) of some well-known resort, local sight, or street scene of England or the United States. They were first used *c*.1850 or sooner. The glass was usually made in Belgium or Germany, and sent to England for the picture to be affixed.

———, *butterfly*. A paper-weight whose central motif is an embedded glass butterfly, either alone or hovering over a single light-coloured flower with green leaves. The butterfly usually has a bluish body, with black head and antennae, and multi-coloured marbled outspread wings; a rare type has folded wings. Some examples have an encircling border of millefiori FLORETS or garlands, and some have a muslin GROUND. Such weights were made by BACCARAT, and occasionally by ST-LOUIS; modern examples are known.

———, *Caithness*. Paper-weights of original design made by CAITHNESS GLASS, LTD, mainly designed by Colin Terris since 1968 and executed by Peter Holmes since 1969. They are mainly made with abstract patterns inspired by the eleven planets and other space themes, but some have floral motifs. Many include powdered coloured glass or non-glass inserts. Most weights are issued in marked limited editions, but some are made one of a kind.

———, *candy*. The same as a scrambled paper-weight (see below).

———, *carpet*. A paper-weight with a carpet GROUND, having interspersed CANES with FLORETS and SILHOUETTES. Weights with carpets of tightly-packed canes arranged in a rare design are highly

paper-weight, basket. Overlay weight, turquoise, cut with six windows showing basket of staves enclosing millefiori and rose canes, Clichy. D. 8 cm. Courtesy, Sotheby's, London.

paper-weight, butterfly. Opaque ground within millefiori circle, Perthshire Paperweights, Ltd. Courtesy, Spink & Son, Ltd., London.

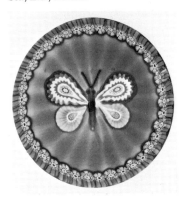

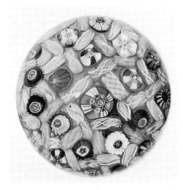

paper-weight, chequer. Florets separated by white *filigrana* stripes, Clichy. D. 8·8 cm. Courtesy, Christie's, London.

paper-weight, crown. Weight with twisted ribbons radiating from central set-up, Baccarat. D. 6·4 cm. Courtesy, Christie's, London.

paper-weight, cut-base. Posy weight with strawberry-cut base, St Louis. D. 7·2 cm. Courtesy, Christie's, London.

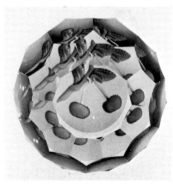

paper-weight, faceted. Cherry paper-weight with overall faceting, Baccarat, c. 1848–60. Courtesy, Spink & Son, Ltd, London.

prized. Weights with carpet grounds were made by BACCARAT and ST-LOUIS.

——, *chequer*. A paper-weight in which scattered or patterned MILLEFIORI or FLORET canes are separated by short lengths of canes (usually of FILIGRANA glass, opaque white or coloured) arranged in a chequered pattern or even irregularly arranged, and sometimes having also an encircling border of such lengths of canes.

——, *circlet*. A paper-weight with a circle of FLORETS, usually found with flower paper-weights (see below).

——, *Clichy*. A paper-weight made at the CLICHY GLASSWORKS. Examples are known from c. 1846 to 1870. Some have a CANE marked 'C' and a very few spell out the full name, sometimes split on two canes. The distinguishing mark of many Clichy weights is the CLICHY ROSE. Most of the millefiori weights have FLORETS arranged to form a design, e.g., loops or garlands, interspersed with florets and PIE-CRUST CANES, or concentric patterns, but some are scrambled. Some have a muslin or moss GROUND, and many have faceting or a star-cut base. There are also miniatures (see below) of the same styles.

——, *cluster*. A paper-weight whose decoration is a cluster of CANES arranged in concentric fashion, repeated several times. The canes in each cluster usually have a similar floral motif, but some are often with SILHOUETTES or are PIE-CRUST CANES.

——, *concentric*. A millefiori paper-weight in which FLORETS are arranged in an overall pattern of concentric circles. The smallest inner circle of some examples was sometimes made from SHAMROCK CANES. The canes are often arranged within a BASKET made of STAVES of two alternating colours.

——, *cross*. A type of close millefiori paper-weight (see below under *millefiori*) that is divided into quadrants by a cross made of like CANES, with a central FLORET. They were made by ST-LOUIS.

——, *crown*. A paper-weight that is hollow and whose decoration is an embedded crown-like pattern of twisted ribbons, usually of opaque white glass alternating with ribbons of various colours, the ribbons starting at the base and flowing upward to meet at a central MILLEFIORI motif set high in the clear CROWN. Glass objects so decorated were also used as door-knobs and as bases for ornaments. They were made by ST-LOUIS.

——, *cut-base*. A paper-weight (of various types) having the base decorated with a cut pattern, usually a STAR, STRAWBERRY DIAMOND CUTTING, or a small chequerboard pattern. Such cutting is usually found on weights that have only a translucent coloured GROUND or are of clear glass without a ground.

——, *ducks-on-a-pond*. A curious type of hollow paper-weight that has from one to three small glass ducks made AT-THE-LAMP and fused to the interior, apparently swimming above the GROUND. A related example has swimming swans, and another has small walking pigs. They were made possibly by BACCARAT or ST-LOUIS.

——, *encased overlay*. A type of overlay paper-weight (see below) that is cut with WINDOWS, PRINTIES, or other design, and then further encased in clear glass. They were made at ST-LOUIS.

——, *faceted*. A paper-weight having parts of the surface decoratively cut, usually in FACETS, PRINTIES or FLUTES. The bottom of such weights is often decorated with STAR CUTTING or HOB-NAIL or STRAWBERRY-DIAMOND pattern. Some examples have the surface cutting overall, down to the base, with graduated facets of hexagonal, square (sometimes termed 'brick') or honeycomb shape, or only on the top or around the lower part of the weight.

——, *flower*. A paper-weight whose principal motif is an embedded individual flower, as distinguished from a bouquet paper-weight (see above). A wide variety of flowers (some imaginative or stylized) is depicted, but usually all have the same type of leaf. Among the most often seen BACCARAT flowers are the primrose, pansy, and clematis; the rose is rare, and the rarest is the lily-of-the-valley. The most usual ST-LOUIS flowers are the dahlia, clematis, camomile (pom-pom), and fuchsia. Few individual flowers are known from CLICHY except the characteristic CLICHY ROSE. Some flowers (e.g., a marguerite) cover the entire piece. The motif is sometimes within a BASKET made of STAVES of different alternating colours. The GROUND of such weights is usually clear glass, but sometimes it is an opaque white muslin ground.

———, *fruit*. (1) A paper-weight whose central motif is an embedded group of fruit, usually placed on a white FILIGRANA basket or cushion. These were a speciality of ST-LOUIS, where the fruits were usually pears, red cherries (usually 2 or 3, but up to 5) and apples, with occasionally other fruits added, e.g., a plum, orange or lemon. A rare BACCARAT example has a cluster of strawberries with a blossom. (2) A type of paper-weight in the form of a piece of fruit (e.g., an apple or pear), usually resting on a clear-glass cushion base with the stem touching the cushion. Examples made at ST-LOUIS, *c*.1841, are of solid glass realistically coloured; they were perhaps inspired by glass replicas of fruit made at MURANO from *c*.1714 onwards. Similar in appearance are fruit weights made by the NEW ENGLAND GLASS CO.; these are of hollow blown glass, made by a Frenchman, François Pierre, by blowing a glass tube of two colours (the colours blended in the blowing) and then fusing the glass fruit to a base. An example of a free-blown solid glass pear without a base was made at VAL-SAINT-LAMBERT in the 20th century.

———, *garland*. A paper-weight in which FLORETS are arranged to form a single garland or two entwined garlands. Some have a central motif, others have added florets interspersed within the curves of the garlands.

———, *hollow*. A special type of paper-weight that, unlike the usual solid weights, is hollow. They are of several varieties. Some have a design impressed into the inside of the glass crown and some have an inner coloured cushion, so that the entrapped air space creates a silvery appearance. They were made in England, possibly at Stourbridge. Types include: crown; ducks-on-a-pond; marbrie; snow.

———, *incrusted*. The same as a sulphide paper-weight (see below).

———, *Kaziun*. Various paper-weights made by CHARLES KAZIUN. They are usually of the millefiori type or the flower type featuring a rose, and some are overlay. They are signed 'K'.

———, *lizard*. A paper-weight whose motif is an embedded glass lizard. In some examples the glass lizard is not embedded within the weight but is gilded and coiled on top of the CROWN of the weight, as made at ST-LOUIS. *See* marbrie paper-weight (below).

———, *macédoine*. The same as a scrambled paper-weight (see below).

———, *magnum*. A paper-weight that is about 2·5 cm. larger in diameter than the average 8 cm. size. The usual large weights are English, made at STOURBRIDGE. French magnums are rare; BACCARAT specimens, dated 1847 and 1848, are of the millefiori type, with the base cut with a star, raised diamonds, or STRAWBERRY DIAMONDS. ST-LOUIS magnums are usually decorated with concentric rings of millefiori florets.

———, *marbrie*. A type of overlay paper-weight (see below) that is hollow (hence blown) and whose decoration, on an opaque white overlay, is trailed loops (usually quatrefoils or quadrants of festoons) that curve inward toward the centre or outward from the centre. The festoons are of one or more colours (blue, red, green), with a central motif of concentric canes. Some have a moulded lizard resting on the top. They were made at ST-LOUIS. The term is derived from the French *marbre* (marble).

———, *medal*. An unusual type of paper-weight whose central motif is a replica of a medal, the medal being made of thick gold leaf and lying flat but as if suspended by enamelled ribbons. Examples include the Legion of Honour and other orders; they were made by BACCARAT.

———, *millefiori*. The most frequently seen type of paper-weight, decorated with slices (set-ups) of many CANES, usually making a pattern. The canes are closely-packed (termed 'close millefiori'), scattered haphazard with intervening clear FILIGRANA glass (termed 'scattered millefiori'), or arranged in rows or circles with fairly even intervals of

paper-weight, flower. Anemone, with white petals on green stalk, Baccarat. D. 7·4 cm. Courtesy, Sotheby's, London.

paper-weight, fruit (1). White *filigrana* cushion ground, Baccarat, *c*.1845–60. Courtesy, Spink & Son, Ltd, London.

paper-weight, fruit (2). Solid glass apple, naturalistically coloured, St Louis, *c*.1851. Private collection.

paper-weight, lizard. Green glass, surmounted by gilded lizard, St Louis, *c*. 1845–60. Courtesy, Spink & Son, Ltd, London.

paper-weight, marbrie. Turquoise swirls, St Louis. Courtesy, Sotheby's, London.

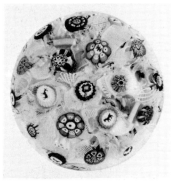

paper-weight, scattered millefiori. Opaque white and coloured canes with florets and silhouettes, Baccarat, marked 'B 1848'. D. 8 cm. Courtesy, Sotheby's, London.

clear glass (termed 'spaced millefiori'). The canes are usually FLORETS but sometimes they show a SILHOUETTE or are shaped like a tiny star, hexagon or geometric figure, or are an ARROWHEAD CANE or a PIE-CRUST CANE. Some millefiori paper-weights are FACETED, some are covered by one or more OVERLAYS with windows. *See* concentric paper-weight, scrambled paper-weight.

——, *Millville*. A paper-weight made since *c*. 1863 by WHITALL, TATUM & CO., and so-called from the location of the factory. The best known are those, usually of pedestal type (see below), enclosing (1) an upright rose motif made *c*. 1905–18 by Ralph Barber (d. 1936), (2) a lily motif made by Emil Stanger or Marcus Kuntz, or (3) a scenic motif made by Michael Kane. *See* JERSEY ROSE.

——, *miniature*. A small paper-weight about 3·5 to 5 cm. in diameter. They are usually similar in pattern to those of normal 8 cm. size, and are generally of the millefiori or flower style, with sometimes a crown type. Most miniatures are from CLICHY, but some are from BACCARAT and ST-LOUIS.

——, *mushroom*. A type of paper-weight whose decoration is a central bundle of CANES that spread out in the form of a mushroom to a domed top supported by a multi-coloured stem. Usually the canes are multi-coloured MILLEFIORI, but sometimes are entirely opaque white, and in rare examples are STAR-DUST CANES. Generally the canes are arranged in five concentric rings, but some examples (usually BACCARAT) have the canes in a random pattern. Sometimes the mushroom is standing in a plain or striped basket. The ST-LOUIS and some BACCARAT weights have an encircling rope (TORSADE) of entwined white and coloured spirals; some ST-LOUIS weights are encircled by a white FILIGRANA ribbon with an added coloured ribbon. Many also have MERCURY BANDS. Some examples are in OVERLAY glass with windows. Alternative terms are 'tuft' and 'pom-pom' paper-weight.

——, *opaline*. A paper-weight that is a rectangular OPALINE glass plaque (about 6 × 8 cm.), and hence without magnification, sometimes shaped like a book. Enclosed in the glass is a single flower or a small flat bouquet. They were made by CLICHY.

——, *overlay*. A highly-prized type of paper-weight that is made with one or more OVERLAYS of coloured glass. There is usually a thin (sometimes hair-line thin) underlay of opaque white glass, then one or two (very rarely three) overlays of coloured glass. The overlays are ground to provide windows (see below, p. 232), one at the top and usually one or two encircling rows of four to six, through which the central motif can be seen. Examples from ST-LOUIS are encased in clear glass after the windows are ground (*see* ENCASED GLASSWARE). The usual colours of the outer overlay are: BACCARAT, turquoise or bright blue, or emerald green; ST-LOUIS, dark blue, emerald-green or apple-green; CLICHY, turquoise or deep pink; one example is known in yellow. At each window the opaque white underlay shows a thin white line.

——, *Pairpoint*. A paper-weight made after *c*.1910 by PAIR-POINT MANUFACTURING CO. The weights were solid glass spheres mounted on a flat-bottomed base and contained glass spirals of glass threads of two colours (white and cobalt blue or ruby red) formed in a vertical ovoid shape surrounded by evenly spaced air bubbles. The base of some bears the engraved name of Carl Banks.

——, *panelled*. A paper-weight in which a number of similar MILLEFIORI CANES are arranged in triangular groups or panels radiating around a central motif, with the canes of the adjacent panels of contrasting colours and motifs. Sometimes the panels are separated by a line of STAVES or of different canes or a ribbon twist.

——, *patterned*. A general term for several types of paper-weight having the canes arranged in a definite pattern rather than in haphazard fashion, e.g., concentric, circlet, garland, panelled.

——, *pedestal*. A type of paper-weight that rests, not on a flat bottom as is usual, but on a small pedestal or a rim at the base that is slightly wider than the body. They were made at the three important French factories and also at Stourbridge. Another term for such weights is '*piédouche*'

——, *pell-mell.* The same as a scrambled paper-weight (see below).

——, *Perthshire.* A paper-weight made by PERTHSHIRE PAPER-WEIGHTS, LTD.

——, *picture.* A rare type of paper-weight whose decorative motif is a painted picture. Examples sometimes have a window on the top.

——, *piédouche.* The same as a pedestal paper-weight (see above).

——, *pinchbeck.* A paper-weight in which is enclosed a decorative piece moulded in high relief and made of pinchbeck (an alloy of copper and zinc used to imitate gold or silver in cheap jewellery; it was invented by Christopher Pinchbeck (1670–1732), a London watch-maker). The weights became popular *c.* 1850, so they were not made by Pinchbeck or his sons; they were made in English and Continental unidentified factories. The design is usually set in a pewter or marble base (sometimes copper, tin, alabaster or cardboard), screwed or cemented to a circular domed magnifying-glass lens which extends laterally beyond the base.

——, *posy.* A paper-weight of which the central motif is a posy or nosegay, being smaller than the motif of the bouquet type (see above). They often have no GROUND but feature a CUT base.

——, *St-Louis.* A paper-weight made by the ST-LOUIS GLASS FACTORY. Some are dated from 1845 to 1849, and rare examples are marked 'SL', alone or with a date. Many are millefiori type (patterned or scrambled), with some CANES showing a silhouette, particularly of dancing figures; there are also some of the flower or subject type (fruits, vegetables, snakes), or crown or overlay. The overlay examples are encased in clear glass and usually have a star-cut or STRAWBERRY-DIAMOND base. There are a few examples of the magnum and miniature types (see above).

——, *salamander.* A paper-weight whose interior decorative motif is a glass salamander. Some examples have a salamander coiled on top of the CROWN.

——, *scattered millefiori. See* above, under *millefiori.*

——, *scrambled.* A paper-weight whose decoration is a mass of different millefiori CANES (some with interspersed pieces of white FILIGRANA) placed in a scrambled haphazard formation. They were made at ST-LOUIS and CLICHY, and also at MURANO. The same technique was also used in making decoration for other objects, such as ink-bottles, bases of vases, etc. Such weights are sometimes also termed 'macédoine', pell-mell', and 'candy'.

——, *snake.* A paper-weight whose motif is an embedded glass snake, usually coiled with its head across its body and lying on a muslin GROUND (by ST-LOUIS) or rock ground (by BACCARAT). The snake usually has mottled coloured markings, e.g., green or red. A few were made by BACCARAT and ST-LOUIS. A related weight has an embedded glass worm, lizard, or Gila monster. *See* animal paper-weight, above.

——, *snow.* A paper-weight that is different from the usual collector's paper-weights in that it is not solid glass but is hollow; it is filled with a liquid in which is a mass of tiny pieces of white substance which, when shaken, float about to give the effect of a snowstorm.

——, *spaced millefiori. See* under millefiori, above.

——, *spoke.* A paper-weight with a design suggestive of the spokes of a wheel, somewhat similar to the swirl type (see below), but with the radiating lines straighter and less twisting. They were made by CLICHY and ST-LOUIS.

——, *Strathearn.* Paper-weights made by STRATHEARN PAPER-WEIGHTS, LTD.

——, *subject.* A broad class of paper-weights of various types, the decoration of which is an embedded motif, made AT-THE-LAMP or hand-made as distinguished from decoration consisting of millefiori CANES (e.g., weights of the flower, bouquet, animal and incrusted types). The process of making such weights involved first MOULDING the motif, then embedding it in a blob of molten glass, and gathering more molten glass until the desired size was obtained, and (if an OVERLAY was desired) covering with a thin layer of coloured glass and grinding WINDOWS.

paper-weight, pedestal. St Louis, marked 'SL 1848'. D. 8 cm. Courtesy, Sotheby's, London.

paper-weight, pinchbeck. Portrait of Isabella of Portugal, dated 1852. Courtesy, Spink & Son Ltd, London.

paper-weight, snake. Pink reptile with green markings, in clear glass, Baccarat. D. 8·3 cm. Courtesy, Sotheby's, London.

paper-weight, sulphide. Portrait of Alfred de Musset on translucent ground, Clichy. D. 8·4 cm. Courtesy, Christie's, London.

paper-weight, window. Mushroom weight with 6 windows, gilt scrollwork, and star-cut base, Baccarat, *c.* 1845–60. Courtesy, Spink & Son, Ltd, London.

——, *sulphide.* A paper-weight whose principal decoration is an embedded SULPHIDE (CAMEO INCRUSTATION). They are very rarely of the OVERLAY type. Examples made at BACCARAT and CLICHY feature portraits of prominent persons, e.g., Queen Victoria or Napoleon, but also show mythological characters and hunting scenes.

——, *swirl.* A type of paper-weight whose decoration is an overall series of embedded spirals of one of several colours, alternating with white spirals, and sometimes having a small central motif, e.g., a MILLEFIORI rosette or a CLICHY ROSE. They were made only at CLICHY.

——, *Tiffany.* A paper-weight made by LOUIS COMFORT TIFFANY that is of heavy glass creating the appearance of enclosed water in which float wavy plants and leaves. Few examples were made. *See* TIFFANY GLASSWARE: *paper-weight.*

——, *tricouleur* (or *tricolore*). A type of bouquet paper-weight in which the central bouquet motif, within green foliage, has at its sides red, white, and blue flowers, the whole on a clear glass GROUND. The style, it is said, was created to commemorate the French Revolution of February 1848.

——, *triple.* A paper-weight made of three superimposed weights of graduated sizes, the smallest at the top; they were made separately and fused together. Only a few examples are known.

——, *tuft.* The same as a mushroom paper-weight (see above).

——, *vegetable.* A paper-weight whose central motif consists of embedded glass vegetables, usually turnips, placed in a white FILIGRANA basket. A few examples were made at ST-LOUIS.

——, *Whitefriars.* A paper-weight made by WHITEFRIARS GLASS, LTD., which has made weights from *c.* 1848 to the present day. Examples are in the traditional form of the French factories, but some modern ones are fancifully shaped, e.g., ovoid, cylindrical, etc., and have original motifs. One example commemorates the coronation of Elizabeth II in 1963. Some have CANES with the date and the insigne of the factory.

——, *window.* A paper-weight made of overlay glass and having openings, usually circular, ground into the overlay so as to reveal the motif within. There is usually a top window and an encircling group of four to six (rarely three) others, but sometimes there are many windows covering the whole weight or the lower part, and occasionally the windows are separated by FLUTES. The overlay is coloured, with a thin underlayer of OPAQUE WHITE GLASS so that the window shows a thin white line. Usually termed an overlay paper-weight.

——, *zodiac.* A type of paper-weight whose decoration is one of the signs of the Zodiac against a coloured ground. They were made by BACCARAT.

parabolico, vetro (Italian). A type of glass developed by Ercole Barovier, *c.* 1957, that has the appearance of a patch-work of pieces in chequerboard pattern, the pieces having alternating vertical and horizontal stripes.

paraison. The term (spelled 'parison' is defined in the *Oxford English Dictionary* as 'Originally – The rounded mass into which the molten glass is first gathered and rolled when taken from the furnace', and still so defined by some writers. However, W. B. Honey and present-day authorities define it as such a GATHER on a BLOW-PIPE after it has been blown into a bubble (calling it sometimes a 'bubble').

Pargeter, Philip (1826–1906). A glass-maker from a Stourbridge glass-making family, his paternal uncle being Benjamin Richardson, of W. H., B. & J. RICHARDSON, who offered a reward of £1,000 for the making of an accurate replica of the PORTLAND VASE after it was smashed in 1845. Pargeter, who was lessee of the RED HOUSE GLASSWORKS at Wordsley, persuaded his cousin, JOHN NORTHWOOD, to work with him to produce such a copy (*see* PORTLAND VASE: *reproductions*), and later to make the MILTON VASE (1878) and three portrait TAZZE in blue CAMEO GLASS which Northwood engraved and dated 1878, 1880, and 1882 (*see* PARGETER-NORTHWOOD TAZZE). Pargeter made CASED GLASS in

BLANKS, to be decorated in CAMEO engraving by others, and also designed glassware and cameo decoration.

Pargeter-Northwood Tazze. A group of three TAZZE made of blue CAMEO GLASS by PHILIP PARGETER and engraved by JOHN NORTHWOOD, with central portraits depicting Isaac Newton (symbolic of Science), John Flaxman (symbolic of Art), and Shakespeare (symbolic of Literature), signed by Northwood and dated respectively 1878, 1880, and 1882. The rims are decorated with a wreath of ivy, hawthorn leaves, and acorns and oak leaves, respectively, and the domed feet with ACANTHUS leaves. The stems are detachable so that the shallow cups may be displayed as PLAQUES. The portraits are from jasper medallions by Josiah Wedgwood. They were on loan to the British Museum from 1959 until 1975, when they were sold on 24 July in London by Sotheby's (Belgravia).

Parker Cut Glass Manufactory. A glasshouse at 69 Fleet St, London, (said to have been started by JEROM JOHNSON) operated by William Parker (fl. 1765–71), a manufacturer and seller of glassware who had a reputation for making large CHANDELIERS of CUT GLASS, the material being supplied by the WHITEFRIARS GLASS WORKS. The chandeliers were in ROCOCO style, and included those made for the Assembly Rooms and Guildhall at Bath, the former marked with his name. The firm supplied glassware to JAMES GILES for decoration at his London workshop.

Parr Pot. Opaque white glass with silver-gilt mounts, probably Netherlands, first half of 16th century. H. 14 cm. Museum of London.

Parr Pot. A JUG with a hinged LID, having wide vertical bands of opaque white (LATTIMO) glass with narrow lines of clear glass between them and having silver-gilt mounts, hall-marked 1546–47, bearing the enamelled arms of William, Lord Parr of Horton (uncle of Henry VIII's sixth wife); it was formerly in the collection of Horace Walpole at Strawberry Hill. A lidded jug of identical glass and shape (but with slightly different mounts) is in the British Museum, attributed probably to the Netherlands, first half of the 16th century; its mounts are hall-marked London 1548–49. It is said that a similar piece is listed in the 1559 and 1574 inventories of Queen Elizabeth I.

part-size mould. A MOULD, either a DIP-MOULD or a small PIECE-MOULD, used to impress a design on a piece of glassware that is later expanded by BLOWING. *See* FULL-SIZE PIECE-MOULD.

partridge eye. *See* OEIL-DE-PERDRIX.

Passglas. Enamelled figures depicting the 'Ages of Man', Germany, first half of 18th century. H. 29 cm. Kunstgewerbemuseum, Cologne.

Passglas (German). A tall BEAKER, cylindrical or CLUB-SHAPED (a type of STANGENGLAS), that is decorated characteristically with equidistant horizontal bands (from 3 to 7) of coloured enamel or trailed rings of glass, and having a pedestal BASE or a KICK-BASE. Such graduated glasses were passed among the beer-drinkers in the German beer-cellars, and each person of the group was, it is said, expected to drink the amount indicated by the next horizontal band on the glass (i.e. to drink exactly to that band or else be required to drink to the next). Often between the bands is found enamelled decoration of pictorial subjects or, rarely, small PRUNTS; examples from Thuringia sometimes have a playing card painted in enamel colours (SPIELKARTENHUMPEN). Such glasses were made in Germany, Bohemia, and the Netherlands in the 16th to 18th centuries. Sometimes a BANDWURMGLAS is called a *Passglas*, but that is manifestly incorrect, as the spiral threads do not make equal divisions.

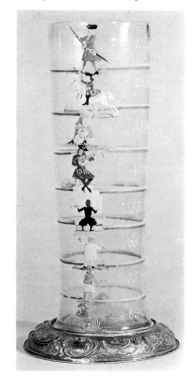

paste jewellery. Imitation gems, usually made of STRASS. It has been said that 'paste gems' were made in the 12th century from LEAD GLASS (which is highly refractive, and therefore rivals the brilliance of natural gems), and that lead glass, used for artificial gems at Birmingham, England, as late as 1836, was known as 'Jew's glass'. The term 'paste', to refer to glass imitations of precious stones, dates from the translation by

Christopher Merret in 1662 of the book by ANTONIO NERI. The use of strass dates from the 18th century.

pastry-mould cane. A type of CANE used in making some PAPER-WEIGHTS. It has a central rod surrounded by other rods forming a scalloped rim. It was used as a central motif or several were scattered in a MILLEFIORI pattern.

patch-box. A small, shallow decorative box in which were kept the black patches which were fashionable cosmetic accessories in the 18th century. They were made of porcelain or glass. The French term is *boîte à mouche. See* PATCH-STAND.

patch-stand. A small circular low stand, about 6·5 cm. in diameter and 2·5 cm. high, with a flat top and a pedestal BASE or footed STEM. Often made of cream-coloured pottery, they were also made of glass. They were used on a dressing table instead of a PATCH-BOX.

pâte-de-cristal (French). Glassware similar to PÂTE-DE-VERRE, but made of powdered glass of finer quality, resulting in greater translucency and resonance. Examples were made by GABRIEL ARGY-ROUSSEAU, FRANÇOIS ÉMILE DÉCORCHEMONT, and, less often, by ALMARIC-V. WALTER.

pâte d'émail (French). *See* DAMMOUSE, ALBERT.

pâte-de-riz (French). A type of OPALINE of greyish colour, so-called at the ST-LOUIS FACTORY from *c*. 1843.

pâte-de-verre (French). Literally, glass paste. A material produced by grinding glass to a powder, adding a fluxing medium so that it would melt readily, and then colouring it (or using coloured powdered glass). Objects were made of it in a mould and the material was fused by FIRING. The varied colouring, suggested by semi-precious stones, was obtained by the positioning of different powdered ingredients in the mould. Some examples were built up into polychrome HIGH RELIEF or BUSTS by successive layers in the mould, and sometimes, after the firing, they were refined by being carved. The process was known in ancient Egypt, and was revived in France in the 19th century, especially by the sculptor HENRI CROS and his son Jean. Other exponents of the technique were ALBERT DAMMOUSE, *c*. 1898, and FRANÇOIS DÉCORCHEMONT, *c*. 1900–30, as well as EMILE GALLÉ and GABRIEL ARGY-ROUSSEAU. It was sometimes called *pâte-de-riz. See* TASSIE MEDALLION.

paten. A plate for the bread in the Eucharist. Some were made of glass conforming to the CHALICE. Examples made at MURANO sometimes have a long projecting HANDLE.

patera (Latin). A Roman vessel similar to a Greek PHIALE, being a small bowl, usually shallow, to pour libations, but having one flat horizontal handle, and a FOOT-RING or sometimes a stemmed FOOT. They were made of SYRIAN GLASS in the 1st/3rd centuries, in forms adapted from silverware. Similar vessels were made at Cologne, in the Rhineland, in the 3rd century, decorated on the bottom with SNAKE-TRAILING in several colours of glass. *See* TRULLA.

pattern-moulded glass. Glassware that has been blown in a mould whose interior has a raised pattern so that the glass object shows the pattern with a concavity on the inside underlying the convexity on the outside, as distinguished from (1) BLOWN-THREE-MOLD GLASS that has the outside convexities opposite the inside convexities, and (2) MOULD-PRESSED GLASS that shows the pattern only on the exterior, with a smooth interior. It is made in a part-size DIP-MOULD or PART-SIZE PIECE MOULD, and later expanded by blowing.

pax. The same as an OSCULATORY.

patch-stand. Hexagonal pedestal stem and folded conical foot, English, *c*. 1740. Ht 8·2 cm. Courtesy, Sotheby's, London.

patera. Greenish glass, Syria, 1st/3rd centuries. D. 16·5 cm. Victoria and Albert Museum, London.

Peach (Bloom) Glass. *See* PEACH BLOW GLASS.

Peach Blow Glass. A type of ART GLASS, made by several United States glassworks, that was intended to resemble the peach-bloom glaze on Chinese porcelain of the reign of K'ang Hsi (1662–1722); *see* MORGAN VASE. One version, made from 1885 by NEW ENGLAND GLASS CO., was of uncased glass and shaded from an opaque cream colour up to deep rose; another, made by MOUNT WASHINGTON GLASS CO., shades in pastel tones from pale blue up to rose-pink. When the latter company sued for patent infringement, it kept the name and the former adopted the name WILD ROSE. Another version, made of CASED GLASS, shaded in darker colours (yellow up to reddish) and had a silky or glossy surface; it was made by HOBBS, BROCKUNIER & CO., and was used for tableware and decorative objects. This last version often has a moulded drapery pattern on the inner layer, termed 'Wheeling Drape'. Another version (called 'Peach Glass') was made by THOMAS WEBB & SONS, of cased glass shading from pink up to deep red, with either an acid-etched (MAT) or a glossy surface, and occasionally with gold decoration; this type was also produced by STEVENS & WILLIAMS and called by them 'Peach Bloom'. An uncased version was made by the BOSTON & SANDWICH GLASS CO., often with a moulded swirl pattern and coloured pink, sometimes with a bail handle resembling a thorny branch. Imitations of all these peach-coloured versions were also made in Bohemia.

peacock-feather glassware. A type of IRIDESCENT glassware developed by ARTHUR J. NASH for LOUIS COMFORT TIFFANY, the characteristic feature of which is glass in a variety of colours embedded in the body in a combed feather pattern and often having inlaid a piece of dark glass to resemble the 'eye' of a peacock feather. The process involved adding to the original PARAISON, at intervals, small amounts of molten glass of various colours and textures, then reheating, enlarging, and manipulating the object so as to achieve the intended form and pattern. The finest examples are vases and bowls made in various shapes.

pear-shaped. Pyriform, as a VASE or TEAPOT made in the form of a pear. The term 'inverted pyriform' is used when the greatest diameter of the piece is in the upper part. *See* BALUSTER.

pearl. A decorative motif in the form of a small attached glass bead such as was applied as an 'eye' to the tip of PRUNTS on the body of some RÖMER, NUPPENBECHER, and GOBLETS. They are found on some glassware made in the southern Netherlands, probably Antwerp, from the second half of the 16th century and made in Germany from the 17th century. The pearls were usually of turquoise blue, set on clear glass.

Pearl Satin-Glass, A type of ART GLASS made by several processes and having different appearances, but basically a pearly lustred surface or lustred stripes. Patents were issued to several companies in the United States and France in the 1880s. The type made by STEVENS & WILLIAMS, LTD, was called by it *Verre de Soie*, and other terms for it were 'Mother-of-Pearl Satin-glass' and 'Pearl Ware'. Some examples of the ware have decoration made by AIR-LOCKS, sometimes with added acid-etching on the outer layer of glass.

pebbled glass. Glass decorated with numerous very small irregularly shaped haphazard spots of glass of bright colours on a differently coloured (often light-blue) background. It was made in France and probably also in Venice, Germany and elsewhere in the 17th century; some such Venetian ware has been attributed to France.

pectoral. A small ornamental tablet to be worn on the breast suspended by a cord or a string of beads, and sometimes counterbalanced by a smaller tablet hanging on the back. In Egypt they were originally worn as jewellery but by the XVIIIth Dynasty (*c.* 1540 BC) they had become

pâte-de-verre. 'Incantation' by Henri Cros, 1892. Ht 33·5 cm. Cleveland Museum of Art (James Parmelee Fund).

amulets and were buried in tombs. They were made of wood and other materials, but some had inlaid glass decoration.

pedestal stem. *See* STEM, *Silesian.*

peg lamp. A type of small OIL LAMP whose bowl has a peg at its bottom that fits into the socket of a CANDLESTICK or other candle holder; when the candle was about to burn out or a substitute was desired, the peg lamp, filled with whale oil, could be inserted into the socket. Some such lamps, made in the United States, were highly decorated and some were of overlay glass.

Pegasus Vase. A CAMEO GLASS vase with blue body and white OVERLAY. It depicts Amphitrite and Aurora, and its FINIAL is in the form of Pegasus. It was made by JOHN NORTHWOOD at the DENNIS GLASS WORKS (then owned by THOMAS WEBB & SONS), hence sometimes called the 'Dennis Vase'. It is signed and dated 1882 by John Northwood on the body and on the cover. The sea-horse handles were roughly carved from solid glass by EDWIN GRICE and were finished by Northwood. It was exhibited before completion at the Paris Exposition of 1878 and is now at the Smithsonian Institution, Washington, D.C.

pegging. The process of pricking an object of MOLTEN GLASS with a tool so as to leave a tiny air-filled hollow which, when covered with molten glass, forms a TEAR as the entrapped air expands as a result of the heat.

Peking glassware. A general type of Chinese glassware made probably at Po Shan, in Shantung Province, for the Imperial Court in the 18th century. It usually was in the form of small objects, e.g. animals or vegetables, in coloured glass of dark blue, cerulean blue, or yellow hues. It has sometimes been called 'Mandarin glass'.

Peking Imperial Palace glassware. Glassware made at the workshops in the Peking Imperial Palace established in 1680, in the Ch'ing Dynasty, by the Emperor K'ang Hsi (1662–1722). This factory was the first in China to produce glass in quantity, making BLOWN GLASS (in the styles of Venice and the Netherlands) under Jesuit guidance, much of which today shows CRISSELLING. It also at a later date made snuff bottles (*see* CHINESE SNUFF BOTTLES) with CAMEO cutting, KU beakers of clear and monochrome-coloured glass, enamelled glassware in KU-YÜEH HSÜAN style, and PENDANTS and other objects in coloured glass imitating jade. Some pieces bear a CHINESE REIGN-MARK.

Peking Imperial Palace glassware. Beaker, *ku*-shaped, of brown glass with crisselling, *c.* 1700. Ht 22 cm. Courtesy, Hugh Moss, Ltd, London.

Pelikan, Franz Anton (1786–1858). The leading member of one of two large families of the same name in Meistersdorf, in Bohemia, who all did glass-engraving in the BIEDERMEIER period. Their subjects were mainly horses and hunting scenes.

Pellatt, Apsley (1791–1863). His father, of the same name, acquired *c.* 1790, the FALCON GLASSWORKS, at Southwark, London. The son became its owner and learned of the French process for making SULPHIDES ('encrusted cameos'); he introduced it in England, obtaining a 14-year patent in 1819. He called his ware 'cameo incrustations' or 'crystallo-ceramie'. His wares included PAPER-WEIGHTS, DECANTERS, ICE-PLATES, CANDLESTICKS, etc. and a well-known SCENT-BOTTLE bearing the portraits of George IV and Queen Caroline. The firm continued in business until after the Great Exhibition of 1851 (*see* CRYSTAL PALACE), at which it displayed examples of its so-called 'Anglo-Venetian' glass and its revived ICE GLASS. Pellatt wrote books on glass-making, published in 1821, 1845, and 1849.

Peloton glass. A type of ART GLASS decorated with applied short thin filaments of glass. It was patented in Bohemia in 1880 by Wilhelm

Kralik of Neuwelt. The process involved dipping a GATHER or a partially shaped piece of glass into a container that held filaments or threads of coloured glass, or throwing such filaments onto the body or rolling the PARAISON on a surface on which the filaments were scattered. The piece was then reheated until the filaments fused, and was then pressed or tooled into the desired form. The body was of clear, coloured or opaque white glass, and sometimes the finished article was acid-treated to produce a satin surface.

pendant. (1) A small ornament worn suspended around the neck. Some were made of coloured glass in Greece, *c*. 1400–100 BC and in Rome in the 1st century AD; they were of various shapes, e.g. hearts, pomegranates, ducks, jugs, etc. Later they were made, in square or rectangular form, in Germany in the 15th century, with metal suspensory mounts. Examples were made in China, of opaque white glass imitating jade, often in sets with conforming necklaces, rings, buckles, and other such objects. (2) A small glass object suspended as decoration from a CHANDELIER, CANDELABRUM, etc. The French term is *pendeloque*.

pennelato, vetro. Group of vases, Venini, Murano.

pen. A pointed instrument for writing or drawing with ink or similar fluid medium. Examples in various styles have been made of glass, especially of so-called NAILSEA GLASS

pen-stand. A glass stand on which to rest a PEN, sometimes of glass. They were made in ornamental form in the mid-18th and 19th centuries.

pennelato, vetro (Italian). Literally painted glass. A type of ornamental glass developed by VENINI, the characteristic feature of which is an overall decoration of irregular swirling streaks of various colours embedded in the clear glass object which is heated, blown, and twisted.

pepper bottle. A small receptacle for ground pepper, having (instead of a pierced cover, as a CASTOR) a cover to which is attached a smalll spoon that fits down into the bottle.

Percival, Sir Thomas. The inventor of a coal-fired furnace, developed before 1611 by making changes in the furnace at Winchester House in Southwark operated by Edward Salter and later leased to EDWARD ZOUCH. In 1611 Percival, with Zouch and others, received a patent granting for 21 years the exclusive right to make glass (presumably window glass) using coal as fuel. Percival joined SIR ROBERT MANSELL in 1615 in forming a new company to exploit the patent, but by 1618 he sold his interest to Mansell.

Perseus-Andromeda plaque. Wheel-engraved decoration by Caspar Lehmann, *c*. 1606. W. 18 cm. Victoria and Albert Museum, London.

perfume sprinkler. *See* SPRINKLER.

Perrot, Bernard (d. 1709). An Italian glass-maker and technologist, originally named Bernardo Per(r)otto, who from 1649 worked with his uncle, JEAN CASTELLO (Castellan) at Nevers and from 1662 at Orléans. He developed the CASTING process for making flat glass panels for windows and mirrors, and was granted a patent for it in 1688; but in 1696 his rights were abrogated by the King in favour of the new royal factory at ST-GOBAIN. Perrot also was granted patents, *c*. 1668–72, for making OPAQUE WHITE GLASS (called *porcelaine en verre*) and for a red-coloured glass (called *rouge des anciens*). He also developed a process (although he was perhaps not the originator) for making cast portraits and small glass figures.

Perseus-Andromeda Plaque. A glass PLAQUE (with WHEEL-ENGRAVING by CASPAR LEHMANN, *c*. 1606), depicting the myth of Perseus rescuing Andromeda from the dragon. It bears, above the scene, cartouches with the crowned initials of the Elector Christian II of Saxony (d. 1611) and of Hedwig, the daughter of Frederick II of Denmark, whom he married in 1602.

Perthshire Paper-weights, Ltd. Bottle with millefiori base and printies on body, and overlay stopper. Courtesy, Spink & Son, Ltd, London.

Pfohl, Karel. Beaker of ruby-flashed glass, Karel Pfohl, Steinschönau, Bohemia, *c.*1850. Museum of Decorative Arts, Prague.

Persian glassware. Glassware made in Persia (now Iran) from the 3rd to 19th centuries and probably some made earlier in the period (550–330 BC) of the Achaemenid dynasty (*see* ACHAEMENIAN BOWL). Although much glassware of the medieval period that has been excavated in Persia may have been made in Iraq, Egypt or Syria and although some pieces found there were made in the 18th/19th centuries in Bohemia, there are many examples of glassware of earlier and later periods that are characteristically Persian in style and have been generally recognized as Persian. The first creative period, the Sassanian period, *c.*226–642, produced blown glass made in the 3rd/4th centuries, and wheel-engraved facet- and linear-cut pieces of the 5th/6th centuries (the latter decorated with shallow concavities in a honeycomb or QUINCUNX pattern), many examples of which were exported. Next was the Islamic period (following the conquest of Islam by the Arabs in the 7th century) when glass-making was revived in the 9th and 10th centuries under the Samanids (819–1004) at Nishapur and Afrasiyab (Samarkand), and when thin glass was skilfully decorated by grinding to leave a design in countersunk-relief; during this period some glassware was decorated with cameo-cutting and some with applied trailing, and perhaps also enamelling and gilding, much of the latter ware having been inspired by the art of China. After the conquest by the Mongols under Tamerlane (Timur) in 1402, little glass was produced in Persia until the industry was revived under Shah Abbas I (1587–1628). Glassware was then, during the Safavid dynasty and later, made at Isfahan and especially at Shiraz until the 19th century, much in Venetian style, it being said that an Italian originally, in the late 16th century, taught the art. In this period glassware (often of deep-blue colour) was made in the form of tall EWERS and perfume SPRINKLERS. Due to the demand for Persian glassware, it was exported in the 17th century to the Far East and, in the mid-19th century, Bohemian glassware, especially cut cased glasses with enamelling and gilding, was exported to Persia via Smyrna.

Perthshire Paper-weights, Ltd. A glasshouse near Crieff, Scotland, started by Stuart Drysdale in 1970 after he had left the VASART GLASS CO. It produces mainly PAPER-WEIGHTS, some of high quality and of original designs in limited editions, as well as inexpensive weights in large quantities. Some have a CANE with the letter 'P' and some bear a date.

pestle. An object, somewhat club-shaped, for triturating a substance in a MORTAR. Some were made of glass, occasionally having handles with spiral twists.

Pettersen, Sverre (1884–1959). A glass-designer engaged in 1928 by HADELANDS GLASVERK; during the 1930s he designed tableware and decorated glassware by the process of SAND-BLASTING.

Peytowe (or Paytowe) family. A family of glass-makers from Normandy who settled near Chiddingfold, Surrey, *c.*1440, intermarried with the SCHURTERRE FAMILY, and continued making glass at least until the arrival of JEAN CARRÉ in 1567, principally making mould-blown and free-blown vessels and window glass.

pezzato, vetro (Italian). A type of ornamental glass developed by VENINI *c.*1951, made of pieces of glass (roughly square) of several colours, fused together into a patchwork-quilt mosaic effect. The process involves laying the pieces side by side, heating them so as to fuse them together into a flat piece, than making a FREE-FORMED object.

Pfohl, Karel (1826–94). A Bohemian decorator of glassware during the 1850–60s. He worked at Steinschönau (Kamenický Senov) and specialized in depicting horses by wheel-engraving, sometimes on a FLASHED glass panel on the body of a BEAKER.

phalera (Latin). In antiquity, a MEDALLION worn on the breast of a horse, and sometimes worn in a leather holder by a soldier as a mark of rank. Some examples were made of ROMAN GLASS in the 1st century AD.

phallus glass. A type of drinking glass in the form of a penis and testicles, with an upcurved funnel-shaped drinking spout above the testicles. Examples were made of greenish clear glass in Germany in the 16th and 17th centuries. Some are of free-formed plain glass with no decoration except a small applied glass ornament, but one example is MOULD-BLOWN with many small oval prunts along the length of the glass. Other German examples of phallus glassware are variously shaped drinking vessels with relief decoration of phallic motifs, made in the 1st century AD at Cologne and in the 15th century at Siegburg. *See* JOKE GLASS.

phallus glass. Green glass, mould-blown, German, 16th century. L. 25 cm. Germanisches Nationalmuseum, Nuremberg.

pharmacy glassware. Various types of glass objects used in a pharmacy. *See* ALBARELLO; BOTTLE; CARBOY; DISPENSARY CONTAINER; JAR; LEECH JAR; PHIAL; VIAL; POISON BOTTLE; RETORT; SPECIE JAR; MEDICINE BOTTLE.

phial. A small glass bottle for ointments, medicines, etc. (1) Some examples were made of ROMAN GLASS, *c.* 100 BC, with encircling trailed decoration. The German term is *Phiole*. (2) Any small glass bottle of modern times used in a pharmacy shop. *See* VIAL.

phiale (Greek). A shallow bowl, with a flat bottom or a FOOT-RING, without handles, used as a wall ornament or to pour libations. The type with a central boss (forming a hollow on the underside) is called a *phiale mesomphalos. See* PATERA; TRULLA.

Philadelphia Glass Works. A glasshouse at Kensington, near Philadelphia, Pennsylvania, founded in 1771/72 to make FLINT GLASS in competition with STIEGEL-TYPE GLASSWARE. It operated successfully until it was sold in 1780 to Thomas Leiper for making bottles and containers, and was again sold in 1800 (among its new owners being Thomas W. Dyott, known for making HISTORICAL FLASKS). *See* DYOTTSVILLE GLASS WORKS.

phalera. Head of Drusus, son of Tiberius, and his sons, bluish glass, first half of 1st century AD. D. 4 cm. Römisch-Germanisches Museum, Cologne.

Phoenix Glass Works. (1) A glassworks in South Boston, Mass., established *c.* 1820–23 by Thomas Caines, an Englishman who went from Bristol to the United States in 1813 to work at the Boston Crown Glass Co. It operated until after 1870, making CUT GLASS, PRESSED GLASS, and possibly LACY GLASS and BLOWN-THREE-MOLD GLASS. (2) A glassworks in Bristol, England, owned by Wadham, Ricketts & Co., which specialized in CUTTING and ENGRAVING in the early 19th century. (3) A glassworks at Pittsburgh, Pennsylvania, founded in 1832 by William McCully (d. 1859) and operated by his associates and successors until after 1868.

photochromic glass. A type of glass invented in 1964 by the CORNING GLASS WORKS which is used in lenses for eye-glasses and sun-glasses. It darkens under ultra-violet radiation, present in sunlight, and clears again when the light source is removed.

pi (Chinese). A circular disk, with a central hole, considered symbolical of Heaven. Glass examples, in imitation of similar jade objects, were made in China during the late Chou and the Han dynasties.

piano feet. Solid glass supports to be placed under the feet of a piano, made by John Davenport, an English glass-maker, by press-moulding (under a patent of 1874).

piano feet. One of a set of press-moulded feet, English, John Davenport, *c.* 1874. Victoria and Albert Museum, London.

piatto (Italian). A PLATE, usually circular and of a large variety of sizes. Italian terms for various sizes are: *piatto reale* (royal plate), *c.* 50 cm. in diameter; *piatto de cappone* (plate for capon), *c.*40 cm.; *piatto da tavagliolo* (napkin plate), *c.*26 cm.; *piatto imperiale*, 65 cm.

piatto da pompa (Italian). Literally, ceremonial plate. A plate that is highly decorated, and usually made to be displayed, placed on a buffet or suspended on the wall. Some have a deep centre well (*cavetto*).

pickler or **pickle jar**. A glass receptacle for serving pickles, usually cylindrical with a flat bottom and a flat-topped STOPPER or COVER.

pictorial flask. *See* HISTORICAL FLASK

pie-crust cane. A type of millefiori CANE whose border is scalloped like a pie crust and which usually has a central raised tiny round rod. Such canes are included in Clichy PAPER-WEIGHTS. Sometimes termed a 'pastry-mould cane'.

piece-mould. A MOULD made of two parts that are hinged together and make glassware in part-size or full-size. They are used for shape only or to make PATTERN-MOULDED GLASSWARE. The part-size piece is later expanded by BLOWING. *See* DIP-MOULD; PART-SIZE PIECE-MOULD; FULL-SIZE PIECE-MOULD.

piédouche (French). Literally, pedestal. A base or foot attached to an object as a supporting pedestal or sometimes as a saucer-like stand under a sauce-boat, jam-pot, etc. Some paper-weights have, instead of the usual flat bottom, such a supporting pedestal-like plinth. *See* PAPER-WEIGHT, *piédouche*.

pier-glass. A tall, narrow MIRROR designed to occupy the pier (wall space) between two windows. The width varied to suit that of the pier, which changed with the styles of different period. In early examples the mirror was in two or more vertically placed sections. Some had attached branches with candle nozzles. The frames on pier glasses followed the styles of mirror frames (*see* MIRROR, *framed*). Many were made in the ADAM style. The French term is *trumeau*. *See* CHIMNEY GLASS; SCONCE.

pig. A fanciful object in the form of a standing pig. Many were made of various types of glass, including clear glass of various colours, speckled glass, and OPAQUE WHITE GLASS, and in various sizes; some were even furnished with wings. Some are solid, some are flasks. They were made as FRIGGERS in England in the 19th century. A similarly shaped piece for use as a FLASK was made of ROMAN GLASS at Cologne; it had a hole at the rear end for pouring.

pigeon-eye pattern. A style of decoration on some ISLAMIC GLASSWARE in the form of a mould-blown overall diaper pattern of contiguous hexagons, diamonds, etc. enclosing raised circles, the centre of each circle being depressed to present an eye-like appearance. Such pieces have been variously attributed to periods from the 7th to the 13th centuries. The term was applied to such patterns by Carl Johan Lamm, the historian of Islamic glass.

piggin. A small glass receptacle used as a dipper for milk or cream. The form is based on wooden pieces made in the form of a small half-barrel with a vertical handle in the form of one tall stave. Some examples, in IRISH glass with cut decoration, were made, *c.*1800. The form is somewhat similar to a Greek KYATHOS.

pilgrim-flask. A flattened gourd-shaped bottle with one or two pairs of lugs at each side through which might be passed a strap whereby it could be slung over the shoulder. Roman in origin, many such flasks have been made of pottery and a few of porcelain, but they also occur since the 15th century (perhaps as ornaments) in glass. As the name suggests, they were originally intended for use by pilgrims to carry drinking-water.

phial (1). Blown Roman glass, *c.*300/100 BC. Ht 9·2 cm. Römisch-Germanisches Museum, Cologne.

pig. Frigger of English speckled glass, 19th century. L. 10 cm. Courtesy, Maureen Thompson, London.

XI *Paper-weights.* Group of French weights, diam. (except top right) 6·0–8·0 cm. ($2\frac{3}{8}$–$3\frac{1}{4}$ in.): (top left) crown type, with twisted ribbons alternating with *filigrana* spirals, St Louis; (top right) magnum type, with standing flowers and black reptile, diam. 11·2 cm. ($4\frac{1}{2}$ in.); (middle left) carpet ground type, with *millefiori* and animal-silhouette canes, Baccarat; (middle right) marbrie type, with turquoise festoons and *millefiori* floret, St Louis; (bottom left) flat bouquet type; (bottom right) fruit type, multi-faceted, containing two cherries, St Louis. Private collection, London.

XII *Roman glass*. Group of objects from the Roman Empire, 1st–4th centuries, including: mould-blown Sidonian types (top left; possibly from Ennion workshop); Alexandrian ribbed bowls; blue œnochoe; early Christian (2nd century) amber bottle; faceted stamnium; and Janus bottle. Courtesy, Sheppard & Cooper Ltd, London.

Some have a metal screw-cap. Alternative terms are 'pilgrim bottle' or 'costrel'. The French term is *bouteille de voyage*; the German is *Pilgerflasche*; the Italian is *fiaschetta da pellegrino. See* FLASK; LENTOID FLASK; MEISTERSTÜCK; HUNTING FLASK.

piling pot. A type of POT (crucible) that was used for melting small BATCHES of artificially coloured glass. It was placed on top of a large covered melting pot in the furnace. Such pots were used before 1662, when they were mentioned by Christopher Merret. Sometimes called a 'jockey pot' or 'monkey pot'. *See* SKITTLE POT.

Pilkington Brothers, Ltd. A British glass-making firm; it is one of the world's largest producers of flat glass and other types of industrial glass, having its main factory at St Helens, near Liverpool. The company was started in 1826 as St Helens Crown Glass Co. by John William Bell and associates; it was renamed Greenall & Pilkington in 1829 and Pilkington Brothers, Ltd., in 1849. In 1773 the British Cast Plate Glass Co. had been formed, with its plant at Ravenhead, near St Helens, to make plate glass, and its factory was acquired by Pilkington in 1901. In 1945 Chance Brothers, Ltd., of Birmingham, became a subsidiary of Pilkington, which has other subsidiaries and factories around the world. The group makes a great variety of industrial glass, including plate glass (made since 1959 by the FLOAT GLASS process), laminated and safety glass, optical glass, FIBRE GLASS, PRESSED GLASS, etc. It maintains an important museum of glassware at St Helens.

pillar. A type of RIBBING where the rib is broad and rounded at the top. *See* PILLAR-MOULDING; PILLAR-MOULDED BOWL.

pillar cutting. A decorative pattern on CUT GLASS in the form of a series of parallel vertical convex ribs. It was used in England on glassware in the REGENCY STYLE. *See* FLUTE CUTTING.

pillar-moulded bowl. A BOWL of depressed hemispherical shape, with no FOOT, characterized by fifteen to thirty closely placed vertical (occasionally slightly slanting) RIBS extending from the bottom to slightly below the RIM. The ribs were produced by the process of PILLAR-MOULDING, which changed over a period of time. The bowls are from 12 to 20 cm. in diameter, and are made of variously coloured ROMAN GLASS, *c.* 100 AD. Sometimes called a 'ribbed bowl'. The German term is *Rippenschale.*

pillar moulding. A style of angular RIBBING that is characteristic of PILLAR-MOULDED BOWLS. The process of production changed from that used on early Roman examples, *c.* 100 AD, to that used on later Venetian pieces. The early ones were MOULD-PRESSED, with interior depressions behind the exterior ridges, then were finished by grinding and sometimes by wheel-polishing or FIRE-POLISHING. W. B. Honey has written that the ribs 'were produced apparently by working up the plastic glass with pincers and other tools while it was on the mould'. Occasionally the ribs are irregularly spaced, and hence Honey has concluded that they 'could not be produced by "twisting in delivery" from a hollow mould as has sometimes been stated'. Later Roman examples were MOULD-BLOWN on a second GATHER, and Venetian ones have additional decoration produced by horizontal TRAILING of threads of glass and melting them in.

pincering. A method of decorating glassware by applying PINCERS to squeeze various pieces of ornamentation, e.g. THREADING, or wings or flame-like protuberances on the stems of some Venetian goblets, especially WINGED GLASSES. Pincered decoration is found on the walls of some Islamic bowls.

pincers. A GLASS-MAKER'S TOOL used for decorating glassware by squeezing various pieces of ornamentation, e.g. THREADING while in a

pigeon-eye pattern. Bottle, mould-blown, Islamic, probably 12th century. Ht 8·4 cm. Wheaton College, Norton, Mass.

pilgrim-flask. Flasks with arms of Sforza and Bentivoglio families honouring a marriage, Murano, late 15th century. Museo Civico Medievale, Bologna.

pine-cone glass. Receptacle made in the *façon de Venise.* Ht 21·5 cm. Kunstsammlungen der Veste Coburg.

pizzo, vetro. Bowl with white opaque decoration, *c.* 1970, Venini, Murano.

Galla Placidia cross. Enamelled portrait on glass, 3rd/4th century. Museo Civico dell'Eta Cristiana, Brescia.

molten state. Some have decorative patterns for impressing a design on an object. *See* PINCERING.

pinched lip. A LIP (of a JUG) of TREFOIL shape. First to be seen on the Greek OENOCHOË, it occurs later on JUGS and EWERS based on classical prototypes.

pinched trailing. A decorative pattern on glass made by TRAILING threads of glass in parallel lines and then PINCERING them together at intervals. Similar to NIPT-DIAMOND-WAIES.

pine-cone glass. A glass vessel whose body is ovoid with overall relief imbricated decoration simulating a pine-cone; it has a stemmed foot and a short neck. They were made in the 16th/17th century in Germany in FAÇON DE VENISE style. Their intended purpose is uncertain; they have been referred to by Walther Bernt, *Altes Glas,* as a POKAL (pl. 62) and as a *Flasche* (p. 59), and elsewhere as a receptacle for smelling salts. Examples are found from Spain. A German term is *Pinienzapfensflasche.*

pink glass. *See* PURPLE OF CASSIUS; CRANBERRY GLASS; SELENIUM; OPALINE. Sometimes there is accidental pink colouring in CRISSELLED glass, especially French.

pink-slag glassware. A type of ART GLASS moulded in ornate patterns, made in various shades of pink by the Indiana Tumbler & Goblet Co., Greentown, Indiana.

pinnacle. A glass ornamental object, tall, thin, and tapering upward, affixed as decoration to a CHANDELIER. Many were FACETED or ornamentally cut.

pipe. *See* TOBACCO PIPE.

pistol flask. A FLASK in the form of a pistol. They were made, usually of coloured glass with enamelled decoration, in Bohemia and Germany (Saxony), two known examples being dated 1615 and 1617, respectively. They were also made of unenamelled glass in MURANO. Many others made of coloured or colourless glass with various kinds of decoration were made at other factories. The flasks usually have a metal stopper to close the end of the barrel of the pistol.

pitcher. A term in general use in the United States for a JUG.

Pitkin Flask. A type of small MOULD-BLOWN flask attributed generally to the PITKIN GLASS WORKS but known to have been made also at other New England and Midwestern glassworks. The flask, with a double-thick lower half, was made by the method called in the United States the 'HALF-POST METHOD'. Some have only vertical or swirled ribbing, but others have a combination of both ribbings; the latter type was made by insertion in a ribbed mould, removing and twisting the object, and then again inserting it in a mould. They were made of amber or green glass, and sometimes blue or amethyst. The shape is ovoid, flattened, and tapering slightly to a short cylindrical neck.

Pitkin Glass Works. A glasshouse founded in 1783 by William and Elisha Pitkin and Samuel Bishop, at East Hartford (later East Manchester), Connecticut, to make CROWN GLASS for windows, but by 1788 it was making bottles and flasks. It received a state monopoly for 25 years in 1783, and operated until 1830. It made free-blown flasks, pattern-moulded flasks, HISTORICAL FLASKS, and CHESTNUT FLASKS, but is best known for the PITKIN FLASK, so that the name 'Pitkin' is now associated mainly with Pitkin flasks, whether made there or elsewhere. *See* GLASTENBURY GLASS FACTORY.

pizzo, vetro (Italian). Literally, lace glass. A type of ornamental glassware developed by VENINI, *c.* 1970, the characteristic feature of which is the presence of irregularly shaped pieces of opaque glass (in a pattern resembling a type of lace having small connected loops and projecting ends of 'threads') embedded in clear glass.

Placidia, Galla, cross. A 3rd/4th century Byzantine jewelled cross (known as *di Desiderio*), bearing, at the centre, an attached FONDO D'ORO with portraits of a family group. The now unidentified group was formerly thought to depict Galla Placidia (daughter of the Roman Emperor Theodosius, 379–395) and her two children. The plaque is inscribed with a name in Greek characters, perhaps that of the artist or the father or the person to whom the plaque was dedicated.

plaque. A flat thin tablet sometimes made of glass, intended to be attached as a wall decoration or for inlaying in furniture. Glass examples were decorated with RELIEF, MOSAIC, ENGRAVING, PAINTING or ENAMELLING. (1) Decorative moulded relief plaques: some with figures of persons or *amorini* or with mythological or religious motifs were made of ROMAN GLASS in the 1st/2nd centuries; some of these may have been used as PANELS. (*See* BONUS EVENTUS PLAQUE.) (2) Mosaic plaques: some were made in Egypt and probably in Italy in the 1st/2nd centuries. (3) Engraved plaques: some were made in Bohemia and Germany in the 17th century. (*See* PERSEUS-ANDROMEDA PLAQUE.) A rectangular plaque (45 × 23 cm.) of crystal glass was made in the late 18th century at LA GRANJA DE SAN ILDEFONSO with a wheel-engraved meticulously detailed view of the Royal Palace; it is signed by Félix Ramos. (4) Enamelled plaques: Some were made of OPAQUE WHITE GLASS in Germany and Italy in the 18th century. (5) Painted plaques: Some, with PAINTING on the back of the glass for protection when displayed, were made of transparent glass. The German terms for plaque are *Platte* and *Plättchen*. *See* MEDALLION; ROUNDEL.

plastic. Susceptible to being readily modelled or shaped, as wax or natural clay before being fired or glass in a molten state. *See* DUCTILE; GLYPTIC.

plate. A shallow table utensil from which food is eaten, usually circular with a ledge and a well or, in some modern styles, a slightly upcurved rim. The size ranges from about 15 to 28 cm. in diameter. Generally made of porcelain or pottery, plates have also been made of glass, either clear glass or OPAQUE WHITE GLASS with enamelled decoration. Some plates of clear glass are decorated with engraved designs, white FILIGRANA patterns, or enamelled designs and sometimes have gilt edges. The French term is *assiette*; the German is *Teller*; the Italian is *piatto* (circular) or *fiamminga* (oval). *See* DISH; SALAD PLATE; CUP PLATE; ICE PLATE; PLATTER; PIATTO; VENETIAN SCENIC PLATES.

plate glass. High-quality, thick SHEET GLASS. It is used for large display windows, mirrors, table tops, etc. *See* ST-GOBAIN GLASS COMPANY.

plate-protector. A circular glass disk to fit in the well of an enamelled porcelain plate so as to protect the decoration during use.

Plated Amberina. A type of cased ART GLASS patented by JOSEPH LOCKE in 1886 and made only at the NEW ENGLAND GLASS CO. It has a creamy OPALESCENT lining, sometimes slightly bluish, cased with an outer layer of AMBERINA glass. All pieces have moulded external vertical ribbing in various patterns. It was issued only in a limited quantity.

platter. A large shallow dish, usually oval, for serving food; it usually has a ledge and sometimes a well. One known example was made of moulded glass in the 1st century AD at Alexandria. *See* PLATE.

Plum Jade Glass A type of ART GLASS developed by FREDERICK CARDER at STEUBEN GLASS WORKS, and made by a complicated process using three layers of glass with an acid-etched design. Two layers of

plaque (1). Ultramarine Roman glass with white decoration, 1st century AD. Ht 3·7 cm. Römisch-Germanisches Museum, Cologne.

plaque (3). Glass plaque (on metal stand) depicting façade of Royal Palace of San Ildefonso, wheel-engraved. W. 23 cm. Museo Arqueologico Nacional, Madrid.

platter. Alexandria, 1st century. W. 27·3 cm. Museo Vetrario, Murano.

amethyst glass were cased around ALABASTER GLASS, then the design was printed in wax ink on the outer layer, and the object was dipped into acid until etched about half-way through this layer. Then the background motif was similarly wax-inked, and the object was again dipped into acid until only a thin layer of amethyst remained. The design thus appears in three shades of amethyst, enhanced by the middle layer of alabaster and the inner layer of amethyst. *See* JADE GLASS.

ply. The term used to designate the number of strands or threads of opaque glass used in making a particular type of TWIST, e.g. up to 20-ply.

pocket-flask. A flat FLASK made to be carried on the person, usually in the hip pocket. It is similar to a PILGRIM FLASK except that it has no LUGS for a carrying strap. It usually has a screw cap, a hinged LID or other form of STOPPER. Early pocket flasks were protected against breaking by being made of heavy glass, sometimes reinforced with added strips of glass; later examples had a leather or metal casing. Some have engraved decoration, or trailing or PRUNTS. Spanish examples were made in the 17th and 18th centuries.

Pokal. Zwischengoldglas, Bohemia, *c.*1730. Ht 20·7 cm. Kunstsammlungen der Veste Coburg.

pocket tumbler. A type of TUMBLER that is blown and flattened to an oval shape so as to fit in a man's pocket. A large number were made in Spain. Some from the early-17th century made in Almería are decorated with trailed threading, and one from a glasshouse in Maria there has a mould-blown inscription 'Ave Maria'. Many were made in the 18th century at LA GRANJA DE SAN ILDEFONSO; they are of transparent cobalt blue or amethyst glass, decorated with COMBED opaque white stripes, with fired gilding, or with white opaque speckling.

poculum (Latin). A drinking-cup, usually of globular or cylindrical shape, with a low funnel-mouth, a rounded bottom resting of a FOOT-RING, and one or two vertical loop or angled handles extending from the rim down to the body. They were often made of pottery but some were made of ROMAN GLASS. *See* SKYPHOS; PTEROTOS CUP.

poison bottle. A type of glass BOTTLE used in a pharmacy, for containing poisons. They are of many shapes, but usually of a distinctive form and with external ridges so as to be recognizable to the touch. Generally the NECK is small, with two tiny holes to regulate the flow of the contents, and is removable for filling or there may be a corked hole in the bottom for filling. A distinctive label is usually attached. The smaller MEDICINE BOTTLE for domestic use is similarly ridged as a safety measure when medicines containing poisonous drugs are prescribed.

Pokal (German). A type of GOBLET having a stemmed foot and a cover (often missing today) with a FINIAL, which was made in Germany from *c.*1680 until the mid-19th century; it was passed among the guests for drinking a toast on a festive occasion, hence sometimes called a *Gesundheitsglas*. Ranging in height from 25 to 60 cm., they were often decorated in BAROQUE style with HOCHSCHNITT and TIEFSCHNITT engraving and with stems having balusters, MERESES, and KNOPS. Some *Pokale* were engraved with elaborate motifs appropriate for a special public or private occasion, but most had simple portraits or scenes with suitable inscriptions. The style of decoration was frequently characteristic of the German region where the glass was engraved. Their function was similar to that of a LOVING CUP. *See* DOPPELPOKAL.

pole-head. A decorative ornament to be placed on the top of a pole such as was carried in and near Bristol, England, by village clubs and guilds. They were usually made of brass or wood, in a great variety of symbolic designs, but a few glass examples are said to have been made of so-called NAILSEA GLASS, being of OPAQUE WHITE GLASS streaked with pink and blue glass. The poles were about 2 m. in height and were decorated with coloured ribbons.

Poli, Flavio (1900–   ). A glass-designer who worked for SEGUSO VETRI D'ARTE, at MURANO, from *c.* 1934. Some of his early glassware is in sculptured form with embedded air bubbles and coloured inlaid glass, but among his most highly regarded pieces, made *c.* 1958, are vases and bowls of heavy clear CASED GLASS in three or four contrasting colours and blown to emphasize simple shapes without added decorative effects. He has also designed CHANDELIERS in *vetro traliccio* (lattice-work glass) and ornamental ware in *vetro astrale* (astral glass, suggestive of forms in outer space), styles introduced by him.

polishing. The process of giving glass pieces a smooth brilliant surface after an object has been cut or engraved. The first step is with a fine-grained stone rotating wheel, then with a finer wheel of lead, wood or cork and the use of putty powder on a brush. Some glass is left partially unpolished to give a contrasting effect. For another process and effect, *see* FIRE-POLISH.

Pollard, Donald (1924–   ). An American glass-designer and painter who became a staff designer at STEUBEN GLASS WORKS in 1950. He made original designs for many important ornamental glass pieces made during the Houghton period at Steuben, the engraving design and the actual engraving being executed by others. He conceived the idea of combining glass with precious metal ornament. *See* MYTH OF ADONIS; CANADA, GREAT RING OF.

polverine. A kind of POTASH from the Levant used as an ALKALI in making glass. It was imported into England for making CRYSTAL GLASS.

polycandelon. A hanging lighting fixture consisting of a flat metal ring suspended by three chains, the ring having apertures to hold a number (3 to 6) of point-tipped CONE-BEAKERS that serve as oil lamps, or possibly cups with a projecting stem or cups shaped like a MASTOS.

polychrome. Strictly, decoration employing more than two colours, but with respect to glass and ceramic ware the term is often applied to ware with only two colours (properly termed 'bichrome') in contrast to MONOCHROME.

pomegranate bottle. A CORE GLASS bottle, in the shape of a pomegranate, and decorated with coloured trailing combed in a festoon or feather pattern. Egyptian, XIXth–XXIst Dynasties (14th/10th centuries BC).

Pomona glass. A type of ART GLASS developed by JOSEPH LOCKE at the NEW ENGLAND GLASS CO. and patented by him in 1885. It is of clear glass enlarged by repeated MOULD-BLOWING and its surface is partially etched and partially stained in amber or rose; it is further decorated with garlands of flowers or fruits in blue (cornflower) staining, with amber stained leaves. There are two versions: 'First grind', made by covering the glass object with an acid resistant substance into which was scratched a multitude of fine lines, then dipping it into acid, leaving etched lines; 'second grind', made by rolling the object in fine particles of an acid resistant substance and then dipping it into acid, leaving a mottled surface. The technique was used for many shapes.

Pompeio, Vincenzo. A glass-maker from MURANO who worked in London at the Savoy Glasshouse of GEORGE RAVENSCROFT and later in Antwerp in the late 17th century.

pontil (sometimes 'puntee' or 'punty'). The iron rod to which a partly-made molten object is transferred from the iron tube (BLOW-PIPE) on which the METAL has been gathered from the melting pot and MARVERED and tentatively shaped by blowing. The final shaping, finishing the neck (if a type of bottle, vase, etc.), attaching of handles and applied ornamentation, etc., are done when the piece is on the pontil.

polycandelon. Bronze, with three cone beakers. Gemeentemuseum, The Hague.

Pomona glassware. Pitcher with etched ground and engraved floral band, New England Glass Co. Courtesy, Antique & Historic Glass Foundation, Toledo.

(Opposite)
Portland Vase. Roman glass, 100
BC–AD 100. Ht 24·5 cm. British
Museum, London.

The pontil can also be used to draw metal to almost any length in the form of a rod or tubing. In order to make the molten object adhere to the pontil, a small GATHERING is first attached to it (sometimes called a 'punty-wad'). *See* PONTIL MARK; GLASS-MAKERS' TOOLS.

pontil mark. A rough mark in the centre of the bottom of a glass object where the PONTIL was attached. On early ware the mark was often pushed upwards into the BASE of the object, making a KICK. On later ware it was generally smoothed by grinding and by *c*. 1850 it was eliminated altogether. (*See* GADGET.) Many old glass objects show a pontil mark (and almost all FORGERIES do) but not a CRUET BOTTLE, from which it was always removed so that the cruet bottle could rest securely on a silver CRUET STAND without scratching the metal. *See* PAD PONTIL MARK; WAFFLE PONTIL MARK.

pony. The same as a WHISKY GLASS.

porcellana contrafatta (Italian). Literally, counterfeit porcelain. A term used in the 16th century for Venetian OPAQUE WHITE GLASS or LATTIMO which has the appearance of white porcelain.

Porcelleinglas (German). Literally, porcelain glass. The German term for OPAQUE WHITE GLASS (MILK GLASS). It was the subject of experiments by JOHANN KUNCKEL, but none of his pieces has been identified. A glasshouse at Basdorf in Prussia (1750 onwards) made such glass, with painted flowers, in imitation of porcelain ware, as did also a glasshouse at Potsdam. *See* BASDORF GLASS.

porringer. A shallow bowl with one or two flat horizontal handles, based on silver and pewter prototypes. They were usually made in English delftware in the second half of the 17th century, but some were made of glass, normally with one handle.

porró. Clear glass with thread-wound neck and trefoil mouth, Catalonia, 18th century. Ht 24·4 cm. Victoria and Albert Museum, London.

porró or **porron** (Spanish). A type of tavern drinking vessel (derived from the wineskin) with a globular or conical body, tapering upward to a narrow NECK and FUNNEL-MOUTH (or TREFOIL mouth), characterized by a long attenuated SPOUT extending from the body at an upward angle, and having no HANDLE. They were (and still are) used for drinking wine by pouring it directly into the drinker's mouth. Examples are known from Catalonia and Valencia from the 17th and 18th centuries. The form is similar to a Spanish CRUET (*setrill*) except that the spout of a *porrón* tapers to the mouth, while that of a *setrill* usually widens a little. *See* CÀNTIR.

port glass. A small drinking glass on a stemmed FOOT, with a BOWL the shape and style of which may vary, but is smaller than that of a SHERRY GLASS. It is intended for drinking port wine.

porte-huilier (French). A type of CRUET-STAND with holders for oil and a vinegar CRUETS. Sometimes called a *porte-vinaigrette*.

Portland Vase. A vase of Greek AMPHORA type, made of cobalt-blue translucent CASED GLASS so dark that it is almost indistinguishable from black except when seen by transmitted light; it has decoration of white opaque glass carved in CAMEO relief. Below each of the two vertical attached handles is a carved head of Pan which separates the encircling frieze into two groups of figures. The bottom is missing (*see* PORTLAND VASE DISK), and it has been suggested by the British Museum that, although the vase might originally have had a flat base, it is more likely that it tapered downwards from below the present white base line to a knob or point. Although its origin is unknown, the vase has been attributed to ROMAN GLASS (1st century BC–1st century AD). The frieze is said to depict the myth of the marriage of Peleus and Thetis, this belief being based largely on the inclusion of a sea-dragon.

History: Now known as the 'Portland Vase', it was originally known as the 'Barberini Vase' because it was in the Palazzo Barberini in Rome

Portland Vase disk. Blue cased glass with opaque white cameo, 1st century BC–AD 1st century. D. 12·1 cm. British Museum, London.

in 1642. It was once conjectured that it had been discovered in 1582 in the Monte del Grasso sarcophagus on the Appian Way near Rome and that it was a CINERARY URN of Alexander Severus (Emperor, 222–235), depicting a scene related to his birth, the figures thought to depict the Emperor and his mother, Julia Mam(m)aea; but such conjectures are now authoritatively discredited. The first record of the vase relates to its being in the possession of Cardinal del Monte (1549–1627) at the Palazzo Madama, Rome; his heirs sold it, c. 1600, to Cardinal Francesco Barberini, nephew of Matteo Barberini (Pope Urban VIII; 1623–44). The vase was again sold, c. 1780, by Donna Cornelia Barberini-Colonna, Princess of Palestrina, to James Byres, a Scottish antiquary resident in Rome. He in 1783 sold it to Sir William Hamilton, British Ambassador to the Court of Naples, who brought it to London and before 1785 sold it to the Dowager Duchess of Portland (from whom it derived its present name). After her death in 1785 it was sold at auction to a Mr Tomlinson who was acting for the Duchess's son, the third Duke of Portland. In 1810 it was lent by the fourth Duke of Portland to the British Museum, which purchased it in 1945. It was reported in 1786 that it had been fractured by the Duchess of Gordon (d. 1812). It was smashed into over 200 fragments in 1845 by William Lloyd; it was expertly restored in 1845 and again in 1948. The vase was lent in 1786 to Josiah Wedgwood (it has been said in consideration of his forgoing a bid for it at the auction) for the purpose of reproducing it in jasper ware.

For a detailed description and history, see D. E. L. Haynes, *The Portland Vase* (1964, 1975). *See* VENDANGE VASE; AULDJO JUG.

Reproductions: (1) A copy of the Portland Vase was made in CASED GLASS by PHILIP PARGETER and was engraved from 1873 to 1876 by JOHN NORTHWOOD; it has a base in the form of the PORTLAND VASE DISK. It is signed and dated 1876. It was on loan at the British Museum from 1959 until 1975 when it was sold at auction in London for £30,000. (2) An earlier copy, in colourless glass with WHEEL ENGRAVING, was made by Pargeter and Northwood, c. 1865, and was possibly a test model. (3) A glass replica was made in 1878 by JOSEPH LOCKE. (4) A series of pottery copies was made in 1790 by Josiah Wedgwood in jasper ware; one example is now in the British Museum. Other copies were made by the Wedgwood factory, the most recent being a series (15 cm. high) made in 1976 in a new Wedgwood colour called 'Portland blue'. (5) A number of less accurate copies have been made by Staffordshire potters. (6) Sixty plaster copies were cast by JAMES TASSIE from a mould made by Giovanni Pichler before the original vase was delivered in 1783 by James Byres to Hamilton. (7) The earliest attempt to make a glass copy was by Edward Thomason and a Mr Biddle at Birmingham Heath Glassworks in 1818, but they failed to complete the white casing; their problem was later solved by Pargeter by first blowing the white outer layer.

Portland Vase Disk. A flat circular disk of blue glass CASED with opaque white glass carved in CAMEO relief with a figure in profile. It formed at one time the base of the PORTLAND VASE, and is believed to have been attached to it before 1642; it is now considered by the British Museum not to be the original base or bottom because (1) the blue colour is paler, (2) the carving is not by the same hand, and (3) the disk has been cut from a larger composition. The disk depicts a Grecian youth wearing a Phrygian cap and is believed to represent Paris, the son of King Priam.

portrait glassware. Glassware, in many forms, decorated in various styles with a miniature portrait. The earliest known examples are medallion portraits engraved through gold leaf behind glass; these are of ROMAN GLASS, from the 2nd to 4th centuries. *See* PLACIDIA, GALLA, CROSS. Later examples include: (1) Portraits enamelled, somewhat crudely, on HUMPEN made in Germany, Bohemia, and Silesia in the 16th–18th centuries. (2) Finer portrait work done by STIPPLING in the Netherlands in the 18th century, and drinking glasses with portraits of contemporary clients engraved in gold behind glass by JOHANN JOSEF

MILDNER at Gutenbrunn and wheel-engraved by DOMINIK BIEMANN at the spa of Franzensbad in the late 18th–early 19th centuries. (3) Glass PANELS with a portrait (a HEAD or a BUST) executed in relief or enamelling; some were made of heavy glass in Venice in the 16th century, e.g. a relief portrait of Andrea Gritti, a Doge of Venice. (4) MEDALLIONS with relief portraits, e.g. TASSIE MEDALLIONS and SULPHIDES. (5) COMMEMORATIVE GLASSWARE with a portrait, e.g. JACOBITE GLASSWARE; KING OF PRUSSIA GLASSWARE; ROYAL OAK GOBLET. (6) MARRIAGE GLASSES and WEDDING CUPS with engraved or enamelled portraits made at Murano (e.g. BAROVIER CUP) and elsewhere (e.g. one portraying Louis XIV, made by BERNARD PERROT at Orléans in the late 17th century).

Poschinger, Ferdinand von (1815–67). A German glass-maker, the son of Benedikt von Poschinger who in 1856 had established a glassworks at Buchenau, in Bavaria, which was inherited by Ferdinand and in turn by his son Ferdinand Benedikt (1867–1921). The factory made many pieces in IRIDESCENT GLASS in the style of ÉMILE GALLÉ and LOUIS COMFORT TIFFANY.

posset-pot. Moulded pot with seal of George Ravenscroft, *c.* 1680. Ht 11·5 cm. Courtesy, Delomosne & Son, London.

posset pot. A vessel for drinking posset (a beverage formerly popular in England, made from hot milk curdled by an infusion of wine or ale, with breadcrumbs and often spiced). They have vertical or curved sides with two vertical loop-handles, sometimes a COVER, and a thin curved SPOUT extending from near the bottom of the piece (so placed to enable the person drinking from the spout to avoid the curdled milk on the surface of the posset). Some examples have a slightly flaring BASE or a low stemmed FOOT. They may also have been used for caudle (*see* CAUDLE CUP) and sometimes for syllabub (*see* SYLLABUB GLASS). They were made in England by GEORGE RAVENSCROFT and others in the 17th century, and some possibly in Venice for the English market. Sometimes called a 'posset glass', a 'spout glass' or a 'spout cup'.

post-horn. Footless drinking glass, Germany, 17th or 18th century. Ht 18 cm. Kunstgewerbemuseum, Cologne.

post-horn. A type of drinking glass made in the form of a post-horn, with a funnel-shaped mouth on which it rests when empty, and a looped stem that brings the other end up to opposite the mouth or sometimes looped again to form a right-angle to the mouth. They were made in Germany in the 17th and 18th centuries. *See* BUGLE; COACH-HORN.

pot. A CRUCIBLE of FIRE-CLAY in which the BATCH of glass ingredients is fused. There are two types: (1) the SKITTLE POT, like an open barrel, and (2) the CROWN POT, with a domed top and a side-opening ('mouth'). The pots, about 4 ft (1·30 m.) high and 3 ft (1 m.) in diameter, are made by hand from clay capable of withstanding intense heat and from which iron particles have been removed magnetically; the bottom, about 5 in. (12·5 cm.) thick, is made in a mould, then the sides are built up slowly from day to day, and the completed pot is subjected to intense heat before being moved to the FURNACE (*see* POT SETTLING) while extremely hot and in controlled temperature conditions. Some pots can contain about a ton of molten glass.

pot arch. A type of furnace in which a POT is fired and from which it must be carefully transferred to the glass-melting FURNACE by the process of POT SETTLING.

pot settling. The process of transferring a POT from a POT ARCH to the glass-melting FURNACE, which must be done while the pot is extremely hot and with no sudden change of temperature.

potash. Potassium carbonate. It is, as the alternative to SODA, the ALKALI ingredient of glass. It was obtained in early Germany and Bohemia by burning beechwood, oak, or certain other timber (*see* WALDGLAS) and in France by burning fern and bracken (*see* VERRE DE FOUGÈRE). The process was to leach the wood ashes, evaporate the lye, and calcine the residue. Other sources of potash were saltpetre (nitrate of potash), refined commercial potassium carbonate ('pearl ash'), and burnt lees of wine. Today much potash is made commercially from potassium chloride. Glass made with potash becomes rigid more quickly after heating than glass made with soda, and it is harder and more brilliant, hence is more suitable for CUTTING and ENGRAVING. It is used in making LEAD GLASS, and is sometimes combined with LIME to avoid CRISSELLING.

potiche (French). A covered JAR, of variable shape and usually without handles; examples are usually circular but some are polygonal. Usually they are of Chinese or Japanese porcelain, but some Spanish glass receptacles were apparently inspired by such pieces.

potichomania. The process of imitating painted porcelain by painting or otherwise decorating the interior of glass vessels, e.g., as on WITCHBALLS made of NAILSEA GLASS. The process involved glueing cut-out pictures on the interior of the vessel, varnishing them, and then sealing them with a chalk or plaster lining that served as a background. The process was referred to by Charles Dickens in 1855.

Potsdam Glass Factory. A glassworks established in 1679 at Potsdam (Berlin), by Elector Friedrich Wilhelm of Brandenburg (d. 1688), in addition to one established in 1674 at nearby Drewitz. It was under the direction, from its founding until 1693, of JOHANN KUNCKEL, who developed the process of making GOLD RUBY GLASS and who also introduced the use of chalk into a BATCH for crystal glass to obviate CRISSELLING. The factory burnt down in 1688, and in 1736 the operation was moved by the successors of Kunckel to ZECHLIN, where it remained a State operation until 1890. *See* BRANDENBURG GLASSWARE.

Potters, Barbara, Goblet. A GOBLET, inscribed 'Barbara Potters, 1602', having a bell-shaped bowl on a tall stem with a hollow urn-shaped KNOP with four lion's-head masks. The bowl and the foot are decorated with floral designs in DIAMOND-POINT engraving. It has been associated with JACOPO VERZELINI, but as it was made after his retirement in 1592, it is now regarded as probably having been made by his former Italian employees at the BLACKFRIARS GLASSHOUSE of his successor, SIR JEROME BOWES. *See* VERZELINI GLASSWARE.

pouch bottle. A drinking vessel of somewhat globular or ovoid shape, with a waisted neck below a wide mouth and having a rounded or pointed bottom. They are shorter and squatter than a BAG BEAKER. The decoration is (1) a horizontal neck spiral in an uninterrupted trail and a zigzag or MERRYTHOUGHT (wishbone) trail on the body, (2) a neck spiral only, or (3) corrugation on the body. Most known examples have been found in Kent (except a few in Scandinavia), and have been attributed to that region in the 7th and 8th centuries.

powder-decorated glass. A type of ART GLASS having a design made of a coating of vari-coloured powdered glass applied to the surface. The process, patented in 1806 by John Davenport, of Longport, Stoke-on-Trent, England, involved applying a paste containing powdered glass, making the design by scraping it away with a pointed tool, then lightly fusing the design onto the glass. Although the patent referred to it as being an imitation of engraving or etching, in fact it resembles neither. The designs often were heraldic insignia and sporting scenes. *See* DAVENPORT'S PATENT GLASS.

powder-glass. A type of SAND-GLASS, consisting of a SET of four such glasses in a compartment-case, to record the quarter-hours by starting the respective glasses at fifteen-minute intervals. It has been said that they were used for timing sermons in the mid-18th century.

powder horn. A flask in the form of an ox-horn, used for holding gunpowder for priming purposes. Some were made of glass in Germany, in the 17th century; they have a metal screw cap or hinged lid.

pouch bottle. Frankish glass, 7th century. Ht 12·3 cm. Statens Historiska Museum, Stockholm.

powder-glass. Four fifteen-minute sand-glasses, mid-18th century. Science Museum, London; Crown copyright.

presentation tumbler. Engraved with portrait of King John III of Poland (1674–96) presented to commemorate his defeat of the Turks at Vienna in 1683, with appropriate inscription in Latin. Bohemia, 1685. Ht 12 cm. Cinzano Glass Collection, London.

pressed glass. Sugar-bowl with cover, colourless flint glass, New England Glass Co., c. 1820. Ht. 24·8 cm. Toledo Museum of Art, Toledo, Ohio.

Powell, Harry J. (1853–1922). A director of the WHITEFRIARS GLASS WORKS, who from c. 1880 contributed much to the development of modern glass in England. He was succeeded by his son Barnaby (1871–1939).

Powell, James, & Sons. *See* WHITEFRIARS GLASS WORKS; POWELL, HARRY J.

Powolny, Michael (1871–1945). A glass-maker who had his own workshop and who designed for J. & L. LOBMEYR and J. Lötz Witwe (*see* LÖTZ GLASSWARE).

preserve jar. A small JAR, made in various shapes and styles, used for serving preserves and jam at table. Many are ornately decorated with cut patterns. They usually have a cover.

Preissler glassware. Glassware tentatively attributed to Daniel Preissler (1636–1733), a HAUSMALER from Friedrichswalde, Silesia, and later Kronstadt, Bohemia, and to his son Ignaz (b. 1670), who are recorded as having decorated glassware in SCHWARZLOT. Ignaz was *Hausmaler* to Count Franz Karl Liebsteinsky von Kolowrat at Kronstadt, 1729–39. The pieces attributed to them (all unsigned) depicted, on early work, landscapes and peasant scenes, but later examples feature CHINOISERIES, MYTHOLOGICAL SUBJECTS, and LAUB- UND BANDELWERK. The work of another unidentified Preissler, who worked at Breslau and may have been related, and who may have decorated glass in *Schwarzlot*, has been included in the attribution 'in the style of the Preisslers'. According to W. B. Honey, this work was 'formerly attributed vaguely to an artist called 'Preussler', possibly Christian Preussler (*see* PREUSSLER FAMILY). For a discussion of some Preissler glassware, *see* Rudolf von Strasser, 'Twelve Preissler Glasses', *Journal of Glass Studies* (Corning), XV 1973, p. 135.

presentation glassware. Glassware in various forms, e.g. a TUMBLER, GOBLET, BOWL, etc., that is decorated to commemorate its presentation to an individual on some auspicious occasion. *See* COMMEMORATIVE GLASSWARE.

pressed glassware. Glassware made by the process of shaping an object by placing a blob of MOLTEN GLASS in a metal MOULD that forms the outside shape of the piece (as in the process of jollying ceramic ware) and then pressing it with a metal 'plunger' or 'follower' to form the inside shape. The resultant piece, termed MOULD-PRESSED, has an interior form independent of the exterior, in contrast to MOULD-BLOWN GLASS whose interior corresponds to the outer form. Pressing was introduced in 1827 in the United States, first by a workman named Enoch Robinson at the NEW ENGLAND GLASS COMPANY, at East Cambridge, Boston, Mass., and then by DEMING JARVES at the rival BOSTON & SANDWICH GLASS CO., at Sandwich, near Cape Cod. The process came into general use in the 1830s and was used for the majority of American glassware by the late 1850s. Pressed glass was produced in England before 1827 (at Birmingham, Stourbridge, Newcastle-upon-Tyne, and Sunderland), and has since been made as a cheap reproduction of CUT GLASS, often in coloured glass and sometimes with pressed FACETING and ridges in LOW RELIEF. Although originally it was used to copy cut glass, its use led to the development of new forms for which the technique was especially suitable, e.g. LACY GLASS, CUP PLATES, etc. The Italian term is *vetro pressato*. *See* MOULD-PRESSED GLASS.

For a discussion of American pressed glassware and the nomenclature of over 300 patterns, *see* Ruth Webb Lee, *Early American Pressed Glass* (1946) and *Sandwich Glass* (1939).

Prestonpans Glasshouse. A glassworks set up at Prestonpans, Scotland. It made bottles, some of gigantic size, said to have had a capacity of 100 gallons. They were made by injecting into the bubble a volatile spirit which, when evaporating due to the heat, expanded and enlarged the bubble.

Preussler family. Glass-makers and engravers in Bohemia and Silesia during the 17th and 18th centuries. In 1658 Georg Preussler, a glass-cutter of the Riesengebirge, in Silesia, became director of a new glasshouse that Friedrich Wilhelm, the Great Elector of Brandenburg, founded in 1653 at Grimnitz, in Silesia. Christian Preussler, a glass enameller who worked at Zellberg, Bohemia, *c.*1680, and whose name is inscribed as donor of a HUMPEN dated 1680, may be the decorator who has been confused with Ignaz Preissler (see PREISSLER GLASSWARE). Wolfgang Preussler started a glasshouse at Schreiberhau in Silesia, and later, *c.*1742, a glasshouse was operated there by Karl Christian Preussler; the latter's ware was inferior to that made at the glasshouses in adjacent Bohemia, from which raw glass was smuggled to the engraving houses in Silesia after Frederick the Great in 1742 forbade the importation of glass into Silesia. A HUMPEN made in 1727, probably in Bohemia, depicts the operation of a glassworks and bears an inscription identifying the members of the Preussler family.

primavera, vetro (Italian). Literally, springtime glass. A type of ornamental glass developed by Ercole Barovier for BAROVIER & TOSO, *c.*1927, the characteristic feature of which is a crackled TRANSLUCENT glass for the body of the pieces, which are decorated with handles and with narrow bands on the base and mouth made of black glass.

printie. A decorative pattern on CUT GLASS, consisting of a shallow concavity, circular or oval, made with a slightly convex cutting wheel. They are generally made in an encircling pattern (sometimes in V-shaped rows) on various glass vessels or on some overlay PAPER-WEIGHTS. The term is said to be derived from the practice of grinding away the PONTIL (PUNTY) mark after the piece has been removed from the iron, and hence sometimes called a 'punty'. Circular examples are also sometimes termed 'ball cut' and oval ones 'thumb marks' or 'finger cut'. Sometimes printies that overlap are termed 'clover cut'; these were made by the NEW ENGLAND GLASS CO. On some glasses with a circle of printies, a small engraved sprig is cut between each pair of printies. *See* CURVED CUTTING; PUNT.

prism. (1) A type of DROP of triangular section. (2) A style of decoration on CUT GLASS; *see* PRISM CUTTING. (3) A rod of glass of triangular section for optical use in the refraction of light.

prism cutting. A decorative pattern (see CUT GLASS: *patterns*) consisting of long straight mitred grooves cut horizontally, in parallel lines, so that the top edges touch. It is usually found on the neck of a jug or decanter in rings of diminishing size, extending upward from the body to the rim. A variation is the 'alternate prism' which is cut, either vertically or horizontally, with small sections of ridges alternating with small sections of grooves.

privateer glassware. Glassware, usually wine glasses, engraved with the representation of a commissioned private sailing vessel in full sail and usually an inscription with the names of the captain and of the ship. They were made of English glass in the 18th century. *See* NAVAL SUBJECTS; SHIP SUBJECTS.

procession cup. A type of STANDING CUP or BOWL that is decorated with an enamelled encircling frieze of a procession of people. These, and similarly decorated TAZZE, have been ascribed to Venetian glass of the middle or last quarter of the 15th century, and are among the earliest examples of MURANO enamelled decoration. They usually have a bowl of cylindrical form, resting on a hollow or knopped STEM with a spreading FOOT. *See* WEDDING CUP; BAROVIER CUP.

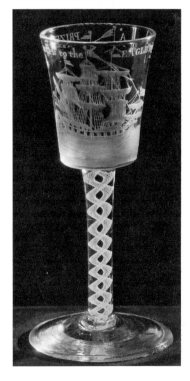

privateer glassware. Engraved decoration on wine glass with bucket bowl and multiple air-twist stem, English, mid-18th century. Courtesy, Derek C. Davis, London.

prunts. Stangenglas with eight vertical rows of nine prunts and internally eight rows of seven prunts, probably Rhineland, early 16th century. H. 26 cm. British Museum, London.

pterotos cup. Moulded bowl, dark-blue glass, Alexandria, 1st century AD. Ht 12 cm. Museo Civico, Adria.

prochoös (Greek). A slender form of OENOCHOË; a small JUG with a tapering globular body, a thin NECK, small MOUTH, high arched HANDLE, and a stemmed FOOT. It was used for pouring wine, and for water for washing the hands before a meal. It was also made in low compressed form. Generally made of Greek pottery, some were made of ROMAN GLASS in the 3rd and 4th centuries.

Prouvé, Victor (1858–1943). A painter from Nancy, France, who was a friend and associate of EMILE GALLÉ, and who was art director of his factory from his death in 1904 until 1913. Prouvé designed the famed Orpheus Vase for Gallé, and in 1892 painted a well-known portrait of Gallé at work.

provenance (or **provenience**). The source of a specimen of glass or other art object; preferably (for clarity), the source as applied to prior ownership or place of discovery, as distinguished from the place of production (the origin).

prunt. A blob of glass applied to a glass object, usually a drinking vessel, as decoration but also to afford a firm grip in the absence of a handle. Prunts are of various forms and sizes, sometimes globular, pointed or irregularly shaped, and are occasionally impressed with a mask (*see* LION'S HEAD PRUNT) or a diaper pattern of droplets (*see* RASPBERRY PRUNT). Usually a large number of prunts were applied to an object, either haphazardly or in a pattern. Prunts were used on ROMAN GLASS and FRANKISH GLASS, on glass made in the *façon de Venise* in the 16th/17th centuries, and on 16th/17th-century German glass (e.g., on a NUPPENBECHER, WARZENBECHER, RÖMER, etc.), and continued to be used until the late 17th century, e.g. by GEORGE RAVENSCROFT in England; some such German glasses have prunts on the interior as well as the exterior (*see* STANGENGLAS). Prunts should be distinguished from small drops of glass in an overall surface pattern, as on some tall *Humpen*.

pseudo-faceted stem. A glass STEM that is PRESSED to resemble cut FACETING. Stems made in this style were made in Hesse and Thuringia and in England (*see* STEM: *Silesian*) in the first half of the 18th century.

pseudo vasa diatreta. Various objects characterized by a surrounding network or other RETICULATED ornamentation of glass that was not cut from the original piece but applied later when the object was reheated. *See* SCALLOP-SHELL CHALICE; CAGE CUP; DIATRETA GLASS.

pterotos cup. A type of SKYPHOS that has vertical handles usually referred to as 'wing-shaped'. It is basically of the same shape as other *skyphoi* (although some have a foot-stand), except that it has two vertical loop (or angled) handles whose upper edge forms a flat thumb-piece level with the rim; the handles are attached midway down the bowl and have at their bottom a curved finger-piece. The cups, of ROMAN GLASS, resemble contemporary silver cups. Some such cups, from the late 3rd century BC to the 1st century AD (with examples from Sidon having an impression made from a coin), have the handles cut from the solid MOULD-BLOWN glass. Others, from *c.* 100 BC to *c.* 100 AD, have attached mould-blown handles (presumably derived from the earlier cut examples). Some bowls of larger size and with a pedestal foot were similarly shaped. It is thought that the term 'pterotos' is derived from the Greek *pteron* (wing), and has been applied to this type of *skyphos* due to the wing-shape of the handles. See *Journal of Glass Studies* (Corning), IX (1967), pp. 23, 33.

pucellas. A glass-maker's tool with two metal spring-arms joined at one end by a curved metal handle. It was formerly used to form the diameter and the shape, inside and outside, of the bowl of a DRINKING GLASS or other circular-shaped piece while it was being rotated on the PONTIL. Later examples, of steel, are called a 'steel jack' or the workmen's term, 'the tool'. The pucellas tended to leave STRIATION marks on the inside and outside of the bowl where it touched, and so was superseded by a wooden-ended tool giving a softer effect. *See* WOODS; TOOLS, GLASS-MAKERS'.

pulegoso, vetro (Italian). Literally, bubbled glass. A type of ornamental opaque glass developed by VENINI *c.* 1928; the characteristic feature is the inclusion of many irregularly spaced air bubbles in the glass and some bursting on the surface, giving a blistered and pitted effect. Sometimes the piece is covered with a FLASHING of smooth clear glass, which is then termed *vetro pulegoso sommerso*. *See* SOMMERSO, VETRO.

pull test. A test to ascertain if two BATCHES of glass, intended to be used to make an object of CASED GLASS, have a compatible coefficient of expansion (if not, the finished object would crack or shatter from any sudden change of temperature). The test used by FREDERICK CARDER at STEUBEN GLASS WORKS involved making a rod of glass from each of two batches, melting them together, and then pulling the resulting molten piece into a thread; the thread was laid on a frame, and if it remained reasonably straight, the two batches were compatible, but if it curled beyond a certain tolerance, one batch was contracting more rapidly than the other, and one had to be corrected. This necessity for COMPATIBILITY is equally important in making CANES for MILLEFIORI glass.

pull-up vase. A type of vase decorated by inserting coloured or white glass threads in the base and drawing them up into the body with a pointed iron hook. The method was developed by JOHN NORTHWOOD at STEVENS & WILLIAMS, LTD, and later Northwood invented a machine to produce the decoration in this style, making even more intricate patterns.

Punch and Judy. A pair of glass FIGURES in the form of these puppet characters, made by John Derbyshire at Manchester, England, 1875.

punch-bowl. A large circular BOWL for serving hot or cold punch. They were usually made of silver, porcelain or pottery, but some were made of glass in the 17th century and later. Early examples may have a short knopped STEM with a domed FOOT; later ones are decorated with CUTTING and ENGRAVING. They are frequently accompanied by a conforming glass PUNCH-LADLE and sometimes by a set of conforming PUNCH-GLASSES. *See* TODDY LIFTER.

punch filler. *See* PUNCH-LADLE; TODDY LIFTER.

punch-glass. A type of drinking glass for drinking punch, sometimes part of a set accompanying and conforming to a PUNCH-BOWL. Some formerly had a conical bowl with a BALUSTER STEM, but since the 19th century, when glassware for punch was introduced, a punch-glass has usually been cup-shaped with a handle but no stemmed foot. It has been said that at one time glass TANKARDS were fashionable for drinking punch.

punch-ladle. A long LADLE for serving punch, often made as an accompaniment to a PUNCH-BOWL. Examples were made in the mid-18th century and later. They were superseded by the TODDY-LIFTER.

punt. A small PRINTIE. Examples of such cut decoration are found in an encircling pattern just below the rim of the bowls of some drinking glasses made in England in the second half of the 18th century.

purlee glassware. A type of glassware so designated in a list, dated 1677, of pieces being sold by the GLASS SELLERS' COMPANY. The precise meaning of the term has not been ascertained, but it has been surmised that 'purlee' referred to a pearl-like surface decoration, or alternatively, by W. A. Thorpe, that it designated a glass trail or thread,

pulegoso, vetro. Vases, *c.* 1930, by Venini, Murano.

punch-bowl. Bowl with separate base, in brilliant style, Libbey Glass Co., 1904. W. 63·5 cm. Toledo Museum of Art, Toledo, Ohio.

probably applied in a circuit (the term being perhaps derived from 'purl', to embroider).

purple glass. Glass coloured purple by the use of oxide of MANGANESE or NICKEL.

purple of Cassius. A crimson-purple colour that was sometimes used to colour glass. The pigment is prepared by precipitating a gold solution by means of chloride of tin and then adding the resultant colloidal gold to the BATCH; the glass so treated became ruby-red when reheated. The process of precipitation was named after Andreas Cassius (d. 1673), but it was first published by the German chemist, Johann Rudolf Glauber (1604–68) in 1659. The pigment was frequently used by German HAUSMALER in the 17th century to colour enamel for decorating faïence, but it was also used by JOHANN KUNCKEL, c. 1679, to colour his GOLD RUBY GLASS. It is also known as *rouge d'or* and 'rose pink', and it is the pigment used in the opaque enamel employed in China in painting *famille rose* porcelain.

putto (Italian). The figure of a small boy, usually nude or nearly so. Since the word *putto* (pl., *putti*) is the Italian for a young male child, the term, although often used loosely and interchangeably with *amorino*, should preferably be applied only to a wingless figure, as distinguished from a winged AMORINO (a word that has affiliations with the Roman Amor, or Cupid).

puzzle jug. A type of drinking glass made in the same spirit as pottery puzzle jugs which have an openwork upper half through which the drink spills unless it is sucked up through a concealed tube from the bottom to an aperture in the rim while the drinker's fingers close the other apertures in the rim. No glass example of the form of the pottery jugs is known, but several variations have been made, e.g., a glass with a central vertical tube surmounted by a glass bird whose beak is the drinking spout, and unless the hole under the bird's tail is closed the contents will spill from it. *See* JOKE GLASS; SIPHON GLASS.

pyramid. An arrangement of SWEETMEAT GLASSES and/or JELLY GLASSES on a SALVER, with a tall sweetmeat glass (sometimes called a 'captain glass') in the centre, surrounded by smaller glasses, and with the salver resting on glasses placed on a larger salver. There were often three such tiers. Such pyramids were customary in England in the 18th century.

Pyrex. The well-known trade name for heat-resistant oven-glass made by the CORNING GLASS WORKS and its world-wide licensees. It is made with BORAX and technically called 'borosilicate glass'. It was developed in 1912 to resist heat-shock that had caused breakage of railway brakemen's lanterns exposed to rain and snow.

Pyroceram. A trade-marked glass-ceramic product developed in 1957 by the CORNING GLASS WORKS. The process involved initially melting and forming ordinary glass, then, by further reheating, transforming the products to closely-grained crystalline ceramics. It has special strength and heat-resistant qualities.

Pyx. In the Christian church, a vessel in which the Host is reserved. They were made of glass, e.g. at Murano. *See* CHURCH GLASSWARE.

pyxis (Greek). A covered box-like receptacle, usually cylindrical. Generally they were made of Greek pottery as a container on the dressing-table for jewels or trinkets. Some were made of blown ROMAN GLASS, in the 1st century BC. A number of covered glass *pyxides* have recently been excavated in Crete; these were apparently made in moulds from fused crushed glass. *See Journal of Glass Studies* (Corning) I, 1959, p. 1.

pyxis. Opaque marbleized Roman glass, 1st/2nd century. Ht 5·8 cm. Römisch-Germanisches Museum, Cologne.

Q

Quartz glass. A type of ART GLASS developed by FREDERICK CARDER at STEUBEN GLASS WORKS. It was made by a process similar to that used for CINTRA GLASS, but was given a crackled effect obtained by briefly plunging a GATHER of molten colourless glass into cold water so as to cause only the surface to shatter. It was reheated so that the crackles were lightly fused, then the PARAISON was heated again and rolled over powdered coloured glass that was pulled into a pattern, and finally it was covered with colourless glass. A satin surface was sometimes given with acid. Depending on the colour of the outer glass layer, the glass was termed 'Rose Quartz', 'Alabaster Quartz', 'Amethyst Quartz', 'Blue Quartz', 'Green Quartz', 'Peach Quartz' or 'Yellow Quartz'. Added decoration was sometimes given by applied glass decoration in relief that was then acid-etched.

Queen's Burmese. *See* BURMESE GLASS.

Queen's design. A decorative pattern on BURMESE GLASS in the form of sprays of flowers. It was used on a tea-set of Burmese glassware made by the MT. WASHINGTON GLASS CO. for Queen Victoria, after which Burmese glassware made by the English licensee, THOMAS WEBB & SONS, was called 'Queen's Burmese Glass'.

Queen's Ivory Ware. A type of OPAQUE WHITE GLASS with a creamy colour, made by SOWERBY'S ELLISON GLASSWORKS, *c.* 1870.

Quezal Art Glass & Decorating Co. A glassworks at Brooklyn, N.Y., founded in 1901 by Martin Bach (d. 1924) and Thomas Johnson, both formerly with LOUIS C. TIFFANY. It made OPALESCENT glassware and other forms of ART GLASS very similar to the FAVRILE GLASS of Tiffany and also the AURENE GLASS of Steuben. The name was trade-marked in 1902, and a second trade-mark, issued in 1907, shows the que(t)zal, a Central American bird with brilliant colourful plumage, which suggested the company's name. After a change of the company's ownership, it closed in 1925, and Martin Bach, Jr., and others went to the VINELAND FLINT GLASS WORKS of Victor Durand, Jr.

quilling. A style of decoration in the form of closely spaced festoons of WHITE OPAQUE GLASS, to be found on some so-called NAILSEA GLASS, and especially glass said to have been made at Alloa, Scotland.

quilt moulded. A decorative moulded style occurring on the bowls of some Venetian wine glasses, the exterior having the appearance of quilting, with diagonal criss-cross ridges enclosing slightly elevated diamond forms.

quilted-cushion glass. A type of ART GLASS made with the appearance of quilting. It was made as CASED glass, the opaque inner layer being mould-blown in a diamond pattern and covered with a thin layer of transparent glass, thus trapping an air bubble within each diamond, after which the surface was given a satin finish with HYDROFLUORIC ACID. Such ware was made by THOMAS WEBB & SONS and by STEVENS & WILLIAMS, LTD., *c.* 1880.

quincunx pattern. A style of decoration with motifs placed in an overall diaper pattern so that groups of five are in the form of the five dots on dice or symbols on playing cards, one central and one at each corner. It was used in the 2nd to 6th centuries and the 9th to 11th centuries on PERSIAN GLASSWARE that is decorated with contiguous (but not overlapping) shallow wheel-engraved circular concavities so arranged. It was also used later in Bohemia and elsewhere.

Quartz glass. Vase designed and made by Frederick Carder for Steuben Glass Works, 1920s. Ht 18·5 cm. Corning Museum of Glass, Corning, N.Y.

quincunx pattern. Flask with circular facets in quincunx formation, Persian, 9th century. Ht 15 cm. British Museum, London.

R

raised diamonds. Tumbler with lower part in raised diamonds and upper with diamond-point engraved decoration, St Petersburg Imperial Glass Factory, *c.*1830. Ht 9·2 cm. Mrs Merriweather Post Collection, Hillwood Museum, Washington, D.C.

raspberry prunts. Römer wheel-engraved by Christoffel Heier, Netherlands. 17th century. Ht 21·5 cm. Rijksmuseum, Amsterdam.

rafraîchissoir (French). (1) A receptacle for cooling, usually a large goblet-type container into which is suspended in a metal mount a smaller receptacle; ice placed in the large part surrounds the inner receptacle and cools its contents. *See* CAVIAR BOWL. (2) A MONTEITH. (3) A cylindrical bowl for cooling a single wine glass.

raised diamond cutting. A decorative CUT GLASS pattern, made with a mitred wheel so that diagonal grooves are close together, with the top edges common, and then crossing these grooves at right angles to a second series of like grooves, resulting in an overall surface of raised or relief four-sided diamonds in pyramid form, each with a sharp apex. Although the pattern is usually diagonal, the grooves are sometimes in a horizontal-vertical grid. The pattern is often found on ANGLO-IRISH GLASSWARE, *c.*1780–1825. *See* HOLLOW DIAMOND CUTTING; STRAWBERRY DIAMOND CUTTING; HOBNAIL CUTTING; CROSS-CUT DIAMONDS.

ramarro, vetro (Italian). Literally, lizard glass. A type of glass, designed by Ercole Barovier for BAROVIER & TOSO, having a green mottled surface.

Ranftbecher (German). A type of low BEAKER with a tapering or waisted body standing on a thick-cut COGWHEEL BASE, often decorated with transparent enamelling and gilding. They were made in Vienna, *c.*1815–28, in BIEDERMEIER STYLE. Their principal decorator was ANTON KOTHGASSER who enamelled some with three tarot-cards, some with scenes of Vienna, and occasionally with only an encircling band of flowers. Some were wheel-engraved in Bohemia and elsewhere.

Raqqa glassware. A type of Syrian Islamic glassware identified with, but not necessarily produced at, Raqqa (Rakka), a city on the Euphrates (in modern Iraq), but possibly made also at Aleppo or Damascus. It is from the earliest Islamic period, *c.*1170–1270, and includes BEAKERS of clear glass with gilding and enamelling, and also inscriptions in various colours, the most characteristic feature being a diaper pattern of dots in white and/or turquoise-blue. *See* EIGHT PRIESTS, GOBLET OF THE; ENAMELLING, ISLAMIC; SYRIAN GLASS.

raspberry prunt. A type of flat circular PRUNT on which there is a relief design, impressed with a tool, that has the appearance of a raspberry. They were used as decoration on several types of German glasses, e.g., a RÖMER, STANGENGLAS or WARZENBECHER, and also on some hollow stems of drinking glasses made by GEORGE RAVENSCROFT, and subsequently by other glassmakers in England. Sometimes called a 'dotted prunt'. The German term is *Beerennuppen*. *See* STRAWBERRY PRUNT.

ratafia glass. A type of drinking glass intended for drinking ratafia (a liqueur flavoured with almonds and fruit kernels). It is a type of FLUTE GLASS, made in two styles: with a very narrow and tall funnel-shaped bowl or with a thin cylinder-shaped bowl having a rounded bottom; both types have a stemmed foot, the stem sometimes having an OPAQUE TWIST. The bowl is sometimes decorated with light FLUTING and/or with ENGRAVED floral designs or border. The capacity was 1 to 1½ fl. oz. Sometimes called a 'flute cordial'.

Ratcliff Glasshouse. A glassworks at Southwark, London, operated by JOHN BOWLES and making FLINT GLASS and CROWN GLASS window panes. In 1678 William Lillington became a partner. Some glasses that bear an applied seal (perhaps inspired by the RAVEN'S HEAD of GEORGE RAVENSCROFT) with the figure of an archer are thought to be from this factory.

Ravenna glass. A type of ornamental glass made by ORREFORS GLASBRUK, developed *c.* 1947 by SVEN PALMQVIST. It is used for making heavy objects, and is tinted in transparent colours; the glass is inlaid with abstract patterns in brilliant hues.

Raven's Head. A seal bearing in relief a raven's head which GEORGE RAVENSCROFT was permitted to use in 1677 on the glass thereafter made by him for the GLASS SELLERS' COMPANY, perhaps in recognition of his work in developing LEAD GLASS and to offer assurance to the buyers as to the quality of the new glass. The emblem was from the coat-of-arms of his family.

Ravenscroft, George (1632–83). The English glass technician who perfected the process of making LEAD GLASS. He was a merchant who dealt in glass in London and Venice, and in 1673 started a glasshouse at the Savoy in London where, after experimenting for eight months, he successfully produced a 'crystalline glass' and in 1674 obtained a patent for seven years. Later in 1674 he made an agreement with the GLASS SELLERS' COMPANY that authorized him to set up a second glasshouse at Henley-on-Thames and provided for the sale to the Company of his entire production of 'crystalline glass'. Due to an excess of POTASH (used to assist fusing), the new glass developed the defect known as CRISSELLING; to counteract this, he added oxide of lead in lieu of some SILICA, and by 1676 he succeeded in making lead glass. To recognize his achievement, and as evidence of its guarantee to the public against crisselling, the Company agreed in 1677 that the new glass should be impressed with a RAVEN'S HEAD seal. Ravenscroft, by notice to the Company, retired from the Henley plant in 1678 and was succeeded there by his brother Francis and by HAWLEY BISHOPP, his associate; he was succeeded at the Savoy in 1682 by Bishopp. The new lead glass was finally perfected in 1681, but in 1968 only 25 extant sealed pieces, including fragments, were recorded. The dates of Ravenscroft's birth and death have, until research reported in 1975, been stated in many books as 1618–81. *See* R. J. Charleston, in *Journal of Glass Studies* (Corning), X (1968), p. 158; Rosemary Rendel, in *The Glass Circle*, No. 2 (1975), p. 65; D. C. Watts, *ibid.*, p. 71.

Ravilious, Eric (1903–42). A glass-designer who worked at STUART & SONS.

Reactive glass. A type of ART GLASS made by LOUIS C. TIFFANY that changed colour when it had been reheated. Some pieces are made with inlaid decoration of many colours, and some are internally lustred, creating a so-called 'under-the-water' effect. The decorative patterns varied, including some known as 'Rainbow', 'Gladiola', 'Morning-glory', and 'Opalescent Optic'. The type of glass was also called 'Pastel Tiffany' and 'Tiffany Flashed Glass'. *See* GOLD RUBY GLASS.

realgar glass. Glass imitating realgar (an almost pure form of disulphide of arsenic, an orange-red mineral poisonous to the touch; it was often used in China to make figures of Taoist deities in the belief that it contained the germ of gold). Such glass was used in China in the 18th century to make figures, CHINESE SNUFF BOTTLES, cups, and vases. The Chinese term is *hsiung hu'ang*.

rectangular bottle. A four-sided bottle of rectangular section, generally similar to a SQUARE BOTTLE. Some have two angular flat reeded handles, or one such handle, extending from the sloping shoulder to just below the mouth.

red glass. Glass coloured red as the result of adding oxide of GOLD, IRON or COPPER to the BATCH. *See* GOLD RUBY GLASS.

Ravenna glass. Bowl by Sven Palmqvist. Courtesy of Orrefors Glasbruk.

raven's head seal. Detail of English lead-glass bowl, George Ravenscroft, *c.* 1765. Victoria and Albert Museum, London.

realgar glass. Snuff bottle, opaque glass mottled orange and yellow, Chinese, 18th century. Ht 6 cm. Courtesy, Hugh Moss, Ltd, London.

reeding. Jug with mould-blown reeding, English (probably by George Ravenscroft at Savoy Glasshouse, *c.* 1676). Ht 27·3 cm. Victoria and Albert Museum, London.

Reformation Beaker. German, probably mid-17th century. Ht 23·7 cm. Corning Museum of Glass, Corning, N.Y.

Red House Glassworks. A glasshouse at Wordsley, near Stourbridge, founded in 1776 by George Ensell. It passed through several ownerships until 1811 when it was leased to Frederick Stuart (b. 1817) and 1885 when it was acquired by STUART & SONS, LTD., of which it is now a part. It was at one time leased to PHILIP PARGETER, under whose direction in 1873 JOHN NORTHWOOD I made his glass replica of the PORTLAND VASE from a blank made there by Daniel Hancox.

red lead. The essential ingredient used in making LEAD GLASS.

Redcliff Backs Glasshouse. A glassworks in Bristol, adjoining a pottery, and said to have been the first glassworks in Bristol to make OPAQUE WHITE GLASS, *c.* 1757. It employed MICHAEL EDKINS between 1762 and 1767.

reducing atmosphere. A non-oxidizing condition in the furnace, rich in carbon monoxide, produced by a smoky fire, that affects the glass in the crucible, e.g., a red colour is produced by the use of copper (cuprous oxide) in the BATCH in conjunction with a reducing atmosphere, in contrast to an oxidizing (smokeless) atmosphere. The effect of such atmospheres was known as long ago as the 7th century BC by the glass-makers of Mesopotamia. Various methods have been used to attain this atmosphere, e.g., the Chinese commonly employed wet wood in the furnace.

reeding. Relief ornamentation in the form of a series of parallel convex reeds. The converse of FLUTING. Also called 'ribbing'. *See* PILLAR-MOULDING.

Reformation Beaker. A HUMPEN (with cover) commemorating the beginning of the Reformation in 1517, depicting seated figures of Martin Luther, Philipp Melanchthon, and two Dukes of Saxony. It was made in Germany, probably in the mid-17th century.

Regency style. An English decorative style, evident in CHANDELIERS and other glassware, which prevailed, strictly speaking, from 1811 to 1830, during that part of the reign of George III when the affairs of the country were in the hands of the Prince Regent, afterwards George IV. In common parlance the term is extended both before and after those dates to an indeterminate extent. The Regency style in England is a version of the French EMPIRE STYLE, with some modifications. *Chinoiseries* were perhaps inspired by the Prince Regent's taste for the subject, to be seen at the Royal Pavilion at Brighton. Generally, the rendering of classical subjects lacked the lightness of the earlier NEO-CLASSICAL and ADAM styles, and is inclined to be pompous and heavy. Egyptian motifs belong to this period, and mark the revival of interest in ancient Egypt which followed Napoleon's campaign there of 1798.

Registry mark. A mark required by the Patent Office from 1842 to 1883 on British manufactured goods using a registered design. It was used on glassware (indicated as Class III) as well as on ceramic and other ware. It is diamond shaped, with a code letter in the corners indicating the year, month, and 'day' (i.e., date) of the design registration, which is not necessarily the date of manufacture of the object. The first series was used from 1842 until 1867, and the second from 1868 to 1883, and thereafter merely serial numbers were used in a continuous sequence. The glass mark appears mainly on PRESSED GLASSWARE. *See* SOWERBY'S ELLISON GLASSWORKS.

Reichsadlerhumpen (German). Literally, Imperial Eagle Beaker. A tall HUMPEN decorated with an enamelled black double-headed Imperial Eagle of the Holy Roman Empire, with its wings outspread and hung with fifty-six shields, in groups of four (the Quaternion System), representing the constituent parts of the Empire (the Ecclesiastical and the Secular Electors being, four each, on the top row, and below, in twelve groups of four each, read vertically, the Landgraves, Margraves, Counts, and others). At the neck of the eagle there is superimposed a crucifix on early examples (*c*.1571, the earliest known example, to *c*.1599) but on others (*c*.1585) there is substituted an orb or (*c*.1670–76) portraits of Emperor Leopold I, flanked by the seven (eight) Electors or by his sons. An inscription reads 'The Holy Roman Empire and its Member States'. On the reverse of some 16th-century examples there is an enamelled 'Brazen Serpent', but later ones have floral designs. The same designs appear in a few instances on a *Stangenglas*, goblet, mug, and bottle, and even on a miniature beaker of MILCHGLAS. Some such *Humpen* have a cover (often missing). The design is derived from a 15th-century woodcut. The glasses were made in Bohemia and Franconia in the late-16th century and also elsewhere in the 17th century. *See* ENAMELLING, *German.*

Reijmyre Glassworks. An important glasshouse in Östergötland, Sweden, founded in 1810, which adopted modern technical methods of production and from 1830 made pressed glass. *See* SWEDISH GLASSWARE.

relief. A style of decoration that projects above the surface of the object to varying degrees, i.e. HIGH RELIEF, MEDIUM RELIEF, and LOW RELIEF. Relief decoration on glass was produced by cutting, wheel engraving, ETCHING, and SAND BLASTING, as well as by MOULDING. Wheel engraving in relief is sometimes termed 'cameo engraving' or 'carving' when executed on CASED GLASS. *See* CAMEO GLASS; HOCHSCHNITT; TIEFSCHNITT; WHEEL ENGRAVING: *relief.*

relief jug. A small JUG or PHIAL (vial) made in the form of an animal or fruit. They were made by Syrian glass-blowers in the 2nd and 3rd centuries AD.

religious subjects. Subjects of a religious nature, ordinarily not found on glassware except on STAINED-GLASS WINDOWS in churches. Although the use of glass objects for Sacramental purposes was banned or discouraged by the Church (although some occurs, e.g., a glass CHALICE and PATEN at the Victoria & Albert Museum), some secular glassware has been decorated with engraved or enamelled religious motifs, e.g., a TANKARD (made in Bohemia, *c*.1684) with an enamelled medallion of the Virgin and Child, and a GOBLET (made in France in the second quarter of the 16th century) with an enamelled crucifix. *See* CHURCH GLASSWARE; BIBLICAL SUBJECTS.

reliquary. A small box, CASKET, goblet-like receptacle, or other object for keeping or displaying a religious relic. Many were made of glass, variously shaped and decorated; some, made especially for the purpose, are tall and cylindrical, with a cover having a finial in the form of a cross; others, adapted to this use, were originally made (in Germany, France and the Netherlands in the 15th and 16th centuries) in the form of a BEAKER or a KRAUTSTRUNK. Some specimens contain a document. Many reliquaries were made at MURANO and HALL-IN-TYROL from the 16th century.

———, *sealed.* A reliquary originally made as a drinking glass (e.g., a WARZENBECHER) but adapted to preserve a religious relic and sometimes a document, with the opening closed with a cover and sealed with wax by the bishop. In some cases a reliquary, or a lipsanotheca, was made from a bottle, the neck of which was broken off and the aperture sealed with wax, as found in Spain.

reproduction. A reasonably close copy of a genuine old glass object, but made without any intent to deceive and so usually having some identifiable differences. Some have been made where the original is unique and known to be a museum piece, e.g. the PORTLAND VASE, the G S GOBLET, the ROYAL OAK GOBLET. *See* FAKE; FORGERY.

Reichsadlerhumpen. Enamelled portrait of Emperor Ludwig I; Bohemia, dated 1683. Ht 18·9 cm. Museum für Kunsthandwerk, Frankfurt.

reliquary, sealed. Warzenbecher with wax seal of Bishop of Brixen, Tyrol, *c*.1500. Ht 9·2 cm. Kunstmuseum, Düsseldorf.

resonance. The quality of returning sound, e.g., the ringing sound when certain glass, particularly lead glass, is struck. It is sometimes considered a test for FORGERY, but it is not invariably so, as it is not only LEAD GLASS that is resonant but also BARIUM GLASS. Resonance in glassware depends on whether the shape of the object permits free vibrations when struck, e.g., a DRINKING GLASS, as distinguished from an object like a DECANTER, but some forms of drinking glass do not resonate.

reticello (Italian). Literally, little network. *See* VETRO A RETICELLO; FILIGRANA.

reticulated. Having a pattern of openwork which forms a net or web (*see* RETICULATED GLASSWARE).

reticulated glassware. Glassware of which the decoration is in the form or has the appearance of network. There are several varieties: (1) glass objects decorated with embedded glass threads forming an overall network or lattice pattern, e.g., VETRO A RETICELLO, or having such a pattern produced by wheel-engraving or cutting; (2) glass objects with an openwork net that is undercut, e.g., the ancient CAGE CUP (VASA DIATRETA) or SITULA PAGANA; (3) glass objects made of glass strands pincered together to form a mesh, e.g., the 19th-century Venetian BASKETS (*see* TRAFORATO); (4) glass objects covered with or blown into a wire netting (*see* NETTED GLASSWARE). Included in group (4) is the 'reticulated glassware' made by LOUIS COMFORT TIFFANY, in which glass (usually opaque jade-green glass) was blown into a network of heavy 'Tiffany green' bronzed wire with wide apertures so that the glass protruded slightly in the interstices; this was a variation of the netted glass made by FRITZ HECKERT and also by SALVIATI.

retort. Pear-shaped retort, with mouth and spout, Germany, 16th century. Ht 9 cm. Kunstgewerbemuseum, Cologne.

retort. An apparatus, for distilling liquids, usually of glass of globular or ovoid shape, characterized by a long thin attenuated spout extending from the upper part obliquely to the receptacle for the distilled liquid. Sometimes there is an orifice for filling the retort. The rounded bottom of the retort rests on a stand over a flame. *See* ALEMBIC.

retorti or **retortoli** (Italian). *See* VETRO A RETORTI; FILIGRANA.

reverberatory furnace. An early type of glass-making furnace, somewhat paraboloid in shape, made so that the heat from the burning fuel was reflected downwards on to the POTS for maximum efficiency. Sometimes called a 'reverberatory kiln' or 'reverberatory oven'. Around the inside were arranged POTS in which the glass ingredients were melted, and above the pots were openings (*bocche*; *see* BOCCA), through which the workmen inserted the rods to take out the molten glass, and some had a small hole (GLORY-HOLE) for reheating a partially made glass object or giving a FIRE-POLISH.

reverse engraving. The same as TRANSFER ENGRAVING.

revolving server. A stand with one or two flat trays that revolve around a central post and on which may be placed several serving bowls, etc. Sometimes called a 'lazy susan'.

rhyton (Greek). A drinking cup, usually in the form of a head of a person or animal, tapering to a rounded or pointed bottom so that it could stand only when inverted and empty. Generally of pottery, some were made of PERSIAN GLASS of the Achaemenid dynasty and of ROMAN GLASS. Some examples were not in the form of a head but somewhat resembled a DRINKING HORN, curving upward to a wide mouth and broad rim, and occasionally having at the tapered end the head of a faun.

ribbed bowl. *See* PILLAR-MOULDED BOWL.

ribbing. The same as REEDING. *See* PILLAR-MOULDING.

ribbon glassware. A type of glassware decorated with bands, narrow or wide or both alternately, of OPAQUE WHITE GLASS (LATTIMO), either vertically, horizontally, or spirally. Such decoration is found on Venetian glass of the 16th and 17th centuries.

Richardson, Henry G., & Sons, Ltd. The company that succeeded W. H., B. & J. RICHARDSON as the owner of a glasshouse at Wordsley, near Stourbridge, England. The partners were Henry Gethin Richardson and his sons Benjamin Richardson III and Arthur Richardson. Henry G. Richardson was the first manufacturer in England of IRIDESCENT GLASS; he also invented a pantograph machine for etching on glass. The business was acquired in the 1930s by THOMAS WEBB & SONS, and it was operated by that firm from its DENNIS GLASSWORKS, until it closed with some glassware with the Richardson label still being made and sold by Thomas Webb & Sons until the late 1960s.

Richardson, W. H., B. & J. The firm that became the owner of the WORDSLEY FLINT GLASSWORKS at Wordsley, near Stourbridge, England, that had been founded in 1720. In 1829 the business was taken over by Thomas Webb I (1804–69) with William Haden Richardson and Benjamin Richardson I, being operated as Webb & Richardson until 1836 when Webb withdrew and was succeeded by a third brother, Jonathan Richardson, the firm then becoming W. H., B. & J. Richardson. In 1842 it acquired the WHITE HOUSE GLASSWORKS from Thomas Webb I. The firm was succeeded some years later by Richardson descendants, operating as HENRY G. RICHARDSON & SONS, LTD. The firm made opaque white glass that was gilded or was painted or printed with pictorial subjects; examples were exhibited at the 1851 Great Exhibition at the CRYSTAL PALACE. In 1857 Benjamin Richardson I received patents for making etched and moulded glassware and for a machine to decorate by threading, and in 1857 another for glassware with AIR-LOCKS in the HOBNAIL pattern. Among the noted artists who worked for the firm were JOHN NORTHWOOD, PHILIP PARGETER, and John Thomas Bott.

Ridley Goblet. A GOBLET of clear purple glass with a flaring mouth and with three equidistant circuits of TRAILED glass around the bowl. It rests on a silver footed STEM connected to a silver encircling band, perhaps to replace damaged parts. It has been said that it was associated with Bishop Ridley of Oxford (d. 1555), but there are no authenticated records.

Riedel family. (1) Josef Riedel (fl. 1830–48), a Bohemian who used URANIUM in making FLUORESCENT greenish and yellowish glass, which he named ANNAGRÜN and *Annagelb* after his wife Anna. (2) Franz Anton Riedel (1786–1844), a glass-engraver in the BIEDERMEIER style and a glass manufacturer. (3) Josef Riedel (1816–94), nephew and son-in-law of F. A. Riedel; he founded in 1849 a glasshouse in the Isergebirge, in Bohemia, making glass for decorators at Kamenický Senov (Steinschönau) and jewellers at Jablonec (Gablonz). He acquired in 1887 the large HARRACHOV GLASSWORKS, founded in 1712 at Nový Svet (Neuwelt), in Bohemia and thereafter made and decorated luxury glassware that was widely exported. (4) Claus Josef Riedel (b. 1925), a descendant who designed glass at the Tiroler Glashütte at Kufstein, Austria, and in the 1950s made tableware and candlesticks with extremely thin stems.

rigaree trail. A decoration of TRAILING, having the threads of glass impressed by a wheel making a pattern of parallel notches.

Rigolleus, Cristalerias. A glasshouse in Argentina that produces glassware of a distinctive nature with thick-walled hand-wrought body and haphazard colour decoration.

rhyton. Clear glass with multi-coloured encrustations, Roman glass, 1st century AD. Museo Civico, Adria.

Riihimäki Glassworks. A glassworks founded in 1910 at Riihimäki, about 70 miles (110 km.) from Helsinki, Finland, and later named Riihimäen Lasi Oy. Its principal production has been container, window, and household glass, but since 1928 it has also made modern ART GLASS. Its early designers were HENRY ERICSSON, ARRTU BRUMMER, and GUNNEL NYMAN. The ware shows the influence of modern Scandinavian functional styles. Recent designers have been Helena Tynell, Nanny Still (b. 1926), Aimo Okkolin (b. 1917), Tamara Aladin (b. 1932), and Ekkitapio Siiroinen (b. 1944). Among its well-known patterns are KUBIIKKI WARE and KORALLI WARE. *See* FINNISH GLASSWARE.

rim. The narrow area adjacent to the EDGE of a vessel, on such objects as a VASE, PLATE or CUP. Rims may be decorated with ENAMELLING, or with applied wrought decoration. *See* ARCADED RIM; STRING RIM.

ring. An ornamental circlet to be worn on the finger. Examples have been made of glass from ancient times. Larger rings were made as BANGLES and ARMLETS. *See* THUMB RING.

ring-beaker. A type of German BEAKER decorated with attached glass loops from which are suspended glass rings. They are similar to the ring-jugs of brown salt-glazed German stoneware made at Dreihausen and Siegburg, also in the 16th century. Rings were similarly suspended from loops attached to GOBLETS and bowls made in Bohemia in the 16th and 17th centuries; some were also made in Venice and Spain in the 17th century. The German term for such ware is *Ringelbecher*.

ring-flask. A type of FLASK of circular shape with a small hole through the centre, and sometimes decorated with PRUNTS having a moulded pattern. They were made in France and Germany in the 17th and 18th centuries.

Rippenglas (German). A type of German glassware, in the form of a FLASK, BOWL or BEAKER, decorated with vertical or diagonal RIBBING, made in the 15th and 16th centuries. Some examples have a footed stem. *See* PILLAR-MOULDED BOWL.

Robinson salt-cellar. A type of SALT-CELLAR in the form of a bireme galley, made of OPALESCENT GLASS by John Robinson, at the STOURBRIDGE FLINT GLASS WORKS, Pittsburgh, *c.* 1830–36.

Robson, William. The first English-born holder of a glass monopoly. (*See* MONOPOLISTS, THE.) He, with William Turner, became assignees of SIR JEROME BOWES and operated the BLACKFRIARS GLASSHOUSE; neither assignee had previously been connected with glassmaking. Turner supplied the capital, and Robson supervised the management and was in sole charge after 1605. Between 1607 and 1611 he was engaged in litigating his rights, and required Bowes to turn over his new patent that commenced from 1607. He was by 1610 dominant in the field of CRISTALLO glass, and prospered until the change of fuel from wood to coal, *c.* 1612. He owned a glasshouse in Ireland from 1603 until 1618. Soon after 1617 he was engaged by SIR ROBERT MANSELL to manage his new BROAD STREET GLASSHOUSE and worked for Mansell for the rest of his lifetime; from *c.* 1618 he managed for Mansell the glasshouse at Newcastle-upon-Tyne.

rocaille (French). Literally, rockwork. A style of ornamentation developed in France in the 18th century and characterized by forms derived from artificial rockwork and pierced shellwork of the period. The term is commonly used in France (together with *style Louis Quinze* and *style Pompadour*) to designate the style called in England and Italy ROCOCO.

rock-crystal. Natural quartz. It is usually colourless (or nearly so) and TRANSLUCENT (or nearly so). Chemically, it is almost pure SILICA.

ring-beaker. Glass rings in attached loops, Germany, early 17th century. Ht 9·5 cm. Museum für Kunsthandwerk, Frankfurt.

Glass-makers from early times sought to imitate it, and carved rock-crystal executed in medieval Egypt, Iraq, and Persia was a source of inspiration for CUT GLASS. Clear rock-crystal was not, however, liked by the Egyptians and this aversion may account for the rarity of clear glass among surviving specimens. Carved rock-crystal pieces were highly prized during the Renaissance (e.g., pieces with metal mounts by Benvenuto Cellini, 1500–71) and later in Bohemia until CASPAR LEHMAN(N) introduced WHEEL-ENGRAVING on hardstones and then on glass (possibly fluxed with POTASH rather than SODA). The name *cristallo*, for glass decolourized with MANGANESE in Venice, was used due to the resemblance of such glass to rock-crystal, but the glass was too fragile for carving in the manner of rock-crystal. The patent granted to GEORGE RAVENSCROFT in England in 1674 was for the production of a crystalline glass 'resembling rock-crystal'. The German term for rock-crystal is *Bergcristal*, the French is *cristal de roche*.

rock-crystal engraving. A style of engraving in which all parts of the engraving are polished, in contrast with ordinary WHEEL-ENGRAVING where the engraved areas were left unpolished. It produced a more brilliant effect, and extended to the whole surface of the object as the area of engraving was enlarged from a mere design. It was developed in England during the 1880s and 1890s, and was extensively used there by THOMAS WEBB & SONS, and especially by FREDERICK E. KNY and WILLIAM FRITSCHE, and also by STEVENS & WILLIAMS, LTD. It was also used in France.

rococo. A style of decoration that followed the BAROQUE style in France, its development having been mainly influenced by Jean Bérain I and Juste-Aurèle Meissonnier, the latter a silversmith. The principal features of rococo are asymmetry of ornament and a repertoire consisting to a considerable extent of rockwork, shells, flowers, foliage, and scroll-work (often termed 'C-and-S scroll', from the typical form adopted). The rococo style was developed in France under Louis XV (1715–74), and it was imitated in Italy, Germany, Austria, and to a lesser extent in England. Rococo was followed by the NEO-CLASSICAL style, popularly termed the ADAM style or the Louis Seize style. A revival of 18th-century rococo ('revived rococo') took place in the 19th century during the reign of Louis-Philippe in France, and it was evident in English decoration of the 1830s. In France rococo is variously termed *rocaille*, *style Boucher*, *style Louis Quinze*, and *style Pompadour*. The term 'rococo' seems to have been of Italian origin, and was first applied to some exaggerated examples of baroque.

As to glassware, the rococo style is mainly evident on glassware made in Silesia (*see* SILESIAN GLASSWARE) after *c*.1725, and especially at Warmbrunn, where ROCAILLE decoration, relief palmettes, and gilt rims were popular (*see* AMBROSIA DISH). In England the change of style to rococo was hastened by the GLASS EXCISE ACT of 1745 which imposed a tax on glass based on weight, so that thereafter the glasses became lighter in weight and style, and decoration by means of engraving, enamelling, and gilding gained popularity. The influence of the 'revived rococo' style was felt in England in the second quarter of the 19th century, when glassware showed profuse cutting and heavy FLUTING, with also radial stars under the bases.

rod. A solid thin cylindrical stick of glass, clear, opaque white, or monochrome coloured, from a number of which a CANE is made or which is used to make FILIGREE decoration. The process of making a rod involves using a GATHERING IRON to take from a POT of molten glass a small GATHERING which is then shaped into a thin cylinder by rolling it on a MARVER; then a PONTIL is attached to the end opposite the gathering iron and the rod is drawn until thin and cut into usable sections. Usually the rod is retained in cylindrical form, but sometimes it is shaped in a mould to give it the cross-section of a star, hexagon, silhouette, lobed figure, etc. The German term for a rod or CANE is *Stab*, the Italian is *bastone*.

Roman glass. Rectangular bottle, green glass showing iridescence, *c.* 100 AD. Ht 16 cm. Kunstmuseum, Düsseldorf.

Römer. Coiled foot and decoration of raspberry prunts, Rhineland, late 17th century. Ht 15·3 cm. Kunstgewerbemuseum, Cologne.

rolling pin. A TOY or FRIGGER in the form of a rolling pin, made in the early 19th century, by many English factories of so-called NAILSEA GLASSWARE. They have a knob at each end for a suspensory cord; one knob has a stoppered hole so that they could be filled with hot or cold water for rolling dough (or to store flour, tea, sugar, or salt). Many are of OPAQUE WHITE GLASS but some are of blue glass. The decoration is often unfired painted mottoes and flowers. They are said to have been popular as gifts ('love tokens') by sailors and possibly as containers for smuggling.

Roman glass. A broad term embracing glass made throughout the Roman Empire, including glass made by Syrians or Alexandrians in Italy, Gaul, and the Rhineland, and even possibly in England. Its extensive production was largely due to the invention of GLASS-BLOWING about the 1st century BC; bowls and other objects were MOULDED, and MOSAIC was produced. The production also included CAMEO GLASS and diminutive GREEK VASES, as well as moulded MEDALLIONS and PANELS. The METAL was generally coloured, and decoration included TRAILING. The period of Roman glass extended from *c.* 100 BC to *c.* 400 AD, by which time the Roman Empire had begun to disintegrate. [*Plate XII*]

Römer (German). A popular German drinking glass which, in typical later form, has a somewhat spherical BOWL, a wide hollow STEM decorated with PRUNTS, a flared FOOT decorated with TRAILING, and sometimes a domed COVER. WALDGLAS examples in the 15th century were short BEAKERS of inverted conical shape, with applied small glass drops on the lower part. In the late-15th and early-16th centuries, the flared MOUTH developed into a hemispherical bowl set on a wide hollow stem with a FOOT formed by a thread wound on a conical core (at first only a few turns, but later a deep threaded band). Later the foot was a hollow blown, around which, in the 18th century, a thread of glass was wound. The applied drops on the stem developed into prunts (often RASPBERRY PRUNTS) or lion's-head masks, and later the drops were sometimes drawn out into points or loops. Some bowls were decorated with enamelling, some with DIAMOND-POINT ENGRAVING. The 17th-century Dutch examples were very large (up to 38 cm. high), with the bowl, made of SODA GLASS, engraved, and the foot plain and folded. The fully developed versions continued to be made in *Waldglas* into the 17th century and later, when they were made of colourless glass. A version was made in England by GEORGE RAVENSCROFT, its stem decorated with raspberry prunts. *Römers* are still made today, but the trailing on the foot is replaced by closely set horizontal raised mould-blown rings of glass. The *Römer* was the usual glass for drinking white Rhine wine. The name may have been derived, not from 'Roman' as once thought, but from the Lower Rhenish word *roemen* (to boast). *See* BERKEMEYER. For their development and many illustrations, see *Journal of Glass Studies* (Corning), X (1968), p. 114, and XI (1969), p. 43.

rope edge. A type of EDGE on some glass bowls or dishes of ancient Rome, *c.* 100 BC–AD 100, made by applying a cable-patterned CANE (usually incorporating a white thread) along the rim; it is found on some MOSAIC ware. A somewhat similar decoration appears on the FOOT-RIM of some Chinese glass bowls from the K'ang Hsi period (1662–1722), where the 'rope' is a twisted single thread that is roughly joined to make a circle. The Roman twisted threads of the rope edge anticipate the later FILIGRANA glassware.

roquetta. A marine plant from Egypt and the Near East, the ash of which was used as a source of SODA, the ALKALI required in making early glass. *See* BARILLA; GLASSWORT.

Rosaline Glass. A type of JADE GLASS that is rose-coloured. It was made by STEUBEN GLASS WORKS, sometimes decorated with engraved designs.

Rosbach, Elias (1700–65). A German glass-engraver who worked independently at Berlin from 1727, being senior master of the glass-engravers' guild in 1735–36. He moved, c. 1742, to work at the glasshouse at Zechlin. He has left several INTAGLIO decorated glasses, done at Berlin, signed with his name in DIAMOND-POINT ENGRAVING.

rose. A decorative motif used in some CANES found in MILLEFIORI glassware (see CLICHY ROSE) and used in some PAPER-WEIGHTS (see JERSEY ROSE; KAZIUN, CHARLES).

Rose-Teinte. A type of ART GLASS that is similar to AMBERINA, tinted from pale amber to rose; it was produced by the BACCARAT GLASSWORKS from 1916 and reintroduced in 1940. It was made with less gold than amberina, and thus was more delicately coloured. Also called 'Baccarat's Amberina'.

rose-water sprinkler. See SPRINKLER.

rosette. (1) A decorative ornament with its elements disposed in a circular plan, as a stylized rose in full bloom; the equally spaced petals sometimes alternate with stylized leaves.(2) A small glass ornament in this form, attached to a metal screw for the purpose of holding a mirror to its frame.

Roubiček, René (1922–   ). A Czechoslovakian instructor and designer of glassware, born at Prague and working at Novy Bor as the supervising artist at the Borske Sklo Glassworks. He has made pieces of free-formed ART GLASS in many original and complicated styles, some with embedded flecks of enamel.

Rouge Flambé glass. A type of ART GLASS developed by FREDERICK CARDER at the STEUBEN GLASS WORKS, and produced c. 1916–17. The colour was inspired by the decoration on Chinese *famille rose* porcelain, and the ware was made in shapes suggested by the Chinese ceramic pieces. The colour (produced by selenium and cadmium sulphide) ranged from rich red to orange or coral pink. The surface is not sprayed, and hence the glass is bright and not IRIDESCENT. It is sometimes erroneously called 'Red Aurene'. Due to strains in the glass, it proved too fragile and few pieces were made.

rouge pot. A small container for rouge. Some examples in ROMAN GLASS of the 1st century AD were ball-shaped with a tiny opening. The German term is *Schminkkugel*. See GLOBULAR UNGUENTARIUM.

round. (1) A pharmacy BOTTLE or JAR cylindrically shaped and ranging usually from 15 to 25 cm. in height. Early examples from the late 18th century were of clear glass, with variously shaped distinguishing stoppers, but later ones were often of coloured glass, usually blue or green. They were used for holding pills or syrups, with a capacity varying from 5 to 60 fluid ounces. They are generally labelled. (2) A tall cylindrical covered jar used by confectioners to hold sweets.

roundel. A circular PLAQUE; sometimes a painting in enamel or gilt on glassware within a circular border or cartouche.

Rousseau, François Eugène (1827–91). A French artist who designed and made glassware and combined many new decorative techniques. He made objects inlaid with glass of various colours and with varied effects, produced by the use of metallic oxides, which were covered with a thin FLASHING of colourless glass, and sometimes finished on the wheel. He was influenced by Japanese decorative styles. The METAL used is often of a champagne colour. His work displayed at the Paris Exhibitions of 1878 and 1884 was highly regarded. He retired in 1885; his pupil and successor, ERNEST BAPTISTE LÉVEILLÉ, carried on his work under the name Léveillé-Rousseau.

rouge pot. Ball-shaped pot, with white trailing and tiny opening, 1st century AD. D. 4·2 cm. Römisch-Germanisches Museum, Cologne.

roundel. Clear glass with gilding and enamelling, Venetian, mid-16th century. W. 39 cm. Victoria and Albert Museum, London.

Rousseau, François Eugene. Vase by Rousseau, Paris, c. 1884. Ht 26 cm. Kunstmuseum, Düsseldorf.

rugiadoso, vetro. Bowl designed by Ercole Barovier, *c.* 1940. Barovier & Toso, Murano.

Royal Flemish glassware. A type of ART GLASS with an acid-finished surface and with raised gold enamel lines separating various elements of the design. The very ornate patterns included Oriental scenes, ducks in flight, and abstract subjects. It was made by the MOUNT WASHINGTON GLASS CO. from *c.* 1890; a patent was issued to Albert Steffin in 1894. Pieces are sometimes marked 'R F'.

Royal Oak Goblet. A MARRIAGE GLASS in the form of a GOBLET with a cylindrical BOWL resting on a knop STEM, made in 1663 to commemorate the marriage of Charles II and Catherine of Braganza. Its decoration in DIAMOND-POINT ENGRAVING shows their portraits, between which is a portrait of Charles II within a wreath of oak leaves on a formal tree inscribed 'Royal Oak'; on the reverse are the engraved Royal Arms of England and the date 1663. It has been attributed to the glasshouse of the DUKE OF BUCKINGHAM, but W. B. Honey has expressed the opinion that it was both made and engraved in the Netherlands (as also certain FLUTES formerly attributed to Buckingham). *See* SCUDAMORE FLUTE. This goblet has been reproduced but with obvious differences negativing any intent to deceive.

Ruby glass. A term applied by FREDERICK CARDER to ruby-coloured glass developed by him at STEUBEN GLASS WORKS, but related to earlier GOLD RUBY GLASS. It was of two types: (1) Glass of a pink colour made by adding to the BATCH 22-carat gold in colloidal form, with sometimes additional varied metallic oxides. The glass was usually produced in solid lumps (called 'sausages') that were yellow until becoming ruby when reheated in the LEHR, and were safe-guarded until needed. (2) Selenium Ruby glass, made by the inclusion of cadmium selenide and zinc sulphide, resulting in a brilliant red colour, sometimes shading to orange; hence it was sometimes called 'Cadmium Ruby Glass'. Both types were used mainly as a thin FLASHING over clear glass or ALABASTER GLASS.

rugiadoso, vetro (Italian). Literally, dew-like glass. A type of ornamental glass, developed by Ercole Barovier for BAROVIER & TOSO, *c.* 1940, which has an overall surface texture and appearance of fine dew. It was used to make bowls, vases, etc. The surface is roughened by fusing small glass splinters to the surface.

ruler. A straight glass rod made for ruling lines on paper. They were about 24 to 25 cm. long and about 1·2 cm. thick. Some were made by the FALCON GLASSWORKS with decoration of twisted MILLEFIORI CANES laid lengthwise.

rullo (Italian; pl., *rulli*). A small circular PANE of glass made in the same manner as CROWN GLASS, having a bull's-eye at the centre and a folded rim. Such panes, placed contiguously in horizontal and vertical rows, were used for windows before large panes were made, and are to be seen in late-15th-century paintings. In the 19th century they were made of coloured glass or multi-coloured FILIGRANA (RETICELLO) glass. The term used at MURANO was *rui*.

rummer. (1) A type of English drinking glass, usually in the 19th-century form of a low GOBLET with a stemmed FOOT, often domed or square, being a gradual evolution from the German RÖMER. The term has been applied to a glass made *c.* 1677 by GEORGE RAVENSCROFT, having a cup-shaped BOWL with a spreading FOOT and a hollow STEM decorated with PRUNTS, but it is questionable whether such early glass should be so termed. Later examples, *c.* 1690, have a knopped stem, sometimes of BALUSTER shape, and a folded foot. In the late 18th century the bowls were of various shapes, often engraved, and the foot often had CUT decoration. The name is a corruption of the German *Römer*, as the early glasses were used for white Rhine wine, and had no connection with rum. (2) A term sometimes used loosely in England for a German *Römer*, and sometimes for an 18th-century green glass with a

Prince Rupert's drop. Greenish-brown glass. L. 5 cm. Challicom Collection, Castle Museum, Taunton. Photo Keith Vincent.

globular bowl and a stemmed and domed foot, occasionally having gilt decoration on the bowl.

Rupert's drop. A curious tadpole-shaped hollow glass object, about 5 cm. long, having a bulbous end tapering to a thin curved tail. They were made, usually of green or yellow-brown glass, by dropping a small blob of moderately-hot molten glass into cold water and leaving it until cooled. They are not affected by a blow on the bulbous end but, if the tail is broken or the surface scratched, the piece, due to different internal stresses, explodes with a noise into fine powder. They were introduced into England from Germany by Prince Rupert (1619–82), nephew of Charles II, who made some in 1661 at Windsor Castle. They were mentioned in Samuel Pepys's *Diary* under date of 13 January 1662 and also by Christopher Merret in 1662, in his translation of *L'Arte Vetraria* by ANTONIO NERI, in describing an experiment by the Royal Society at Gresham College. They are also called '[Prince] Rupert's ball' or 'tear'.

Rüsselbecher (German). Literally, trunk, or proboscis, beaker. The same as a CLAW-BEAKER; also called a 'trunk-beaker'.

Russian glassware. Glassware made, since the 11th century, at many glassworks in Russia. The early ware included beads, jewellery, goblets, pilgrim flasks, and a type of beaker known as a STOPKA. Such ware was made at glassworks at Kiev, Kostroma, and elsewhere (*see* NOVO-GRUDOK GLASSWARE). Recent finds revealed glass objects made in the 3rd–14th centuries. During the 17th century there were two known glasshouses: (1) at Dukhanin, near Moscow, making mediocre greenish flasks and panes; it received a licence in 1634, issued to Julius Koiet of Sweden, and was opened after his death by his son Anton; and (2) the IZMAILOVSKII GLASSWORKS, near Moscow, opened in 1669, making tableware; it closed between 1706 and 1725. In the second decade of the 18th century, state-owned factories were opened at JAMBURG (now Kingisepp) and Zhabino, but both closed in the 1730s. During the 18th century several privately-owned glasshouses were founded (including the MAL'TSEV GLASS FACTORY, the BACHMETOV GLASS FACTORY, and the MINTER GLASS FACTORY) and also the ST PETERSBURG IMPERIAL GLASS FACTORY and the factory of H. V. LOMONOSOV. By 1812 there were over 140 private glasshouses, but only a few made fine-quality glassware; later, from 1850 to 1870, many more glass factories were started and old ones expanded. Glassware of many varieties was made, including colourless crystal glass, coloured glass, opaque white glass, and CASED GLASS, with decoration in engraving, enamelling, and transfer printing. After *c.*1900 much glassware was made in the ART NOUVEAU style of EMILE GALLÉ. Today the factory at Gus-Krusthal'nii is the leading Russian producer. *See* HEDWIG GLASSES; SMALTO; FURNITURE. For a discussion of Russian glassware, factories, and styles, *see* B. A. Shelkovnikov, in *Journal of Glass Studies* (Corning), II (1960); VI (1964); VIII (1966); IX (1967).

Russian glassware. Tumbler with engraved inscription in Russian, Jamburg, 1720. Hermitage Museum, Leningrad.

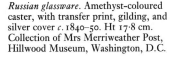

Russian glassware. Amethyst-coloured caster, with transfer print, gilding, and silver cover *c.*1840–50. Ht 17·8 cm. Collection of Mrs Merriweather Post, Hillwood Museum, Washington, D.C.

S

sack glass. A type of drinking glass formerly used for drinking sack (originally any strong dry white wine, but in the 17th century the term was applied in England to any wine imported from southern Europe, and especially to sherry from Spain; the name was derived from the French *vin sec*, dry wine). The glass is similar to a SHERRY GLASS used in the 19th century when the term sack had ceased to be used. Some sack glasses ordered from Venice and so designated by ALESSIO MORELLI had a straight-sided cylindrical bowl and a short stem of inverted baluster form between two MERESES.

Saint-Amans, Pierre-Honoré Boudon de (1774–1858). A French nobleman who in 1818 secured a patent for making cameos enclosed in glass, in the manner of cameo-incrustations (SULPHIDES).

St-Cloud Glassworks. A glassworks established at St-Cloud, west of Paris, in 1783 by Philippe-Charles Lambert under the name of Verrerie de la Reine, it being formerly under the patronage of Marie Antoinette. It was transferred in 1787 to Mont-Cénis, near Le Creusot. It was one of the first French glasshouses to make glassware in the English style. It also made LUSTRES and PLATE GLASS. *See* CREUSOT, LE, GLASS FACTORY.

St-Gobain Glass Company. The French national glass company for making PLATE GLASS and MIRRORS, founded in 1693 and still operating today. In 1653 a glassworks had been established at Tourlaville, in Normandy, by RICHARD-LUCAS DE NEHOU, which in 1667 united with a glassworks established in 1665 by JEAN-BAPTISTE COLBERT in the Faubourg St-Antoine in Paris, with a monopoly to make mirrors. A rival company had been set up in 1688 by Abraham Thévart at Tourlaville, under the direction of Nehou's nephew, LOUIS-LUCAS DE NEHOU; this company, under the patronage of Louis XIV, started in 1693 a company at St-Gobain in Picardy under the title of Manufacture Royale des Grandes Glaces. The Tourlaville, Picardy, and Paris companies were combined in 1695 under the name Manufacture Royale des Glaces de France. The Tourlaville operations were given up in 1787, but the Picardy company thrived to become the French national glass company, known since 1830 as St-Gobain Glass Co. Its monopoly for making cast-plate mirrors lasted over 100 years, until the founding in England in 1773 of a plate-glass works now owned by PILKINGTON BROTHERS, LTD.

St-Louis Glass Factory. A glassworks (named today 'Compagnie des Cristalleries de St-Louis' and now one of the most important in France), founded in 1767, at St-Louis, near Bitche, in the Münzthal, Lorraine, on the site of a glassworks that existed from 1586. A decree of Louis XV in 1767 conferred on it the name 'Verrerie Royale de Saint-Louis'. It made at first glass comparable to Bohemian glassware, but in 1781 it independently discovered the process of making crystal glass similar to English LEAD GLASS. Its success was attested in 1782 by the Académie des Sciences, and in 1784 by a decree of the State Council. Several changes of ownership ensued, and from 1791 to 1797 it was owned and directed by AIMÉ-GABRIEL D'ARTIGUES. In 1829 it came into new ownership and its name was changed to 'Compagnie des Verreries et Cristalleries de St-Louis'. Being in Alsace-Lorraine (which was part of Germany from 1871 to 1918), some of its ware was then signed 'St-Louis-Münzthal' or 'd'Argental' (*see* ARGENTAL) and sometimes 'Arsale' or 'Arsal'. From 1825 onwards the factory has collaborated with the BACCARAT GLASS FACTORY, and in 1932 they jointly acquired the LE CREUSOT GLASS FACTORY (formerly the ST-CLOUD GLASS-

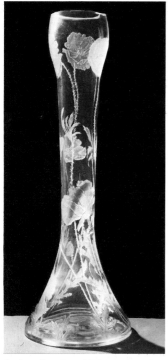

St Petersburg Imperial Glass Factory. Art Nouveau vase, with cipher of Tsar Nicholas II and date 1914 on the base. Ht 56·4 cm. Collection of Mrs Merriweather Post, Hillwood Museum, Washington, D.C.

WORKS). From the mid-19th century the factory produced OPALINE and wheel-engraved and acid-etched glassware with decoration of flowers and landscapes. After surviving periods of closure during the War of 1870 and the two World Wars, it has grown to become to-day one of the leading makers of crystal tableware in both traditional and modern styles. It is known for its 18th-century paper-weights (*see* PAPER-WEIGHT, *St-Louis*) and from 1952 it has revived the making of weights in classic style, including SULPHIDES and OVERLAYS.

St Petersburg Imperial Glass Factory. The State glasshouse originally established by Peter I (the Great) (Tsar, 1682–1725) in the early 18th century at St Petersburg (now Leningrad). After the State glasshouses that had been founded in the second decade of the 18th century at Jamburg (now Kingisepp), near St Petersburg, and nearby Zhabino had been ordered to be closed by Peter I in the late 1730s (and after he had previously ordered closed the other State glasshouses near Moscow), a new State glass factory was started, *c.* 1743, in St Petersburg, which operated until the 1770s. In the mid-18th century another State glasshouse was built which in 1777 was leased to Prince Potemkin who soon moved it to Ozerki, near St Petersburg; in 1792, after Potemkin's death, it reverted to the State. It became known as the St Petersburg Imperial Glass Factory, and at the end of the 18th century it was administered together with the Imperial Porcelain Factory. These State factories, under Peter II (Tsar, 1727–30), Anna Ivanovna (Tsarina, 1730–40), Elizabeth Petrovna (Tsarina, 1741–62), and Catherine the Great (Tsarina, 1762–96), made articles principally for the Court and the aristocracy, especially crystal glass goblets and large chandeliers and girandoles made at Ozerki. During the 19th century, the St Petersburg factory made fine drinking vessels of uncoloured glass with engraved and enamelled decoration, the motifs often being the heraldic eagle and the Imperial monogram, and also some CASED GLASS in two or three colours and much SMALTO. In 1862 unsuccessful efforts were made to sell the St Petersburg factory, and economies had to be introduced. Before 1890 it was moved to the site of, and combined with, the Imperial Porcelain Factory until it was closed during the 1917 Revolution. *See* RUSSIAN GLASSWARE.

Sala, Jean (1895–   ). An artist-craftsman who made STUDIO GLASS in Paris. He was the son and pupil of Dominique Sala, a Spanish glass-maker who settled in Paris. He worked in a greenish-blue light METAL imperfectly made, with many bubbles. He made a collection of glass FISH for the aquarium at Monaco.

Sala, Jean. Fish designed and made by Sala, *c.* 1912–25. L. *c.* 20 cm. Institut Océanographique de Monaco.

salad dish. A medium-size plate, sometimes of glass, for salad. A modern version is crescent-shaped, with a low vertical rim, to fit alongside the dinner-plate at the table.

salt-cellar. One of a pair with mark of cipher of Tsar Alexander III, St Petersburg Imperial Glass Factory, late 19th century. Ht 3·8 cm. Collection of Mrs Merriweather Post, Hillwood Museum, Washington, D.C.

saliva. A defect in glassware in the form of a group of small bubbles resulting from insufficient expulsion of air from the GATHER or premature cooling of the gather.

salt. (1) A SALT-CELLAR. The addition of the word 'cellar' came about as a result of the corruption of the French *salière*. (2) The term is sometimes applied to an ALKALI used in making glass.

salt-cellar. A small shallow bowl for holding table salt. It is usually circular, oval or boat-shaped. Most glass examples have a flat base or rest on a stemmed foot, but some English ones, dating from the first half of 18th century, have three claw feet surmounted by a lion's-head mask or a gargoyle. Early English salt-cellars were large, up to 30 cm. high, and were for communal use at table (called 'standing salts' or 'great salts'); in the 17th century individual ones were provided for each diner (called 'SMALL SALT' or 'TRENCHER SALT'). From *c.* 1780 English and Irish examples were decorated with cutting. Some French salts are square or rectangular, with two wells. Many Italian ones are large, some having a

salver. Venetian glass, *c.* 1708. D. 32 cm. Royal Collection, Rosenborg Castle, Copenhagen.

Salviati glassware. 'Abstract Element', designed by M. Weichberger, 1969. Salviati & Co., Murano.

sand-blasted glass. Plate designed by Ladislav Oliva, Podebrady Glassworks, Czechoslovakia, 1970. Oliva Collection, Museum of Decorative Arts, Prague.

figure mounted on the rim at each end. Some from Bohemia are decorated in the manner of ZWISCHENGOLDGLAS. Some were made in pairs or sets. The French term is *salière*, the Italian is *saliera*, and the German is *Salzfässchen*. *See* SALT; IRISH GLASSWARE: *salt-cellars*; ROBINSON BOAT SALT.

salver. A large circular glass dish or shallow bowl usually having a pedestal FOOT, but sometimes with a baluster STEM or a Silesian stem (as distinguished from a silver salver, which is a flat tray for presenting letters or visiting cards). Sometimes JELLY GLASSES or SWEETMEAT GLASSES were placed in a circle on a glass salver (and occasionally another smaller similarly laden salver was placed on the glasses as a dinner-table CENTRE-PIECE, with a large sweetmeat glass in the centre, the complete arrangement being sometimes called a 'pyramid'). A salver was also sometimes used as a *présentoir* for offering a glass of wine. Salvers were made and used in England in the late 18th century. They were also made at MURANO; *see* ALZATA.

Salviati & Co. A glasshouse at MURANO that makes a wide variety of ornamental glassware and tableware, especially coloured, CASED, and MOSAIC glassware, as well as enamelled and gilded ware. A glassworks founded at Murano *c.* 1859 by Antonio Salviati, a Venetian glass-maker, was purchased in 1886 by three Englishmen who formed a British holding company, Salviati & Co., Ltd, the name of which was changed in 1871 to 'Venice & Murano Glass & Mosaic Co., Ltd (Salviati & Co.)'. In 1877 Antonio Salviati withdrew to found his own glassworks at Murano, and the original factory became known as 'Venice & Murano Glass Co.'. The Salviati glassworks has been owned from 1920 by Maurizio Camerino (1859–1930) and later by his sons Renzo (b. 1904) and Mario; Renzo Tedeschi (b. 1923), the grandson of Maurizio, is the present general manager and partner.

San Ildefonso. *See* GRANJA DE SAN ILDEFONSO, LA.

sanctuary lamp. A type of OIL LAMP made in Spain in the 16th century, of cylindrical form, slightly taller than wide, and with a horizontal disk RIM. They often are decorated with ENAMELLED insignia of religious orders. *See* HANGING LAMP.

sand. The most common form of SILICA used in making glass. It is an impure silica, taken from the seashore or preferably from inland beds whose sand is more readily ground and is freer from impurities. It should have a low content of iron and other impurities, and must first be well washed, heated to remove carbonaceous matter, and screened to obtain uniform small grains. The early Venetians used crushed pebbles from the River Sile, and later from the River Ticino. Early English silica came from local sands during the 16th and early 17th centuries, then from crushed FLINT and sand in the second half of the 17th century, and from purer sands in the 18th century and later; the sand came from the Sussex Weald and the Isle of Wight and from near Maidstone (Kent) and King's Lynn (Norfolk).

sand-blasting. The process of decorating glassware by projecting fine grains of sand (or crushed flint or powdered iron) at high velocity by means of a special tool on to the surface of the glass, masking the part outside the design by a stencil impervious to the sand. A wide range of effects can be obtained by varying the quality of the sand, the force of the jet, and the duration of the blast. The design is left in a greyish MAT finish. The process was invented by Benjamin Tilghman, a Philadelphia chemist, in 1870. It was first used mainly for large glass panels, but it has been used for decorative effects by modern glass-makers (e.g., RENÉ LALIQUE), combining it with other processes. Very original patterns with deep cutting have been designed by LADISLAV OLIVA, *c.* 1959–60, in Czechoslovakia.

sand-core glass. *See* CORE GLASS. The term 'sand core', often used, is a misnomer, as it is now the authoritative opinion that sand was not used as the core but probably mud, with straw, perhaps, as a binder. *See* D. B. Harden, in *Masterpieces of Glass*, 1968, p. 12 note.

sand-glass. An instrument for measuring time, in the form of a reversible device made of two vertically connected glass vials from the upper of which a quantity of sand runs down through a small neck into the lower (similarly shaped and sized but inverted) vial; when all the sand has passed through, the device can be turned over and the process repeated. The clear-glass blown vials are set into a frame of wood or other material. The basic sand-glass measures an hour (hence the general term usually applied to all such devices is an 'hour-glass'), but small sand-glasses measure shorter periods of time (such as the LOG GLASS used on ships, set for 14 to 28 minutes) and large ones measure longer periods (*see* POWDER GLASS). In early examples a metal disk or diaphragm with a tiny hole was placed where the necks meet (the two vials being joined by an applied glass thread); in the 18th century the two vials were joined by the sand-glass-maker with a sealed-in metal bead, and later the metal was dispensed with and the device was made in one glass piece with a small passage for the sand. After the specially selected and treated sand was inserted, the large open end of one of the vials was sealed. Sand-glasses were imported into England from Germany, the Low Countries, and Venice in the 16th century and until *c*.1610–20 when glass vials made by SIR ROBERT MANSELL became the main source of supply to the English sand-glass-makers.

sand-moulding. The process of moulding a glass LINER for a silver vessel by carving a wooden block to fit exactly into the vessel, then forcing the block into a bed of damp sand, withdrawing it so as to leave an exact depression, and then blowing in molten glass. After cooling, the glass is removed, and the marks left by the sand eliminated.

Sandwich glass. (1) A general term used in the United States for PRESSED GLASS, being derived from its extensive production by the BOSTON & SANDWICH GLASS CO., but applied to such glass also made by other glasshouses. (2) Double-wall glass with inner decoration of engraved gold-leaf. *See* ZWISCHENGOLDGLAS.

sanfirico (Italian). Same as ZANFIRICO.

Sang, Andreas Friedrich (fl. 1719–60). A German glass-engraver born at Ilmenau, in Thuringia; he is recorded as working at Erfurt in 1719, at Weimar from 1720 to 1744 (where he was Court Engraver in 1738 to Duke Ernest Augustus of Saxe-Weimar), at Ilmenau until 1747, and later in the Netherlands. He was the father of JOHANN HEINRICH BALTHASAR SANG and probably the brother (or father) of JACOB SANG. Some of his glasses are signed 'Sang' or 'A F Sang' and are dated.

Sang, Jacob (d. 1783). An accomplished glass-engraver who worked at Amsterdam. He was probably the brother (or son?) of ANDREAS FRIEDRICH SANG. Some signed examples of his work, with the added word 'Amsterdam', are known. He engraved English glass from Newcastle, although the covers of some pieces were possibly from the Netherlands. He decorated glasses (some signed and dated) commemorating births, betrothals, and marriages. It has been said that Jacob Sang is not the same person as the Simon Jacob Sang who advertised CUT GLASS and ENGRAVED GLASS in Amsterdam in 1753. *See* R. J. Charleston, *Journal of Glass Technology*, XLI (1957), p. 242.

Sang, Johann Heinrich Balthasar (fl. 1745–55). A German glass-engraver who worked at Brunswick and was Court Engraver from 1747. He was the son of ANDREAS FRIEDRICH SANG.

Sargon Alabastron. An ALABASTRON of green transparent glass, decorated with an engraved lion, and, in cuneiform characters, the name

sand-glass. Dismantled sand-glass, showing separate vials and diaphragm, *c*.18th century. Science Museum, London; Crown copyright.

Sang, Jacob. Marriage goblet with arms of Artois and Flanders, Newcastle glass, decorated by Jacob Sang, *c*.1745–50. Ht 22·7 cm. Courtesy, Maureen Thompson, London.

Sargon alabastron. Clear dark-green glass, *c.* 722–705 BC. Ht 8 cm. British Museum, London.

Sarpaneva, Timo. Glass sculpture blown in burned wooden mould, designed by Sarpaneva. Ht 60 cm. Iittala Glassworks, Finland.

of Sargon II of Assyria (722–705 BC). It is of thick glass and is said to be the earliest surviving glass vessel not made over a CORE. It has been said, by D. B. Harden of the British Museum, to be ground and polished from a raw block of glass (the interior showing spiral grooving); but it has also been suggested that it was made in a mould and that the interior markings are from grinding after it had cooled. The surface is weathered and appears whitish with a greenish IRIDESCENCE. It was excavated at Nimrud, near Nineveh, and was regarded by W. B. Honey to be more likely of Mesopotamian than of Assyrian origin, of the late 8th century. *See* COLD CUTTING.

Sarpaneva, Timo (1926–   ). A Finnish designer in several media, including glassware. He has been designer at the IITTALA GLASSWORKS from 1950. He has been recognized for his glass sculptures as well as his original designs for tableware. He has developed a new technique for glass sculptures made in wooden moulds that are allowed to burn during blowing, giving a special surface texture, sometimes emphasized by being rolled in sawdust while hot. In 1967 he produced a large glass panel, about 4·40 m. by 9·60 m., made from over 500 pieces of glass.

Sartory, Franz (fl. 1799–1841). A Viennese painter of landscapes on porcelain at the Royal Vienna Porcelain Factory who also decorated with TRANSPARENT ENAMELLING a few pieces of glassware.

Sassanian glassware. Pre-Islamic glassware made in Persia and Mesopotamia under the Sassanian Emperors, *c.* 226–642, when glass-making there was revived after a lapse following the Achaemenid dynasty, *c.* 550–330 BC. The Roman techniques were perhaps brought in by craftsmen from conquered territories. Some glass was coloured and MOULD-BLOWN, the finest examples being in CUT GLASS with faceted decoration, such as pieces with faceted overall decoration in QUINCUNX PATTERN. More elaborate types of cut decoration are found on Islamic ROCK-CRYSTAL, and these may have been used on glassware, although examples have not yet come to light. *See* PERSIAN GLASSWARE; MESOPOTAMIAN GLASSWARE.

satin glass. Glass with a satin finish. It is decorated with HYDRO-FLUORIC ACID by a process that is the reverse of ETCHING, i.e., the design is covered and protected, and the acid attacks only the background surface. Some examples are of OVERLAY glass, with a pattern of air bubbles trapped between an opaque body and a transparent coloured overlay that is given a satin finish. *See* VERRE DE SOIE.

satirical glassware. BEAKERS and other glassware decorated with subjects of a satirical nature, e.g., a Netherlands beaker, dated 1604, with a scene in DIAMOND-POINT ENGRAVING of a confrontation between Christ on a donkey and the Pope on a caparisoned horse, with satirical Dutch verses.

sauce boat. A boat-shaped vessel, usually with a HANDLE at one end and a pouring LIP at the other, for serving sauce or gravy. The vessel usually rests on a spreading base or on short legs, and sometimes has a COVER. Some have a lip on both ends and two side-handles. Generally made of porcelain or pottery, some have been made of glass. In design, 18th-century glass sauce-boats are very commonly based on that of contemporary silver. They must be distinguished from a BOURDALOU. The French term is *saucière*. *See* CREAM-BOAT.

sauce bottle. A small BOTTLE shaped like a DECANTER, but sometimes with a pouring LIP, for serving sauces and condiments. Some are inscribed with the name of the contents.

sauce dish. A circular dish (about 11·5 cm. wide) with a flat bottom and slightly curved sides, for serving sauce or preserves. They were made in the United States especially of LACY GLASS by the BOSTON & SANDWICH GLASS CO. *See* NAPPIE.

save-all. A hollow cylindrical object that slips over or into the nozzle of a CANDLESTICK and which spreads out at the bottom or the top to form a circular disk for catching the wax drippings. They are made of silver for silver candlesticks but also of glass for glass candlesticks. Their main function is to permit the dripped wax to be cleaned off simply by removing the save-all without disturbing the candlestick, or especially a CHANDELIER or a SCONCE. Also called a 'sleeve'. *See* DRIP PAN; BOBÊCHE.

savon, boule de (French). Literally, soap-bubble. A colour of some OPALINE glass; it is of an iridescent soap-bubble hue (sometimes referred to as *savonneuse*).

Savoy Vase. A vase of thick blue glass in the same form as the AMIENS CHALICE, and similarly showing CRISSELLING. It was formerly attributed to the Savoy Glasshouse of GEORGE RAVENSCROFT, *c*. 1674–75, and has been described as made of a low lead METAL; however, no authority for its lead content has been cited, and its attribution to Ravenscroft is questionable. It is in the Toledo (Ohio) Museum of Art.

Saxon glassware. Glassware made in Saxony, Germany, from the Middle Ages, in the forest regions, mainly decorated with enamelling in the style of neighbouring Bohemia and Franconia; clear glass was decorated with borders of white enamel dots on a gold ground, together with polychrome motifs. Later the main centres were Dresden and Weissenfels. From 1606 to 1608 CASPAR LEHMAN(N) worked in Dresden after being forced to move from Prague. From 1610, Caspar and Wolfgang Schindler were glass-engravers at Dresden, and from 1635 Georg Schindler. In 1712 there were nineteen glass cutters and engravers in Dresden; a number of glasshouses were founded by Freiherr Walther von Tschirnhaus upon instructions of Augustus the Strong, with craftsmen from Bohemia. From 1717 to 1744 JOHANN CHRISTOPH KIESSLING was Court Engraver at Dresden. At Weissenfels, Court glass-engravers were JOHANN NEUMANN from 1680 to 1697, under the patronage of Duke Johann Adolf I of Saxe-Weissenfels, and, later, JOHANN GEORG MÜLLER until 1783. Much enamelled glassware was made from *c*. 1610, including examples of the HOFKELLEREIGLAS and the HALLORENGLAS, and also goblets made at Dresden decorated with horizontal oval FACETING and engraved portraits, as well as some glassware with white FILIGRANA decoration.

scabbard fitting. A small decorative attachment for a scabbard, made in glass in China during the Han dynasty (206 BC–AD 220).

scale cutting. A decorative pattern on CUT GLASS, found on the stems of some English drinking glasses in the form of overlapping scales making an IMBRICATED design.

scale faceting. A style of IMBRICATED decoration on glassware with FACETING in the form of overlapping scales resembling an enamelled SCALE GROUND. It was used on the STEMS of some English drinking glasses.

scale ground. An IMBRICATED diaper pattern composed of overlapping scales. It was used extensively as a ground pattern on Worcester porcelain, *c*. 1760–85, and sometimes as a moulded ground pattern on St-Cloud procelain early in the 18th century. It is found on enamelled glassware from Murano dating from the late 15th and early 16th centuries (it has been suggested that it was copied from mosaics); this island usage may account for the origin of the fish-scale pattern, but the exterior walls of timber houses in Austria and Germany are also

satirical glassware. Beaker with diamond-point engraving of Christ and the Pope, with satirical inscription, dated 1604. Ht 26·5 cm. Rijksmuseum, Amsterdam.

similarly covered with overlapping wooden shingles, producing the same effect. *See* SCALE FACETING; CESENDELLO; STANDING CUP.

scallop-shell chalice. A CHALICE of Roman (Rhenish) glass, dating from the 4th century. It has a tall cylindrical bowl resting on a stemmed FOOT, the bowl being decorated in PSEUDO-VASA DIATRETA style, with attached vertical columns, each ornamented with three scallop shells. Some similar chalices have attached columns without the shells.

scallop-shell decoration. A decorative relief motif often found on glassware. It was used on SHELL FLASKS made of ROMAN GLASS and Venetian glass, and also on a SCALLOP-SHELL CHALICE made at Cologne in the 4th century.

scavo, vetro (Italian). Literally, excavated glass. A type of ornamental glassware, developed by CENEDESE, of which the characteristic feature is an overall surface rough MAT finish, predominantly grey but with splashes of various colours. It is produced by an application of powdered minerals fused into the surface of the glass, and made into many forms, e.g., bowls, animal figures, etc.

scenic decoration. Transparent glass enamelled on the reverse with scene of St Mark's Square, Venice. Museo Vetrario, Murano.

scenic decoration. Opaque white glass enamelled on the surface with scene of Doge's Palace, Venice, 18th century. Museo Vetrario, Murano.

scenic decoration. Decoration on glassware depicting scenic views, usually well-known buildings and localities. (1) It was executed in enamelling on the surface of OPAQUE WHITE GLASS drinking glasses and panels, e.g., the glasses decorated by SAMUEL MOHN and ANTON KOTHGASSER. Large PANELS decorated in Venice in the 18th century depict colourful local scenes, with numerous figures, gondolas, etc., and Venetian scenes are also shown on enamelled plates (*see* VENETIAN SCENIC PLATES). (2) Such scenic decoration also was done by engraving, e.g., the views near Warmbrunn by CHRISTIAN GOTTFRIED SCHNEIDER. (3) It was also done in variously coloured paste depicting scenes in low relief and fused to the exterior of the glassware; this was done in the mid-17th century, attributed possibly to Germany.

scenic paper-weight. *See* PAPER-WEIGHT, *British (Victorian)*.

scent barrel. A type of SCENT BOTTLE made in the form of a barrel resting horizontally and having a bung-hole in the centre of the side. Some were made of glass.

scent bottle. A small bottle for perfume, made in a wide variety of forms and many materials, including gold, hardstones, enamelware, porcelain, terracotta, and also glass. They have a tiny orifice to retard the flow of the perfume, and usually a ground glass STOPPER (sometimes also a metal screw cap). They have been made of glass in Egypt from the 6th century BC, and also of Islamic and Roman glass, and later in many countries of Europe and in China. Some have decoration in MILLEFIORI technique, some Bohemian examples are in CASED GLASS, and some made in England are of OPAQUE WHITE GLASS or blue or green glass, some with embedded SULPHIDES and some enamelled. Some are double bottles for two perfumes. In the early 20th century fanciful commercial scent bottles were made by RENÉ LALIQUE for Coty and by ÉMILE GALLÉ for Guerlain. Some scent bottles are provided with silver or gold MOUNTS. Sometimes called a 'perfume bottle'. The French term is *flacon*; the German is *Flakon*. *See* SMELLING BOTTLE; COLOGNE BOTTLE; SCENT BARREL; SPRINGEL GLASS.

sceptre. A term that has been applied to a rod (about 18 cm. long) made of opaque grey glass in which is embedded lengthwise a bronze wire, but whose intended purpose is unknown; such pieces are decorated with a combed feather pattern. The wire was fused into the rod to prevent breaking, suggesting that the pieces preceded Egyptian glass that was not fragile, and so may have been made in Palestine (where they have been found), *c.* 1000 BC.

Schaper, Johann (1621–70). The first south German HAUSMALER, who was a painter of STAINED-GLASS WINDOWS and created a new style of decorating glass vessels, working both in SCHWARZLOT and in GRISAILLE, with the same technique of scratching out details with a needle. He also invented some TRANSPARENT ENAMELS. Born in Hamburg, he worked in Nuremberg from 1655 and was in Ratisbon (Regensburg) in 1664. He decorated a number of low cylindrical BEAKERS, usually with three flattened hollow ball-feet; glasses of such shape and decoration are sometimes referred to as a 'Schaper glass'. The motifs were often landscapes and battle scenes. He also decorated some TANKARDS. Some of his pieces are signed and dated.

Schaper glass. A type of BEAKER, enamelled in SCHWARZLOT in the style of JOHANN SCHAPER, which is of low cylindrical form and usually has three flattened hollow ball feet (*see* BUN FOOT). Beakers in such form were also decorated by others with WHEEL-ENGRAVED designs and some were also decorated in *Schwarzlot* in the 19th century.

Scheuer (German). A type of low drinking glass, the bottom half being hemispherical and the upper part being a short cylindrical NECK with a wide MOUTH. They are usually decorated with PRUNTS, one of which is elongated to serve as a handle. They were made in Germany in the 15th century. *See* DOPPELSCHEUER.

Schiesshausglas (German). Literally, shooting-lodge glass. A type of drinking glass, made in the 17th century for use by members of a shooting lodge, and having decoration depicting the emblem of the lodge. Examples are of conical shape with a pedestal foot and have a cover with a finial. They were made in Saxony, some examples being dated 1678.

Schindler, Caspar. A glass-engraver who, with Wolfgang Schindler, worked in Dresden from 1610. Georg Schindler was also a glass-engraver there from 1635. *See* SAXON GLASS.

Schlangenfadenglas (German). Glassware decorated with applied threads of glass trailed on in serpentine patterns and then pressed flat and notched. It was made in the 2nd and 3rd centuries in the Rhineland. *See* SNAKE THREADING.

Schlangenpokal (German). Literally, snake goblet. A type of WINGED GLASS in the form of a POKAL with a snake STEM. Examples were made in the FAÇON DE VENISE in Germany and the Netherlands in the 17th century. *See* SNAKE GLASS.

Schmelzglas (German). Literally, enamel glass. As the word *Schmelz* referred, in 17th/18th century German parlance, to opaque glass, it has been later applied in Germany (according to W. B. Honey, 'inappropriately' and 'pointlessly') to AGATE GLASS, which is solid opaque glass of several blended colours, and not decorated with enamelling.

Schmidt, Hans Wolfgang (fl. 1676–1710). A glass-engraver of Nuremberg who depicted battle scenes, ruins, and hunting scenes. He engraved drinking glasses made in Bohemia with solid stems and some with hollow knop STEMS. Several pieces by him are signed. *See* HOHLBALUSTERPOKAL.

Schneider, Christian Gottfried (1710–73). An engraver of glassware who was the master-engraver among the many engravers at WARMBRUNN in Silesia. He combined HOCHSCHNITT and INTAGLIO engraving, usually with motifs derived from Augsburg prints of pastoral scenes and of allegories, but often from his own sketches of local scenes. His work is unsigned. *See* AMBROSIA DISH.

Schneider, Cristallerie. A glasshouse founded at Epinay-sur-Seine in 1908 by Charles Schneider (1881–1962) and his brother Ernest, the

Schaper glass. Beaker enamelled in black (*Schwarzlot*), probably by Johann Schaper, Nuremberg, *c.*1665. Ht 10 cm. Kunstmuseum, Düsseldorf.

Scheuer. Dark bluish-green glass, Rhineland, 15th–16th century. Ht 6·2 cm. Kunstgewerbemuseum, Cologne.

Schneider glassware. 'Galaxie', decorative ball with air bubbles, 1970. Cristallerie Schneider, Paris.

former having studied the art of glass decoration under EMILE GALLÉ and DAUM FRÈRES. The original commercial production was changed in the 1920s to decorative glassware of the ART NOUVEAU style, featuring CASED GLASS of two or three colours, sometimes decorated by means hydrofluoric acid. Pieces were signed 'Schneider', 'Le Verre Française' or with a filigree design, the rarest pieces being signed 'Charder'. The artistic direction passed in 1948 to Robert Schneider, son of Charles. The factory was moved in 1962 to Lorris (Loiret).

Schouman, Aert (1710–92). A Dutch painter and engraver who was a follower of FRANS GREENWOOD in the use of STIPPLING. He was born in Dordrecht and worked mainly there but also in The Hague. His engraving was predominantly on English glass, and almost all of his known examples are signed. *See* Wilfred Buckley, *Aert Schouman*, (1931).

Schraubflasche (German). Literally, screw flask. A type of FLASK having a threaded neck for screwing on the cover.

Schröder, C. C. (fl. 1760s). A Dutch glass-engraver of Delft, who decorated flat panes by wheel-engraving with portraits, sometimes after 17th-century prints. Some pieces bear his signature or initials.

Schulze, Paul (1934–). An American glass-designer who joined the design department of STEUBEN GLASS WORKS in 1961. He created many designs for important ornamental glassware, often in abstract forms, during the Houghton period at Steuben. In 1970 he succeeded JOHN MONTEITH GATES as design director.

Schurman(n), A. F. à (1730–83). A Dutch engraver of Amsterdam who did work in STIPPLING in conjunction with HATCHING. His dated glasses are known from 1757 until 1780.

Schurman, Anna Maria van (1607–78). A Dutch engraver who decorated glassware with flowers and insects lightly engraved, together with calligraphic inscriptions (*see* CALLIGRAPHY).

Schurterre family. A family who came from Normandy to England, *c.*1343, and settled near Chiddingfold, Surrey, where LAURENCE VITREARIUS had been making glass. By 1367 John Schurterre leased the property of JOHN ALEMAYNE, and began making glass, which his descendants continued for several generations. Various agents, including Alemayne, obtained orders for some of the STAINED-GLASS WINDOWS for St Stephen's Chapel, Westminster, and St George's Chapel, Windsor, as well as for some Oxford colleges, and it may be that much of that glass was made by the Schurterres, including 'Peter Shortere', *c.*1400.

Schusterglas (German). Literally, shoe-maker glass. A general term for various forms of caricature HEAD-FLASK, so-called because one well-known example depicts a misshapen head that is said to have been that of Nero's fool who had been a shoe-maker. They were made of ROMAN GLASS in the 1st/3rd centuries AD.

Schusterglas. Roman head-flask, blown in two-piece mould, 1st/3rd century. Römisch-Germanisches Museum, Cologne.

Schwanhardt, Georg (the Elder; 1601–67). A master of wheel-engraving on both crystal and glass, polished and unpolished. Born in Nuremberg, he worked from 1618 with, and succeeded to the engraving privilege of, CASPAR LEHMANN at Prague; after Lehmann's death in 1622 he returned to Nuremberg to continue his work there. He engraved landscapes and BAROQUE scrollwork, supplemented by diamond-point engraving; he was the first to use (1632) on the reverse of a glass object a polished DIMINISHING LENS. He was succeeded by his two sons, HEINRICH SCHWANHARDT and Georg (d. 1676), and three daughters, Sophia, Maria, and Susanna, all of them accomplished glass-engravers. About fifty pieces by Georg Schwanhardt the Elder have been identified, some of them signed, but no piece signed by any other member of the family is known. *See* STURZBECHER.

Schwanhardt, Heinrich (d. 1693). A son of GEORG SCHWANHARDT, who was born and worked in Nuremberg. He was the first known exponent of ETCHING as a means of decorating glass, *c*. 1670, which technique was not generally practised until the 19th century. An etched glass dated 1686 is attributed to him; the only wheel-engraved glass attributed with reasonable certainty to him is dated 1681. He used acid on the background of his etched pieces, so that the smooth and brilliant figures contrast with the mat ground. He also decorated with inscriptions in CALLIGRAPHIC style.

Schwartz, Samuel (1681–1737). A glass-engraver from Thuringia, possibly a pupil of GOTTFRIED SPILLER. He applied, unsuccessfully, to become Court Glass-Engraver at Gotha in 1713. He worked independently for several patrons until 1731, when he was appointed Court Glass-Engraver to Prince Günther I of Schwarzburg-Sondershausen at Arnstadt, Thuringia. One GOBLET made by him in 1730 is the only signed work by him known to survive, although another signed piece dated 1712 has been recorded.

Schwarzlot (German). Literally, black lead. Decoration in TRANSPARENT ENAMEL in black linear style, inspired by engraving, principally used by German HAUSMALER during the second half of the 17th century and the first half of the 18th, especially on porcelain. In many cases IRON-RED is additionally employed, and a little gilding is often added. Its use on glass was introduced at Nuremberg, by JOHANN SCHAPER and his followers, particularly JOHANN LUDWIG FABER. It was also practised by *Hausmaler* in the early 18th century in Bohemia and Silesia (particularly by DANIEL PREISSLER) and at Nuremberg (by ABRAHAM HELMHACK). The designs consisted principally of landscapes, battle scenes, and mythological subjects, but towards the beginning of the 18th century they included LAUB- UND BANDELWERK and CHINOISERIES, principally by Preissler. Sometimes details of the decoration were scratched out with a needle. Reproductions were made in the 19th century.

Schwinger, Hermann (1640–83). A glass-engraver of Nuremberg who was a pupil of Hans Stefan Schmidt; he decorated BEAKERS and GOBLETS with hollow-knopped STEMS, using WHEEL-ENGRAVING and signing some of his work. His motifs were pastoral landscapes, architectural subjects, and bacchanalian scenes, with added CALLIGRAPHIC inscriptions.

sconce. A type of wall-bracket with one or more projecting arms for candle-sockets. Some are made with a mirror or a PLAQUE, and some have figures which serve as the candle-holders. The candle NOZZLE usually has a DRIP-PAN. Such pieces are known from Venice in the early 17th century, and were copied in England before the 18th century. Some examples are about 1 metre high. Glass sockets for candles attached to a PIER-GLASS or CHIMNEY-GLASS are not classified as sconces. Sometimes called a 'wall sconce'; the French term is *bras de cheminée*. The term 'sconce' was used as early as 1715 to designate a sort of bracketed candlestick; it is now sometimes loosely used to refer to a detachable nozzle of a CANDLESTICK, especially one with a DRIP-PAN attached.

Scottish glassware. Glassware made in Scotland from *c*. 1610 until the present day. The first glasshouse was established by SIR GEORGE HAY in 1610, as Wemyss; it was transferred to SIR ROBERT MANSELL in 1627. Two competing companies were started in 1617. After the use of wood for glass-making was banned in 1615, glasshouses using coal were opened in 1628 at Leith, the port of Edinburgh, mainly to make green bottles, and *c*. 1664 the public was forbidden in Scotland to buy bottles not made there. In 1682 Charles Hay, a relative of Sir George Hay,

Schwarzlot. Goblet with *chinoiserie* decoration, attributed to Ignaz Preissler, *c*. 1725–30. Museum of Decorative Arts, Prague.

established a glasshouse at Leith, as did others soon after, and by 1777 there were seven there, making bottles and drinking glasses, as well as FRIGGERS. In 1750 a glassworks was started at Alloa, making bottles with QUILLING, and also friggers. In the second half of the 17th century the PRESTONPANS GLASSHOUSE was founded. Another glasshouse set up at Leith in 1864 has become WEBB'S CRYSTAL GLASS CO., making tableware. The North British Glass Works was started c. 1864 at Perth, becoming later the MONCRIEFF GLASSWORKS, maker of MONART GLASS. For makers of modern PAPER-WEIGHTS, see CAITHNESS GLASS, LTD; PERTHSHIRE PAPER-WEIGHTS, LTD; STRATHEARN GLASS, LTD; YSART, PAUL. See also EDINBURGH CRYSTAL GLASS CO.

screw cap. A type of COVER (usually metal, but occasionally glass) that has interior threading to screw on and fit securely over the MOUTH of a vessel, such as a CASTOR, SNUFF BOTTLE, SCENT BOTTLE, spirit flask etc.

screw-joined glassware. Objects of glass with a STEM made in two parts, the upper part being threaded to screw into a KNOP on the lower part. They were made in Bohemia and Germany. The threads were cut by a specialized craftsman called a *Schraubmacher*.

scroll flask. A type of flattened FLASK of MOULD-BLOWN glass made in the United States, shaped generally like a heart with point upward, having a flat base and a short cylindrical neck, and decorated with several elaborate relief scrolls following the outline of the flask. There is often further decoration or inscriptions within the scrolls.

Scudamore Flute. A FLUTE GLASS in the form of the EXETER FLUTE, bearing the engraved Royal Arms of England and also the arms of the Scudamore family. It belonged to the Earl of Chesterfield. The arms are enclosed within an engraved wreath and below them is an engraved festoon of fruit. It has been formerly attributed to the English glasshouse of the DUKE OF BUCKINGHAM, but this has been disputed by W. B. Honey in favour of the idea that it was made and engraved in the Netherlands, as also in the case of the EXETER FLUTE and ROYAL OAK GOBLET. It is sometimes referred to as the 'Chesterfield Flute'.

sculptured glassware. A type of STUDIO GLASSWARE that is shaped by the artist from a solid block of molten glass and carved by him without BLOWING or MOULDING. An Italian term is *a masello*. See ALFREDO BARBINI.

Sea Chase, The. An ornamental crystal-glass LUMINOR depicting a crescent cluster of five dolphins emerging from a crescent wave. The flat base, made in the form of a wave, rests above a concealed light that illuminates the wave and travels up the piece to illuminate the dolphins. It was designed by LLOYD ATKINS for Steuben Glass, Inc., in 1969, and engraved by Peter Schelling.

sea-horse bottle. A type of small SMELLING BOTTLE having a flattened body with its lower part curled to one side in a snail-like form. They are free-blown and usually decorated with applied trailing. Some are of colourless glass with coloured trailing. They were made in the United States in South Jersey and New England, and some were probably made in England or Scotland.

seal, bottle. A seal bearing a monogram or design made of glass, usually MOULDED, and affixed to a BOTTLE to identify (1) the owner by showing his name, initials, or coat of arms, and sometimes his address; (2) a tavern; or (3) in a few cases, a factory, to show the source of the contents. Over 800 different such seals have been recorded. See SEALED GLASSWARE.

seal, fob. A seal bearing a monogram or design made in INTAGLIO for imparting an impression on sealing wax; the seal itself is mounted on a

'Sea Chase'. Designed by Lloyd Atkins for Steuben Glass Works, 1969. Ht 27 cm. Collection of Mrs Gratia R. Waters. Courtesy, Steuben Glass, New York.

shank, about 2·5 cm. to 5 cm. long, having a small ring for suspension. Some seals made in England, *c*.1750 and later, are of coloured glass. Some Chinese examples have a figure of a lion surmounting the handle, usually square in section. The glass shanks for some seals made in the late 18th century have decoration with a colour TWIST, some with a coloured corkscrew twist, and some made in the mid-19th century have MILLEFIORI florets. *See* SEAL, LETTER.

seal, letter. A seal (*see* SEAL, FOB) mounted on a shank for use at a desk for sealing letters or documents. The shank is about 6 cm. long and some have been made of glass; examples made at the ST-LOUIS GLASS FACTORY are of coloured glass with white FILIGRANA decoration. The French term is *cachet*.

sealed glassware. (1) Glassware bearing a seal identifying the glassworks that made it, e.g., glassware with the RAVEN'S HEAD seal. (2) BOTTLES with a bottle seal affixed (*see* SEAL, BOTTLE).

seam mark. A slight narrow ridge on a glass object which indicates that it has been made by MOULDING. The seam appears where the joins in the parts of the mould have permitted molten glass to seep during the process of formation, and is indicative (together with the usually accompanying blunted relief decoration) of a much-worn mould. On well-made pieces the seam marks are usually smoothed away by grinding or by FIRE POLISHING.

sea-plants, ashes of. *See* KELP; ROQUETTA; BARILLA; GLASSWORT. *See also* FERN ASH.

section. The outline of an object as it would appear if bisected by a plane, either horizontally or vertically.

seeds. An undissolved tiny bubble of gas or air that rises to the surface while melting the ingredients of glass, especially in an inadequately heated old-style FURNACE. If not removed, seeds show as tiny specks in the finished glass object. They do not usually occur in a modern furnace with heat control, but when they do they must be skimmed off, particularly in making high-quality crystal glass.

Seguso Vetri d'Arte. A glassworks at MURANO which has been making ART GLASS since *c*.1934; the pieces are often free-blown and partially FLASHED with coloured glass, including an amethyst shade. Wares designed by FLAVIO POLI have been produced here.

Selbing, John (1908–   ). A glass-designer with ORREFORS GLAS-BRUK from 1927. Although essentially a photographer specializing in portraying glass objects, he also created original designs, e.g., vases in the form of bubbles resting on a hollow conical pinnacle.

selenium. A metallic agent used in modern times to colour glass. When used alone it produces a pink colour in SODA GLASS, and hence is used as a decolourizer in making bottles to neutralize the natural green from the iron impurities. When used in making LEAD GLASS, it produces an amber colour. When combined with cadmium sulphide, it produces a brilliant sealing-wax red, e.g., as used by FREDERICK CARDER at STEUBEN GLASS WORKS to make a type of RUBY GLASS.

Selenium Ruby glass. *See* RUBY GLASS (2).

Senses, The Five. A group of allegorical figures representing the five Aristotelian senses: hearing, smell, taste, touch, and sight. They have been depicted as engraved and enamelled decoration on glassware made in Germany and Bohemia in the 18th century.

serpent-stemmed glass. *See* SNAKE GLASS.

sealed bottle. Green glass with kick-base, seal dated 1721, English. Ht 15·9 cm. New Orleans Museum of Art (Billups Collection).

service. Decanter from Prince of Wales service, featuring swirling pattern with wheel-engraving; unknown factory, *c*.1810–20. Ht 23 cm. (without stopper); weight 8 lb. By gracious permission of Her Majesty the Queen.

serving bottles (decanter bottles).
English. Ht 27 cm.; 20·6 cm. Courtesy,
W. G. T. Burne, Ltd., London.

serrated. Notched, or resembling the teeth of a saw, e.g., the form of the
RIM of some English glass vessels.

service. A set of glass tableware, including both hollow-ware and
flatware, made *en suite*. During the 18th century drinking glasses were
made of varying shapes for particular wines and beverages, and also of
varying sizes, and soon various other table objects were made *en suite*,
e.g., bowls, decanters, dishes, etc. The range of objects of the same
design made by a particular factory was so vast that a buyer would select
only a few of his own choice, and likewise the range of shapes, sizes, and
decorative patterns was so great as to offer a wide selection. Thus a
service of glassware differs from one of ceramic ware, where the
complete service is often practicable for the owner's use. *See* SETTING.

——, *Duke of Wellington.* A very large service consisting of many
types of pieces that was acquired (by donation or purchase) by the first
Duke of Wellington, *c.* 1820, and said to be of IRISH GLASSWARE, but of
unidentified factory. It is now kept divided between Apsley House,
London, and Stratfield Saye, near Reading.

——, *Hull.* A service of over eighty pieces, reputedly IRISH GLASS-
WARE, all completely decorated with PRISM CUTTING. The tradition
is that it was seized and taken off the English frigate *Guerrière* by Captain
Isaac Hull after his U.S. frigate, the *Constitution*, had defeated the
English ship on 19 August 1812; it was donated by a descendant of
Captain Hull to the Wadsworth Atheneum, Hartford, Conn.

——, *Prince of Wales.* A service of glassware, mainly in the Royal
Collection at Windsor Castle, bearing the engraved insignia, three
feathers, of the Prince of Wales. It includes many DECANTERS and
WINE-GLASSES made *c.* 1810–20, presumably for the future George IV
while he was Prince Regent. It is not known which factory made it. The
glasses have a wide CUSHION PAD and an ornamented FOOT.

serving bottle. The same as a DECANTER BOTTLE.

servitor. The principal assistant to the GAFFER on the team (CHAIR or
SHOP) that makes objects such as wine-glasses.

set. More than two objects of the same or similar form, suited to each
other and intended to be used together as a complete unit, but not
essentially identical. An extension of a PAIR.

setrill. The Spanish term for an oil CRUET. Those made in Catalonia,
15th century, have a globular BODY, a tall thick NECK, a loop HANDLE,
and a long thin curved SPOUT. *See* PORRÓ (PORRON).

setting. Pieces of glassware for one person at a dinner table, including
GOBLET, WINE GLASSES, FINGER BOWL, SALAD DISH. *See* SERVICE.

set-up. A sliced section from a CANE, as used in making MOSAIC and
MILLEFIORI glass.

Seuter, Bartolomäus (1678–1754). An artist in Augsburg, Germany,
to whom some enamelling on glassware, *c.* 1700, has been attributed on
the basis of its resemblance to his painting on some faience. The
enamelling depicts naturalistic flowers with birds, in bright colours, and
was done on some small SCENT BOTTLES.

Sèvres-type vase. A vase of English coloured glass decorated in the
19th century by an unidentified painter (possibly an artist who worked
in a porcelain factory) in the style of some Sèvres porcelain vases, e.g.,
with a polychrome portrait within a gilt cartouche on a *bleu-de roi*
ground with a diaper *œil-de-perdrix* pattern. They were made probably
at Bristol, *c.* 1840. Comparable vases were made at Stourbridge, *c.* 1840,
with decoration simulating that on some Derby porcelain vases, e.g.,
with exotic birds painted within a gilt cartouche on a blue ground. *See*
ENAMELLING: *English.*

Sèvres-type vase. Blue glass with gilded
oeil-de-perdrix ground and enamelled
portrait (probably by J. T. Bott), 1847;
vase probably by W. H., B. & J.
Richardson, Stourbridge. Ht 32·5 cm.
Dudley Metropolitan Borough
(Stourbridge Council House
Collection).

shade. A glass protective cover for a clock or ornament. They are large, usually cylindrical but sometimes conical, with a domed top and often a decorative FINIAL. They were made in France from the 18th century for use in protecting NEVERS FIGURES, and later in England where they were popular in the Victorian period. *See* DOME.

shaded glass. Glass that has a variation of colours blending into one another. It is found usually as the interior layer of an object of CASED GLASS. The process involved taking a GATHER of glass of one colour and blowing it slightly so as to make thinner the end away from the BLOWING IRON; the bubble was then dipped into a BATCH of another colour to the extent of covering only the thin lower part, then blown to develop a glass of two equal layers, with blended colours. Other methods are called 'striking' and 'die away'. Such shaded glass was used by FREDERICK CARDER at STEUBEN GLASS WORKS to make his GROTESQUE GLASSWARE.

shaft. The column of a CANDLESTICK between the nozzle and the base. They were made of glass in many styles, solid or hollow, and ribbed, faceted, knopped, etc. Some for a FIGURE-CANDLESTICK were made in ornamental form, e.g., in the shape of a caryatid or a dolphin.

shaft-and-globe bottle. *See* WINE BOTTLES, *English*.

shamrock cane. A type of CANE used in making a paper-weight that is made with a silhouette of a shamrock, hence usually green. They were used by BACCARAT (*see* PAPER-WEIGHT, *concentric*).

shank glass. A drinking glass that is made in three parts by attaching to the BOWL a separately made STEM, and then attaching the FOOT. Also called a 'stuck-shank glass' or a 'three-part glass'. *See* DRAWN SHANK; STUCK SHANK.

sheared glass. A type of glass object (e.g. the bowl of a WINE-GLASS) which, after being blown by the glass-blower, is sheared from the BLOW-PIPE while hot. The rim is then made round and smooth by reheating. Sometimes called 'driven glass'. *See* CRACKING-OFF.

shears. A glass-maker's tool used in the process of trimming a piece of glassware in the course of production, e.g., the RIM of a drinking glass when it is attached to the PONTIL by the base or held by a GADGET. *See* TOOLS, GLASS-MAKERS'.

sheet glass. Glass made in large flat sheets with parallel plane sides, by various methods that have changed with time, by CASTING, by the BROAD GLASS process, the BICHEROUX PROCESS, the LUBBERS PROCESS, and since 1959 the FLOAT GLASS process. *See* PLATE GLASS.

shell dish. A type of dish in the form of a scallop shell; examples were made of ROMAN GLASS in the 3rd century AD. Later examples have the shell resting on a stemmed FOOT; these were made in Bohemia of ROCK-CRYSTAL and other hardstones in the 17th century and of glass in the 18th century. *See* AMBROSIA DISH.

shellfish beaker. A type of BEAKER of ovoid form, usually with four small feet, decorated with two or more encircling rows of applied hollow glass figures in the form of shrimp or other shellfish. They were made at Cologne in the 4th century. The German term is *Konchilienbecher*. *See* DOLPHIN BEAKER.

shell flask. A type of FLASK (AMPHORISKOS) decorated with a MOULD-BLOWN replica of a scallop shell, identical on both sides. Some have merely a short neck and small mouth, but others, with a stemmed FOOT, have a tall neck, a disk mouth, two vertical side-handles, and two small loops at or below the rim for a carrying strap. Often the seam indicates

shade. Glass cover, French, 18th century. Victoria and Albert Museum, London.

shellfish beaker. Colourless glass, with encircling hollow glass shellfish. Cologne, 4th century. Ht 13 cm. Römisch-Germanisches Museum, Cologne.

shell flask. Double-moulded flask, green Roman glass, *c.*300 AD. Ht 19 cm. Römisch-Germanisches Museum, Cologne.

ship. Colourless glass, with coloured glass fittings and glass fibre simulating sea. Collection of Mr and Mrs S. T. Painter, Somerset. Photo Keith Vincent.

where the two identical sides were joined. They were made of ROMAN GLASS in the Rhineland in the 1st to 3rd centuries, and were copied at MURANO in the 16th–17th centuries. For the same technique, *see* GRAPE-CLUSTER FLASK.

shepherd's crook. A tall staff used by a shepherd, with a curved handle. An example in black glass was made, presumably as an ornament, probably in Persia, in the 19th century. *See* WALKING STICK.

sherry glass. A small drinking glass, intended for drinking sherry; it has a stemmed FOOT, with a BOWL which may vary in shape and size but is usually slightly larger than that of a PORT GLASS. *See* SACK GLASS.

ship. A small model of a rigged sailing vessel. They were made in England, in the region of Bristol, in the 19th century, by craftsmen working AT-THE-LAMP. They are usually of uncoloured glass with details of coloured glass and with GLASS FIBRE as the sea. *See* NEF.

ship candlestick. A CANDLESTICK with a circular base that is especially wide to assure stability at sea. They were made in England in the early 19th century. *See* DECANTER, *ship*.

ship glass. An English drinking glass with a wide thick foot and a short stem, intended to provide stability at sea. They had a wide variety of bowl shapes and stems.

ship-in-bottle. A miniature replica of a sailing ship within a glass bottle. The traditional process of making such objects was to make a small wooden hull with masts hinged to it and strings attached to the masts in such a way that, when the masts were pulled upright thereby, the strings remained as part of the rigging; the surplus string was burnt off or attached to the bottle with wax, and then the bottle was sealed with wax. Modern objects of this type are made with a glass ship inserted in the bottle by processes undisclosed by the makers.

ship subjects. Decoration on glassware, engraved or enamelled, depicting ships. Glassware, wheel-engraved in the 18th century, e.g. Dutch and Russian, shows various types of ships, e.g., sailing vessels, whalers, etc. Some rare bowls of BEILBY GLASS, *c.*1765, are decorated with an enamelled sailing ship. *See* PRIVATEER GLASSWARE; NAVAL SUBJECTS.

Shirley, Frederick S. A glass technician with MT. WASHINGTON GLASS CO. who in the 1880s received patents for several types of ART GLASS, including CROWN MILANO glass and BURMESE GLASS. He also received licences in 1886 to use the trade-name 'Peach Blow' (*see* PEACH BLOW GLASS) and 'Peach Skin'.

shoe. A decorative object made in the form of a lady's shoe. They were made of glass in NAILSEA GLASSWARE and also in Murano (with LATTIMO stripes), France, and Spain. *See* BOOT GLASS; SLIPPER.

shop. The same as CHAIR, the team that makes glass objects such as wine-glasses.

shot glass. (1) A glass vessel, to be filled with lead shot, to support and clean quill pens. Some were made by ST-LOUIS with a ribbed waisted bowl set on a base in the general form of a PAPER-WEIGHT of the millefiori or crown type. (2) A small drinking glass similar to a WHISKY GLASS, so-called mainly in the United States where they are used for serving a single measure (shot) of whisky.

shoulder. The bulge just below the neck of a VASE, BOTTLE or similar vessel.

sidone, vetro (Italian). A type of ornamental glass developed by Ercole Barovier for BAROVIER & TOSO, *c.* 1957; it is characterized by square and rectangular pieces of OPALESCENT glass fused in a chequerboard pattern, each piece having a narrow coloured centre stripe in the opposite direction from that of the adjacent pieces. *See* MURRINA.

sidone, vetro. Group designed by Ercole Barovier, *c.* 1957. Barovier & Toso, Murano.

Sidonian glassware. Glassware possibly made from before the 15th century BC at Sidon (and Tyre) on the Syrian coast, which are considered, with sites in Mesopotamia, to be perhaps the earliest glass-making centres. It is said to have been made of fine sand from the River Belus. (*See* NATRON.) Some MOULD-BLOWN vessels, stamped on a pad by the handle with the word 'Sidon', have provided the basis for ascribing to Sidon the first glass made by the process of BLOWING, in the 1st century BC. Sometimes the name of the glass-maker was also stamped (*see* GLASS-MAKER'S MARK), which is said to have been a sort of trade-mark. The glass was thin, not TRANSPARENT and colourless; the early untreated glass was often pale bluish-green, later yellowish, but was also deliberately coloured purple, blue, opaque white, etc. The vessels were mould-blown FLASKS, and occasionally JUGS, BEAKERS, BOWLS, and MERCURY BOTTLES. The Sidonian glassware was widely distributed by the Phoenician traders, as Sidon (now Saida, in Lebanon) was the capital of Phoenicia until it ceased to be a nation by the 1st century BC. The Syrian glass-makers migrated to Italy and elsewhere to establish glass-making centres. *See* SYRIAN GLASSWARE.

Siebel, Franz Anton (1777–1842). A German *Hausmaler* of porcelain who probably painted glassware in TRANSPARENT ENAMELLING, *c.* 1825, at Lichtenfels-am-Main, in Upper Franconia, No signed piece is known, but his daughter Klara is recorded as having painted some glassware.

siege. The floor of the glass FURNACE on which the POTS sit. *See* EYE.

sifter. The same as a CASTOR.

Silesian glass. Glassware made in Silesia, which was part of Poland from the 11th century, then from 1526 part of Bohemia under the Austrian Hapsburgs until 1742, after which different provinces came under Austria, Prussia, and Poland, and are now parts of Czechoslovakia and Poland. The principal glass-making centres were in the Riesengebirge, a mountainous area adjoining Bohemia, and in the Hirschberger Tal, and included the glassworks of the Counts Schaffgotsch at Petersdorf, and those at Schreiberhau (from mid-14th century to the 19th), and Friedrichsgrund. Glassware was decorated by engravers at WARMBRUNN. *See* ENGRAVING, *Bohemian-Silesian*.

silhouette. A representation of the outline of an object, figure or portrait filled in with a monochrome colour, usually black. (1) They were used in the EMPIRE period as enamelled decoration on some glassware, e.g., by SAMUEL MOHN, and by glass-decorators (e.g., J. S. MENZEL) at WARMBRUNN, a spa in Silesia where there were many glass-engravers and where pieces were decorated especially for the numerous visitors. Miniature MEDALLIONS with silhouettes in black were inserted in the body of GOBLETS. The decoration was on the reverse of the glass. (2) They were made of glass in a mould and inserted in some CANES used in making MILLEFIORI glass. The process of making these involved carving the shape on the inside of an iron mould, pouring in molten dark-coloured glass, removing the silhouette on a PONTIL, covering it with clear or opaque glass, and drawing it out to make a long thin cane, to be sliced and set in molten glass. Such silhouettes were made to represent a variety of animals, a butterfly, or dancing figures, and several different types are often found in a single PAPER-WEIGHT, surrounded by FLORETS.

silhouette cane. A type of CANE decorated with the silhouette in black, white, or a colour, of an animal (e.g. horse, dog, stag, bird), a butterfly, a

ship subject. Goblet with engraved decoration and inscription, Jamburg Factory. Ht 20 cm. Hermitage Museum, Leningrad.

flower, or often a dancing person or devil. Modern examples have the signs of the Zodiac. They have been made by BACCARAT.

Silhouette glassware. A type of stemmed tableware of which the stem was made of black glass in the form of various animals. It was developed by A. DOUGLAS NASH while he was with the A. Douglas Nash Corporation, but not produced until after he had left and joined the LIBBEY GLASS CO. at Toledo. *See* STEM, *figure*.

silica. A mineral that is one of the essential ingredients in making glass. The most common form of silica used in glass-making is SAND (an impure silica) taken from the seashore or inland beds. The Venetians used white pebbles from the rivers. In England at least as early as *c.* 1650 calcined and powdered flints were used. (*See* FLINT GLASS.) Pure silica can be melted to form glass, but its fusion requires such a high temperature that it is commercially impracticable; but it is so used for making certain laboratory glass that is subject to high temperatures.

sillabub glass. *See* SYLLABUB GLASS.

silver. A metallic agent used in colouring or decorating glass or in backing mirrors. It has been used by various techniques: (1) Silver leaf was used sometimes, with or instead of gold leaf, in decorating ZWISCHENGOLDGLAS. (2) Silver nitrate has been used in the backing of mirrors and in making SILVERED GLASS. (3) Silver sulphide has been used to produce STAINING of a deep-yellow colour, e.g., in medieval STAINED-GLASS WINDOWS and some glassware made by ANTON KOTHGASSER, and also to develop a straw-yellow colour in LUSTRE PAINTING. (4) Silver has been used to make SILVER-ELECTROPLATED GLASS. (5) Silver foil has been used in making SILVERIA GLASSWARE. The presence of a small amount of silver in the METAL of the LYCURGUS CUP may help to produce the DICHROIC effect.

silver-cased glassware. A vessel of glass blown as a lining into a silver case with openings in a decorative pattern through which the glass projects in bulges. Such ware was known as *opus interrasile* and was made of ROMAN GLASS in the 1st century AD. *See* METAL-CASED GLASSWARE.

silver-electroplated glass. A type of ART GLASS having an electro-deposit of silver on the surface. One process, patented in 1889 by Oscar Pierre Erard and John Benjamin Round for STEVENS & WILLIAMS LTD, involved the use of a special flux containing silver, which was then mixed with turpentine to form a wash; the wash was painted on the glass to make the desired design and the piece was fired, then placed in a solution of silver through which an electric current was passed to deposit the silver. The style was popular from *c.* 1890 until *c.* 1920.

silver shape. The form of glassware copied or adapted from a silver prototype, e.g., a SAUCE BOAT.

silvered glass. A type of decorative glassware having an overall silver appearance. It was made by applying a solution of silver nitrate to the inside of a double-walled vessel by pouring it in through a hole in the base, which hole was then tightly sealed with a glass disk to protect the silver from discolouring through oxidation. Sometimes such ware was engraved or was FLASHED with coloured glass which was then cut to show the silver inner layer. A patent for the silvering process was taken out in 1849 by F. Hale Thomson of London, and some disks are impressed with his name as patentee or that of E. Varnish & Co., as licensee. The process was also patented in 1855 by William Leighton (*see* LEIGHTON FAMILY) of the NEW ENGLAND GLASS CO., whose mark appears below the seal on some pieces. Other companies in England, France, and Belgium produced this type of glass between 1855 and 1885. It is sometimes called 'mercury glass'.

silver-cased glassware. Cobalt-blue beaker in silver case, Roman glass, 1st century AD. Ht 9·3 cm. British Museum, London.

silvered glass. Beaker with engraved vine decoration, probably by James Powell & Sons, London, *c.* 1857. Ht 12·4 cm. Victoria and Albert Museum, London.

silveria glassware. A type of ART GLASS having a silvery appearance due to silver foil having been embedded between two layers of clear colourless or coloured glass. Some examples are further decorated with irregular vertical trails of glass, usually green and some with small leafy attachments. Sometimes the silver has become oxidized if CRISSELLING has affected the outer layer or if air otherwise penetrated between the layers. The process was developed *c.* 1900 by JOHN NORTHWOOD II, and the ware was produced by STEVENS & WILLIAMS, LTD. Some examples are marked 's & w', with the word 'England' or a small fleur-de-lis. The glass is to be distinguished from SILVERINA GLASS made by STEUBEN GLASS WORKS.

Silverina glass. A type of ART GLASS developed by FREDERICK CARDER in the 1920s at STEUBEN GLASS WORKS, but using a technique previously known. It was made by picking up mica flakes on the PARAISON and covering them with a layer of clear coloured glass. The flakes vary in size and density, and the silvery colour from the mica is affected by the colour of the outer layer. The flakes are combined with trapped air bubbles. The glass is different from SILVERIA GLASS made by STEVENS & WILLIAMS, LTD.

Simm family. A family of Jablonec (Gablonz), in Bohemia, which included a number of highly skilled glass-engravers.

simpulum (Latin). A small (10/11 cm. high) Roman LADLE for dipping wine for libations. They have a circular bowl from one side of which extends vertically a short flat handle. Examples in ROMAN GLASS, from shortly before 79 AD (the year in which Vesuvius erupted), were found in 1961 in the ruins of Herculaneum. They are related to the Greek KYATHOS.

simulated cameo. An inexpensive type of glassware made in Bohemia with the design in white enamelling to simulate true CAMEO GLASS with a carved white OVERLAY.

sinumbra lamp (Literally, without shadow lamp). A type of OIL LAMP by the use of which a shadow, if not destroyed, is rendered imperceptible by the formation of the vessel. A lamp shown in Apsley Pellatt's *Memoirs*, with a globular glass body on a flat bottom and a short neck, called there a 'sinumbra lamp', is in fact a type of oil lamp known as an 'astral lamp', and the sinumbra lamp, as patented in France in 1820 by George Philips, of London, has a completely different form, being of metal with a globular reservoir (FONT) for oil, resting on a metal stand.

siphon glass. A type of JOKE GLASS in the centre of the bowl of which is a vertical rod extending upward; placed over this rod is a wider tube extending downward from a hollow stag-like figure whose mouth reaches outside the rim of the glass and from which the user sucks the contents. Such glasses were made in Germany and Russia in the 16th/18th centuries. The French term for a glass involving difficulty when drinking is a *verre à surprise*.

Sit, (Buena)ventura (d. 1755). A Catalan glass-maker, who started as a workman under Juan Goyeneche, the first to produce good-quality colourless glass in Spain. In 1728, with the encouragement of Queen Isabella, he founded the royal glassworks at LA GRANJA DE SAN ILDEFONSO, near Segovia. He specialized in making MIRRORS (from 1738 under the patronage of Philip V), and also (until 1740) in vessels in the Catalan traditional style.

site. A flat plate of glass made of CRISTALLO and used in the making of mirrors after being ground, polished, and silvered. They were made in Venice and exported by 1570 to mirror-makers in Antwerp, Rouen, and England.

simpulum. Pale-green with white trailing, Roman glass, 1st century AD. Ht 11·3 cm. Wheaton College, Norton, Mass.

siphon glass. Tumbler with stag, German, 17th century. Ht 12·6 cm. Kunstsammlungen der Veste Coburg.

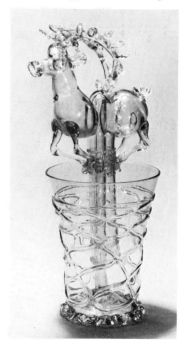

situla. Violet-blue glass, with wheel-engraved *intaglio* decoration of Bacchanalian scenes, probably Byzantine, 6th/7th century. Ht 20·3 cm. Treasury of St Mark's, Venice.

Situla Pagana. Yellow-green glass with *diatreta* decoration, probably Italian, 3rd/4th century. Ht 25 cm. Treasury of St Mark's, Venice.

situla (Latin). A bucket-shaped receptacle, usually for holy water, having a flat base and two small loops or lugs near the rim for attaching an overhead HANDLE. Generally Roman examples were made of bronze or terracotta, but some Roman and Byzantine examples were made of glass, in the 3rd–5th centuries. They are about 20 cm. deep. The Italian term is *secchio. See* SITULA PAGANA; ASPERSORIUM; BUCKET.

Situla Pagana. (1) A SITULA of greenish ROMAN GLASS with a swinging overhead HANDLE and decorated in VASA DIATRETA technique. The upper half has an encircling hunting frieze and the lower half a reticulated decoration of tangential arches in cage style. It was made probably of Italian workmanship, 3rd/4th century AD, and came into the Treasury of St Mark's Cathedral, Venice, when Venice conquered Byzantium (hence often called the *Situla di San Marco*). (2) Another similar *situla*, of thick violet-coloured glass, probably Byzantine, 5th-century, is also at St Mark's; its decoration consists of WHEEL-ENGRAVED Bacchanalian scenes in INTAGLIO.

skittle pot. A type of POT (crucible) in which the ingredients for making glass were fused until the invention of the CROWN POT. They were also used for making coloured glass. It was tall and cylindrical, and open at the top.

skyphos (Greek). A drinking cup, hemispherical or semi-ovoid, having a low or medium-height bowl with sides curving inward toward a flat base or a FOOT-RING and having two loop side-handles (horizontal or vertical) attached at or just below the rim. The glass forms were adapted from *skyphoi* made of Greek pottery. The term *skyphos* has been used to include the type with wing handles, sometimes referred to as a PTEROTOS CUP. *See* POČULUM.

slag glass. A type of glass that is streaky like marble, due to the inclusion in the METAL of some waste slag, e.g. silicate skimmed off molten steel from the furnaces (as was done in making 'ironstone china', a type of opaque stoneware). It was usually made by PRESS-MOULDING inexpensive ware, some having a bright-blue colour, some purple or black. It was made in England *c.* 1875–90 by several factories in the north-east, e.g., SOWERBY'S ELLISON GLASSWORKS, at Gateshead-on-Tyne, Henry Greener & Co., at the Wear Glass Works, in Sunderland (its mark: a lion with a halberd), and George Davidson & Co., at the Teams Glass Works, at Gateshead-on-Tyne (its mark: a lion on a battlement). It continues to be made, with the same mark, by the Davidson company. *See* TYNESIDE GLASSWARE.

sleeve. The same as a SAVE-ALL.

slicker. The same as a LINEN SMOOTHER.

slipper. An object in the form of a lady's flat slipper, made by flattening a glass bottle lengthwise, and decorated on the upper part with SNAKE THREADING. Examples were made in the Rhineland in the 3rd century AD. *See* SHOE.

small salt. A small SALT-CELLAR, introduced in England in the 17th century when it became customary to provide each diner with an individual one, in lieu of the large communal salt-cellars ('great salt' or 'standing salt'). They were usually about 2·5 cm. high and 5 to 8 cm. wide. Some early examples had a hemispherical bowl on a pedestal base, but later ones followed contemporary silver patterns.

smalt. A deep-blue pigment prepared by fusing together ZAFFER, potassium carbonate, and SILICA, and grinding the resulting product to a powder. It is used in colouring BLUE GLASS. Smalt was imported into England from Saxony until native deposits of cobalt were found.

smalto (pl., *smalti*; Italian). Literally, enamel. A piece of opaque coloured glass made in a thin slab (called *smalto in pan*) and used in quantity to make MOSAICS and to decorate glassware, as well as mirror-frames, tables, and consoles, and in Russia to inlay floors. It was itself sometimes decorated with gilding. It was made in Russia at the ST PETERSBURG IMPERIAL GLASS FACTORY and also at the studio of M. V. LOMONOSOV. It was also used as source of jewellers' enamelling colours. *See* GLASS 'CAKES'.

smelling bottle. (1) A small glass bottle, with a ground stopper, to contain an aromatic volatile salt (basically ammonia) for inhaling by ladies to avoid faintness or headaches, related to a SCENT BOTTLE but not intended to contain perfume. (2) A term used in the United States for a type of scent-bottle made there which was FREE-BLOWN or MOULDED in a variety of shapes and colours. Many were of STIEGEL-TYPE GLASSWARE, but they were also made by the BOSTON & SANDWICH GLASS CO. and other glasshouses. *See* COLOGNE BOTTLE; SEA-HORSE BOTTLE.

Smith, William. A glass-decorator who migrated to the United States after twelve years at factories in Birmingham, England. He worked for the BOSTON & SANDWICH GLASS CO. from 1855, and then formed his own decorating firm for gilding and enamelling glass lamps. His sons, Henry and Alfred, were trained by him as decorators and operated the design department for the MOUNT WASHINGTON GLASS CO. from 1871 until they withdrew to start a decorating shop in New Bedford, Mass., using imported BLANKS which they decorated with landscapes, Oriental scenes, flowers, and fruits. Their mark was a lion rampant within a shield or the words 'Smith Bros.'.

smoke bell. A bell-shaped shade with a ring handle, made to hang over an oil lamp or a candle to catch the soot and thus protect the ceiling.

snake. A decorative object, of English glass, used on LUSTRES and CANDELABRA, in the form of a curved, tapering, faceted ornament, pointing upward and attached to the central SHAFT.

snake glass. A type of WINGED GLASS of which the stem was made by manipulating glass rods (enclosing twisted coloured opaque threads) into a convoluted pattern resembling a snake and so elaborate that there was no independent stem. Sometimes the FINIAL on the cover was similarly made. They were made in the FAÇON DE VENISE in Germany and the Netherlands in the late 16th and 17th centuries, and copied in the late 19th century at EHRENFELD, near Cologne. Sometimes the rods were further embellished by clear or coloured pincered CRESTING along the outer edge, and sometimes the pattern has at the top a stylized eagle's head. The German term is *Schlangenglas*, the Italian is *calice a serpenti*. *See* THREADING.

snake trailing. A type of decoration made of applied trailed glass threads in a twisting haphazard serpentine pattern. Some BOTTLES, BOWLS, and JUGS so decorated have been found in the Near East and are considered to be SYRIAN GLASSWARE; others found in the Rhineland were probably made at Cologne in the 2nd and 3rd centuries AD. The trailing is usually coloured, but some of the Near Eastern examples have self-coloured trails. On some examples from Cyprus the 'snake' is given a triangular head and the trailing is flattened. When such decoration is essential to an object, e.g., a WINGED GLASS, it is preferably called 'snake threading'. *See* MEISTERSTÜCK.

snowflake glass. A type of glass made in China in the 18th century; it is a colourless glass with white INCLUSIONS and bubbles that produce, probably unintentionally, a snowstorm effect. It was used for some CHINESE SNUFF BOTTLES of CAMEO GLASS with a coloured OVERLAY.

skyphos. Brownish glass with angled handles, Roman glass, 1st century AD. Ht 7·8 cm. Römisch-Germanisches Museum, Cologne.

snake trailing. Patera with decoration in blue and yellow on underside, Roman, 3rd century AD. Römisch-Germanisches Museum, Cologne.

snowflake glass. Chinese snuff bottle with red cameo overlay, 18th century. Ht 7 cm. Courtesy, Hugh Moss, Ltd., London.

snuff bottle. A small phial for holding snuff, generally similar to a SCENT BOTTLE but with a larger mouth and fitted with a tiny spoon attached to the cover. They are predominantly of Chinese production. *See* CHINESE SNUFF BOTTLE.

soda. Sodium carbonate. It is used, alternatively with POTASH, as the ALKALI ingredient of glass. It serves as a FLUX to reduce the fusion point of the SILICA in making glass. In early Egypt soda was perhaps derived from naturally occurring NATRON, but later it was derived from the calcined ASHES of marine plants, i.e. GLASSWORT (the prepared salt called *barilla* when exported from Spain and *roquetta* when exported from Egypt) and KELP (used in Normandy, Scandinavia, and England). When supplies of soda became insufficent for their needs, Germany, Bohemia, and France substituted potash (*see* WALDGLAS and VERRE DE FOUGÈRE), produced by burning wood of certain trees or ferns and bracken. Venetian and Venetian-style glass continued to be made with soda. Today both soda (produced chemically) and potash are used in making glass. *See* SODA GLASS; WATER GLASS.

soda glass. Glass of which the ALKALI constituent is SODA, rather than POTASH, e.g., EGYPTIAN GLASS, ROMAN GLASS, VENETIAN GLASS, and SPANISH GLASS, as contrasted with WALDGLAS. In England soda glass was made until the use of LEAD in 1676. Much soda glassware is in the FAÇON DE VENISE style, and so is difficult to distinguish as to origin. Soda. glass is slightly brownish, yellowish or green-grey, and, unlike lead glass, lacks resonance. It is light in weight and remains PLASTIC longer than glass made with potash, and hence is easier to manipulate, as in the making of Venetian glassware. Soda glass also contains some LIME, the basic proportions being SILICA 60%, soda 25%, and lime 15%; hence it is often called 'soda-lime glass'.

sommerso, vetro (Italian). A type of decorative glassware developed by VENINI, *c.*1934, similar to CASED GLASS or FLASHED GLASS, with a covering layer of glass, of varying thickness, over the ornamented under-layer, the latter often having a design in MURRINA, or *vetro pulegoso* (*see* PULEGOSO, VETRO), the outer glass being colourless or of a different colour from the inner.

South Jersey type glassware. Glassware made in the style and tradition of that made at the factory of CASPAR WISTAR, being of the same technique and decoration, but not of any assignable factory or area, or even always indigenous to the region, some coming from New England and New York, and even the Midwest. It was made from *c.*1740 until the mid-19th century; that made until *c.*1825 is referred to as 'Early South Jersey Glass'. It was made of BOTTLE GLASS evidencing sturdiness and individuality, and often was made as OFF-HAND GLASS (FRIGGERS); the pieces were FREE-BLOWN, not MOULDED, and were ornamented by manipulation with TRAILING and PRUNTS. Much of the glass was coloured, often aquamarine with opaque white. Characteristic features were the LILY-PAD DECORATION and the SWAN FINIAL.

South Netherlands glassware. Glassware made in the southern portion of the Low Countries (now Belgium), including glassware made in Antwerp from *c.*1549 to *c.*1629 in Liège from *c.*1569 to *c.*1611 and from *c.*1626 to *c.*1700, and in Brussels from *c.*1623 to *c.*1700. *See* BELGIAN GLASSWARE; VAL-SAINT-LAMBERT GLASS CO.

Sowerby's Ellison Glassworks. A glasshouse at Gateshead-on-Tyne in north-east England, which operated in the second half of the 19th century. It made inexpensive PRESSED GLASSWARE in great quantity, including SLAG GLASS and colourless, coloured, and opaque glass – all in a great variety of objects. It also produced a type of deep-blue SPANGLED GLASS which it called 'Blue Nugget'. The pieces bear a factory mark (a peacock's head) and sometimes a REGISTRY MARK. In

XIII *Standing cup*. Green glass with enamelled and gilded decoration, Murano, late 15th century; ht 17 cm. (6¾ in.). Victoria and Albert Museum, London.

XIV   *Steuben glassware*. Group of Gold Aurene objects: disk, diam. 26·5 cm (10½ in.), 1920s; iridescent vase with *millefiori* and green trailed decoration, ht 27·6 cm. (10⅞ in.), *c*. 1905–10; early iridescent vase (imitating Roman glass), ht 12·6 cm. (5 in.). Private collection; Smithsonian Institution, Washington, D.C.; private collection. Courtesy, Steuben Glass, New York.

1883 it was licensed by NEW ENGLAND GLASS CO. to make pressed
AMBERINA. *See* TYNESIDE GLASSWARE.

spa glass. A type of TUMBLER decorated and sold in the 18th/19th
centuries at such fashionable spas and health resorts as WARMBRUNN, in
Silesia, and Franzensbad (*see* DOMINIK BIEMANN), sometimes decorated
with the engraved portrait, from life, of one of the visitors, together with
some flowery inscription, and sometimes with local views.

spangled glass. A type of ART GLASS with an OVERLAY of glass of
variously tinted colours, covering an inner layer of clear glass embedded
with flakes of mica. It was developed by William Leighton, Jr., at
HOBBS, BROCKUNIER & CO. in 1883 and used for various objects, but
mainly glass baskets. Sometimes they are further decorated with applied
glass flowers and leaves. *See* CLUTHA GLASS; CLUTHRA GLASS.

Spanish glass. Glassware made in Spain from the Roman and medieval
eras, but mainly that of great artistic merit made in the 16th and 17th
centuries in many glassworks throughout the country, principally in
Catalonia (Barcelona), Castile (Cadalso), and Andalucia (Seville and
Granada), and at the Royal Factory at LA GRANJA DE SAN ILDEFONSO.
Spanish glass shows Moorish influence and later that of Venice, and to a
limited extent of Bohemia and England, but local styles were developed
in the forms and the decoration. STAINED-GLASS WINDOWS for
churches were made from the 13th to the 16th centuries. Later
important production included mirrors, chandeliers, and many
indigenous varieties of jugs, vases, drinking glasses, and pieces made
and decorated AT-THE-LAMP (*see* ALMORRATXA; CANTÍR; PORRÓ;
SETRILL). Coloured, colourless, and opaque glass was made, with
decoration mainly in white FILIGRANA (LATTICINO) and with
intricate embellishments in PINCERED work and TRAILING. Some early
Barcelona pieces were decorated with enamelling, as was some of the
later glassware from San Ildefonso. In the period from 1918 to 1940,
some glassware was designed by JOSÉ MARIA GOL, of Barcelona. For a
discussion of Spanish glass, its makers, types, and forms, *see* Alice
Wilson Frothingham, *Spanish Glass* (1964).

sparking lamp. A type of small low OIL LAMP with an opening at the
top of the FONT for the wick, which is usually inserted through a metal
collar inserted in, or screwed to, the rim, or occasionally merely held in
place by a wire. They are popularly said to be so-called because courting
('sparking') swains, *c.* 1820, were supposed to measure their visit by the
duration of the light. They were made in the United States in various
forms, sizes, and styles of decoration; the main types were the WINE
GLASS LAMPS, with a stemmed foot like a wine-glass, and the examples
with an attached loop-handle.

spearhead. The point attached to a spear, sometimes made of a solid
piece of imported glass and chipped to a sharp point and jagged edges by
aborigines in Central Australia. Arrowheads were similarly made of
glass.

Spechter (German). A term first applied by Johann Mathesius, pastor
of the Joachimstal, Bohemia, in a sermon published in 1562 at
Nuremberg, to certain unidentified and undescribed glasses attributed
by him to the Spessart region of Germany, an area in HESSE that had
made glass from the early 15th century. The term was repeated in 1884
by Carl Friedrich as applicable to a tall glass (similar to a STANGENGLAS)
decorated with an exterior pattern of chequered spiral trailing and
attributed to the Spessart region, and this identification has been
repeated and accepted for almost a century (as recently as 1966).
However, Hugh Tait, of the British Museum, in 1967 authoritatively
attributed a number of glasses in such style to Antwerp (*see* CHEQUERED
SPIRAL-TRAIL DECORATION) and has asserted that the term *Spechter*
cannot be attributed to any identifiable form of glassware nor to any

snuff bottle. Semi-opaque white glass
with enamel decoration, Chinese,
Peking Palace Workshops, with
enamelled decoration by palace artists
and Ch'ien-lung four-character mark,
1736–95. Ht 4·7 cm. Courtesy, Hugh
Moss, Ltd, London.

spearhead. Chipped from solid glass
piece, Central Australia. Science
Museum (Lovatt Collection), London.

Spielkartenhumpen. Passglas decorated with playing card, Thuringia, 1734. Ht 24·5 cm. Museum für Kunsthandwerk, Frankfurt.

known glassware made in the Spessart region. *See* Hugh Tait, 'Chequered Spiral-trail Decoration', *Journal of Glass Studies* (Corning), IX (1967), p. 96.

specie jar (or 'specie glass' in the 17th century). A large cylindrical (sometimes globular) glass JAR used in a pharmacy, originally to hold drugs but in the second half of the 18th century for window display; it was filled with coloured water (or sometimes made of coloured glass), with a hollow glass STOPPER similarly filled (or coloured). Later the jars were used for storage. Some had a narrow neck and mouth, others (usually cylindrical) a wide mouth. After *c.* 1830 they were larger, made in three sizes, and were usually decorated with a painted coat-of-arms. Sometimes a light was placed behind them so that they would serve as a condensing lens (*see* WATER LENS).

Spiegelmonogramm (German). The same as MIRROR MONOGRAM.

Spielkartenhumpen (German). A HUMPEN, usually a PASSGLAS, decorated with one or more enamelled representations of playing cards. Usually the glasses were made in complete sets including all 52 cards. They were made in the 17th and 18th centuries in Thuringia, and were probably inspired by the playing-card industry at nearby Altenburg. Some examples may be of Saxon origin. Sometimes called an *Altenburg Passglas*.

spiked kick. A KICK that projects upward into a BOWL (e.g., a MAIGELEIN) in the form of a spike, coming to a point.

spill holder. A type of VASE, usually cylindrical, used for holding spills (wood splinters or paper tapers for obtaining a light from a fire and lighting a pipe, a candle or a lamp). Those made of glass in the United States usually have a heavy bucket-shaped bowl on a short footed STEM. They are also termed a 'spill vase', 'spill pot', 'spill box', or 'quill holder'. *See* SPOON-HOLDER.

Spiller, Gottfried (1663?–1728). An engraver of glass from Silesia who was a nephew and a pupil (from 1675) and a partner (from 1683) of MARTIN WINTER at Berlin and with whom he founded (under a petition made in 1687) a water-powered engraving workshop in conjunction with the glass-factory at Potsdam. When Winter died in 1702, Spiller became Royal Glass Engraver. With the aid of the Berlin water-power mill, he did outstanding work in HOCHSCHNITT and TIEFSCHNITT, making some BEAKERS with figures usually in *Tiefschnitt* (sometimes 7 mm. deep in glass 10 mm. thick) and with encircling border decoration of acanthus or laurel in *Hochschnitt*. He often depicted *putti* and MYTHOLOGICAL SUBJECTS on covered GOBLETS and BEAKERS, some made of GOLD RUBY GLASS by JOHANN KUNCKEL. He worked on glass made at Potsdam (this has often deteriorated due to CRISSELLING). His most celebrated piece is the ORPHEUS BEAKER.

spindle. A round stick, used in hand spinning, that is tapering toward each end, with a notch or catch at one end to attach the yarn; the spindle is twirled, usually by a movement against the right leg, to twist the yarn as it is drawn from a bunch of wool or flax held on the distaff. Spindles were made of glass in Spain in the 18th and 19th centuries, sometimes in the form of a tapering rod with an ornate handle.

spindle whorl. A pierced weight, usually thicker at the centre than at the top and bottom; it was attached to a primitive SPINDLE, before the advent of the spinning-wheel, as an aid to keep the spindle turning and in manipulating the thread. Such pieces were made of various materials, including some made of glass in China and elsewhere.

spittoon (or **cuspidor**). A receptacle in which to spit; usually globular-shaped with a wide flaring rim, but sometimes with a high large

FUNNEL-MOUTH and sometimes with a short neck and a small funnel-mouth. Although generally made of porcelain or pottery, some were made of glass, e.g., (1) in India by Persian craftsmen in the Mogul period in the 16th century, (2) probably in Persia in the medieval period and in the 18th and 19th centuries, (3) in Venice in the 19th century for export to Persia, and (4) in China, at the PEKING IMPERIAL PALACE workshops in the Ch'ien Lung period (1736–95). The Chinese term is *cha-tou*.

Spitzbecher (German). A type of CONE-BEAKER of FRANKISH GLASS, in the form of a tall thin inverted cone. They were decorated on the lower part with diagonal TRAILING in the 5th/7th centuries and with vertical trailing in the 5th-early 6th centuries; the upper part, toward the MOUTH, was decorated with horizontal trailing. Like the STURZBECHER, they had to be inverted to be put down when empty, to 'ward off the evil eye'.

splayed (or **spreading**) **base**. The BASE of a vessel that flares outward toward the bottom.

split. A decorative pattern on CUT GLASS in the form of a thin vertical oval, pointed at both ends. It is made with a mitred disk. Several ovals were arranged in an encircling band (especially in the Adam period; *see* ADAM STYLE) but sometimes they are found forming star patterns or other designs. *See* COMBED FLUTING.

Spon Lane Glassworks. A glassworks at Smethwick, near Birmingham, England, owned from 1824 by ROBERT LUCAS CHANCE. From 1830 he developed there improved methods for making BROAD GLASS. In 1851 it produced 300,000 panes of glass for the CRYSTAL PALACE. It employed GEORGES BONTEMPS from 1848 to 1884.

spoon. An implement for serving, stirring or eating, consisting of a shallow bowl, usually oval or circular, attached to a horizontal handle the length and style of which is variable. They were made of ROMAN GLASS; later examples of VENETIAN GLASS in the 16th century usually have RETICELLO decoration. The Italian term is *cucchiaio*, the German is *Löffel*, the French is *cuiller*. *See* MEDICAL SPOON; MEASURING SPOON; CUTLERY.

spoon-holder. A glass receptacle in the form of a GOBLET with a large deep ovoid bowl resting on a stemmed FOOT. They usually have a longer stem than a SPILL HOLDER and often have a scalloped rim. They have been made of PRESSED GLASS in many patterns at various United States factories.

sporting glassware. Various glass vessels, usually drinking vessels, e.g. BEAKERS, RUMMERS, or GOBLETS, that are decorated with engraved scenes depicting sporting activity, e.g., horse-racing, hunting, shooting, etc. English examples have been made from *c*.1760. Some enamelled HUMPEN are decorated with sporting scenes.

spot mould. A type of MOULD used to make a PAPER-WEIGHT or other solid glass object that is decorated with a pattern of regularly spaced air bubbles. It is a sectional mould with small spike-like protuberances in the interior, and when the PARAISON (sometimes already with an interior decorative motif) is placed in it and the sectional parts of the wall are tightened around it and the spikes are forced into the glass, small cavities are created so that, when the piece is then CASED, air bubbles are trapped and formed into a pattern.

spout. The tubular protuberance through which the liquid contents of a vessel are poured. Sometimes the term is loosely applied to a pinched or beak-shaped pouring LIP. Spouts appear in many shapes, and after *c*.1850 are usually MOULDED and attached to the BODY. *See* BEAK; POSSET POT.

Spiller, Gottfried. Covered goblet, wheel-engraved by Spiller, Potsdam, *c*.1700. Victoria and Albert Museum, London.

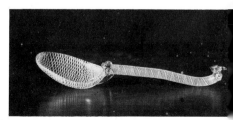

spoon. Spoon with *reticello* decoration, Venice, 16th century. Museo Vetrario, Murano.

spout cup (or spout glass). *See* POSSET POT.

spreader. An ornamental disk with a central hole that slips over the nozzle of a CANDLESTICK. Some serve to catch the wax drippings and are removable for cleaning (*see* DRIP PAN); but when the candlestick also has a SAVE-ALL above the spreader, it serves, especially on a SCONCE or CHANDELIER, as the piece to which DROPS are attached.

springel glass. A type of SCENT-BOTTLE used for scattering scent about a room. Also called a 'casting bottle'.

sprinkler. A glass vessel of various forms but all characterized by a shape that causes the contents to pour slowly, or in drops as needed for perfume-sprinkling. (1) Persian examples from the 18th/19th centuries, probably made at Shiraz (often of dark-blue glass), and examples made in Venice for export to Persia, have a globular body and a tall, vertical, curving, thin neck with a wide mouth that has a long pouring lip pointed upward. Their function is not certain, but the shape has led to their being generally regarded as perfume-sprinklers, perhaps an adaptation of the slow-pouring KUTTROLF (some related pieces even have three or four thin tubes between the base and the bowl); but A. Nesbitt has curiously called such an object a 'fanciful *bouquetière*'. The form was copied by TIFFANY in FAVRILE glass. (2) Another type made at Shiraz has a globular (sometimes cylindrical) body (resembling an EWER) with a long wide vertical neck having a FUNNEL MOUTH for filling and a long curved SPOUT extending from the side of the body, the mouth of the spout having a pincered tip with a tiny hole to reduce the flow. These usually have a small handle attached to the shoulder of the body and to a decorative ring encircling the lower part of the neck; they rest usually on a wide circular folded foot with a double-knopped short stem. These were made in the 18th/19th centuries. (3) A Syrian type (sometimes called an OMOM) has a globular, often flattened, body and a tall, thin, vertical attenuated neck. (4) For the Spanish form, *see* ALMORRATXA. All these forms are sometimes referred to as 'rose-water sprinklers' or 'perfume sprinklers'. The German terms are *Sprenggefäss* or *Rosenwasserflasche*. *See* HEAD FLASK; HELMET FLASK.

sprinkler. (1) Blue glass, Shiraz, Persia, 19th century. Ht 31·8 cm. Museum für Kunsthandwerk, Frankfurt. (2) Shiraz, 18th century. Ht 18·8 cm. Kunstgewerbemuseum, Cologne.

spun glass. An obsolete term for GLASS FIBRE. Recently it has been used very loosely to designate glass fibre drawn from molten glass by a centrifugal process.

spy-hole. A narrow opening in the covering of the BOCCA through which, during the process of melting the BATCH, impurities could be removed and samples taken for testing. *See* BYE-HOLE; GLORY-HOLE.

square bottle. A four-sided bottle, usually of thick green or blue Roman glass, made in the 1st and 2nd centuries AD. They generally have a cylindrical neck with an extended rim and a handle that is REEDED (*see* CELERY HANDLE) and angular, extending from just below the mouth to the sloping shoulder. The handle extends only slightly beyond the perimeter of the bottle, and thus does not interfere with ease of packing and storing such bottles. The bottles were MOULD-BLOWN or were free blown and then flattened on the sides. They sometimes have relief decoration, and are sometimes decorated on the bottom with a Greek or Latin inscription. The size varies greatly (from 6·4 cm. to 40 cm. in height), and the capacity of bottles of equal size varies depending on the thickness of the glass in the walls and on the shape of the shoulder; and the proportion of width to height also varies. They are related to the RECTANGULAR BOTTLE, the HEXAGONAL BOTTLE, and the octagonal bottle, but are not to be confused with the square JARS with a wide mouth or with the MERCURY FLASK. The German term is *Vierkantflasche*. *See* MALTESE CROSS BOTTLE. For a discussion of Roman square bottles, *see* Dorothy Charlesworth, *Journal of Glass Studies* (Corning), VIII (1961), p. 26.

Stab. The German term for a CANE.

Stachelglas (German). Literally, prickly glass. A type of bottle, made in the Rhineland, decorated with scattered small PRUNTS, not so densely placed as on a KRAUTSTRUNK.

stacking glasses. Drinking glasses so designed (in both shape and size) that they may be placed safely and effectively in a vertical stack, the foot and stem of an upper glass resting in the bowl of the next lower glass. The height of the foot and stem should approximately equal the height of the bowl, so that the bottom of the bowl of each glass is only minimally above the rim of the lower bowl. A very striking set of such glasses was designed by TAPIO WIRKKALA for IITTALA GLASSWORKS, of Finland, c. 1965.

staff head. A knob decorating the upper end of a staff. An example produced by COLD CUTTING a raw block of dark-blue glass was made in Assyria in the 8th–7th century BC. A small hole has been ground out to insert the staff.

Stag Head, The. A glass ornament designed by Georges Chevalier, 1952, for the Cristalleries de Baccarat (see BACCARAT), weighing 20 kgs. Examples were presented to President de Gaulle, Leonid Brezhnev, and other heads of State.

stained-glass window. A decorative window, usually found in churches, made of pieces of COLOURED GLASS or glass that has been FLASHED or ENAMELLED, fitted into channelled lead strips, and set in an iron framework. The oldest are said to be in Augsburg cathedral, c. 1065, and the most renowned examples are in the cathedrals at Chartres and Le Mans, and in the Sainte-Chapelle, Paris. Medieval windows were made of small pieces of glass, but the size of the individual pieces has grown over the years. The usual process of production involves first the making of a cartoon with the design, then cutting the glass into shapes to fit according to the colours needed, then painting the pieces, leading them together, soldering and cementing to secure them, and fixing copper ties, with which to secure the panels to the iron cross-bars of the window. Glass coloured with metallic oxides was used, and also flashed glass. Sometimes details were defined with GRISAILLE, a brownish enamel made with iron oxide and fused on to the surface of the glass. The finest windows are those of the 13th and 14th centuries, made with coloured glass; those of the 16th century and later were made by painting with enamel colours on mainly colourless glass panels. The art declined but was revived in the 19th century with designs by Sir Edward Burne-Jones (1833–98) executed by William Morris (1834–96), e.g. in Birmingham Cathedral; by the American, John La Farge (1835–1910) and others; and in the 20th century by Marc Chagall (1889–   ). A modern development is the use of 'slab glass', cemented together instead of leaded, with the surface decorated by chipping. See Lee, Seldon and Stephens, *Stained Glass* (1976).

staining. The process of colouring the surface of ANNEALED glassware by the use of brushed coloured pigments that do not colour the entire body but merely sink into the surface, leaving it almost smooth. It was used on early ISLAMIC GLASSWARE but the technique employed has not been rediscovered, although some use was then made of silver stain and that technique has been used in medieval Europe (see SILVER). A later method of staining glass was developed by FRIEDRICH EGERMANN, using silver chloride to produce a yellow stain in 1820 and a red stain in 1840; this was intended as a cheap substitute for FLASHED GLASS, and the pieces were usually decorated with engraving or etching. The process was again used in England in the second half of the 19th century; the usual colours were brownish yellow (amber) and deep ruby, lightly fired, and the decoration was by engraving that revealed the thinness of the staining. The process was used only rarely in the United States.

Stachelglas. Greenish glass with prunts, Rhineland, 15th/17th century. Ht 7 cm. Römisch-Germanisches Museum, Cologne.

'*Stag Head*'. Crystal ornament, designed by Georges Chevalier, 1952. Cristalleries de Baccarat.

stacking glasses. Set designed by Tapio Wirkkala, 1970; made by Iittala Glassworks, Finland.

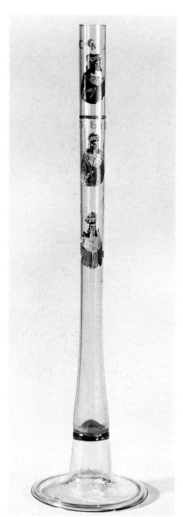

Stangenglas. Enamelled portraits of Electors, Germany, 17th century. Ht 42·5 cm. Kunstmuseum, Düsseldorf.

stamnium (Latin). A type of BOTTLE with a cylindrical BODY and a tall thin NECK having a disk mouth. The characteristic feature is that the two handles extend from the angle of the shoulder vertically upward and curve inward to be attached just below the level of the mouth. They were made of ROMAN GLASS in the 3rd–4th centuries.

stamnos (Greek). A type of storage JAR used in ancient Greece. It is ovoid in form, with a high shoulder, short NECK, wide MOUTH, and two horizontal loop-handles attached at the shoulder, and is sometimes provided with a COVER. It is similar to an AMPHORA of the neck type. Generally made of Greek pottery, some were made of ROMAN GLASS.

standard. A term used in the United States for the STEM of any glass.

Ständehumpen (German). Literally, profession glass. A type of HUMPEN with enamelled decoration depicting persons in various (4 or 5) professions. They were made in Franconia in the 17th century.

standing bouquet. A decorative motif in a paper-weight (*see* PAPER-WEIGHT, *bouquet*), in which the bouquet of glass flowers and leaves is three-dimensional and is vertical in relation to the base. *See* FLAT BOUQUET.

standing bowl. A large glass bowl (usually about 12 to 16 cm. high), with sometimes a slightly everted rim and standing on a wide hollow spreading FOOT. Examples occur in colourless glass decorated with an enamelled formalized diaper pattern, or in OPAQUE WHITE GLASS so decorated. Some were made in Venice in the 15th century, decorated as a WEDDING CUP. They are said to be for serving fruit or sweetmeats. A forerunner of such bowls may be the ISLAMIC GLASS bowls made in Syria in the early 14th century which stand on a wide hollow cylindrical stem attached to a splayed, slightly domed foot; this type has been called a 'footed bowl'. *See* STANDING CUP.

standing cup. A large glass cup (usually about 16 to 20 cm. high) with a somewhat cylindrical bowl resting on a wide hollow (sometimes RIBBED) spreading FOOT. Of the same general style are some GOBLETS with a tapering or conical bowl resting on a wide hollow spreading foot, with occasionally a KNOP in the stem, and sometimes a domed COVER. These are Venetian, from the 15th or early 16th centuries. They are often of coloured glass (blue, green, rarely sapphire), with enamelled decoration, often white dots and scale diaper patterns, and occasionally a procession or medallions with portraits or figures (a rare example has a Biblical scene). Such cups are of the same category as the STANDING BOWLS, but are usually taller and less massive. *See* PROCESSION CUP; BOLOGNA GOBLET; BAROVIER CUP.   [*Plate XIII*]

Stangenglas (German). Literally, pole glass. A tall narrow cylindrical BEAKER, usually with a pedestal BASE or a KICK BASE, made in a number of forms and with various styles of decoration. They are usually about 20 to 30 cm. high, but some are up to 42 cm. Early examples, of FRANKISH GLASS, 5th and early 6th centuries, are decorated with applied drops of glass (*see* KEULENGLAS), but later ones, made in Germany in the 15th to 17th centuries, are (1) smooth, with some spiral trailing or enamelled decoration, or (2) covered over the entire surface with PRUNTS (some RASPBERRY PRUNTS, some pointed prunts), sometimes with the prunts on the inside as well as the exterior of the glass. Many are of greenish or brownish WALDGLAS, probably from Germany but some from the Netherlands; some of nearly colourless glass are from Saxony, and others are Venetian of the 16th century, made for the German market, or early German or Bohemian somewhat in the *façon de Venise* and with enamelled decoration. The Italian terms are *bicchiere a canna* or *a tubo*. *See* KNOPFBECHER; SPECHTER; PASSGLAS; BANDWURMGLAS.

Stankard, Paul J. (b. 1942). An American maker of PAPER-WEIGHTS, working at Mantua, New Jersey. He specializes in weights with motifs of flowers, herbs, and other subjects from nature. All are dated and marked, and some are in limited editions.

star-cut base. The BASE of a PAPER-WEIGHT, ICE PLATE or other glass object decorated in the manner of a star FOOT.

star cutting. A decorative pattern on CUT GLASS, popular in the 18th and 19th centuries, in the form of a star of various styles, cut into the sides ('barrels') of some Irish DECANTERS and on the bottom of almost all CRUETS and many PAPER-WEIGHTS and other glass objects. The earliest form of the star was 6-pointed or 8-pointed, made by mitred grooves radiating from a central point and tapering toward the extremities, but variations were made with 12, 16, 24, or 32 points. Complicated variations are the BRUNSWICK STAR and the JEWEL STAR. Imperfect cutting results in the radiating grooves being of unequal length or not meeting precisely at the centre.

star-dust cane. A type of CANE used in making a PAPER-WEIGHT. It is made of a central ROD surrounded by six white rods to form a star shaped section or a corrugated shape, sometimes with a coloured star or dot in the centre, and is sometimes cased with coloured glass. They were used at BACCARAT to make a GROUND or sometimes with other canes as a centre motif.

star-holly glassware. A type of ART GLASS made of moulded OPAQUE WHITE GLASS with relief encircling bands of holly leaves and having the body of the object of a solid blue or green colour brushed on around the design, with a MAT surface. It was intended to simulate Wedgwood jasper ware. It was made in the early part of the 20th century by the Imperial Glass Co., Bellaire, Ohio.

stave. A flat, narrow strip of glass, coloured or opaque white, used in the decoration of some paper-weights. In one type (*see* PAPER-WEIGHT, *swirl*) long strips are used in a swirling pattern, and in others (*see* PAPER-WEIGHT, *millefiori*) short pieces are placed adjacent to the design (e.g., the concentric or flower type) and disposed vertically so as to form a kind of basket enclosing it.

Stein. A thick, heavy drinking glass, similar in form to a MUG, having a looped handle and, in addition, a hinged LID. The capacity is usually approximately one pint, but may be up to two pints. Such glasses have long been popular in Germany and elsewhere on the Continent for drinking beer.

stem. The part of a vessel, especially a drinking glass, that unites the BOWL to the BASE or FOOT. It may be long or short, and of various styles. The styles of stems on English WINE GLASSES changed gradually, from *c.* 1680 to after *c.* 1777, as to the inclusion of various KNOPS and TWISTS and the use of CUTTING, and characteristic features serve to identify the period of production. Many of the changes resulted from changes in the taxation on glass under the Glass Excise Acts (*see* EXCISE ACTS) of 1745 and 1777.

TYPES OF STEM

——, *air twist.* A stem having as its decoration an air twist (*see* TWIST, *air*).

——, *applied.* The same as STUCK SHANK.

——, *baluster.* A stem on a drinking glass or other vessel, basically in the form of the upright supports of a balustrade, somewhat pear-shaped but swelling slightly outward toward the narrow end. When the bulbous portion is in the lower part, it is known as a 'true baluster'; when it is in the upper part, it is known as an 'inverted baluster'. Sometimes there are two connected baluster forms, with either large ends facing or small ends facing. Drinking glasses made with such stems are known in England as balusters. *See* BALUSTER, LIGHT; BALUSTER, HEAVY; BALUSTROID.

——, *balustroid.* A type of stem on an English WINE GLASS, developed *c.* 1725–60 from the earlier balusters (*see* BALUSTER, HEAVY).

stem, hollow baluster. Detail of goblet with three hollow knops, Belgium, 17th century. Museum für Kunsthandwerk, Frankfurt.

stem, faceted. Stem of goblet, English, *c.* 1785. Courtesy, Delomosne & Son, London.

stem, figure. Goblet with blue prancing horse, Venice, 17th century. Ht 28·2 cm. Museum fur Kunsthandwerk, Frankfurt.

stem, knopped. Bobbin knops, with central beaded knop. Courtesy, Sotheby's London.

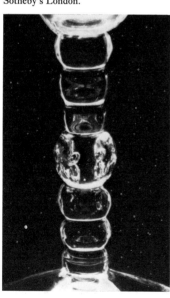

The stem is tall, thin, light, and most examples have either (1) a simple or double of triple KNOP, or (2) an inverted baluster shape (see above) with one or more knops. See BALUSTROID.

——, capstan. A type of stem usually found on English ALE GLASSES, c. 1800–30, somewhat in the form of a ship's capstan, with a cylindrical spool-like stem, a COLLAR between the stem and the bowl, and a larger ring above the foot. Also called a 'spool stem'.

——, cigar-shaped. A hollow circular stem tapering inward slightly toward the bottom. It is found on some Venetian TAZZE of the 16th/17th centuries, and was a common type made by SIR ROBERT MANSELL. See FUSIFORM.

——, collared. A stem decorated with an encircling COLLAR that is plain, faceted or VERMICULAR.

——, colour twist. See TWIST, colour.

——, composite. A stem of an English drinking glass that combines a section of AIR-TWIST and a section of plain clear glass, the latter often in the form of a KNOP. They date from between c. 1745 and c. 1765.

——, cut. A stem on a drinking glass that is decorated with CUTTING or FACETING. Before 1777 such cutting was unusual and elongated cut diamonds were used to save time and cost, but thereafter the diamonds became more equilateral, and cutting on the stems became frequent, and extended on to the BOWL. Some stems were given a series of encircling vertical connected plane surfaces (called a 'flat cut stem') or had SCALE FACETING or HEXAGONAL FACETING, and the faceting extended on to the KNOPS.

——, drawn. The same as DRAWN SHANK.

——, faceted. A stem that is decorated by FACETING which may be in various styles. Some examples have encircling vertical connected plane surfaces. Some have, in diaper style, diamond facets, higher than wide, or connected hexagons; others SCALE FACETING or shell faceting. Sometimes a single KNOP is in the centre of the stem, with the faceting continuing over it. Such stems are a late type on English glasses, c. 1750 to c. 1810. See flat cut stem, below.

——, figure. A stem, usually of a GOBLET, in the form of a figure, e.g., a prancing horse standing on his rear legs and supporting the bowl on its head; such pieces were made at MURANO, with the horse in colourless or coloured glass or in spattered colours. Some goblets have a stem in the form of a human figure, e.g. an elaborate goblet of Venetian glass, 17th century, of which the stem is a figure of a hunchback. See SILHOUETTE GLASSWARE.

——, flat cut. A stem that has encircling vertical connected plane surfaces. Sometimes the FACETING is waisted, sometimes bulging; and sometimes the edges of the facets are notched.

——, hollow. A type of stem (on an English drinking glass) that is in the form of a hollow cylinder closed at the ends by the joining of the BOWL and FOOT, respectively. It must be distinguished from the plain stem (see below) with an elongated TEAR. There are no KNOPS but sometimes an occasional swelling of the cylinder midway between bowl and foot. Such stems are rare, and were made c. 1745–60.

——, incised twist. A stem having decoration in the form of an incised TWIST. Some such stems have a KNOP. They date from c. 1750 to c. 1775.

——, inverted baluster. See baluster stem, above.

——, knopped. A stem on a drinking glass that is decorated with one or more KNOPS, which may be in various styles, either alone or in conjunction with a TWIST or FACETING, and also sometimes with COLLARS or MERESES. The twists or faceting usually continue unchanged through or over the knops. The knop may be solid or hollow, and some of the latter variety have a coin inserted (see COIN GLASS).

——, laddered. A stem of inverted baluster shape (see above), on which there are four vertical columns of small cut squares, each making a ladder-like pattern; between each column there is cut decoration.

——, opaque twist. A stem whose decoration includes any one of a variety of TWISTS made of OPAQUE WHITE GLASS in the form of THREADS or TAPES. They succeeded air-twist stems (see above) in favour, and continued until c. 1777, whereupon cut stems (see above), succeeded them. Examples have a simple single-spiral twist, then a gauze twist, a corkscrew twist or a double-series twist.

——, *pedestal*. *See* Silesian stem, below.

——, *plain*. A stem without any decoration in the form of KNOPS or TWISTS, but sometimes with a TEAR.

——, *quatrefoil*. A stem on a drinking glass that is of inverted BALUSTER shape and is usually hollow (sometimes solid), with the wider part pinched into four vertical lobes, with sometimes the sides of the lobes impressed with HATCHING. Examples are recorded in some fragments of wine glasses bearing the raven's head seal of GEORGE RAVENSCROFT.

——, *rudimentary*. A type of stem that is plain and very short, usually only a KNOP between the BOWL and the FOOT, or a plain stem with a small knop or MERESE or COLLAR.

——, *shouldered*. *See* Silesian stem, below.

——, *Silesian*. A type of MOULDED stem, made in England and Germany, that is ribbed and shouldered. Early examples, from *c.*1715, are four-sided; next they were six-sided and, from *c.*1725–30, eight-sided, and then 'debased' with rounded shoulders. They were not used often after *c.*1740 on WINE GLASSES, but more so on heavier TAZZE, CANDLESTICKS, and SWEETMEAT GLASSES. Sometimes an inscription ('Long live King George') or a crown, sceptre, or the letters 'G R' in moulded relief, appear on the shoulders. After *c.*1725–30 some Silesian stems on wine-glasses are combined with one or two knops, but knops are not found on sweetmeat glasses. Such stems were made from *c.*1715 to *c.*1765; they have no connection with Silesia, but were first made (except for a possible early example from Venice) in Thuringia or Hesse, whence George I came to the throne of England in 1714, and they were possibly so-called to honour him. They are sometimes called a 'pseudo-faceted stem', a 'pedestal stem' or a 'shoulder stem'.

——, *straw*. The same as drawn stem or DRAWN SHANK.

——, *stuck-shank*. The same as STUCK SHANK.

——, *tear*. A stem that has embedded within it in an air bubble in the form of a TEAR, sometimes elongated or twisted.

——, *wire*. A stem that is tall and very thin (about 7 mm. in diameter) that is made by being drawn from a TRUMPET-SHAPED bowl of a drinking glass. It is found mainly on a TOASTING GLASS, appropriately, as it was an 18th-century custom for the user to snap the stem after offering a toast, and hence few examples survive.

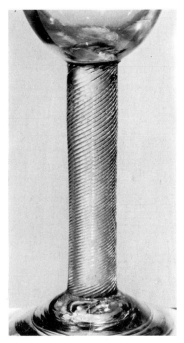

stem, incised. English. Courtesy, Derek C. Davis, London.

stem-gatherer. A member of the team (CHAIR or SHOP) assisting the GAFFER in making objects such as wine-glasses.

stem-maker. A member of the team (CHAIR or SHOP) assisting the GAFFER in making objects such as wine-glasses.

stem, Silesian. Stem of wine glass, English, *c.*1745–50. Courtesy, Delomosne & Son, London.

stem-work. Intricate patterns of glass THREADING that constitute the entire STEM of some drinking glasses without any vertical supporting column, e.g. the stems of WINGED GLASSES and threaded stems suggestive of serpents and dragons on some glasses in the *façon de Venise*.

Stengelglas. The German term for a glass with a stemmed FOOT, e.g. a GOBLET, POKAL, etc.

step cutting. A style of CUTTING in the form of adjacent horizontal angular REEDING. It is found on some English and Irish vases, jugs, celery glasses, etc.

step-stemmed goblet. A type of GOBLET characterized by a stem consisting of a series of about six thick adjacent graduated rings, which diminish in size toward the foot. They were made, sometimes of coloured glass, in Bohemia, *c.*1850.

step cutting. Piggin, Georgian, *c*. 1820.
Courtesy, Delomosne & Son, London.

Steuben Glass Works. The leading American maker of ornamental glassware (now called Steuben Glass, Inc.). It was founded as Steuben Glass Works, at Corning, New York, in 1903 by FREDERICK CARDER, together with Thomas G. Hawkes and Mr & Mrs Willard Reed, for the purpose of making ornamental glass; it was acquired in 1918 by the CORNING GLASS WORKS to become its ART GLASS division. In 1932 the production manager, Robert J. Leavy, created a new transparent METAL of great brilliance that became the crystal glass used in later productions. During his 30-year regime Carder created new types of glass of many colours and surface effects and produced ware in innumerable forms. By 1933, years of financial loss led to a reorganization under ARTHUR AMORY HOUGHTON, JR (a great-grandson of the founder of the Corning Glass Works) with the intention of changing from coloured glass to exploiting the new crystal glass by creating new forms and using engraved decoration. Steuben Glass, Inc., was formed, with Houghton to direct policies. He promptly appointed JOHN MONTEITH GATES as Chief Designer and the sculptor SIDNEY B. WAUGH as first designer; in 1936 GEORGE THOMPSON took charge of a newly created Design Department. International recognition soon followed. From the beginning of this period tableware was made, emphasizing elegance, with some services specially created for individual customers. However, ornamental pieces of imaginative design and with skilled engraved decoration came to be Steuben's forte. In 1937 the company produced a series of pieces designed by 27 contemporary artists, in 1954 a group designed by 20 British artists and sculptors, and in 1956 designed by Asian artists. Generally it has relied on its own staff designers, including Gates, Waugh, and Thompson, and later LLOYD ATKINS, DONALD POLLARD, PAUL SCHULZE, JAMES HOUSTON, and DON WIER. In 1963 it created a series inspired by 31 commissioned works of poetry. A recent development is the use of gold ornament made by leading jewellers to accent the crystal-glass piece, e.g., the MYTH OF ADONIS. Among outstanding pieces produced in the Houghton period are the GAZELLE BOWL, the VALOR CUP, the TROUT AND FLY, the GREAT RING OF CANADA, and the SEA CHASE. Unique pieces have been created for Presidential presentation (from Truman to Ford), e.g., MERRY-GO-ROUND and FOREST SPIRES. In 1973 Houghton retired and was succeeded by Thomas S. Buechner. [*Plate XIV*]

Stevens & Williams, Ltd. A glasshouse at Brierley Hill, near Stourbridge. Its origin dates back to the Moor Lane Glass House at 'Briar Lea Hill' owned by a Mr Honeyborne, whose daughter married in 1740 John Pidcock, a nephew of the last of the Henzey family in Stourbridge who came from Lorraine with JEAN CARRÉ. In the early 19th century it was leased to Joseph Silvers, whose daughters married respectively William Stevens and Samuel Cox Williams, both of whom became principals in the firm in 1847, changing its name to Stevens & Williams, Ltd. From 1932 KEITH MURRAY was engaged as designer, but the company continued to make table services of LEAD GLASS in well-established CUT GLASS styles, first known as BRIERLEY GLASSWARE and, since receiving its first Royal Warrant in the reign of George V, as 'Royal Brierley'. Tom Jones was appointed chief designer in 1956. One of the firm's specialities for many years has been COMMEMORATIVE GLASS-WARE for royal and international events. *See* ALEXANDRITE; SILVERIA GLASSWARE.

Stiegel, Henry William (1729–85). A native of Cologne, Germany, who settled in the United States in 1750 and whose American father-in-law, a wealthy ironmaster, financed him in 1763 to start a glassworks at Elizabeth Furnace, Pennsylvania. He started another in 1765, and a third, the American Flint Glass Works, in 1769, both at Manheim, Pennsylvania, employing English and German craftsmen. He at first made bottles and window glass, but later SODA-LIME GLASS, LEAD GLASS, and COLOURED GLASS, especially blue and amethyst, usually MOULD-BLOWN (often with the daisy pattern – *see* DAISY GLASSWARE); he also decorated glassware with enamelling and wheel-engraving. The

Steuben glass. 'Flying Eagle', designed by Paul Schulze, 1975. Ht 36·7 cm. Steuben Glass, New York.

ware was of good quality (comparable with European glassware) but, due to extravagant management and competition from imported glassware, the factories were sold in 1774. *See* STIEGEL-TYPE GLASSWARE.

Stiegel-type glassware. Glassware made in the style and tradition of that made at the factories of HENRY WILLIAM STIEGEL, being of the same technique and decoration, but not of any assignable factory or area, or even indigenous to the region (Pennsylvania), some coming from as far away as Ohio. The ware was skilfully made, of clear and coloured glass (some LEAD GLASS), with decoration in varied styles, including WHEEL ENGRAVING, ENAMELLING, and PATTERN MOULDING, following English and Continental styles. Styles included DAISY GLASSWARE and SUNKEN PANEL glassware. Some similar imported glassware is indistinguishable from the American products. Glassware of the Stiegel type made at Zanesville, Ohio, by craftsmen from Manheim, Pennsylvania, has been called 'Ohio-Stiegel'.

stippling. The process of decorating glass by striking a diamond or hardened steel point against the glass, so that the image desired is produced by many tiny shallow dots ('stipples'), but sometimes by short lines, indented in the surface of the glass. The graded dots provide the highlights; the intensity varies with their closeness, and the shadows and background result from leaving the polished glass untouched (producing an effect similar to that of mezzotint engraving, which was invented in 1643). The technique of stippling was introduced *c.* 1621 in Holland by ANNA ROEMERS VISSCHER, but it was first used as the exclusive means of decorating by FRANS GREENWOOD, *c.* 1720. Only Dutch engravers, mainly amateur artists, used stippling in the 18th century (*see* DAVID WOLFF, G. H. HOOLAART, AERT SCHOUMAN, JACOBUS VAN DEN BLIJK). Much of the glass used for stippling came from Newcastle, England, it being more suitable for the medium. In the 19th century, other Dutch stipplers were ANDRIES MELORT, DANIEL HENRIQUEZ DE CASTRO (d. 1863) and E. Voet. In England stippling has been revived since 1935 by LAURENCE WHISTLER and is now also practised by his three children and by other decorators.

stirring rod. A thin rod of glass for stirring; such rods, of varying length (from 15 to 24 cm.), are sometimes twisted and have the ends flattened (sometimes one is pointed). They were made of ROMAN GLASS in the 1st/2nd centuries AD.

stirrup cup. A drinking vessel which is an adaptation of the ancient RHYTON. It is without HANDLE or FOOT, and must rest (inverted) on its rim when not in use. They were made in glass as well as in porcelain, pottery, and silver, in many forms, such as various animals. One pair made in Potsdam, early 18th century, has FINIALS in the form of the head of a nun and a monk respectively. (Strictly, the term refers to the final drink taken by a mounted rider or huntsman about to depart, but it has come to apply also to this special type of drinking vessel which was sometimes used on such occasions.) The stirrup cup has sometimes been termed a 'hunting-glass' or a 'coaching-glass'. *See* DECANTER, *stirrup*.

Stölzle Glasindustrie. An Austrian company that has had for many years a number of glass factories throughout Austria making a wide range of industrial and household, and some ornamental, glassware. It was founded by Carl Stölzle in 1843. Its main factories are at Köflach and Nagelberg.

stone. A speck of foreign matter, sometimes found in a BATCH of MOLTEN GLASS, which must be removed, especially in the making of high-quality crystal glass.

stopka (Russian). A type of BEAKER that is tall and cylindrical, slightly tapering toward the bottom, and having a flaring mouth and thin FOOT

Stiegel-type glass. Flask with spiral ribbing. Henry Ford Museum, Dearborn, Mich.

stippling. Bowl of goblet with stippled decoration by Frans Greenwood, 1722. Rijksmuseum, Amsterdam.

stoppers. Assorted decanter stoppers, late 18th/early 19th centuries. Courtesy, Delomosne & Son, London.

stopper, bull's eye. Irish glass, *c.* 1820. Courtesy, Delomosne & Son, London.

stopper, hollow. Club-shaped plain decanter, and stopper with open-end hollow peg, *c.* 1795. Courtesy, Delomosne & Son, London.

RING. They were made at several Russian factories in the 11th and 12th centuries.

stopper. A matching piece that fits into the MOUTH of a vessel (e.g., a DECANTER or CRUET) to close it and prevent the contents from evaporating. They are made in various shapes, sizes, and styles. Some, made *c.* 1750, are ground to fit more securely, and a few screw into the NECK of the vessel. Often the stopper for a decanter was lost or broken, and hence replacements are frequent, some being good REPRODUCTIONS of the original.

TYPES OF STOPPER

———, *bull's-eye.* A stopper in the form of a vertical flat disk with a circular depression in the centre that is sometimes a DIMINISHING LENS. Some have the outer area plain, others have cut or moulded ridges radiating from the eye to the rim. They were made in the 19th century. Sometimes called a 'target stopper'.

———, *conical.* A stopper that is conical in shape, but not as high as a spire stopper (see below). Some are 'conical-faceted', with overall faceting.

———, *disk.* A type of stopper that came into use after the spire stopper (see below). It is flat and circular, with its edges sometimes plain but more often scalloped or notched.

———, *grid.* A stopper that is vertically flat, and has ridges vertically on one side and horizontally on the other side, so that, seen together, the ridges form a grille. Such stoppers were made of PRESSED GLASS.

———, *hollow.* A stopper that is hollow and, when open at the bottom, convenient for a quick nip.

———, *initialled.* A stopper that is decorated with an initial to correspond with the name of the contents of the vessel, as also painted on the vessel. Sometimes the full name appears on the stopper. *See* BOTTLE LABEL.

———, *lapidary.* A stopper in the form of a circular knob, FACETED over its entire surface. It is most often found on a square decanter (*see* DECANTER, *whisky*).

———, *lozenge.* A type of stopper that came into use after the disk stopper (see above), to which it was similar but more elongated into a lozenge form. Some have FACETED decoration near the edge and some are of scalloped or other decorative shape. They were used mostly on decanters with long sloping shoulders.

———, *mushroom.* A stopper that is mushroom-shaped with a flat or domed top, and usually having moulded ribbing radiating on the top from the centre to the rim. Some have cut decoration, but most are moulded. They were made after the disk and lozenge types (see above), *c.* 1780–1825, usually of ANGLO-IRISH GLASS. On some the mushroom rests on a plain short shank, but sometimes on a globular knop.

———, *pinnacle.* The same as a spire stopper (see below).

———, *spire.* A stopper with a high point, early examples being cone-shaped and uncut, but later ones almost entirely cut with FACETING or with hollows forming a shallow-diamond pattern. They are sometimes found on whisky DECANTERS. They are an early form, made *c.* 1750, shortly after stoppers first came into use for decanters. Also called a 'pinnacle stopper'.

———, *target.* The same as a bull's-eye stopper (see above).

stoup, holy-water. A small stoup or font for holy water. Often made of faience, they have been made of glass in France (called a *bénitier*), Spain, and the Netherlands in the late 17th and the 18th centuries, but apparently not in England. They are usually made to be hung on a wall.

Stourbridge Flint Glass Works. A glasshouse at Pittsburgh, Pennsylvania, founded in 1824 by John Robinson; its name was changed in 1830 to J. & T. Robinson, and in 1836 to Robinson, Anderson & Co. It made FLINT GLASS, and OPALESCENT GLASS, and later PRESSED GLASS and green glass. The factory closed in 1845. *See* DOOR KNOB; ROBINSON SALT CELLAR.

Stourbridge glassware. Glassware made in England at or near Stourbridge, Worcestershire, first made by the descendants of the emigrant glass-makers who came from Lorraine with JEAN CARRÉ, *c.*1567, initially to the Surrey-Sussex region and then moving to the Stourbridge area. In 1618 Paul Tysack had a glasshouse there. At first only window glass was made but by the 18th century fine crystal glass was being made. Among the factories there today are STEVENS & WILLIAMS, LTD; STUART & SONS, LTD; THOMAS WEBB & SONS; and WEBB, CORBETT, LTD.

strain-cracking. Deterioration in a glass object having the appearance of tiny internal cracks in the wall of the object caused by stresses and strains due to faulty ANNEALING.

strass. A brilliant LEAD GLASS used in the manufacture of artificial gems. It is a borosilicate of potassium and lead, with small quantities of alumina and arsenic. When uncoloured it is used for imitation diamonds, being TRANSPARENT and very refractive, although much softer; the addition of metallic oxides and salts produces coloured strass to imitate most known precious stones. It is named after Georges-Frédéric Strass (1701–73) who was born at Wolfisheim, near Strasbourg, where he learned to make jewellery and artifical gems; he moved to Paris in 1724, and there invented, from 1730 to 1734, new techniques for making imitation precious stones (imitation diamonds having been known in Paris in the 17th century). He ceded the making of such ware in 1752 to Georges-Michel Bapst (1718–70), husband of his niece. Such artificial gems are now made mainly in Czechoslovakia, Austria, and France. Many writers and standard dictionaries have attributed the invention to a Joseph Stras(s)(er) of Vienna, but no record of the existence of such a person is known, and the legend about him has been discredited. *See* Hans Haug, 'Les Pierres de Strass', *Cahiers de la Céramique*, No. 23 (1961), p. 115. *See* PASTE JEWELLERY.

Strathearn Glass, Ltd. A glasshouse founded in 1963 to take over the business of VASART GLASS, LTD., after a major interest in it had been acquired by the Scotch whisky distillers, Wm. Teacher & Sons, Ltd. The operations and personnel were moved from Perth in 1964 to a new factory built at nearby Crieff. It continued to make PAPER-WEIGHTS, some in new designs and bearing the Company's initial and the date; it also produces objects in coloured glass and the 'Glenarty' range of heavy clear glassware. In 1968 Stuart Drysdale, who had reorganized Vasart Glass, Ltd., left with some of the personnel and in 1970 started a new factory, PERTHSHIRE PAPERWEIGHTS, LTD., making only paper-weights.

strawberry diamond cutting. A variation of RAISED DIAMOND CUTTING, made by leaving an uncut space between the diagonal grooves so that a flat area results instead of pointed diamonds, then cutting the flat areas with cross-hatched grooves to make them into a group of (often 16) low-relief diamonds. It is sometimes erroneously called a 'hobnail pattern'. A further and more complicated pattern is made by making the original grooves in pairs (or threes), so that a new ridge is made between the rows of strawberry diamonds. A simplified variation, called 'chequered diamonds', has the flat areas cut in groups of only four small diamonds instead of the usual sixteen. *See* FLAT DIAMOND CUTTING.

strawberry prunt. A type of PRUNT that is slightly convex and has relief decoration, impressed with a tool, resembling a strawberry. Examples appear on glassware made probably in England at the glasshouse of the DUKE OF BUCKINGHAM, *c.*1660–63. *See* RASPBERRY PRUNT.

striae. Undulating or cord-like markings on glassware, caused by variation of temperature in the furnace or by the unequal density of the materials used. *See* STRIATED DECORATION.

Strathearn Glass, Ltd. Paper-weight, double overlay millefiori type, with six windows. Courtesy, Strathearn Glass, Ltd.

striated decoration. An intentional style of decoration on glassware in the form of a series of parallel lines in the METAL, often lines that are curved. It occurs on English ware of the late 18th and early 19th centuries. *See* STRIAE.

striking. The process of reheating glass after it has cooled, in order to develop a colour or an opacifying agent, e.g. in making GOLD RUBY GLASS.

string rim. A ring of trailed glass encircling the NECK of a WINE BOTTLE or DECANTER BOTTLE just below the MOUTH, to facilitate the attaching of a string or wire to secure the cork in the bottle; it was used before the introduction of the close-fitting 'flush cork'.

stringing. The same as TRAILING.

striped glassware. (1) Hollow glassware decorated with parallel stripes of opaque white or coloured glass, sometimes spiralling around the piece as a result of twisting it after glass rods had been embedded in the body. Such glassware was made in the 19th century, and a patent for a special process was issued in England in 1885 to William Webb Boulton, of Boulton & Mills, at the Audman Bank Glasshouse. (2) The term is sometimes applied to VETRO A RETORTI, and particularly in the United States to glassware in that style made by NICHOLAS LUTZ of the BOSTON & SANDWICH GLASS CO.

Strömberg, Edvard (1872–1946). A Swedish glass technician who from 1917 to 1918 was manager of the KOSTA GLASSWORKS, and from 1918 to 1928 was technical director at the ORREFORS GLASBRUK. From 1928 he and his wife Gerda (d. 1960) directed the EDA GLASSWORKS. In 1933 he and his wife founded their own glasshouse, Strömbergshyttan, in the Småland glass-making region of Sweden, where they made colourless and sometimes slightly tinted glass, in forms that she designed, depending for their effect on the graceful lines and varying thickness of the material. In 1953 Gunnar Nylund (b. 1904) joined the glasshouse.

Stuart (or Stewart), Charles Edward, glassware. Glassware decorated with the portrait of Charles Edward Stuart (1720–88), the grandson of James II and known as Bonnie Prince Charlie or the Young Pretender. He led the Jacobite movement to restore his father until defeated at the battle of Culloden (1746). His portrait is found on some JACOBITE GLASSWARE, usually wheel-engraved but sometimes an enamelled BUST, decorated either in three colours (red, blue, white) or in multiple (i.e. five) colours (blue, green, yellow, red, white). The portrait is from an engraving by Sir Robert Strange.

Stuart & Sons, Ltd. A glasshouse at Stourbridge, started in 1881 by Frederick Stuart (1817–1900), who had been associated with the glass industry since 1828. It has been continued until the present by his descendants. It formerly occupied the premises used by the RED HOUSE GLASSWORKS and the WHITE HOUSE GLASSWORKS, now superseded by a new glasshouse. Although originally it made coloured glass and chandeliers, it has since World War II made only crystal glass. Among its designers have been Ludwig Kny and later ERIC RAVILIOUS and other artists; since 1948 the chief designer has been G. John Luxton.

stuck shank. The STEM of a drinking glass that is made separately from the BOWL and is attached to the bowl, in contrast to a DRAWN SHANK. In this method the glass is made in three parts, the bowl, the stem, and the FOOT. *See* SHANK GLASS.

studio glassware. A class of glassware made by artist-craftsmen rather than by glass-makers in a factory, the ware being made and decorated, and usually signed by the designer or under his immediate supervision.

Sturzbecher. Inverted glass with silver whistle, Netherlands, 17th century. Ht 22·6 cm. Museum für Kunst und Gewerbe, Hamburg.

Some studio artists developed a new formula for the glass, so that it can be melted at a temperature low enough to be practicable in the average studio. In the 19th century there were MAURICE MARINOT, EUGÈNE ROUSSEAU, and JOSEPH BROCARD, and in the 20th century are HARVEY LITTLETON and DOMINICK LABINO. In 1962 the Toledo (Ohio) Art Museum sponsored a group to work in this manner.

Sturzbecher (German). Literally, somersault glass. A type of drinking glass that has a STEM but no FOOT, so that when it is emptied it must be stood inverted on the table (like a RHYTON or a Siegburg stoneware *Sturzbecher* or an English STIRRUP CUP). Early examples are of FRANKISH GLASS, made *c*.400–*c*.700; later ones (sometimes double) were decorated by GEORG SCHWANHARDT at Nuremberg in the mid-17th century. *See* DOPPELSTURZBECHER; SPITZBECHER; WHISTLE GOBLET.

stylized. Modified, as a decoration, from the natural appearance of an object to a more abstract pattern or conventionalized style of expression.

sugar-bowl. A BOWL, sometimes with a cover, for serving sugar. Such bowls were often made in porcelain or pottery, but some examples were made of glass, including OPAQUE WHITE GLASS. They are often cylindrical, globular or pear-shaped, resting on a flat bottom, spreading BASE, a knop-stemmed FOOT, or small bun feet, and with or without two side handles. They were sometimes made *en suite* with a TEA SERVICE.

sugar-castor. A CASTOR, usually of baluster shape or cylindrical shape, with a domed COVER (which was sometimes screwed to the container) pierced for the sprinkling of sugar. Sometimes called a 'sugar dredger'. The French term is *saupoudrière* or *saupoudroir*.

sugar-crusher. An ornamented rod, *c*.10 cm. long, with a larger flattened end for crushing sugar in a glass of toddy; in form it is similar to a PESTLE. They were made in England in the 19th century. Also called a 'toddy-stick'.

suite, en (French). Shaped and decorated in like fashion to other pieces to form a SET, e.g. a dinner service or a tea service.

sulphide. A silvery-looking opaque MEDALLION enclosed in transparent glass. The process had been attempted *c*.1750 in Bohemia, but was successfully introduced in France by a Frenchman, BARTHÉLEMY DESPREZ, at Paris; in the last quarter of the 18th century he made fine cameos and medallions of prominent people in a white porcellanous material. In the early 19th century he enclosed them in transparent glass that, because of a thin layer of air under the glass, gave them a silvery appearance. These were set into plaques of glass or were framed to hang on the wall or were inserted in the sides or bottoms of GOBLETS, BEAKERS, PAPER-WEIGHTS, and other objects. In England, APSLEY PELLATT made similar ware which he called 'cameo incrustations' or 'crystallo-ceramie', and obtained a patent for the process in 1819. Such ware was also made at BACCARAT, and CLICHY, and other French factories, and was copied in Bohemia and Germany, *c*.1830. It was also produced *c*.1814 by BAKEWELL'S at Pittsburgh, Pennsylvania, with profile portraits of prominent citizens. Examples depicting prominent persons of the past and present are being made at BACCARAT. *See* PAPER-WEIGHT, *encrusted*.

sunburst pattern. A decorative pattern, on some FLASKS of MOULD-BLOWN glass made in the United States, in the form of a sun surrounded by radiating rays, often with the points of the rays within a plain or scalloped oval border. Sometimes the rays almost meet at the centre, sometimes the centre consists of a plain oval, concentric rings or other decorative device.

sugar-bowl. Opaque white glass with enamel decoration, Bohemia, *c*.1790. Museum of Decorative Arts, Prague.

sulphide. Scent bottle with sulphide by Apsley Pellatt, English, *c*.1820–30. British Museum, London.

Sunderland Bridge glassware. Tumbler with engraved scene and printies. English, 1797–1800. Courtesy, Derek C. Davis, London.

sweetmeat glass. Panel-moulded bowl with dentil rim, opaque white twist stem, and domed and folded foot, *c.*1706. Ht 17 cm. Courtesy, Delomosne & Son, London.

Sunderland Bridge glassware. English glassware (mainly drinking vessels such as TANKARDS, RUMMERS, and TUMBLERS, but also other objects, e.g. ROLLING PINS) with engraved decoration depicting the famous Iron Bridge over the River Wear (actually the Wearmouth Bridge, not the Sunderland Bridge) which was opened in 1796. The view, usually inscribed 'Sunderland Bridge', generally shows a ship (sometimes more than one) sailing under the bridge, and the decoration sometimes includes Masonic or other emblematic subjects or sentimental dedicatory inscriptions and verses. They were made *c.*1797–1840.

sunken panel. A decorative style in the form of adjacent or close vertical depressions in the form of arches encircling a receptacle, made by mould-blowing a single gather of METAL or sometimes a second gather. It is found on American glassware, some of which may have come from the glasshouse of HENRY WILLIAM STIEGEL. *See* PANELLING.

swag. A decorative pattern in the form of a FESTOON, often in relief.

swan finial. A type of FINIAL, so-called even though it more nearly resembles a duck or chicken. It is often found on the cover of a sugar-bowl.

Swedish glassware. Glassware made in Sweden in the 16th to 19th centuries, generally copied from German, Bohemian and English forms and styles, and modern glassware made in the 20th century. In 1676 Kungsholm Glassworks was founded in Stockholm for making glassware in Venetian style, but from *c.*1700 it produced ware in German and Bohemian styles, with engraved decoration, until it closed in 1815. KOSTA GLASSWORKS was founded in 1742, REIJMYRE GLASSWORKS in 1810, and BODA GLASSWORKS in 1864. Today the leading factories are KOSTA-BODA and ORREFORS GLASBRUK. Over one hundred small glassworks have operated in the glass-making district near Växjö in Smaland, of which over thirty-five are still producing. Sweden is noted for glassware called 'Swedish Modern' made from 1917, especially pieces of original design and of artistic ornamental styles, as well as superior tableware.

sweetmeat glass. A tall stemmed receptacle, used in England in the late 17th to early 18th centuries, for serving (toward the end of a meal) various kinds of sweetmeats, probably 'dry sweetmeats', such as chocolates, nuts, cachous, candied or dried fruits, etc. (as distinguished from 'wet sweetmeats', such as trifle, etc. eaten with a spoon and more likely served in JELLY GLASSES placed on a SALVER around the central sweetmeat glass). (*See* PYRAMID.) Sometimes the central sweetmeat glass held an orange (and was called an 'orange glass'). Sweetmeat glasses were made in a variety of forms and styles, but often the wide low bowl was of DOUBLE-OGEE shape, the foot shaped or DOMED, and many had a Silesian STEM or, more rarely, an opaque twist stem, and often there was CUT decoration. The rim of the bowl was sometimes flanged or ornamented, e.g., scalloped or serrated, or with LOOPED TRAILING; the type with a plain level rim, suitable for use as a drinking glass, has sometimes (perhaps because of the wide low bowl) been erroneously identified as a 'champagne glass'. From the mid-18th century, sweetmeats were sometimes served in a larger DESSERT GLASS or in small bowls suspended from an ÉPERGNE. A small sweetmeat glass is usually termed a COMFIT GLASS.

sweetmeat stand. A large composite piece to be used as a table CENTRE-PIECE, consisting of a stand or EPERGNE from which hang several small dishes for sweetmeats.

swizzle stick. A thin glass rod with some type of enlarged end that is used to stir a short drink to remove its effervescence; it was originally used to stir swizzle (a sweetened alcoholic drink made with rum) but in

XV *Venetian glassware*. Group of 16th-century glass objects (left to right): blue vase with *vetro a retorti* decoration; plate with enamelled decoration showing the coat-of-arms of a Medici Pope; standing cup with enamelled and gilded decoration of grotesques; dragon-stem goblet, ht 26·2 cm. (10¼ in.). Corning Museum of Glass, Corning, N.Y.

XVI *Zwischengoldglas* (gold sandwich glass). Beaker of transparent colourless glass in two layers, the inner one flashed with red, the outer one decorated on the inside with engraved gold- and silver-leaf flowers, and on the outside with vertical faceting; Southern Germany, *c.* 1700. Courtesy, Cinzano Collection, London.

modern times (although deprecated by œnophiles) to stir champagne. *See* TODDY STICK.

sword. A full-size replica of a sabre or other type of sword, usually made of clear glass with an ornamented guard and grip. They were made as FRIGGERS in England from the late 18th century to the mid-19th century, and were used in glass-makers' processions in Bristol and Newcastle.

sword guard. A small slitted protective object to be fitted around the blade of a sword. They were sometimes made of glass in China, during the Han Dynasty (206 BC–AD 220).

sword slide. A grooved ornament to slide over an iron sword. They were sometimes made of glass in China, imitating jade, in the Han Dynasty (206 BC–AD 220).

syllabub (or **sillabub**) **glass**. A vessel for drinking syllabub (a beverage formerly popular in England, made from fresh cream whipped to a froth, with added sherry, ratafia, and spices). The early practice, *c*. 1677, was to drink it from a POSSET POT. From *c*. 1725 special glasses were introduced (sometimes called 'whips' from 'whipt-syllabub'), similar to the posset pot but without the spout and with only one handle or none. Then, *c*. 1770, they became similar to the JELLY GLASS, some being engraved with an 'S' to distinguish them.

Syrian glassware. Glassware made in Syria from the early SIDONIAN GLASSWARE until the end of the period of ISLAMIC GLASSWARE in 1402. The art of glass-blowing was first introduced in Syria, *c*. 100 BC and, possibly after a period of freely-blown glass, the process of MOULD-BLOWN GLASS was developed. Syrian glass includes that attributed to RAQQA, ALEPPO, and DAMASCUS. *See* SYRO-FRANKISH GLASS; CHINESE-ISLAMIC GLASSWARE.

Syro-Frankish glassware. A type of ISLAMIC GLASSWARE of uncertain origin, once attributed to Venice but more recently ascribed to Syrian craftsmen working at an unidentified Frankish Court during the period of the Crusades, 1260–90. It includes a group of twenty-five enamelled BEAKERS such as the ALDREVANDINI BEAKER. *See* HOPE BEAKER.

sweetmeat stand. English, *c*. 1770. Ht 66 cm. Courtesy, W. G. T. Burne Ltd, London.

sword guard. Chinese, Han Dynasty (206 BC–AD 220). British Museum, London.

T

table-bell. *Filigrana*, with silver mount, Venice, *c.* 1600. Ht 20·5 cm. Kunstsammlungen der Veste Coburg.

tankard. Blue and white enamelling depicting 'Virgin and Child', Bohemia, dated 1684. Museum of Decorative Arts, Prague.

table-bell. A bell with an ornamental FINIAL rather than a vertical handle, such as on a HAND-BELL. Some were made in Venice or in the *façon de Venise* from the 17th century. An unusual table-bell is in the form of a DOUBLE WINE GLASS, with the clapper hanging in the lower glass so that the bell tinkles when the glass is tipped for drinking. The German term is *Tischglocke*.

table-lustre. The same as a LUSTRE (1).

table-ornament. The same as a CENTRE-PIECE.

taker-in. The apprentice on the team (CHAIR or SHOP) assisting the GAFFER in making a glass object, who in particular takes the finished object to the LEHR.

Tanagra-style figures. Replicas or adaptations of terracotta Tanagra figures, made of PÂTE-DE-VERRE by such French glass-makers as GABRIEL ARGY-ROUSSEAU, GEORGES DESPRET, and ALMERIC-V. WALTER. The original figures, made of terracotta, were moulded in several centres in Greece and at colonies in Italy and Asia Minor, the examples of highest quality being those discovered in 1874 in the Tanagra district of Boeotia; these date from the latter part of the 4th century BC onwards, and represent men and women, and gods and goddesses (the former usually in costumes of the period), the usual height being about 17 to 20 cm.

tankard. A drinking vessel with a handle on one side, a plain circular RIM, and usually a hinged metal LID. Some have a detached COVER. The prescribed capacity of an old tankard was about a quart (approx. 1·4 litres); anything smaller should be called a MUG. Generally they were of stoneware, but some were made of glass, e.g. in Venice. *See* BURGHLEY, LORD, TANKARD; TANKARD, COVERED.

tankard, covered. A TANKARD that has, instead of the usual hinged metal LID, a separate glass COVER. English examples were made in the early 18th century.

tape. A thin flat strip of glass, such as is used in making certain stems (*see* STEM, *twist*).

taperstick. A holder for a taper or thin candle, usually with a flat or domed circular BASE and a candle-socket or NOZZLE of small diameter. It is infrequently accompanied by a snuffer. A taper was used mainly for lighting candles or pipes, but some tapersticks may have been intended to be used as a small TEA-CANDLESTICK on a tea-table. They were usually made in the style of contemporary CANDLESTICKS.

Tassie, William (1735–99). A Scottish sculptor, resident briefly in Dublin and later in London, who, with his nephew William (1777–1860) and their successor (from 1840), John Wilson, made from 1766 reproductions of engraved gems in white and coloured glass paste, and later made TASSIE MEDALLIONS. In 1769 he was asked by Josiah Wedgwood to make casts for CAMEOS and INTAGLIOS to be made in jasper ware. In 1775 he produced a catalogue of 3,000 items. Later, *c.* 1781–82, he cast in plaster sixty copies of the PORTLAND VASE from moulds made by Giovanni Pichler before James Byres sold the original to Sir William Hamilton. He was commissioned by Catherine the Great of Russia before 1783 to produce cameos for her.

Tassie medallion. A MEDALLION made by WILLIAM TASSIE, or his son William, or by their successor John Wilson. Tassie first made reproductions of ancient engraved gems and later, from *c.*1766, medallions in white and coloured glass paste, a finely powdered potash-lead glass which was softened by heating and pressed into a plaster-of-Paris MOULD. The moulds were made from his own original wax relief portraits in profile. The medallions were made from the mould in opaque white glass paste; some pieces are reliefs mounted on glass or other materials. The medallions are from 2 cm. to 11 cm. in height, and 2 cm. to 9 cm. in width; the reliefs are larger. The subjects are Royalty and famous persons, usually contemporaries. Some bear an impressed signature or a 'T'. *See* PÂTE-DE-VERRE; PAPER-WEIGHT, *Tassie*; SULPHIDE.

tazza (Italian). (1) A type of ornamental CUP that has a wide flat or shallow BOWL which rests on a stemmed FOOT. It may be with or without HANDLES. *See* TIERED TAZZA; STANDING BOWL. (2) The Italian term for a cup.

tea-bottle. The same as a TEA-CADDY.

tea bowl. A small cup without a handle, made and used in the Far East for drinking tea, and commonly made in many European ceramic factories in the 18th century. Examples in OPAQUE WHITE GLASS with enamelled decoration were made in China, sometimes with a saucer EN SUITE. Many such bowls were made in Europe, probably Bohemia and South Germany, in enamelled opaque white glass or in white glass speckled with blue or manganese (sometimes with green).

tea-caddy. A covered or stoppered container for dry tea, used on the tea-table in conjunction with a teapot and tea-kettle. It is often of rectangular section, with sloping curved SHOULDERS and a short cylindrical NECK; some have a metal hinged LID. In England the tea-caddy (or 'tea jar') was sometimes erroneously called a 'teapoy'. Other terms are a 'tea cannister' or a 'tea bottle'. Glass examples were made in clear glass and in OPAQUE WHITE GLASS; often they are in pairs, inscribed 'Bohea', 'Green', or 'Hyson'.

tea-candlestick. A type of CANDLESTICK that is smaller than the usual 25 cm. height, being from about 12 cm. to 18 cm. high and having a thin NOZZLE. Their purpose is uncertain, either: (1) to be placed on a tea-table, to burn for a short time; or (2) to serve as a TAPERSTICK, for holding a lighted taper with which to light candles or melt sealing wax. The style changed with the styles of candlesticks.

tea service. A service, usually for one or more persons, for serving tea. It usually consists of a TEAPOT, TEA-CADDY, MILK-JUG, CREAM-JUG, SUGAR-BOWL, and CUPS and SAUCERS. They were made of pottery or porcelain in large quantities and of many styles, and infrequently of glass.

tea-vase. A type of covered TEA-CADDY or tea-jar that is ovoid in shape, and has a small mouth intended for pouring out dry tea leaves rather than spooning them.

teacup. A CUP, usually of hemispherical form, with one HANDLE and accompanied by a SAUCER, used for drinking tea. Its capacity is about 4 fluid ounces. Generally made of porcelain or pottery, they have been made in glass in many countries and of several types of glass. *See* CUP PLATE; TEA BOWL.

teapot. A covered vessel made in a great variety of shapes, styles, and sizes, intended for the brewing and serving of tea. It usually has one SPOUT and, on the opposite side of the vessel, one HANDLE. The spout is usually attached near the bottom of the BODY so as to drain the infused

taperstick. Hollow fluted staff, with domed and terraced foot, English, mid-18th century. Ht 21 cm. Courtesy, Sotheby's, London.

tea-caddy. Cut diamond pattern and flutes, with ormolu mounts, English, *c.*1810. Courtesy, Delomosne & Son, London.

teapot. Gold ruby glass with silver mounts, Johann Kunckel, Potsdam, *c*.1700. Ht 12 cm. Kunstgewerbemuseum, Cologne.

tear. Engraved wine-glass with drawn shank stem showing enclosed tear, English, Ht 21·5 cm. Courtesy, Derek C. Davis, London.

tea without disturbing the floating tea leaves. The pot also has a strainer of small holes at the junction of the spout and the body to prevent tea leaves from being poured into the cup with the liquid. Many teapots are made in fanciful forms. Generally made of porcelain or pottery, some have been made of glass, including FACETED clear glass, GOLD RUBY GLASS, and opaque white or blue glass.

tear. An air-bubble in the shape of a tear-drop, encased in the STEM of some drinking glasses or other glassware, or occasionally in a FINIAL or a solid KNOP. It was originally produced accidentally, but later made deliberately as decoration. The process involved PEGGING (i.e. pricking a small depression into the stem with a metal tool) and then covering the depression with MOLTEN GLASS to entrap the air and form a bubble. When the stem was drawn out, the bubble became tear-shaped (inverted pear-shaped). Occasionally the stem was slightly twisted to twist the bubble. Some later glassware had the tear in the BOWL or in KNOPS in the stem. *See* TWIST, *air*.

tear bottle. A type of UNGUENTARIUM, for containing perfumes and unguents, as depicted on ancient wall-paintings. They have been frequently found in Roman tombs (as have many types of *unguentaria*) and, on the groundless supposition that they were intended for the tears of the mourners, the type has been called a 'tear bottle' or lachrymatory. They are of various shapes but generally have a small body and a tall narrow neck; exaggerated examples have a very long neck (about three to four times the diameter of the body) or have no bowl but just a tall cylindrical neck resting on a small flat circular base, the neck being about 15 cm. to 30 cm. high. *See* CANDLESTICK UNGUENTARIUM.

Temple St Glassworks. A glassworks at Bristol, England, founded by LAZARUS JACOBS, father of ISAAC JACOBS. It is said that he employed MICHAEL EDKINS from 1785 to 1787 to do ENAMELLING and GILDING. It specialized in dark-blue glass, some being signed by Isaac Jacobs, and also engaged in CUTTING and ENGRAVING.

ten-diamond-mould glass. A type of glassware, made at Zanesville, Ohio (*see* ZANESVILLE GLASSWARE) and elsewhere in the United States, in a mould making a pattern of ten roughly formed diamonds in a lateral chain encircling the object, with a row of ten vertical ribs toward the bottom.

tesserae. Small pieces, usually roughly square, of glass or other suitable hard material used in the making of MOSAICS. The term is derived from a Greek word meaning four-sided. In the history of glass, the earliest source of tesserae is uncertain, but fragments found at Torcello, near Venice, indicate that they were made in the 7th century. Many were made, from the 9th to the 14th century, at nearby Murano for the decoration of churches in Venice and possibly those at Ravenna; these tesserae are of coloured glass or are of gold. The process involved cutting cakes of glass and embedding the pieces in cement rather than fusing them together. *See* OPUS SECTILE.

tessuto, vetro (Italian). A type of ornamental glass developed by VENINI with numerous narrow vertical stripes of white opaque glass. It is similar to FILIGRANA glass but the white threads are flattened to make stripes or narrow bands, sometimes twisted and manipulated to create an effect resembling moiré material.

test-tube unguentarium. A type of UNGUENTARIUM that is of thin and high cylindrical form with a rounded bottom. They were made of ROMAN GLASS in the 4th century AD.

thermometer. An instrument for measuring temperature, usually a closed tube with a bulb at one end containing mercury, or alcohol, or other substance which expands or contracts with changes of temperature and indicates the changes in relation to graduated markings on the tube. Early examples were made of FLORENTINE GLASSWARE in the late 16th/early 17th centuries.

thistle glassware. A drinking glass or CELERY GLASS, made in the form of a thistle, of various sizes. Each piece has a stemmed FOOT and a waisted bowl of which the lower half is hemispherical (sometimes diamond-cut or moulded) and above it is a funnel-shaped mouth representing the flower head.

Thompson, George (1913–). An American glass-designer who joined STEUBEN GLASS WORKS upon the formation of its design department in 1936 and has been a senior staff designer. He designed many important pieces of ornamental colourless glassware as well as vases, bowls, candlesticks, and tableware, throughout the Houghton era at Steuben.

thread. A thin strand of glass, circular in section, such as is used in making some twist STEMS, FILIGRANA decoration, etc. *See* TRAILING; THREADING.

threading. (1) The process of attaching glass THREADS as independent decoration, to be distinguished from TRAILING where the threads are applied to the surface of the object. A typical example is the threaded STEM of a WINGED GLASS. Sometimes the terms are used interchangeably, especially when the threads are incidental decoration on the surface of an object rather than an essential part of it. *See* STEM-WORK. (2) The process of drawing glass threads through molten glass as a method of decoration.

thread-wound bottle. A type of BOTTLE that is decorated with an opaque-glass thread spiralling equidistantly over the entire body, starting from the base and continuing upward to the neck; the pattern is not made by one continuous thread but is renewed a number of times, as is apparent from the several points of thickening at the joins. Such pieces were made of ROMAN GLASS, 1st century AD. Similar decoration is sometimes found on the neck of some bottles.

three-ring decanter. A type of DECANTER with three glass rings encircling its NECK. They were made in many sizes (from small to larger than a magnum) and styles from *c.* 1780 until recent times. The early ones were of light METAL, the later ones heavy. *See* IRISH GLASSWARE, *decanters*.

thumb glass. A drinking glass having circular indentations on the sides to permit it to be firmly held between the thumb and fingers; some tall drinking glasses, up to 35 cm. in height, were provided with three hollow indentations on each side. The German terms are *Daumenhumpen* and *Daumenglas*.

thumb print. A style of decoration on PRESSED GLASS, made in the United States, in the form of an overall diaper pattern of oval-shaped shallow depressions arranged in encircling rows. It was used on plates, flips, pitchers, mugs, bowls, compotes, etc. Variations of the pattern include the 'almond thumb print' (slightly pointed at upper end) and 'diamond thumb print' (enclosed in diamonds made by diagonal stripes). When used on a TUMBLER, the depressions served as diminishing lenses through which to see those on the opposite side of the glass.

thumb ring. A broad ring originally designed to be worn by an Oriental archer to protect his thumb when drawing the bow-string, but later worn generally. They were made of bone, jade, ivory, and also of glass in China during the K'ang Hsi period (1662–1722).

tessuto, vetro. Lamps with opaque white decoration. Venini, Murano.

thread-wound. Clear blue glass bottle with opaque white threads. Roman glass, 1st century AD. Ht 12·4 cm. Wheaton College, Norton, Mass.

tiered goblet. Goblets with engraved landscapes, gadrooning, knop-stems, and mereses, Bohemia, *c.* 1680. Museum of Decorative Arts, Prague.

Thuret, André (b. 1898). A French glass-designer who made massive pieces of glass AT-THE-FIRE and some pieces of sculptured glass. He was influenced by the ART DECO style of MAURICE MARINOT.

Thuringian glassware. Glassware made at several glasshouses in Thuringia, Germany, the earliest being at Lausnitz (1196) and Suhl (1350), but especially at Weimar, Arnstadt, and Gotha. A glassworks at Tambach, in nearby Franconia, was owned by Duke Bernhard von Sachsen-Weimar (1604–39) who became a patron of glass-making in Thuringia; he employed ELIAS FRITZSCHE (fl. 1630) as engraver. At Weimar ANDREAS FRIEDRICH SANG was the Court engraver from 1738 to 1744. At Arnstadt, Alexander Seifferd (1660–1714) – who was the son of Anton Seifferd, Court Engraver, *c.* 1650–60, to Carl XI Gustavus, at Stockholm – was Glass Engraver from 1685 to 1714; here also HEINRICH JÄGER probably engraved from 1715 to 1720, and SAMUEL SCHWARTZ from 1731 to 1737. At Gotha CASPAR CREUTZBURG engraved *c.* 1689 and GEORG ERNST KUNCKEL from 1721 to 1750. At Altenburg examples of the PASSGLAS were enamelled with representations of playing cards, the making of which was a local industry (*see* SPIELKAR-TENHUMPEN). At Lauscha, near Coburg, a glasshouse existed from 1595, doing enamelling; here also were made (and still are made) glassware AT-THE-LAMP, BEADS, TOYS, etc. At Ilmenau technical pharmaceutical glassware was made in the 19th century.

Tiefschnitt. Literally, deep carving. The German term for INTAGLIO, i.e. WHEEL-ENGRAVED decoration on glass where the design is cut below the surface of the glass in the reverse of RELIEF. It is the opposite of HOCHSCHNITT. The technique was introduced by CASPAR LEHMANN at Dresden and Prague, *c.* 1605, and was developed at Nuremberg by GEORG SCHWANHARDT and his followers. Sometimes the engraving was polished, but sometimes only partially so for greater effectiveness. *See* MATTSCHNITT.

tiered goblet. A double GOBLET, with the COVER of the lower large one being the base of a smaller one. Some were made in Bohemia in the late 17th century.

tiered tazza A double TAZZA, with a small one set atop a larger one.

Tierhumpen (German). Literally, animal beaker. A tall HUMPEN decorated with enamelled scenes humorously depicting various animals, e.g. small animals playing musical instruments. They were made in Germany and Bohemia in the late 16th and 17th centuries. *See* ENAMELLING, *German.*

Tiffany, Louis Comfort (1848–1933). The leading exponent of the ART NOUVEAU style in glassware in the United States. Inspired by the work of EMILE GALLÉ, he produced a great variety and quantity of ART GLASS of original techniques and styles of decoration, much in coloured CASED GLASS and IRIDESCENT GLASS. His success was due to his artistic capabilities and the technical skill of his associate, ARTHUR J. NASH. He was not a glass-maker, but he designed glassware and closely supervised its production at his glassworks by his large and skilled staff. He was the son of Charles L. Tiffany, the prominent New York jeweller. He studied painting in 1866 in New York and in 1869 in Paris, where he became acquainted with the Art Nouveau movement, and later became acquainted with Siegfried (generally known as Samuel) Bing (1838–1905), whose shop, *Le Salon d'Art Nouveau* (the eponym of the new style), was later the European outlet for Tiffany and other Art Nouveau products. After his return to New York, he entered the glassware field, first designing decorated windows for churches and aristocratic homes, and also lighting fixtures and lamps. In 1885 he established a glass factory at Corona, Long Island, New York, which continued to produce glassware until 1928. His success was established at the Chicago World's Fair in 1893 and the Paris Exposition of 1900. After several changes of name of his companies, he withdrew from active participation in 1919. In addition to glassware, he was engaged in making other products in Art Nouveau style which he distributed from 1900 through his Tiffany Studios, located in New York City. For Tiffany and the Nash family and many illustrations of glassware made by them, *see* Albert Christian Revi, *American Art Nouveau Glass* (1968).

——, *factories*. Tiffany's first connection with glass-making was with Louis C. Tiffany & Associated Artists, which he formed upon his return to New York from Paris in 1879, mainly to do interior decorating. At this time he did experimental work at the Heidt Glassworks in Brooklyn, developing the glassware that he later called FAVRILE GLASS. In 1885 he established in Brooklyn his first glassworks, Tiffany Glass Co., which mainly made decorative glass windows and did interior decorating. In 1892 he moved from Brooklyn, and founded at Corona, Long Island, New York, Tiffany Glass & Decorating Co., for making blown and decorative glassware; a division of the company, called Tiffany Furnaces, was operated by ARTHUR J. NASH under the supervision of Tiffany. In 1898 Tiffany founded, with Nash and others, Stourbridge Glass Co. at Corona; its name was changed in 1902 to Tiffany Furnaces, Inc. In 1900 Tiffany formed Tiffany Studios to offer interior decorating services and to sell Tiffany products in New York City. In 1919 Tiffany and Nash withdrew from active participation and the operations were divided between: (1) Tiffany Furnaces, whose assets were sold to A. DOUGLAS NASH, thereafter operating as A. Douglas Nash Associates until he formed in 1920 Louis C. Tiffany Furnaces, Inc.; and (2) Tiffany Studios, which was thereafter headed by Joseph Briggs until its liquidation upon his death in 1938. In 1928 Tiffany withdrew his financial support and use of his name from Tiffany Furnaces, and A. Douglas Nash formed his own company, A. Douglas Nash Corporation, which continued until its failure in 1931.

Tiffany glassware. Glassware in a vast variety of forms and styles made by the successive companies of LOUIS COMFORT TIFFANY. The ware, mostly in ART NOUVEAU style, was of coloured CASED GLASS and IRIDESCENT glass, the best-known among them being FAVRILE GLASS, decorated with iridescent effects produced with metallic lustre and featuring naturalistic motifs. None of the Tiffany decoration was by enamelling, but was only in the shape, texture, and colours of the glassware itself and the lustre finish. Some of the pieces are signed with Tiffany's initials or his full name. Forged signatures have appeared on imitations made by several glassworks. *See* AGATE GLASS; CYPRIOTE GLASS; DIATRETA GLASS; LAVA GLASS; TIFFANY LAMP; TIFFANY WINDOW.

——, *aquamarine*. A type of glassware made by Tiffany from 1913, with embedded decoration of aquatic plants or marine life. It was a light-green glass, to simulate sea-water. The pieces (usually bowls, vases, DOOR-STOPS, PAPER-WEIGHTS) were generally heavy and thick, with a slight depression in the top to suggest that the object was filled with water. Only a few examples were made.

——, *carved glass*. A term applied by Tiffany to any type of his glassware (colourless, coloured or lustred) that was decorated by cutting or engraving, including relief and intaglio.

——, *damascened*. A type of FAVRILE GLASS developed by Tiffany, *c.*1910, with stripes of gold and of coloured lustre arranged in a wavy pattern to simulate the watery appearance of damascened steel. It was also made later by A. DOUGLAS NASH.

——, *Favrile fabrique*. A type of PRESSED GLASS that has the appearance of fabric with numerous vertically or horizontally arranged folds. It was used exclusively in panels of lamp-shades, but a circular shade of one piece of such glass was invented by LESLIE NASH in 1913. Also called 'linenfold glass'. *See* FAVRILE GLASS.

——, *jewels*. Glass objects made to be set and worn as jewellery or to decorate various glass objects. They included glass 'scarabs' in the form of a beetle, made in several sizes and colours, some iridescent.

Tierhumpen. Humpen with enamelled encircling scene of animals playing music, Upper Franconia, dated 1683. Ht 17·6 cm. Kunstsammlungen der Veste Coburg.

Tiffany glassware. Vase, iridescent opaque white glass with inlain vertical white stripes and horizontal peacock-blue bands, *c.*1950; mark 'L C TIFFANY'. Ht 21·5 cm. Courtesy, Sotheby's Belgravia, London.

Tiffany glassware: Favrile fabrique.
Lamp-shade by Leslie Nash, *c.* 1913.
Galerie des Arts Décoratifs, Lausanne.

Tiffany paperweight glass. Vase with interior decoration of red poppies and green leaves, and lined with gold lustre. Ht 15·2 cm. Courtesy, Sotheby's Belgravia, London.

——, *mosaic.* MOSAICS composed of glass TESSERAE, some iridescent, made to be embedded as decoration in a wall or in a copper ornamental disk, and sometimes used to embellish lampshades and other objects.

——, *Paperweight Tiffany glass.* A curiously termed type of ART GLASS, developed by Tiffany, and occurring in a great many forms, colours, and styles of decoration, usually with floral patterns. It was made by embedding multi-coloured glass making the design in an inner layer of glass and then covering it with another layer that is usually given an iridescent surface. The term 'paper-weight' has also been applied in the United States to such glass made by STEUBEN GLASS WORKS and others; the use of the term derives probably from the similarity of the decoration to that of some Tiffany paper-weights (see PAPER-WEIGHT, *Tiffany*), not because the glass itself is especially light in weight.

——, *tiles.* Squares of glass (from 2·5 cm. to 10 cm. wide, and 1·2 cm. deep) to be embedded in screens and to decorate mantelpieces and walls. They were made by pouring molten glass into open moulds, then pressing to form several different patterns. Some are iridescent. Some rare ones, called 'turtleback', have a convex surface with a central depression; these were used to decorate shades of lighting fixtures. Such tiles were made by Louis C. Tiffany & Associated Artists, from 1879.

——, *vetro di trina.* A type of glassware, rarely made, *c.* 1910, and so-called by Tiffany, that purported to simulate Venetian VETRO DI RETICELLO but which greatly differed by having the criss-crossed opaque white threads much farther apart so as to have large interstices, and being without embedded air bubbles (perhaps due to MARVERING after the threads were applied). It sometimes was given a lustred surface.

Tiffany lamp. A type of table-lamp or hanging lamp made by LOUIS COMFORT TIFFANY, of which the shade is characterized by having been constructed from a large number of variably shaped pieces of multi-coloured glass, in an arrangement which often imitates floral or other natural forms. The table-lamps have a bronze base in the form of a tree trunk. A well-known example is the 'wistaria lamp', from a design by Curtis Freshel in 1904, simulating a wistaria vine, with a verdigris-bronze tree-like base and a shade resembling naturalistically trailing wistaria having an irregular bottom line. Many types of such shades were made, including the 'Apple Blossom', 'Lotus Blossom', 'Dragon Fly', and 'Spider Web'. Related but of different form is the 'Lily lamp' with individual lily-shaped shades suspended from separate branches.

Tiffany window. A decorated window with coloured glass designs depicting floral subjects, landscapes, biblical stories, copies of paintings, or abstract motifs, made by LOUIS COMFORT TIFFANY. In such windows the coloured glass was first cut to conform to the intended design, and the leading was afterwards shaped to fit the glass.

Tille, Christoph (fl. 1685). A German wheel-engraver who followed GEORG GONDELACH, from a factory at Dessau, to work at the glasshouse at Drewitz, near Potsdam.

tin oxide. A chemical used to opacify glass in making OPAQUE WHITE GLASS, and also used in making tin-glaze for pottery. It is also called 'stannic oxide'.

tint. A colour tinge due to an ingredient in the BATCH as distinguished from a colour resulting from the deliberate addition of a mineral oxide or other colouring agent.

titanium. A metallic agent used in modern times in colouring glass, producing hues of yellow-brown.

toasting glass. A tall thin WINE GLASS with a small conical BOWL, a very thin STEM, and a stemmed FOOT; its capacity is about 2 to 4 ounces.

They were made in England in the second quarter of the 18th century, and used for drinking a toast.

toast-master's glass. A glass, sometimes tall and thin, having a BOWL with a thick wall and bottom so as to give a visually deceptive impression of the contents, its capacity being only ½-oz to ¾-oz. They were made in England from the middle of the 18th century.

tobacco flask. The same as a SNUFF BOTTLE. Some made of CASED GLASS with cut CAMEO decoration have been attributed to China in the 19th century, and some of smooth coloured glass have been ascribed to Japan in the 18th century. The German term is *Tabaksfläschchen*.

tobacco pipe. An ornamental object in the form of a smoking pipe. Some, of coloured glass, were made in the Nailsea, Bristol, and Stourbridge areas. Probably they were made for display, especially for display in the shop-windows of tobacconists. Some are of fantastic shape, such as one with a series of six hollow globes along the stem. The type with a very long stem and small bowl was known as a 'church-warden's pipe'. Examples with a long twisted STEM and ornately decorated were made in Spain in the 18th century.

toddy glass. A type of drinking glass intended for drinking hot toddy, tall and made with some thickness. They usually have no STEM, and often have WHEEL-ENGRAVED decoration and an inscription. The same as a FLIP. *See* TODDY PLATE.

toddy lifter. A hollow vessel, about 15 cm. long, for transferring punch from a bowl to a glass. They are shaped somewhat like a small DECANTER with a long thin NECK, bulbous at one end and with a hole in each end. The bulbous end was dipped into the punch-bowl until the object filled (capacity about a wine-glass), when a thumb closed the opposite hole until the punch was released into the glass. Some examples have, at the middle of the neck, a collar as a finger grip. They are said to have been invented in Scotland, and were made in England and Ireland in the 18th and early 19th centuries. The usual decoration is CUTTING. Sometimes called a 'punch filler'. *See* WINE SIPHON.

toddy plate. A type of plate (about 11·5 cm. wide, larger than a CUP PLATE) made in the United States, some of LACY GLASS, some of PATTERN-MOULDED GLASS, for use in connection with a glass of hot toddy, possibly as a rest for the hot glass or as a plate on which to rest the spoon. *See* TODDY GLASS.

toddy stick. (1) A small spatula for stirring hot toddy; examples were made of glass in England in the Victorian period. (2) A term sometimes applied to a SUGAR CRUSHER.

toilet set. A group of objects made for use on a lady's toilet-table, including powder-bowl with cover, toilet-water (cologne) bottle with stopper, jar for bath-salts, etc. Some examples in modern style glass have been designed by KEITH MURRAY for STEVENS & WILLIAMS, LTD, *c.* 1935.

toilet-water bottle. A small BOTTLE, with a STOPPER or CAP COVER, for holding toilet-water. They were accessories to a toilet-table. Many were made of glass, in various styles, including some of OPAQUE WHITE GLASS or coloured glass, with ENAMELLED DECORATION. *See* COLOGNE BOTTLE.

Toledo, Ohio, glass companies. Four leading glass manufacturers located in Toledo, Ohio, that make a wide variety of industrial glass: (1) OWENS-ILLINOIS, INC.; (2) Libbey-Owens-Ford, Inc. (until 1968, Libbey-Owens-Ford Glass Co.), which resulted from a merger in 1930 of Libbey-Owens Sheet Glass Co. (organized by Edward Drummond

Tiffany lamp. Wistaria pattern, multi-coloured glass, *c.* 1900–05. Ht 68 cm. Courtesy, Sotheby's Belgravia, London.

toast-master's glass. Ratafia glass with deceptive bowl, English, *c.* 1760. Ht 18 cm. Courtesy, W. G. T. Burne, Ltd, London.

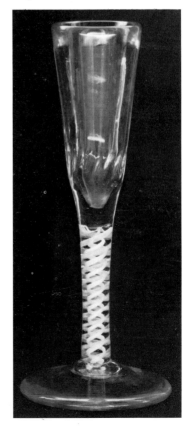

toddy lifters. Irish, late 18th/early 19th century. Ht 19·5 cm. Collection of Phelps Warren, New York City.

trailed circuit. Lead glass decanter jug with encircling chain-trail, and with gadrooning at bottom, George Ravenscroft, *c.* 1685. Ht 21·5 cm. Courtesy, Delomosne & Son, London.

Libbey and Michael J. Owens) with Edward Ford Plate Glass Co. (formed by Edward Ford in 1898 in Toledo); (3) Owens-Corning Fiberglass Corp., established jointly in 1938 in Toledo by Owens-Illinois, Inc., and CORNING GLASS WORKS of Corning, New York, specializing in fibrous glass products; and (4) Johns-Manville Fiber Glass, Inc. (now a subsidiary of Johns-Manville Corp.), formed after the purchase in 1958 by Johns-Manville of L-O-F Glass Fibers Co., which had itself been created in 1955 by the merger of Glass Fibers, Inc. (founded in Toledo in 1944) with the fibre-glass division of Libbey-Owens-Ford, Inc. that had been established in 1951.

tools, glass-makers'. The various instruments used by the glass-making CHAIR (team) to develop and shape an object, including the BLOW-PIPE, PONTIL, GATHERING IRON, PUCELLAS, SHEARS, CLAPPER, WOODS, PALLET, BLOCKING WOOD, PINCERS, BATTLEDORE, LIPPER; CRIMPER.

topographical glassware. Glassware decorated with scenes of known buildings or views of cities, usually by engraving. Such style was popular in Germany, Vienna, and Bohemia (*see* MOHN, SAMUEL; MOHN, SAMUEL GOTTLOB; KOTHGASSER, ANTON). It is also found in English glassware; among the known examples are scenes of Alnwick Castle, Hereford Cathedral, Windsor Castle, Sunderland bridge, etc.

Torcello glassware. Glassware excavated and thought to have been made on the island of Torcello, north of Venice, in the 7th and 8th centuries. Excavations in 1961/62 unearthed TESSERAE and fragments of drinking vessels, MOSAIC ware, etc.

torsade. A twisted ring of OPAQUE WHITE GLASS or COLOURED GLASS, or both, that forms part of the decoration of some paper-weights; it is seen encircling the motif (*see* PAPER-WEIGHT, *mushroom*; *upright bouquet*), especially on examples from ST-LOUIS. Some torsades consist of a gauze twist within a spiral twist (*see* TWIST).

tortoise-shell glass. A type of ART GLASS having the appearance of tortoise-shell, with brown mottling enclosed between two layers of glass with a glossy finish. One process, developed by Francis Pohl, of Silesia, and registered in 1880, involved breaking into small pieces several bubbles of blown glass of different shades of brown; then a PARAISON of clear glass was blown into a globular shape, and cut in half, and into it was inserted another semi-globular piece of glass that had been rolled onto the pieces of brown glass and MARVERED, and then the two pieces, enclosing the brown glass, were blown together and formed into the desired shape. Such glass was made in the United States by BOSTON & SANDWICH GLASS CO. and others, and in Europe.

toy. An English term used in the 18th century, with respect to glassware, for small objects or trifles made for amusement or as souvenirs. Similar to toys in porcelain made at Chelsea and Meissen, these were small objects such as étuis, scent-bottles, thimbles, and other such elegant trifles, and also miniature animals and small figures, particularly such as were popular in Spain and in peasant markets.

traforato (Italian). Literally, perforated. A style of decoration, used mainly on glass BASKETS, made in an openwork pattern in the form of wicker basketry. The openwork was not actually perforated, as English pierced earthenware, but was made by pincering parallel horizontal strands of clear glass. Sometimes termed *a traforo*.

trailed circuit. A decorative style featuring several closely spaced applied glass THREADS forming a continuous pattern.

trailing. The process of applying threads of glass as a decoration, as on the BODY, HANDLE, or FOOT of a vessel. It is done by laying or winding softened threads on to a glass object. The threads may be laid in a pattern (*see* SNAKE TRAILING), drawn into a pattern (*see* NIPT-DIAMOND-WAIES), impressed with a patterned wheel (*see* RIGAREE TRAIL), or drawn into chains (*see* CHAIN TRAILING), or pincered or notched as added decoration. The process was used on ROMAN GLASS, and also in medieval and later times. Other uses were winding threads around the base of a RÖMER or around the exterior of a PASSGLAS or a BANDWURMGLAS. The process is to be distinguished from THREADING.

transfer engraving. A polychrome coloured picture on the back of flat glass. Such pictures were made in England from *c*.1670 (the earliest recorded example is a portrait of Charles II) until the Victorian era. The glass used was very thin, usually CROWN GLASS, in which slight ripples are apparent. The process was very difficult: first, the back of the glass was coated with Venetian turpentine which was allowed to dry and harden; then a black-and-white mezzotint was soaked in water, drained and then spread carefully (leaving no air bubble) on the prepared glass and, after being allowed to dry, was again soaked with water and the paper carefully rubbed off with the fingers until only the ink but no pulp remained. Thereafter the adhered ink picture was painted with transparent oil colours, usually flat washes that left the shading to the mezzotint, so that the painting was in some cases done by amateurs. The pictures were mainly portraits, commemorative motifs, rural and indoor scenes, and allegorical subjects. They were not signed (the only recorded signed example being signed 'mezzotinto engraver'). They were never varnished before or after painting, as that would have detracted from the desired translucency; occasionally a small amount of oil was used to facilitate the removal of the paper. There are many FORGERIES made by improper processes. Other terms are 'glass picture', 'transfer print', or 'reverse engraving'. *See* PANEL; PLAQUE.

transfer print. The same as TRANSFER ENGRAVING, but in no way connected with TRANSFER PRINTING.

transfer printing. The process of decorating ceramic ware (and later glassware) by inking an engraved copper plate with an ink prepared from one of the METALLIC OXIDES, and then transferring the design to paper which, while the pigment was still wet, was pressed onto the ware, leaving the desired imprint; this was subsequently fixed by FIRING. The process was probably invented *c*.1753. Its use in England to decorate glassware in the manner of ceramic ware was started *c*.1809 by GEORGE BACCHUS & SON, and it was also used by HENRY G. RICHARDSON & SONS, at Stourbridge. It was also used on Russian glassware, *c*.1840. Sometimes the design was printed in monochrome on OPAQUE WHITE GLASS, and sometimes printed in outline and filled in by polychrome painting. *See* TRANSFER ENGRAVING.

translucent. Permitting the passage of light but in a diffused manner so that objects behind cannot be clearly seen. *See* TRANSPARENT; OPAQUE.

transmutation colour. A colour that changes hue when seen by transmitted light. Glass coloured turquoise-blue by the use of COPPER oxide in a REDUCING ATMOSPHERE shows as red when so viewed; *see* FAIRFAX CUP. The LYCURGUS CUP in reflected light appears pea-green, but is wine-red when viewed by transmitted light.

transparent. Having the quality of transmitting light without diffusion so that objects behind can be distinctly seen. *See* TRANSLUCENT; OPAQUE.

transparent enamel. An ENAMEL that is TRANSPARENT, as distinguished from the earlier and more customary OPAQUE ENAMEL. It was used principally on STAINED-GLASS WINDOWS, and its use on glass vessels was originally by artists who decorated such windows, such as JOHANN SCHAPER, who *c*.1650 invented some transparent enamels, and

traforato. Basket of pincered colourless glass, Venice, 17th century. Museo Vetrario, Murano.

transfer printing. Tumbler with portrait of Tsar Nicholas I, St Petersburg Imperial Glass Factory, second half of 19th century. Ht 8·5 cm. Hermitage Museum, Leningrad.

trembleuse. Cup and saucer of aventurine glass, Venetian, 17th century. Kunstsammlungen der Veste Coburg.

triple bottle. Greenish glass, with reeded handle, late 2nd or 3rd century. Ht 22 cm. British Museum, London.

Trivulzio Bowl. Cage cup of *diatreta* glass (network blue, inscription emerald green). Ht 12 cm. Museo Archeologico, Milan.

SAMUEL MOHN, GOTTLOB SAMUEL MOHN, and ANTON KOTHGASSER, who after *c*.1810 reintroduced the use of such enamels. *See* Jerome Strauss, *Journal of Glass Studies* (Corning), VI (1964), p. 123.

tray. A flat receptacle, varying in size and shape, with a low vertical, sloping, or curved rim. Some glass examples have a low stemmed FOOT.

trefoil. A form or an ornamentation which has three equal foils or lobes. *See* PINCHED LIP.

trellis pattern. A moulded pattern in the form of crossed diagonal raised lines forming small diamond-shaped depressions.

trembleuse (French). A cup and saucer, the saucer having a vertical projecting ring into which the cup fits; the ring may be of considerable height and is occasionally pierced with ornamental designs. Sometimes a deep well was used instead of the vertical ring. The trembleuse was made in these forms to prevent spillage of the contents caused by a shaky hand. Saucers of this type were usually made to accompany a coffee-cup, but they were also employed with a chocolate-cup, teacup or bouillon cup. Usually made of porcelain, some are made of glass.

trencher salt. A type of SALT-CELLAR without feet that rests flat on the table. They are circular, oval or rectangular, and are made with one, two or three wells. They were originally made to be placed alongside the diner's trencher flat plate.

Trichterbecher. The German term for a CONE-BEAKER.

Trichterpokal (German). A type of POKAL with a funnel-shaped BOWL.

trick glass. *See* JOKE GLASS; PUZZLE JUG; SIPHON GLASS; YARD-OF-ALE. Some trick glasses are an adaptation of the yard-of-ale, having a funnel neck set above a large globular bowl, so that the contents of the bowl squirts out when the glass is tipped for drinking.

trionfo (Italian). Literally, triumph. A very elaborate CENTRE-PIECE, made of many different pieces arranged together to adorn a large ceremonial dining-table. They were made of Venetian glass in the 18th century, especially by GIUSEPPE BRIATI, and often represented a formal garden scene, with lawn, trees, fountains, columns, statues, balustrades, flowers, and vases. Especially ornate examples were called *grande trionfo da tavolo.*

triple bottle. A BOTTLE that is divided internally into three compartments, each with its own MOUTH or SPOUT. Some were made of ROMAN GLASS in the 2nd/3rd centuries. Some are globular-shaped with an angled handle on one section. *See* MULTIPLE BOTTLE.

Trivulzio Bowl. A CAGE CUP of VASA DIATRETA type, the network of which is of blue glass and the inscription emerald-green glass. It came from the Trivulzio Collection, Milan, and, according to W. B. Honey, it 'has been absurdly called "Nero's cup" '.

trombone. A glass ornament in the form of a trombone. Such pieces were made in Venice in the 18th century. *See* MUSICAL INSTRUMENTS; HORN.

trophies. A decorative motif depicting so-called trophies of various types, used as decoration on porcelain and also on glassware. The trophy was originally the arms of a beaten enemy hung by the Greeks on an oak tree to commemorate victory. This led to the motif of martial trophies, including helmets, shields, axes, and drums. Later, other motifs were used and arranged in ornamental form: amatory trophies, with doves, bows and arrows, and quivers; musical trophies, with violins, flutes, etc.; and horticultural trophies, with beehives, watering-pails, and wheel-barrows. The Italian term is *trofei*.

Trout and Fly, The. An ornamental piece made of crystal glass by STEUBEN GLASS, 1966. It depicts a leaping trout taking a Royal Coachman fly (the fly is of 18-carat gold). The body of the fish is CASED, with air bubbles trapped by indentations made in the inner glass before casing. It was designed by JAMES HOUSTON.

trulla (Latin). A LADLE or dipper in the form of a bowl with a solid horizontal handle. The bowl was usually shallow (about 4 cm. deep), but some were deeper. They were made of Roman glass, 1st century AD. *See* PATERA.

Trümper, Carl Ludwig (d. 1753). A relative of JOHANN FRANZ TRÜMPER and a glass-engraver at Cassel (Kassel) in Hesse.

Trümper, Johann Franz. The son of the brother-in-law of FRANZ GONDELACH, whom he succeeded as a glass-engraver at Cassel (Kassel) in Hesse.

Trümper, Johann Moritz (1680–1742). A glass-engraver from Cassel (Kassel), in Hesse, who went to Potsdam in 1713 and in 1718–19 held the lease of the Potsdam factory.

trumpet neck. The NECK of a vase or other vessel that is flared outward toward the MOUTH.

trunk-beaker. *See* CLAW-BEAKER.

tube. A thin hollow stick of glass, as contrasted with a solid ROD or CANE. It is made by forming an air bubble in a PARAISON of molten glass, and then blowing it and drawing it out to the desired thinness.

tubular receptacle. A tall thin cylindrical glass receptacle of uncertain purpose. They were made of Mesopotamian or Persian glass in the 5th/6th centuries, usually with overall wheel-cut decoration of a network of diamond or oval facets. An example at the British Museum, with the top broken off and missing, has been thought there to have been a scent or unguent container, but a complete example in Copenhagen, with a ground rim (as for a cover), has been thought there to have been a document container.

tulip glass. A medium-size drinking glass with a tulip-shaped bowl rounded toward the bottom and resting on a stemmed foot. The rim of the bowl is sometimes everted but usually is curved slightly inward. The term is also applied to a modern type of CHAMPAGNE GLASS of which the tall thin bowl is almost cylindrical. *See* FLUTE.

tumbler. A drinking glass without a stem, foot or handle, and having a flat base on which it rests. It is usually of circular section, cylindrical, waisted, barrel-shaped, or with sides that taper slightly inward toward the base. The cylindrical form was called in England in the 18th century a 'water-glass'. They are of various sizes and styles, but they do not have a flared mouth. Finely decorated examples of tumblers were made by JOHANN JOSEPH MILDNER, including some in ZWISCHEN-GOLDGLAS with MEDALLIONS in the base. Early examples in FRANKISH GLASS, 4th to 7th centuries, were made with a convex base so that they could not stand erect when filled. *See* BEAKER; POCKET GLASS.

tureen. A circular or oval serving bowl, usually with a cover, for soup or vegetables. They were made in a variety of shapes, sizes, and decoration in porcelain and pottery, but some were made of enamelled OPAQUE

Trout and Fly. Ornament designed by James Houston for Steuben Glass Works, 1966. Ht 24 cm. Steuben Glass, New York.

Trichterpokal. Covered funnel-bowl goblet, Thuringia or Saxony. Ht 23 cm. Kunstsammlungen der Veste Coburg.

tureen. Opaque white glass covered tureen with stand, German, in *façon de Venise.* Ht 14·4 cm. Kunstsammlungen der Veste Coburg.

turn-over rim. Fruit (or salad) bowl with faceted rim and pillar-moulded oval base, Irish, *c.* 1780/90. Courtesy, W. G. T. Burne Ltd, London.

Tuthmosis Jug. Core glass, with coloured trailed decoration, Egyptian, 15th century BC. Ht 8·8 cm. British Museum, London.

WHITE GLASS. They sometimes have an accompanying stand. The French term is *soupière.*

turn-over rim. A type of RIM found on BOWLS and SALT-CELLARS of IRISH GLASSWARE; the rim is curved outward and downward.

Tuscan glassware. Glassware made in many towns in Tuscany where production was well developed by the 16th century, but declined toward the end of the 18th century after the last of the Medicis. The ware included mainly bottles and flasks, but also STAINED-GLASS WINDOWS. *See* FLORENTINE GLASSWARE.

Tuthmosis glass. A glass vessel of CORE GLASS bearing the cartouche of Tuthmosis (Thothmes) III (*c.* 1504–1450 BC). There are three known glass vessels with this cartouche, presumably made to honour this pharaoh who is said to have founded the glass industry in Egypt. One is in the British Museum (a JUG of opaque light-blue glass decorated with coloured TRAILING and with powdered coloured glass fired on but not fused, which W. B. Honey and others have said was 'painted'); the second is in the Ägyptische Staatssammlung, Munich (a CUP of inverted-bell or lotus-bud shape, of light-blue glass with decoration of yellow and dark-blue COMBED DECORATION); and the third is in the Metropolitan Museum, New York (a jug, inverted bell-shaped, of undecorated turquoise-blue glass). These are all of EGYPTIAN GLASS, and are among the earliest known glass vessels, it having been said that glass vessels were not made before *c.* 1500 BC.

Twelve Months, The. A decorative ALLEGORICAL SUBJECT in which twelve enamelled figures with dress and attributes appropriate for the month that each represents are depicted, usually with the name of the month inscribed. Such motifs appear on glasses decorated at Nuremberg and in Bohemia, *c.* 1680–1770.

twists. Decoration in the STEMS of English drinking glasses produced by twisting a rod of glass in which were embedded air bubbles or THREADS or TAPES of opaque white or of coloured glass. It dates from *c.* 1735, and was popular from the 1740s until the 1760s. It has been said that there are over 150 varieties of twists in different forms and combinations. For photographs of numerous twists, *see* L. M. Bickerton, *Eighteenth-Century English Drinking Glasses* (1971). *See* PLY.

TYPES OF TWIST
———, *air.* A twist in which is embedded one or several twisted columns of air. Its origin and development are essentially English. It was one of the earliest styles of decorating STEMS, dating from *c.* 1735 to *c.* 1770. It was produced by two basic methods: (1) making a TEAR in a ROD of glass, then drawing the rod until it became very thin and twisting it to make the column of air spiral; or (2) making a pattern of circular holes or flat slits in the top of a glass rod, covering them with molten glass, and drawing and twisting the rod to make a spiral pattern of air (used to make multiple series twists, see below). In early examples the twist was irregularly formed and spaced, but later the threading was uniform; it was sometimes carried down from the BOWL of a two-piece glass. Another process involved placing several rods containing elongated tears into a cylindrical mould with grooves on its interior surface, then covering them with molten glass and, after withdrawing the mass on a PONTIL, attaching another pontil and twisting until the desired pattern was developed. In some examples one twist is concentrically within another (made by repeating the process with the mould, thus twice twisting the inner rod into a tighter twist); these are called 'double series air twist' or even 'triple series air twist'. Occasionally an air-twist stem includes one or more KNOPS, and the twist continues unbroken (but sometimes slightly distorted) through the knops.

——, *cable*. A type of twist produced by a single CABLE, extending vertically or spirally, or by two cables intertwined. It is sometimes encircled by other twists in a double-series (see below) stem.

——, *colour*. A twist in which at least one spiral thread is coloured, usually opaque but sometimes translucent. Sometimes two colours are combined, but more often there is at least one thread of opaque white with a coloured thread. Occasionally, there is also an air twist (see above) with a coloured thread, and very rarely with a coloured and also a white thread (see mixed twist, below). The coloured twists are rare in England, dating from *c*. 1755 to *c*. 1775, but are found on Continental glasses. The colours usually found are red, blue, green, pink, brown, and (very rarely) canary yellow. *See* also tartan twist, below.

——, *corkscrew*. A twist produced where a single air twist or an opaque TAPE is twisted in corkscrew fashion, or two such tapes are intertwined. Sometimes the tapes are edged with a glass thread of one or two colours. (*See* mercury twist, below).

——, *cotton*. A general term for all types of twists made of OPAQUE WHITE GLASS.

——, *double series*. A type of twist with another (air or opaque) twist encircling it, produced by making the first twist and then adding more molten glass and repeating the process, thus twisting the inner rod into an even tighter twist. Sometimes several intertwined outer twists are used, or one or more heavy spiral threads, or a multi-ply band of threads encircles the inner twist. The two types are abbreviated for reference purposes as DSAT (air) and DSOT (opaque) respectively. Occasionally KNOPS are found on a double series twist, but none is known on a multiple series twist (see below). *See* also single series twist, below.

——, *enamel*. A twist produced by using threads of opaque white glass or of opaque coloured glass embedded in clear glass (as in the method of making CANES of MILLEFIORI type) and twisting and manipulating them.

——, *gauze*. A twist produced with opaque THREADS wound to create a rope effect, similar to a cable twist (see above), but with the threads finer and loosely wound. It is found singly or in intertwined pairs, or groups of four and as part of a double series twist (see above).

——, *incised*. A decoration on the exterior of a STEM in the form of closely spiralled grooving running the whole length of the stem. In early pieces the grooving is rather coarse, but a few later ones have fine-drawn twists. Some such stems have a KNOP. They date from *c*. 1750 to *c*. 1775.

——, *lace*. A twist produced from numerous fine opaque threads arranged in a corkscrew fashion, but very loosely so as to present a lace-like effect.

Twelve Months, The. Tumbler with engraved decoration, Nuremberg, *c*. 1680–1700. Ht 14·8 cm. Kunstsammlungen der Veste Coburg.

twists. Stem formation in English wine glasses. Courtesy, L. M. Bickerton, Worthing.

double series air twist : pair of corkscrews outside air column.

double series opaque twist : close multi-ply spiral band outside pair of heavy spiral threads.

double series opaque twist : 12-ply spiral band outside gauze column.

double series opaque twist : band of 3 spiral threads outside spiral gauze.

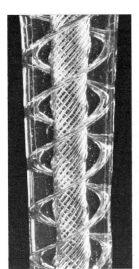

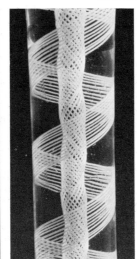

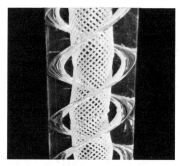

mixed twist: pair of air spirals outside opaque gauze.

mixed twist: four opaque heavy threads outside multiple spiral air gauze.

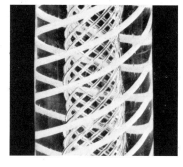

multiple spiral air twist

multiple spiral opaque twist

single series opaque spiral cable twist

single series opaque lace twist outlined

———, *mercury.* A type of air twist (see above) made with a clear METAL of high lead content so as to afford extra brilliance, and produced by using two flat pins to prick the top of the ROD, so that when it was drawn out the effect was of a wide corkscrew, giving the appearance of being filled with mercury. The corkscrews are sometimes tight and of almost round section, but the true mercury twist is more open. Sometimes one or more corkscrews, alternating with each other, encircle an inner twist. *See* corkscrew twist, above.

———, *mixed.* A type of twist produced by combining one or more air twists (see above) and one or more opaque white twists (or sometimes one or more opaque white threads). Sometimes a coloured twist (see above) is combined with an air twist, called a 'coloured mixed twist'.

———, *multiple series.* A combination of one twist with another (air or opaque) twist encircling it and a third twist encircling the second. They were produced by making a pattern of holes in the top of a rod, covering these with molten glass, drawing and twisting, and then adding more molten glass and repeating the process two or more times, using an air twist or an opaque twist (abbreviated for reference purposes as MSAT and MSOT respectively). *See* also double series (above) and single series (below) twists.

———, *outlined.* A type of twist having an outside thread of a colour to contrast with the white opaque twist.

———, *single series.* A type of twist with one column of spiral decoration, either an air twist (see above) or an opaque twist surrounded by clear glass, with no outside encircling twist. These are sometimes abbreviated for reference as SSAT and SSOT. *See* double series twist, above.

———, *tape.* A twist made of a strip or tape of opaque glass instead of threads of glass. Sometimes tapes of different widths are combined in a spiral twist, and sometimes they are part of a double-series twist (see above).

———, *tartan.* A type of twist in the form of several interlaced spirals of different colours, e.g., red, blue, and green, often enclosing a white spiral core.

———, *thread.* A twist made of a single thread of opaque glass, sometimes in a vertical column or in a spiral, and usually in conjunction with another type of twist in a double-series twist (see above).

———, *triple.* An extension of a double series twist (see above).

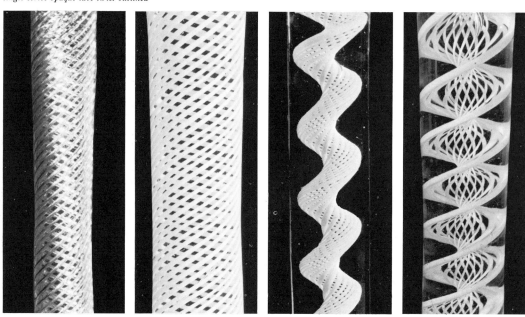

Tyneside glassware. Glassware (apart from that made at Newcastle-upon-Tyne in the 17th/18th centuries), mainly inexpensive PRESSED GLASS, made in the 1880s in great quantity at several factories in northeast England. It was made by SOWERBY'S ELLISON GLASSWORKS, by Greener & Co. at the Ware Glass Works in Sunderland (its mark, a lion with a halberd), and by George Davidson & Co., at the Teams Glass Works at Gateshead-on-Tyne (its mark, a lion on a battlement). Other small factories made like ware. *See* NEWCASTLE GLASSWARE.

Tyrian glass. A rare type of ART GLASS developed by FREDERICK CARDER at STEUBEN GLASS WORKS, *c*. 1916–17. The body of the pieces shows two colours: a pale blue-green when the piece is first fired but which, after reheating in a controlled atmosphere, changes to purple. Some examples show both colours, blending from greenish-turquoise at the top to purple toward the base. Some have added decoration of applied leaves and trailed threads of gold AURENE GLASS smoothed into the surface; sometimes the Aurene glass was sprayed AT-THE-FIRE with stannous chloride to provide an IRIDESCENT sheen. The name, given by Carder, was suggested by the imperial purple fabrics of Tyre. All pieces, except a few experimental ones, are signed.

Tyrian glass. Vase designed by Frederick Carder for Steuben Glass Works, 1917: Gold Aurene trailed decoration, neck pastel green, lower part purple. Ht 26·7 cm. Corning Museum of Glass, Corning, N.Y.

U

ultra-violet test. A test employing ultra-violet radiation, used to identify some glassware and ceramic ware, which produces a fluorescence with a fairly characteristic colour. The reaction depends on the presence of fluorescent elements in the body of a piece. The method is sometimes used to seek to distinguish PAPER-WEIGHTS made by BACCARAT, CLICHY, and ST-LOUIS, but the result is not always conclusive. *See* LEAD TEST.

undercutting. The process of decorating glass in HIGH RELIEF by cutting away part of the glass between the BODY of a piece and its decoration, leaving an intervening open space. *See* VASA DIATRETA.

underglaze decoration. A very rare type of decoration on glassware, involving the use of applied coloured threads beneath and fused to an outer layer of clear glass. A unique example is a fish-shaped hollow receptacle of Egyptian glass from the late XVIIIth Dynasty (1567–1320 BC) which has blue glass spots and lines forming a pattern applied to CORE GLASS before it was covered with transparent glass; it is in the Brooklyn Museum. *See* Elizabeth Riefstahl, in *Journal of Glass Studies* (Corning), XIV (1972), p. 11.

undine. A glass vessel for use by a physician, to irrigate a part such as an eye or a nasal passage. It is globular, with an opening near the top for filling and opposite the opening a projecting thin curved tube for irrigating.

unfired painting. *See* COLD COLOURS.

unguentarium. A small receptacle for toilet preparations (e.g. oil, scent, kohl). They were made of glass in numerous sizes (usually from 3 to 16 cm. high) and forms. Some examples made of CORE GLASS are in the form of various types of GREEK VASES (e.g. ARYBALLOS, ALABASTRON, LEKYTHOS, AMPULLA) and have COMBED DECORATION. Others of the Roman period were of coloured or colourless glass and were of spherical or conical shape, with a slender neck. Sometimes two are joined together, and some are in fanciful shapes, e.g. pigs, fish, etc. Many have a high overhead loop-handle or loops at the side; *see* HANGING UNGUENTARIUM. Examples were made of ROMAN GLASS from the 1st century AD and later of ISLAMIC GLASS in the 5th/8th centuries. Many such objects were deposited in Roman tombs, mistakenly thought to have been used to hold tears of the mourners, and were sometimes called 'tear bottles' or 'lachrymatories'; the German term is *Salbgefäss*, the French *lacrymatoire*. Also called a balsamarium. *See* DOUBLE UNGUENTARIUM; CANDLESTICK UNGUENTARIUM; GLOBULAR UNGUENTARIUM; TEST-TUBE UNGUENTARIUM.

Union Glass Co. A glasshouse established in 1851 at Somerville, Mass., by Amory and Francis Houghton, who left the company in 1864 when they acquired the Brooklyn Flint Glass Co. from JOHN GILLILAND. The Union company continued under a Mr Dana and was prosperous *c.* 1870/85, making PRESSED GLASS, SILVERED GLASS, and BLANKS. After the death of Dana, *c.* 1900, it was continued by his son-in-law, Julian de Cordova, and thereafter made ART GLASS until it closed in 1924. It is best known for its KEW BLAS GLASSWARE.

uranium. A metallic agent used in colouring glass, producing a fluorescent yellowish-green or greenish-yellow colour. When it is mixed with antimony, topaz and amber colours result. *See* URANIUM GLASS.

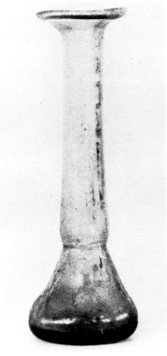

unguentarium. Clear Roman glass, 1st/2nd centuries AD. Museo Civico, Adria.

uranium glass. A type of glass produced by the use of URANIUM in the BATCH. It is yellowish-green (ANNAGRÜN) or greenish-yellow (*Annagelb*). It was developed by JOSEF RIEDEL, and was later produced by GEORGES BONTEMPS at CHOISY-LE-ROI. *See* VASELINE GLASS.

urinal. A receptacle, appropriately shaped for male or female use, to hold urine, but in the past used for uroscopy. Early examples, from before *c.* 1500, were shaped like a pouch, but later types were globular with a vertical neck, and those for male use were globular with a horizontal neck at the top or the modern shape of hemi-ovoid with a flat bottom and uptilted neck. It has not been recorded where or when urinals were first made, but they are mentioned in texts as early as 1300. Glass examples have been made in England since the 16th century. They are still being made, of colourless glass, for invalid and bedroom use. *See* BOURDALOU.

urn. A classical form of various shapes and styles, but usually its BODY is globular or ovoid and curved inward toward a stemmed foot, and the mouth is smaller in diameter than that of the body. The urn often has two side-handles. In classical times it had many uses: for holding liquids, for ornament, for holding lots to be drawn, and for containing the ashes of the dead; but all vessels for the last purpose are not to be classified as urns, despite popular usage. Although urns are generally made of porcelain, they are also made of glass.

ushabti (also *shabti, shawabti*). A type of ancient Egyptian mummiform tomb-figure, generally made of EGYPTIAN FAIENCE but very rarely of glass, which was deposited in the tomb with the mummy of the deceased, intended to serve his spirit in the after-world. The earliest datable examples of glass sculpture are two such pieces, probably made at the Royal Factory at Thebes, during the XVIIIth Dynasty, in the reign of Amenhotep II (1436–1411 BC); these are of opaque blue glass, said to have been moulded and then tooled. Most *ushabtis* bear inscriptions in hieroglyphs. They are usually about 10 cm. high, but range from 2 cm. to 90 cm., the early glass examples being from 17·5 cm. to 38 cm. Forgeries are known. *See Journal of Glass Studies* (Corning), II (1960), p. 11.

unguentarium. Hanging double unguentarium, Syracuse, 4th century AD. Ht 18 cm. Museo Vetrario, Murano.

V

vacuum tube. A sealed glass tube with the contained gas exhausted to a pressure low enough to permit the passage of electric discharges between metallic electrodes projecting into the tube from the outside.

vacuum tube. Made by Dr H. Grissler, 1876. Science Museum, London; Crown copyright.

Valor Cup. Design by John Monteith Gates for Steuben, 1940; replica from Gates Collection. Ht 38 cm. Courtesy, Steuben Glass, New York.

Valor Cup, The. A large two-handled covered URN designed in 1940 by JOHN MONTEITH GATES and made by Steuben Glass, Inc., for the British War Relief Society to commemorate the Battle of Britain. It bears the Royal Coat-of-Arms. It was made by John Jansson and engraved by JOSEPH LIBISCH. A replica was presented to Gates by the Steuben company upon his retirement.

Val-Saint-Lambert, Cristalleries du. The largest maker of crystal glass in Belgium. It was established in 1825 at Seraing-sur-Meuse, near Liège, on the site of the Val-Saint-Lambert Cistercian monastery that was founded in 1202 and abandoned in 1797. The monastery was bought in 1825 by François Kemlin (1784–1855), formerly a leading glass-maker at the VONÊCHE GLASSWORKS. Kemlin, with Auguste Lelièvre (1796–1879), also formerly with Vonêche, built a new glass furnace that was started in 1826, operating with English craftsmen. Kemlin was succeeded as manager by Lelièvre from 1838 to 1863, by Jules Deprez from 1863 to 1889, by Georges Deprez from 1894 to 1908, and by Marcel Fraiport from 1908 to 1940. The artistic director from 1926 was Joseph Simon (1874–1940), succeeded by his former assistants, CHARLES GRAFFART (1893–1967) from 1942 to 1958, and René Delvenne (1902–72) from 1958 to 1967. The factory first made crystal glassware in the English styles, but later produced ART GLASS in the ART NOUVEAU and ART DECO styles. It made some PAPER-WEIGHTS in the mid-19th century. It is now one of the world's leading makers of cut and engraved glassware. In 1879–83 it acquired three subsidiaries near Liège and Namur, and in 1972 a plant in Canada. It is under the control of the Société Générale de Belgique. For a detailed history and many illustrations, *see* Joseph Philippe, *Le Val-Saint-Lambert* (1974).

vandyke rim. A type of RIM that on the bowl of some objects is scalloped (usually in TREFOIL pattern) and on the foot of some drinking glasses is in the form of a series of sharply pointed motifs, either cut or engraved. The term is derived from the scalloped lace collars in portraits by Sir Anthony Van Dyck (Dyke) (1549–1641).

vasa diatreta (Latin). Glass vessels, the body of which is surrounded by a network or other openwork pattern of glass that is attached by small glass struts. The Latin name was given by J. J. Winckelmann on his mistaken theory that the DIATRETARII in Roman times did only such RETICULATED work, whereas they were in fact glass-cutters generally, as distinguished from the VITREARII who were the glass-makers and glass-blowers. For examples of *diatreta* work, *see* CAGE CUP; LYCURGUS CUP; SITULA PAGANA; TRIVULZIO BOWL.
   Process of production. The process of making cage cups and other examples of *vasa diatreta* has been disputed by many authorities:
   (1) The inner receptacle and the outer network (or other outer relief-decorated part) were made independently and then fused or soldered together. This view has been discarded.

(2) The double-walled vessel was made by inserting the separately made inner glass into the outer glass without fusing them together, and then pushing through the heated outer wall tiny glass rods to connect the two walls, after which the outer wall was ground to form the cage or other design. This view has now been discredited.

(3) The outer wall was cut and undercut from a single cold mass (possibly mould-blown), leaving the cage (or other outer relief-decorated part) standing free from the inner receptacle except for the small connecting glass struts. This view of a lapidary process, of meticulously grinding a thick mass of clear or flashed glass so as to leave the tiny struts that connect the cage with the inner body, was held by W. B. Honey and is now authoritatively considered confirmed by observation of grinding marks on the outer wall of the inner receptacle and of the same direction of air bubbles in adjacent parts of the inner receptacle and the outer relief work.

vasa murrhina. (1) The Latin term applied to MURRHINE GLASS-WARE. (2) A type of ART GLASS, made in the United States, that has an inner layer of coloured glass streaked with flakes of vari-coloured metal fused into the glass and covered with a FLASHING of clear glass. It was developed *c*.1880/85 by Dr Flower at the then idle plant of the CAPE COD GLASS CO. *See* MURRINA.

Vasart Glass, Ltd. A glasshouse founded in 1948 at Perth, Scotland, by the father and brothers of PAUL YSART after they left MONCRIEFF GLASSWORKS. It made vases sold as 'Vasart Glass', and made PAPER-WEIGHTS (mainly with MILLEFIORI decoration) that often formed part of various utilitarian objects. In 1960 it was reorganized by Stuart Drysdale, a lawyer, and in 1963 a major interest in it was purchased by the Scotch whisky distilling firm of Wm. Teacher & Sons, Ltd. The business and personnel were transferred in 1963 to STRATHEARN GLASS, LTD., which moved to a new factory built in 1964 at nearby Crieff.

vase. Covered cut glass vase, English, *c*.1815. Courtesy, Derek C. Davis, London.

vase. A type of vessel, made in many forms, styles, sizes, and SECTIONS, but usually rounded and taller than it is wide. It is used mainly for ornamental or flower-display purposes. Some have an ornamental COVER. *See* CANDLESTICK VASE; SÈVRES-TYPE VASE.

vase à oreilles (French). Literally, vase with ears. A type of VASE, introduced at Sèvres in the 1750s, of baluster shape, supported by a stemmed and spreading base. It has a short neck and two scroll-handles arising from the sides of the mouth and curving down to the shoulder; the handles have a profile resembling the human ear, hence the name. Such vases were usually made in porcelain, but some Bohemian examples were made of glass, including some miniatures in OPAQUE WHITE GLASS with enamelled decoration.

vase-candlestick. A type of CANDLESTICK resting on a heavy base, of silver or other metal, to provide stability. Some have, instead of a columnar stem, a glass urn-shaped bowl (supported on metal legs) from the rim of which are suspended pendent LUSTRES or chain festoons. They were sometimes made of porcelain or pottery but some were of CUT GLASS. They were made from *c*.1777 to the late 19th century. *See* CANDLESTICK VASE.

vase with enclosed bouquet. Fixed interior polychrome floral bouquet, Venice, *c*.1708. Ht 34 cm. Royal Collection, Rosenborg Castle, Copenhagen.

vase lustre. A type of VASE, decorated with LUSTRES and made of coloured glass, for decorating a mantelpiece, and usually made in pairs. They are in the form of a tall GOBLET or CHALICE, usually having a scalloped RIM and, hanging around the rim, a circle of ICICLES that tinkled in a breeze. They were made in Bohemia in the Victorian period, often with opaque coloured FLASHING that was cut to show the transparent glass beneath. Many had oil GILDING. Copies were made in England.

vase with enclosed figure. Clear glass with internal ovoid bluish glass pedestal supporting figure (probably porcelain) of boy, Venice, 16th/17th century. Ht 22·5 cm. Victoria and Albert Museum, London.

vase with flower stalk. Vase with crested handles and flower stalk, Venice, c.1708. Ht 43·5 cm. Royal Collection, Rosenborg Castle, Copenhagen.

vase with enclosed bouquet. A large vase having within the bowl a glass floral bouquet attached to the bottom. They were made in MURANO in the 17th/18th centuries, and in Shiraz in the 19th century. *See* FLOWER-ENCLOSING FLASK.

vase with enclosed figure. A type of vase which has inside the glass body a figure, e.g. a Venetian vase of the 16th/17th century which has inside the body an upright glass bulb upon which sits the figure (probably of Saint-Cloud porcelain) of a boy.

vase with flower stalk. A VASE having projecting upward through the mouth a tall flower stalk, complete with leaves and flowers, also in glass. They were made in MURANO in the 17th/18th centuries and also in south Germany in the 18th century.

vaseline glass. Glass made with a small amount of URANIUM, which imparts a yellowish-green or greenish-yellow colour and, according to some, a greasy appearance like vaseline, e.g. glass known as ANNAGRÜN or 'Annagelb'.

vaso fazzoletto (Italian). Literally, handkerchief vase. Same as HANDKERCHIEF BOWL.

Vaupel, Louis F. (1824–1903). A master glass-engraver who worked at the NEW ENGLAND GLASS CO. at Cambridge, Mass., for 32 years. He was born at Schildhorst, Germany, and at 12 worked with his father in a local glasshouse. In 1836 the family founded a glassworks at Breitenstein, where he became a glass-blower and engraver. In 1850 he migrated to the United States and in 1853 became engraver at the New England Glass Co., remaining until he retired in 1885. Thereafter he did some wheel-engraving at his home until c.1890. His engraving was in the Bohemian style, and he often made commemorative or presentation pieces. His early work was on uncoloured glass, but later he engraved coloured flashed glass, mainly ruby coloured. A few examples are signed.

Vauxhall. Site (then called Foxhall) of a glasshouse in London operated by the DUKE OF BUCKINGHAM from c.1663, for which he had the sole privilege to make 'mirror plate'. It was managed from 1671 to 1674 by JOHN BELLINGHAM.

vegetable dish. A dish, usually oval and having a wide rim, accompanied by a flat-topped cover similarly shaped which, when its detachable handle was removed, could be used to form a pair of such dishes. They were derived from silver forms, and examples made in the United States were sometimes of LACY GLASS.

Vendange Vase. A large AMPHORISKOS of blue glass, with CAMEO decoration in white in the manner of the PORTLAND VASE, but with deeper relief. The white decoration depicts scenes with Bacchanalian *putti*, curiously placed not at the sides, but beneath the handles, with embellishments of vines and doves, and (below) an encircling scene with sheep. Beneath the flat bottom there is a base-knob, unlike the Portland Vase. It was found at Pompeii, and is thought to have been made in Rome. It is sometimes called the 'Pompeian Blue Vase'. *See* AULDJO VASE.

Venetian ball. An object, usually globular, composed of a collection of fragments of CANES of various patterns and colours compressed together without any design and then pressed into a transparent glass bubble which is fused to the ball by sucking out the trapped air. They were made from c.1840. Such objects are not PAPER-WEIGHTS, but some paper-weights have been made by the use of such fragments, and are termed 'scrambled', 'macédoine' or 'pell-mell'.

Venetian glassware. Glassware made both in Venice from *c*.450 AD when glass-makers from AQUILEIA fled there and were soon joined by others from Byzantium, and in MURANO from before 1292 to the present day. By 1268 there was a guild of Venetian glass-makers which imposed strict regulations, including a ban on the divulging of trade secrets outside Venice. By 1292, when Venetian glassware was well-established and highly regarded, a fear of furnace fires led to an edict ordering the many glasshouses to move to the nearby island of Murano, where most Venetian glass continues to be made today. The early glassware consisted mainly of beads, jewellery, mirrors, and window glass. Venice exported glass from *c*.1300, but its importance followed the development, by the use of MANGANESE, of CRISTALLO in the 15th century. This SODA GLASS was thin and fragile, not suitable for engraving, and this prompted decoration during production, such as the use of TRAILING and THREADING, especially the making of wine glasses with elaborate threading for stems, e.g. WINGED GLASSES. Glass-making was at its height in the 16th and 17th centuries, and Venetian products were copied throughout Europe (as glassware *à la façon de Venise*, difficult to distinguish from Venetian ware); but the industry deteriorated in the 18th and 19th centuries, when Venice lost its dominant commercial position and as the more durable POTASH GLASS and LEAD GLASS of the north superseded the fragile soda glass. Venice excelled in making coloured glass (blue, green, purple) and AGATE GLASS, and in the development of FILIGRANA, MILLEFIORI, MOSAIC, and ICE GLASS. Only a small amount of decoration was done by engraving, but enamelling and gilding were important styles in the 15th/16th centuries. The Venetian reputation was revived by the late 19th-century and 20th-century production of glass at Murano by SALVIATI, BAROVIER, VENINI, CENEDESE, ALFREDO BARBINI, and others, alongside the making of large quantities of beads and souvenir glassware for tourists. *See* ENAMELLING, *Venetian*.   [*Plate XV*]

Venetian scenic plates. A series of OPAQUE WHITE GLASS (LATTIMO) plates painted in iron-red monochrome with 24 different views of Ven'ce. For many years it had been thought that a set of 24 such plates was decorated with views from engravings published by G. B. Brustolone after Canaletto; they had been brought to England from Venice by

Vaupel glassware. Chalice of ruby flashed glass by New England Glass Co., wheel-engraved by Louis Vaupel, 1872. Ht 22·8 cm. Courtesy, Antique & Historic Glass Foundation, Toledo, Ohio.

Vendange Vase. Blue glass vase cased with opaque white glass, with encircling cameo carved scenes. Ht 30 cm. Museo Nazionale, Naples.

Horace Walpole and were sold at the Strawberry Hill sale (7 May 1842) to various buyers. The existence of duplicates later showed that there were at least two such sets, but in 1959 it was established by Robert J. Charleston that: (a) there were three such sets (there being three plates with the same scene in some cases); (b) they had been made in or soon after 1741 (before the Brustolone engravings, which were not published until 1763); and (c) they had been painted (probably at the AL GESÙ GLASSWORKS of the MIOTTO family at MURANO) from 15 engravings of the Grand Canal after Canaletto by Antonio Visentini, c. 1735–37, and from nine other Venetian scenes from original etchings by Luca Carlevaris made in 1703. For a full discussion and illustrations of such plates, *see* Robert J. Charleston, 'Souvenirs of the Grand Tour', *Journal of Glass Studies* (Corning), I (1959), p. 63.

Venini & Co. A glassworks at MURANO founded in 1921 by Paolo Venini (1895–1959), of Milan and Giacomo Cappellin (1887–   ) of Venice. Cappellin had previously ordered for his shop in Milan glassware to be made in Murano by Andrea Rioda (d. 1921), in forms and styles of Venetian glass of the 16th century. In 1921, just before Rioda died, Venini and Cappellin took over the Rioda glassworks, engaging as designer Vittorio Zecchin (1878–1947), and continuing to make glassware inspired by old paintings. In 1926 the firm dissolved, Cappellin withdrew, and Venini & Co. was formed, reviving the production of MILLEFIORI and FILIGRANA glassware, and developing many new techniques, especially with original surface treatments. Although Venini designed, he also engaged outside designers, e.g. the Finnish glass-maker TAPIO WIRKKALA, the sculptor Napoleone Martinuzzi (from 1928), the Swedish potter Tyra Lundgren (from 1934 to 1940), and Salvador Dali. The factory is now owned by Venini's daughter and her husband, Ludovico Diaz de Santillana. Among the new styles developed by Venini are: VETRO BATTUTE, VETRO COMPOSTO, VETRO CORROSO, VETRO GROVIGLIO, VETRO MEMBRANO, VETRO PENNELATO, VETRO PULEGOSO, VETRO SOMMERSO, VETRO TESSUTO, VETRO PEZZATO. In the 1950s Venini introduced his well-known HANDKERCHIEF BOWL (*vaso fazzoletto*) made of VETRO A RETORTI (ZANFIRICO glass).

vermicular. (1) A decorative pattern in the form of innumerable painted lines with tiny curves, said to suggest the appearance of having been eaten by worms or having been covered by their tracks. The pattern was created at Sèvres for a porcelain ground decoration in the 1750s, and was adapted as enamelled or gilt decoration on Bohemian glassware in the 19th century. (2) A wavy clear glass thread TRAILED to encircle a COLLAR on a STEM or the neck of a BOTTLE or JUG.

Verona glass. A type of ART GLASS that is clear with overall decoration made by covering the design with wax and then spraying the object with acid, leaving the design on the glass when the wax was removed. It was made by the MT WASHINGTON GLASS CO.

verre de fougère (French). Literally, fern glass. A primitive type of greenish glass produced in France after the Roman era, the ALKALI ingredient of which was provided by POTASH made from burnt bracken (fern), just as burnt beechwood was used in Germany during the same period for making the potash used in WALDGLAS. This type of glass was produced in forest areas where this source of potash was readily available, such as the Dauphiné in south-eastern France, and where a large quantity was made in the 14th century. The term is often used interchangeably with *Waldglas* as the types of glass were similar.

Verre de Soie. Literally, silk glass. (1) A type of ART GLASS developed by FREDERICK CARDER at STEUBEN GLASS WORKS soon after 1903 and made until 1918. It has a LEAD GLASS (FLINT GLASS) body with a silky surface made IRIDESCENT (ranging from slightly tinted to rainbow-hued) by a spray AT-THE-FIRE of stannous chloride, with thin radiating

Venetian scenic plate. Opaque white glass with scene in red enamel 'The Arsenal', from etching by Luca Carlevaris; probably Al Gesù Glassworks, Murano, c. 1741. D. 22·7 cm. British Museum, London.

lines on the metallic surface where the glass expanded upon reheating after being sprayed. Such pieces are further decorated with TRAILING, PRUNTS, or engraving, and sometimes with a turquoise-coloured glass ring applied around the rim. The ware includes vases, stemware, etc. Such glass was originally termed 'Flint Iridescent', and was referred to at the factory as 'VDS'. *See* AQUA MARINE GLASS. (2) A term applied to PEARL SATIN-GLASS made by STEVENS & WILLIAMS, LTD., in 1885–86.

verre de tristesse (French). Literally, glass of sadness. A type of glass vase made by ÉMILE GALLÉ, from *c.* 1892. The vases were intended to be buried with the dead, although this practice was not sanctioned by the Church. Several examples, in dark brown (blackish) glass, with carved decoration, are known.

verre doublé (French). The same as FLASHED GLASS or CASED GLASS. It was so-called by EMILE GALLÉ.

verrerie parlante (French). Literally, speaking glassware. Glassware made by EMILE GALLÉ from 1884 that bears on the piece a quotation from a poet (e.g., Mallarmé) or prose writer (e.g., Victor Hugo) that inspired the form and decoration.

Verzelini, Jacopo (1522–1616). A Venetian glass-maker, who, after working twenty years in Antwerp, was brought to England in 1571 by JEAN CARRÉ to manage his CRUTCHED FRIARS GLASSHOUSE. When Carré died in 1572, Verzelini took over the factory. In 1574 he was granted by Elizabeth I an exclusive licence for 21 years to make CRISTALLO glass in the Venetian style, and was expected to teach glass-making to the English. After the factory was destroyed in 1575 by fire (suspected of being caused by competitors) Verzelini rebuilt it, having become naturalized and so entitled to own property. He thrived for 17 years (becoming known as 'Mr Jacob'). The factory was the first in England to use the technique of DIAMOND-POINT ENGRAVING, making commemorative glasses engraved probably by ANTHONY DE LYSLE. Verzelini retired in 1592, and the remaining three years of his licence were taken over by SIR JEROME BOWES. After his retirement, the Crutched Friars Glasshouse was managed by his sons, Francis and Jacob, who became involved in litigation with SIR JEROME BOWES and his assignees; they went to prison in 1598. *See* VERZELINI GLASSWARE.

Verzelini glassware. Glass GOBLETS generally attributed to the CRUTCHED FRIARS GLASSHOUSE of JACOPO VERZELINI, of which there are only nine known surviving pieces; eight, dated from 1577 to 1586, are decorated with DIAMOND-POINT ENGRAVING, one having added gilding; the ninth, dated 1590, is decorated only with gilding. The engraving is generally thought to have been done by ANTHONY DE LYSLE, or in the case of the BARBARA POTTERS GOBLET, by workmen at the Verzelini factory after it was taken over by SIR JEROME BOWES. *See* DIER GOBLET; G S GOBLET; GEARES, WENYJFRID, GOBLET; CLUNY GOBLET. The other five goblets, respectively, bear the initials and dates: RB : IB 1577; AT : RT 1578; AF 1580; KY 1583; MP : RP 1586. *See* LORD BURGHLEY TANKARD.

vesica pattern. A style of engraved decoration used mainly on DECANTERS made at Cork, Ireland, from *c.* 1783. It is made by pairs of incised intersecting curved lines forming a connected series of horizontal flat pointed ovals joined directly together or with small diamond-shaped figures between them. Sometimes the ovals enclose an 8-pointed incised star. There are usually six such ovals engraved around the widest part of the decanter. *See* IRISH GLASSWARE, *decanters*.

vetro. The Italian term for glass. Numerous types of ornamental glass have been made at MURANO, especially in the 20th century with names locally adopted, e.g. *vetro pezzato*; these are usually listed here under the name of the qualifying term, e.g. *pezzato, vetro*.

Verre de Soie. Bowl with spiral trailing and turquoise prunts, designed by Frederick Carder for Steuben Glass Works, *c.* 1916. Ht 26·5 cm. Smithsonian Institution, Washington, D.C. Courtesy, Corning Museum of Glass, Corning, N.Y.

Verzelini glassware. Goblet of soda glass by Verzelini, engraved probably by Anthony de Lysle; initials 'AT : RT' and date 1578. Ht 21 cm. Fitzwilliam Museum, Cambridge. Courtesy, Delomosne & Son, London.

vetro a reticello. Drinking glass with bowl in *vetro a reticello*, blue foot and knopped stem, Venice, 16th century. Ht 18·7 cm. Kunstsammlungen der Veste Coburg.

vetro a reticello. Magnified detail showing crossed opaque white glass threads and entrapped air bubbles. From *Venetian Glass at Rosenborg Castle* by Gudmund Boesen (1960).

vetro a fili (Italian). Literally, threaded glass. A type of glass decorated with opaque white or with coloured threads of glass, or both, embedded in clear glass in continuous lines without any crossing of the threads (as in VETRO A RETORTI or VETRO A RETICELLO), the lines being in a spiral or helix pattern (e.g., on plates) or in a spiral or volute pattern (e.g., on vases). It is a type of FILIGRANA (FILIGREE) decoration (sometimes called LATTICINO when the threads are all white).

vetro a reticello (Italian). Literally, glass with small network. A type of glass decorated with opaque white or with coloured threads of glass, or both, embedded in clear glass in the form of criss-cross diagonal threads making an overall diamond lattice network, as originally made at Murano. It is a type of FILIGRANA (sometimes called LATTICINO) decoration. Terms formerly used at Murano are *a redexelo* and *a redexin*. There are three separate varieties, depending on whether it is made with fine threads, coarse threads, or fine and coarse threads running in opposite directions. As the threads protrude slightly, tiny air bubbles (sometimes microscopic) are entrapped within each criss-cross diamond, the size depending on the process of production; but on some few rare examples there are, instead of the air bubbles, pairs of thin wavy lines running in only one direction between the rows of white threads. The style has been used on vases, bowls, jugs, etc., and also on plates (where the network often becomes distorted toward the centre or the edge of the plate). The German term is *Netzglas*, and a useful English term would be 'network glass', as distinguished from 'lace glass'. *See* VETRO A RETORTI; VETRO DI TRINA.

  Method of production. Several processes have been described by writers, and in Murano, according to local glass-makers, several methods have been used, methods (1) and (2) below being most authoritatively stated:

  (1) A bulb (or cylinder) of blown glass in a mould, on which have been picked up, on a GATHER, parallel threads of glass (almost always white opaque threads) running diagonally in one direction (resulting from twisting the PARAISON after gathering the threads) and which is then blown into another similar bulb (or cylinder) with threads running diagonally in the opposite direction; the two bulbs (or cylinders) become fused together. The difficulty arises from the necessity of having the threads on each bulb exactly equidistant so that they will form equal and similar diamonds (except for distortions toward the extremities of the glass).

  (2) A bulb of glass is made, as described above, with diagonal threads, and then half of the bulb is bent into the other half and fused, thus making the criss-cross threads. This method assures equal spacing, but is a difficult process.

  (3) A bulb of glass is made, as described above, with diagonal threads, and then the glass-blower sucks on the tube and collapses the farther half into the nearer half, making a double-wall with criss-crossed threads.

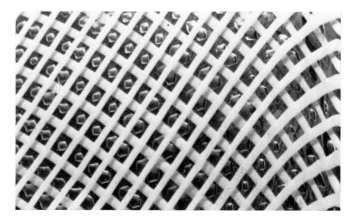

(4) A bulb of glass is made, as described above, with diagonal threads, then it is cut and twirled to make, by centrifugal force, a flat plate; another such plate is made with threads in the opposite direction, and then the two plates are fused together. This method also presents the difficulty of obtaining equal spacing on both plates.

In all these methods there develop, within the interstitial spaces made by the crossed threads, small (sometimes barely perceptible) air bubbles, but occasionally they are elongated into thin wavy lines, and the enclosed areas sometimes have unequal sides. After the piece of glass with the crossed threads is made, it is reheated, then blown and manipulated into various forms, e.g., a vase, plate, jug, etc.

vetro a retorti (or **retortoli**) (Italian). Literally, glass with twists. A type of glass decorated with twisted threads of glass embedded in clear glass, as originally made at MURANO. Today, it is usually termed in Venice and Murano ZANFIRICO or *sanfirico*. The term is applied to a style of decoration with parallel adjacent CANES (vertical or spiral) of glass having embedded threads in various intricate patterns (as in the STEMS of some English wine glasses) and made by flattening the canes and fusing them together. It includes decoration made with opaque white or with coloured glass threads, or both, embedded in the fused canes. It is included in the broad general term FILIGRANA (FILIGREE), and also in LATTICINO when only white threads are used. The style has been used for plates, bowls, vases, etc., and sometimes for PAPER-WEIGHTS. *See* VETRO A RETICELLO; VETRO DI TRINA; STRIPED GLASS.

vetro di trina (Italian). Literally, lace glass. A term that has been loosely used for various types of FILIGRANA glass. It has been said recently by Luigi Zecchin, of MURANO (*Journal of Glass Studies* (Corning), XVII (1975), p. 113), that the term has only a 'romantic' origin, and he has informed the author that it is not used today in Murano, and hence should be abandoned. It has been applied to VETRO A RETICELLO by Apsley Pellatt, London (1849), Paul Perrot, United States (1968), and several other writers, but other authorities have used it to refer merely to LATTICINO glass 'of intricate pattern' without distinguishing which type of FILIGRANA was intended, e.g., W. B. Honey, London (1946), and Hugh Tait, London (1968), while Astone Gasparetto, Murano (1958) and Prof. Giovanni Mariacher, Venice (1961), have omitted it from their dictionaries of glass, and Gudmund Boesen, Rosenborg Castle, Copenhagen (1960), also avoids it, using merely *vetro a reticello* and *vetro a retorti*. As these two terms suffice to designate adequately the two styles, we deem it preferable that any further confusion should be avoided by following the Italian experts and abandoning '*vetro di trina*' as superfluous and of no historical significance, especially as use can also be made of the corresponding English terms 'network glass' and 'lace glass' for these two types of filigree (*filigrana*) glass.

vetro a reticello. Two-part bowl, Venice, 16th century. Ht 10·8 cm. Seattle Art Museum, Wash. Gift of Mrs John C. Atwood Jr.

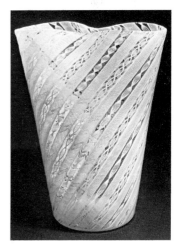

vetro a retorti. Vase made by Venini, Murano.

vetro a retorti. Magnified detail showing parallel canes with twisted designs. From *Venetian Glass at Rosenborg Castle*, by Gudmund Boesen (1960).

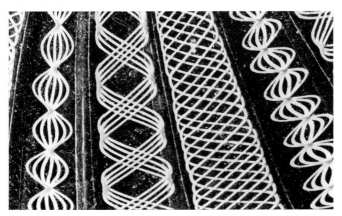

vial. A small glass bottle for liquids, usually medicines. The same as 'phial'. In Ireland it designates a type of SCENT BOTTLE that is oval and flat, with a small glass stopper enclosed within a screw-on silver cap.

Victory cup. A type of broad cylindrical cup from early Roman Imperial days, moulded with laurels and inscriptions commemorative of victory. These were forerunners of the CIRCUS BOWL.

vinaigrette. (1) A glass CRUET for holding vinegar. *See* PORTE-HUILIER. (2) A small glass bottle to contain scented vinegar, formerly intended to be used by ladies to avoid faintness.

Vineland Flint Glass Works. A glasshouse founded in 1897 at Vineland, New Jersey, by Victor Durand, Jr. (1870–1931) who, with his father, leased for it in 1925 the Vineland Glass Manufacturing Co. (established in 1892). Durand commenced making ART GLASS under the supervision of Martin Bach, Jr., who had formerly been with the QUEZAL ART GLASS & DECORATING CO.; it first copied Quezal glassware but soon created new colours and designs. After Durand died, the company was merged into the KIMBLE GLASS CO. in 1931, and the making of art glass was in general discontinued. The factory made many varieties of lustred glassware, including: a yellow glass called 'Ambergris'; glassware with overall trailing of fine glass threads, called 'Spider Webbing'; 'Peacock feather glassware', a flashed glass with a feather pattern; and 'King Tut' glassware, with an overall pattern of pulled threads. It also made, under Durand, many lamps with glass shades of various shapes and colours. Its ware was signed 'Durand' in script, sometimes with a superimposed 'V'.

violin flask. A BOTTLE in the shape of a violin, made in the United States in the 19th century, of MOULD-BLOWN glass, in many sizes and colours. They were often decorated with emblems and inscriptions.

violin flask. Mould-blown flask with stars. Henry Ford Museum, Dearborn, Mich.

Virtues, The Twelve Cardinal. A decorative allegorical motif depicting the Virtues (e.g. Faith, Hope, Charity, Humility, etc.). It was used as enamelled decoration on German HUMPEN. Sometimes the opposite motif was used, e.g. depicting vices such as laziness. Such *Humpen* were made in the 16th and 17th centuries, probably in Franconia.

vision aids. Various devices with one or two glass lenses made to assist an individual's short-range vision. They were first made in Italy from *c.* 1286. Their use increased after the invention of printing in 1436, and especially from the 16th century. The lenses were then made principally in Germany and the Low Countries, and were imported into England until *c.* 1630 when lenses were being made (and exported) by SIR ROBERT MANSELL and sold by him to the English spectacle-makers, who had a guild from 1629. The devices were made in many forms and styles in different periods, including: *nose spectacles* (spectacles with two lenses but no side-arms, having a bridge to rest on the nose); *temple spectacles* (spectacles with two lenses having tight-fitting side-arms to which were attached rings to press against the temples); *wig spectacles* (spectacles with two lenses and having long, or hinged double-length, side-arms at the end of each of which was a ring for threading a ribbon to be tied at the back of a wig); *quizzer* (single lens on a short frame, to be suspended by a ribbon; mainly for fashionable use); *pivoted eye glasses* (two lenses that fold over each other on a frame); *lorgnette* (two lenses that fold alongside, or into, a long ornamental handle); *fan spectacle* (one lens set at the pivot of a lady's fan); *monocle* (single lens, to be held in place by an eyelid); *spectacles* (two lenses with a nose bridge and two arms to be placed over the ears; early ones had straight arms, later ones curved arms bending around the ears); *pince-nez* (two lens with no side arms; to be clipped to the nose by a spring); *spy glass* (short telescopic lens for one eye); *opera glass* (two medium-strength convex lenses in an adjustable frame), *bifocals* (spectacles having each glass made up of two different lenses, for close and distant vision); *contact glass* (single lens without any frame and fitting snugly over the eye-ball).

Long-distance and scientific vision aids (e.g. binoculars, microscopes, telescopes) are beyond the scope of this book. *See* DIMINISHING LENS; MAGNIFYING LENS; WATER LENS.

Visscher, Anna Roemers (1583–1651). A scholar and poet of Amsterdam who, as a talented amateur glass-decorator, introduced in the second quarter of the 17th century the style of decorating glass with CALLIGRAPHY. She also decorated, before the time of FRANS GREENWOOD, a small number of pieces in STIPPLING, outlining the design in linear fashion and using stippled dots to produce light and shade. She mainly decorated BEAKERS and RÖMERS in DIAMOND-POINT ENGRAVING with calligraphic inscriptions, accompanied by representations of naturalistic flowers, fruit, and insects copied from contemporary prints. Some of her pieces are signed.

Visscher, Maria Tesselschade Roemers (1594–1649). The younger sister of ANNA ROEMERS VISSCHER, who also decorated glass with CALLIGRAPHY.

vitrearii (Latin). The glass-makers and glass-blowers of ancient Rome, as distinguished from the DIATRETARII, who were the cutters of glass.

Vitrearius, Laurence. A glass-maker, probably from Normandy, who established a glasshouse at Dyers Cross, near Chiddingfold in Surrey, before 1240. Laurence (the cognomen means 'glass-maker') supplied *c.*1240 window glass for King Henry III's Chapel in Westminster Abbey, probably from coloured glass he had imported from the Continent. His son, William le Verrir (also meaning glass-maker), and possibly his grandson, are said to have continued the business, making window glass and small bottles and flasks. *See* WEALDEN GLASSWARE.

vitrification. The process of changing certain materials into glass, or a glassy substance, by heat and fusion, such as occurs when SILICA (sand) and an ALKALI are heated to a sufficient temperature. *See* DE-VITRIFICATION.

Vitruvian scroll. A classical style of decoration in the form of a repetitive series of convoluted scrolls resembling stylized waves. Sometimes called the 'Greek wave pattern'. *See* CANOSA BOWL.

volunteer glass. A type of drinking glass, usually a RUMMER, engraved with inscriptions inspired by the British and Irish volunteers in the war with France in 1793, and hence made in the early 19th century.

Vonêche Glassworks. A glassworks established as the *Cristallerie de Vonêche* in 1778 in the Ardennes region near Namur, Belgium, under authority granted by Empress Marie Thérèse. After ten successful years, the problems created by the French Revolution in 1789–90 would have closed the factory if it had not been then purchased in 1802 by AIMÉ-GABRIEL D'ARTIGUES, who had operated the ST-LOUIS GLASS FACTORY in Lorraine. He operated it successfully with two assistants, François Kemlin and Auguste Lelièvre, until the French sovereignty ended in 1815 and the company lost its markets. Its operations were transferred in 1816 to the *Verreries de Sainte-Anne* at Baccarat which d'Artigues acquired. Kemlin and Lelièvre then founded the glassworks of VAL-SAINT-LAMBERT. *See* BACCARAT GLASS FACTORY.

W

wafer dish. A small flat glass dish on a stemmed FOOT (somewhat like a small TAZZA) on which to keep wafers (small flat adhesive disks, used as a seal on letters).

wafer stamp. A small columnar object similar to a LETTER SEAL except that the end of the shank has no INTAGLIO design but is smooth, to be used to press a glued wafer stamp to seal a letter or document. The shank was about 2·5 to 9 cm. long, and was sometimes rectangular, enclosing a single or double SULPHIDE (some being made by APSLEY PELLATT), embedded in a coloured ground.

waffle pontil mark. A type of PONTIL MARK made by attaching to the PONTIL, before it was attached to a PARAISON on a BLOWING IRON, a small GATHER of molten glass which was then pressed into a criss-cross pattern. The resulting 'waffle' allows the pontil to be separated from the glass object more easily and it remains on the object. Such marks appear on some STEUBEN glass.

Wagenfeld, Wilhelm (1900–). A German glass-designer who has worked at the glassworks of the Württembergische Metallwarenfabrik at Geislingen, and at the glassworks of the Rosenthal Porcelain Co.

waisted. Of smaller diameter at the middle than at the top or the bottom, and having sides that form a continuous inward curve, e.g. of some TUMBLERS or of the BOWLS on some drinking glasses.

Waldglas (German). Literally, forest glass. A primitive type of glass (greenish, yellowish or brownish) produced in the glasshouses (WALDGLASHÜTTEN) in the forests of Germany during the Middle Ages; the ALKALI ingredient of such glass was provided by potash from the ashes of burnt beechwood and other woods, just as bracken (fern) was used in France during the same period for the making of potash used in VERRE DE FOUGÈRE. The glass, produced in the forest areas where such source of potash was readily and plentifully available, was thick, robust, and primitive in form, and the pieces, usually mould-blown, were decorated with PRUNTS and TRAILING. The term is often used interchangeably with *verre de fougère* as the types of glass were similar; *Waldglas* is sometimes called 'green glass'. The natural colouring, though formerly little appreciated, has recently become recognized as an element of attractiveness and has been commercially imitated in modern glass.

Waldglashütte (German). A forest glasshouse such as existed in Central Europe from the early Middle Ages, when the glass-makers moved from time to time as they exhausted the local wood supply. The term no longer applied when the glass-workers settled in one place, especially later when they came under the patronage of a local ruler. The term *Hütte* (hut) is said to be derived from the peaked roof of such temporary quarters. Each year the craftsmen, who were organized into guilds, worked in the forests from Easter until St Martin's Day (11 November), while their families lived in the villages which grew up around the glasshouses.

walking-stick. A glass replica of a walking-stick, of various styles, generally about 1·20 to 1·40 m. high. Related pieces are those made as a shepherd's crook. They were sometimes made of OVERLAY glass with coloured inner glass encased in clear glass; but more often they were hollow, and decorated by being filled with confections ('hundreds-and-

Warmbrunn. Pokal with inserted silhouette portraits by J. S. Menzel, 1805. Ht 14·1 cm. Staatliche Kunstsammlungen, Kassel.

thousands' in haphazard mixture or in colour stripes) or lentils. They
were made in England in the first half of the 19th century. Some were
used in glass-makers' processions in Bristol and Newcastle, but it has
been said that they were also hung in homes as a spell against disease,
and were cleaned daily out of superstition, and that a broken one was
regarded as a bad omen. They are usually attributed to so-called
NAILSEA GLASSWARE; but some were made in the United States.

wall-pocket. A flower-vase made to be attached to a wall, often loosely
called a 'cornucopia' because most wall-pockets are of this shape.
Frequently made in ceramic ware, there are English examples in glass.
Sometimes called a 'wall vase'.

Walter, Almeric-V. (1859–1942). A French glass-maker who worked
at Sèvres and Nancy, producing objects in PÂTE-DE-VERRE, particularly
sculptured three-dimensional figures of lizards, frogs, crabs, and fish,
and also female forms; these were sometimes placed on the rim of a
bowl. He also made bowls and vases with coloured abstract patterns. He
had worked for DAUM FRÈRES at Nancy from 1908 to 1914, before having
his own studio after the First World War, making a variety of objects,
often trays. Some pieces signed by him also bear the name of a
collaborating designer, e.g. Henri Bergé. He also made copies of 18th-
century figures and TANAGRA-STYLE FIGURES.

Walter, Almeric. Bowl in *pâte-de-verre,*
c. 1925, signed 'A. Walter, Nancy', W.
17·8 cm. Kunstmuseum, Düsseldorf.

Walton, George (1867–1933). An English designer in many fields and
an interior decorator. He designed some CLUTHA GLASS for James
Couper & Sons of Glasgow between 1896 and 1898.

Walzenhumpen (German). Literally, cylinder beaker. A type of
HUMPEN decorated with the coat-of-arms of a member of the nobility.

Wanne (German). Literally, tub. A regenerative tank furnace (*see*
FURNACE, TANK) developed in 1856 by Friedrich and Karl Wilhelm
Siemens for making steel and adapted to glass-making by their brother
Hans at Dresden by 1860, hence known as the 'Siemens furnace'. The
glass ingredients were placed in one large tank and the molten glass
flowed out and was later remelted in pots. It was heated by gas. During
the 1870/80s its use spread world-wide for the mass production of flat
glass and bottles, but less so for tableware and luxury glass which,
needing smaller quantities, continued to be prepared in pots.

Warzenbecher. Thick *Waldglas* with
raspberry prunts and engraved
inscription, Germany, 1659. Ht
13·2 cm. Kunstmuseum, Düsseldorf.

Wapping Glass-house. A glass-house at Wapping (London) in which
the Duke of York (brother of Charles II), who later succeeded to the
throne as James II, is said to have had an interest and which, according
to an advertisement of 1684, used a mark with a lion and coronet. No
specimen of its glassware has been recorded.

Warmbrunn. A fashionable spa in the Hirschberger Tal in Silesia
where, in the 18th century, visitors during the season had their portraits
engraved on SPA GLASSES. The leading artist was CHRISTIAN
GOTTFRIED SCHNEIDER; there were six independent engravers there in
1733 and forty by 1743. Portraits in silhouette were made JOHANN
SIGISMUND MENZEL. Here FRIEDRICH WINTER established in the late
17th century a workshop for decorating CAMEO engraved glassware in
exuberant BAROQUE style, and also set up a retail shop for his wares.

Warzenbecher (German). Literally, wart beaker. A very heavy and
thick cylindrical glass vessel shaped like a TUMBLER, with flat bottom
and having the sides rounded at the top. They have on the lower two-
thirds of the exterior an overall decoration of scattered PRUNTS of
irregular shapes and sizes or diagonal rows of RASPBERRY PRUNTS,
above which there is usually an engraved inscription in German – 'Fill
me up and drain me/Throw me down/Pick me up/And fill me again!' –
thus confirming their use as a drinking vessel rather than as a candle-
holder, as would be suggested by their durable quality (whence they

have been termed by E. Barrington Haynes an 'unbreakable glass'). They were made of German WALDGLAS in the 16th/17th century. The type has sometimes been called *Kerzenbecher*. Some specimens have been adapted to serve as a reliquary (*see* RELIQUARY, *sealed*).

waster. Glassware that was defective in manufacture or that became out-of-fashion and hence unsaleable and was discarded, usually in a place known as a 'waster heap'. There are fewer wasters in glassware than in ceramic ware because those in glass were re-used as CULLET. When found on glassworks sites, wasters are valuable and often conclusive evidence for the identification and dating of wares, but some caution is necessary because wares from elsewhere which had been broken were often discarded with the rest. *See* DOCUMENTARY SPECIMEN; FRAGMENT.

water glass. A glassy substance with a high content of SODA, sometimes unintentionally produced when molten glass in the CRUCIBLE had too much SILICA and became thick and viscous, a fault which was sought to be remedied by the addition of more soda. The substance can be dissolved in water as a viscous syrupy fluid, and is used commercially for fireproofing, as a sealant (e.g., for preserving eggs), and for many other purposes. Sometimes called 'soluble glass'.

water-glass. (1) A drinking-glass of cylindrical form with an expanded foot, made from *c.*1680 to *c.*1770. The term was applied in England in the 18th century to a glass distinct from a TUMBLER, which was a glass with sides slanting slightly inward to the base. (2) A term formerly sometimes applied to a FINGER-BOWL.

water-jug. (1) A type of JUG with a globular body and a flaring neck and a lipped rim, and having one loop-handle opposite the lip. Such jugs occur in English and Irish glass, *c.*1780–1820, and sometimes have engraved decoration. (2) A jug made as part of a WATER-SET.

water-lamp. A type of lamp in the form of a large ovoid glass vase on a flat base and having a constricted neck widening to an everted mouth. It served as a vase, but also as a lamp-base when electric fixtures and a shade mounted on an insulated metal unit were inserted in the mouth after the vase had been filled with water. The water enhanced the optical effect (as with a WATER LENS) and provided greater stability. Examples were designed by FREDERICK CARDER at STEUBEN GLASS WORKS in the 1920s.

water lens. Used with candle as a lace-maker's lamp; constructed from a 17th-century description. Science Museum, London.

water-pot. Semi-opaque white glass with enamel decoration. Ch'ien Lung mark and period, 1736–95. Ht 5·8 cm. Percival David Foundation of Chinese Art, London.

water lens. A glass globular receptacle, usually with a vertical neck and a spreading base, which, when filled with water, was used as a condensing lens to concentrate the light from a nearby light source, e.g. an oil lamp or a candle. They were used by lace-makers, engravers, and others who required good illumination for close work. Sometimes several such objects were placed around a single light source to accommodate a group of workers. It was sometimes called a 'condenser glass', and sometimes erroneously a 'LACE-MAKER'S LAMP' or 'engraver's lamp'. A clear glass globe on a stemmed foot, made in Germany and dated 1677, and designated a *Studierlampe*, may have been intended to be filled with water and used as a lens for a lamp.

water-pot. A type of small Chinese pot for holding the water used to dissolve ink for writing or painting. Usually made of porcelain, some were made of glass in the reign of Ch'ien Lung (1736–95), decorated with enamelling in KU-YÜEH HSÜAN style.

water-set. A set of vessels for serving drinking water, usually consisting of a JUG, one or two TUMBLERS, and often a TRAY, all decorated EN SUITE. Some examples have all the units of glass.

Waterford glassware. Glassware made at the Waterford Glass House at Waterford, Ireland; the glasshouse was founded in 1783 by George and William Penrose, local merchants who financed its operation and brought in JOHN HILL, a glass-maker from Stourbridge, to take charge. Hill left in 1786 and JONATHAN GATCHELL, a protégé, succeeded him and became a partner in 1799 (when the Penroses retired); he managed the firm until his death in 1823, and was succeeded by George Gatchell until 1851, when the factory closed due to the high Irish tax on glass. It was the only glasshouse in Waterford in the period, and its products account for the fame of the Waterford name. The ware was of a wide variety (including DECANTERS and large FRUIT BOWLS), heavy in weight, deeply cut on the wheel (as contrasted with CORK GLASSWARE), and colourless. The long-held claim that Waterford glass has a blue tint has been authoritatively repudiated. No so-called 'Waterford chandelier' has been authenticated as made by this factory, although that in Waterford town-hall may well have been. A few decanters marked on the bottom 'Penrose Waterford' were presumably made before 1799; such pieces have three neck-rings, a wide pouring lip, and an ovoid body, with the bases usually encircled with moulded COMBED FLUTING.

A new company, Waterford Glass Ltd, opened in 1951; its modern plant has a world-wide export business in cut-glass ware.

Watteau subjects. Pastoral scenes, taken from or inspired by French engravings of paintings by Antoine Watteau (1684–1721), used as enamelled decoration on glass, especially on OPAQUE WHITE GLASS of the ROCOCO period.

Waugh, Sidney B. (1904–63). An American sculptor who was appointed by ARTHUR A. HOUGHTON, JR. in 1933 as first designer of Steuben Glass, Inc. He designed forms for moulded and cut glass pieces, and also made designs for engraved decoration, e.g. the GAZELLE BOWL (1935), the Zodiac Bowl (1935), and the MERRY-GO-ROUND (1947). *See* STEUBEN GLASS WORKS.

Wealden glassware. Glassware made in England near Chiddingfold (in the Weald of Kent, Surrey, and Sussex) by glass-makers from France who migrated in the 13th to 15th centuries and settled in this well-wooded region, and also by some English craftsmen who joined them. (*See* VITREARIUS, LAURENCE; SCHURTERRE FAMILY; PEYTOWE FAMILY.) They were joined, *c.*1567, by glass-makers from Lorraine, France, under JEAN CARRÉ. From *c.*1580 they began dispersing to Hampshire and the West of England, and then to the Stourbridge area, Newcastle-upon-Tyne, and other parts of England. The METAL made by the Wealden glass-makers was similar to that of late ROMAN GLASS; it was first light-green in colour but later a darker green like WALDGLAS. The ware was MOULD-BLOWN, decorated with TRAILING and some was WRYTHEN. When the Proclamation of 1615 prohibited the use of wood for industrial purposes, the Wealden glass industry came to an end.

wear-marks. Markings on the FOOT or BOTTOM of a drinking glass, resulting from use, normally scratches running in many directions. When such marks are placed on a late-period glass with intent to deceive, they tend to run in one direction through having been produced by rubbing against an abrasive surface.

weathering. Changes on the surface of glass (caused by exposure to adverse conditions) which appear as dulling, FROSTING, IRIDESCENCE, or decomposition. It is to be distinguished from CRISSELLING. *See* Roy Newton, 'Weathering of medieval window glass', *Journal of Glass Studies* (Corning), XVII (1975), p. 161.

Webb family. A family of leading glass-makers in and near Stourbridge, England, descended from John Webb (1774–1835) who, with John Shepherd, operated as Shepherd & Webb, in the early 1830s, the WHITE HOUSE GLASSWORKS. His son, Thomas Webb I (1804–69),

water-set. Decanter, mug, and tumbler, with inserted opaque white medallions. St Petersburg Imperial Glass Factory, *c.*1810–20. Hermitage Museum, Leningrad.

Waterford glass. Cut glass honey jar with raised diamond pattern, 18th century. National Museum, Dublin.

with William Haden Richardson and Benjamin Richardson, operating as Webb & Richardson, took over the WORDSLEY GLASSWORKS in 1829, and after inheriting the White House Glassworks he withdrew from Webb & Richardson to operate it as Thomas Webb & Co; he then transferred it to W. H., B. & J. RICHARDSON in order to form his own firm, THOMAS WEBB & SONS, to operate from 1837 Platts Glasshouse and then in 1856 to build DENNIS GLASSWORKS. He was succeeded in 1863 by his sons Thomas Wilkes Webb II (1837–91) and Charles Webb, and from 1869 by another son, Walter Wilkes Webb; after Thomas Webb II died, the firm was continued by Walter and Charles Webb until Walter's death in 1919, when the business was sold. Thomas Webb III (1865–1925), a son of Thomas Webb II, became in 1897 a partner in Thomas Webb & Corbett, Ltd. (later WEBB, CORBETT, LTD.) to acquire from W. H., B. & J. Richardson the White House Glassworks.

Webb, Corbett, Ltd. The owner of glasshouses making high-quality household and ornamental glassware. Its original glasshouse, known as the WHITE HOUSE GLASSWORKS, at Wordsley, near Stourbridge, England, was conveyed by W. H., B. & J. RICHARDSON in 1897 to Thomas Webb & Corbett, Ltd. (Webb being Thomas Webb III (1865–1925), son of Thomas Webb II and grandson of Thomas Webb I). The name of the firm was changed in the 1930s to Webb, Corbett, Ltd. The company's operations, after a fire in 1913, were moved to the present site at Coalburnhill, near Stourbridge. In 1906 it took over another glasshouse at Tutbury, Staffordshire, and in 1969 the company became a unit of the Royal Doulton Group.

Webb, Philip (1831–1915). An architect and artist who was commissioned by WILLIAM MORRIS in 1859 to design glasses embodying his principles of form and decoration, derived from the Arts and Crafts Movement, a precursor of the style later to be known as ART NOUVEAU. The glasses, functional and simple in shape, were made by James Powell & Sons, at the WHITEFRIARS GLASS WORKS, London, which subsequently made other glassware for Morris to designs by T. G. Jackson.

Webb, Thomas, & Sons. The company now operating a large glasshouse near Stourbridge, England, making high-quality household and ornamental glassware. Its foundation was laid when John Webb (1774–1835) became connected in the early 1830s with the WHITE HOUSE GLASSWORKS, at Wordsley, near Stourbridge. In 1835 Thomas Webb I (1802–69) inherited the glasshouse and, in order to devote his efforts to it, withdrew in 1836 from Webb & Richardson that was operating the WORDSLEY FLINT GLASSWORKS. Soon thereafter he withdrew from Thomas Webb & Co. that was operating the White House Glassworks. He then formed his own firm, Thomas Webb & Sons, which operated from 1837 to 1856 Platts Glasshouse (which had been making glass at Amblecote, near Stourbridge, for over 100 years under Henry Henzey and his son), and then from 1856 a new glasshouse that it built in 1855 at nearby Dennis Hall, Amblecote, known as DENNIS GLASSWORKS. On Webb's retirement in 1863 he was succeeded by his sons, Thomas Wilkes Webb II (1837–91) and Charles Webb, who were joined in 1869 by another son, Walter Wilkes Webb. In 1919 Dennis Glassworks was acquired by WEBB'S CRYSTAL GLASS CO., LTD., which, together with Thomas Webb & Sons (with no longer any Webb in the firm), became in 1964 a part of a group owned by Crown House, Ltd. Thomas Webb & Sons had in 1886 acquired a licence to make BURMESE GLASS, and it became noted for its CAMEO GLASS so that much glass of that type is still referred to as 'Webb's Cameo Glass'. It also produced in the 1880s ALEXANDRITE, PEACH (BLOW) GLASS, IRIDESCENT GLASS, SATIN GLASS, and other types of ART GLASS. Among the noted engravers at Dennis Glassworks in the second half of the 19th century were FREDERICK KNY, WILLIAM FRITSCHE, and FRANZ JOSEPH PALME; JULES BARBE was an enameller there.

wedding cup. Standing bowl, opaque white glass, enamel decoration with medallions of marriage couple, and scale pattern of blue and red gem-like dots, Venetian, late 15th century. Ht 12·7 cm. National Museum, Prague.

Webb's Crystal Glass Co., Ltd. A company formed in 1919 to take over the DENNIS GLASSWORKS at Amblecote, near Stourbridge, from THOMAS WEBB & SONS; soon thereafter it acquired the Edinburgh & Leith Flint Glass Co. (whose factory at Norton Park was closed in 1974 when the business was moved to Penicuik). In the 1930s it acquired WORDSLEY FLINT GLASSWORKS. The business of all these glasshouses was in 1964 acquired, along with Dema Glass Co., Chesterfield, by Crown House, Ltd.

wedding cup. A large STANDING CUP (or BOWL) made and decorated to commemorate a marriage. Venetian examples of the 15th century usually depict the portraits of the couple and are often decorated with an enamelled encircling procession or allegorical scenes. They are usually in the form of a bowl (about 12 to 16 cm. high) with slightly everted rim, and standing on a tall hollow spreading FOOT. They have been made of transparent blue or green glass or of OPAQUE WHITE GLASS, an example of the latter having portraits of the couple and formalized diaper-pattern decoration. *See* BAROVIER CUP; STANDING CUP.

wedding glass. *See* MARRIAGE GLASS; WEDDING CUP.

Wedgwood glassware. Cut glass wares made by two subsidiary companies of Josiah Wedgwood & Sons, Ltd: (1) Wedgwood Glass, at King's Lynn, Norfolk, which acquired in 1969 King's Lynn Glass, Ltd (founded 1967); (2) Galway Crystal, Galway, Ireland, founded in 1967 and acquired in 1974. The head designer is Ronald Stennett-Willson. The wares include drinking glasses, decanters, and ornamental objects of full-lead CRYSTAL glass, both clear and coloured, and solid glass zoomorphic and fruit PAPER-WEIGHTS, etc.

well. The depressed central portion of a PLATE or DISH that is within the LEDGE or marli. The Italian term is *cavetto*.

Well Spring Carafe. A CARAFE of ovoid form on an extended base and having a high neck with an undulating mouth. It is of clear glass, decorated with green enamelled thin leaves encircling the body and extending upward from the base, and with a circle of flowers around the shoulder. The piece was designed by Richard Redgrave (1804–88) for Henry Cole's Summerly's Art Manufacturers, and made by John Fell Christy at Stangate Glass Works in Lambeth, London, 1847.

welt. The small strip of glass at the edge of a folded FOOT that is turned usually under, but sometimes over, to strengthen the foot and reduce the risk of chipping. It occurs frequently on stemmed drinking glasses.

Wennerberg, Gunnar Gunnarson. A Swedish painter who designed cased-glass vases for KOSTA GLASSWORKS from 1898 until 1903, following the style of EMILE GALLÉ.

Werther, C. A. F. A Canon of Cologne who did DIAMOND-POINT ENGRAVING in the style of, but less notably than, Canon AUGUST OTTO ERNST VON DEM BUSCH.

West Lothian Glassworks. A glasshouse founded at Bathgate, Scotland, by Donald Fraser (d. 1869); it operated from 1866 to 1887, and two dated tumblers and other pieces are recorded. After the death of Fraser, the factory was taken over by Wilson & Sons and was a thriving business in 1871, making all types of domestic glassware, some with engraved fern patterns. Later it was acquired by James Couper & Sons and the name was changed to the Bathgate Glass Co.

Westphalia Treaty Humpen (German). A type of HUMPEN with enamelled decoration commemorating the Peace of Westphalia (1649). The scene shows the Holy Roman Emperor joining the hands of the King of France and the Queen of Sweden, surrounded by kneeling ecclesiastics and laymen, all beneath a celestial group, and with

Wedgwood glassware. Salad bowl with faceting, printies, fan cutting, and jewel stars; Galway Crystal, 1976. D. 25·2 cm. Courtesy, Wedgwood, London.

Westphalia Treaty Humpen. Clear glass *Humpen* with enamelled decoration, Ht 31·5 cm. Museum für Kunst und Kulturgeschichte, Lübeck.

wheel engraving, surface. Bowl with floral design, English, mid-18th century. Courtesy, Derek C. Davis, London.

Whitefriars glassware. Decanter in heavily textured 'Glacier' glass, designed by Geoffrey P. Baxter, 1968. Ht 26 cm. Whitefriars Glass, Ltd.

inscribed biblical quotations and a prayer of thanks. It is based on an engraving by Mathäus Rembold in Ulm, *c.* 1648/49. The glasses, which have a domed cover with a high FINIAL, were made in Franconia, 1649–55. *See* HISTORICAL GLASSWARE.

wetting-off. The same as CRACKING-OFF.

whale-oil lamp. A type of OIL LAMP using sperm or whale oil; it was made in the form of a tall table lamp, with a large FONT resting on a base or pedestal, and having a 'burner' (made of tin and cork) atop the font and fitted with a tin cap through which the wick passed into the font; above the 'burner' was a glass shade to protect the flame, sometimes ornamented with a decorative pattern.

wheel engraving. The process of decorating the surface of glass with pictorial or formal decoration or inscriptions by the grinding action of a wheel using disks of various materials and sizes with an abrasive in a grease applied to the wheel as the engraver holds the object against the underside of the rotating wheel. The process developed from the use of the lapidary's wheel to grind CRYSTAL, precious stones, and CAMEOS. The technique is similar to that of CUTTING glass, but involves more skilful and artistic workmanship. After the glass is engraved, it is usually POLISHED, but sometimes some areas are left unpolished. The technique was used in medieval Egypt, Syria, and Persia and medieval Rome. The modern use of wheel engraving for decorating glass was developed by CASPAR LEHMAN(N) at Prague, *c.* 1590–1605, and it became popular after the thin VENETIAN GLASS, unsuitable for RELIEF engraving, was largely replaced by the thicker and more durable Venetian CRISTALLO, Bohemian potash-lime glass, and English LEAD GLASS (FLINT GLASS). The technique of wheel engraving in RELIEF (HOCHSCHNITT) is sometimes loosely termed 'cameo engraving' or 'carving'. See below for various types of engraving: *intaglio*; *relief*; *surface.* Sometimes all types were used on the same piece.

——, *intaglio.* The technique of decorating the surface of glass by wheel engraving that incises and carves below the surface, known as INTAGLIO or *Tiefschnitt.* The technique was introduced by CASPAR LEHMAN(N) at Prague and Nuremberg, *c.* 1590–1605, and was developed at Nuremberg by GEORG SCHWANHARDT and his followers, including some of the same decorators who used wheel engraving for relief decoration (*Hochschnitt*) – see below; leading exponents were FRANZ GONDELACH, GOTTFRIED SPILLER, MARTIN WINTER, and HEINRICH JÄGER.

——, *relief.* The technique of decorating the surface of glass in RELIEF by wheel engraving that cuts away the ground so as to leave a raised design and that carves the design in the manner of CAMEO relief. The result is sometimes then polished or may be left MAT. The work was done by Bohemians and Silesians toward the end of the 17th century, and also at Cassel and Berlin. Leading exponents of the technique were MARTIN WINTER, FRIEDRICH WINTER, GOTTFRIED SPILLER, and FRANZ GONDELACH. The technique was revived in the 19th century on overlaid FLASHED (CASED) GLASS, especially by KAREL PFOHL and J. & L. LOBMEYR. The process is sometimes termed 'carving' or 'cameo relief'. The German term is HOCHSCHNITT.

——, *surface.* The technique of decorating glass by wheel engraving, but only slightly below the surface, e.g., INSCRIPTIONS and some, usually sketchy, ornamental designs.

whips. *See* SYLLABUB GLASS.

whisky glass. A small drinking glass without a STEM, in the form of a small TUMBLER. Its capacity is usually 2 to 3 ounces. It is intended for drinking neat (straight) whisky. Sometimes called a 'pony' or (in the United States) a 'shot glass'.

whisky tot. A small conical decanter-type receptacle for serving a dram of whisky (some are of double size). The glass body, plain or engraved, is provided with a silver lid having a thumb-rest. They were made in the early 20th century by John Grinsell & Sons, Ltd., of Birmingham, with the mount made by that firm or by Barker Elis Silver Co., Ltd., Birmingham.

whistle goblet. A type of GOBLET that has attached to it a metal whistle, said to be used when the drinker desired a refill. Such glasses were generally not true goblets, as they had no foot and were inverted and stood on the rim when empty. The whistle was attached to the end of the stem, and in some cases may have been added at a date later than that of the glass. Such pieces were made in Venice, Germany, and Netherlands, in the 17th century. A related object is a goblet having attached to the stem a metal rattle. *See* STURZBECHER.

Whistler, Laurence (1912–   ). An English independent glass-engraver (also a writer and poet) who has achieved a high reputation for point-engraving (stipple and drill). He began the present revival of this technique in England in 1935, using first a diamond, later carborundum. His work has passed through three broad phases, first inscriptional and emblematic (often on old glass), then after World War II architectural, and recently pictorial, with imaginary landscapes and symbolic content. Since the War he has had his own shapes blown by WHITEFRIARS GLASS WORKS and LEERDAM, and has worked on some glass by STEUBEN. Some pieces are engraved both outside and inside. In the last twenty years he has also provided church windows as well as panes for private houses. His work is usually signed 'LW' with the date. *See* Whistler, *The Image on the Glass* (1975), with 80 illustrations of his work.

Whitall, Tatum & Co. A glassworks at Millville, New Jersey, that specialized in making glass PAPER-WEIGHTS. It was founded in 1806 and had various names and owners until 1849 when it became Whitall Bros. & Co., and 1857 Whitall, Tatum & Co., and later became a unit of the Armstrong Cork Co. It made perfume and apothecary bottles.

White Flint Glasshouse. A glassworks at Bristol, England, that made much OPAQUE WHITE GLASS decorated with enamelling. Its proprietor in 1752 was Jacob Little, and his successors were Little & Longman (1762–67) and Longman & Vigor (1767–87). The ware was in the style of Chinese porcelain, including covered oviform vases and trumpet-mouth beakers, often made as GARNITURES.

White House Glassworks. A glasshouse originally at Wordsley, near Stourbridge, England, built in 1825 and leased in the early 1830s to John Webb (1774–1835) in partnership with John Shepherd, the firm being called Shepherd & Webb. After Thomas Webb I (1804–69) inherited it in 1835 and Shepherd then retired, Webb decided to devote himself to the White House, and withdrew in 1836 from the firm of Webb & Richardson that operated the WORDSLEY FLINT GLASSWORKS; when Shepherd died in 1842, the firm became Thomas Webb & Co. Thomas Webb I soon conveyed it in 1842 to W. H., B. & J. RICHARDSON, who transferred it in 1897 to Thomas Webb & Corbett, Ltd. (Webb being Thomas Webb III (1865–1925), the son of Thomas Webb II and grandson of Thomas Webb I), the name of which was changed in the 1930s to WEBB, CORBETT, LTD. The company's operations, after a fire in 1913, were moved to the present site at Coalbournhill, near Stourbridge.

White Mills Glassworks. A glasshouse at Honesdale, Wayne County, Pennsylvania, which operated from 1862 until 1921. *See* DORFLINGER GLASSWORKS.

Whitefriars Glass Works. A glass factory founded in London before 1680 near the Temple, between Fleet St and the Thames, in a former Carmelite monastery (hence its name). It was acquired, after several ownerships, by James Powell in 1834, and as James Powell & Sons it

Whistler, Laurence. Bowl with stippled engraving, 1976. D. 25·3 cm. Courtesy, Laurence Whistler.

Wiegel, Albert. Glass block with engraved portrait of Empress Auguste Viktoria, signed 'Albert Wiegel', Kassel, 1899. Ht 21 cm. Staatliche Kunstsammlungen, Kassel.

wig. Glass fibre wig, Antonio Lera, Milan, *c*.1740. Nationalmuseet, Copenhagen.

wig-stand vase. Opaque white glass with enamelling, France, *c*.1715–30. Ht 23 cm. Courtesy, Christie's, London.

made from 1860 some glassware commissioned by William Morris and designed by PHILIP WEBB that was the forerunner of English glass of the ART NOUVEAU style. From the 1850s it made coloured flat glass for stained-glass windows and TESSERAE for MOSAICS. From *c*.1874, under the direction of HARRY J. POWELL, hand-blown glass was made, in a style largely of Roman and WALDGLAS derivation. The company moved in 1923 to Wealdstone, in Middlesex, and in 1962 resumed its former name as Whitefriars Glass, Ltd. It features hand-cut LEAD GLASS, including such glass in several colours, making bowls, decanters, and tableware; a heavily textured line of tableware is known as 'Glacier glass'. It has made PAPER-WEIGHTS since *c*.1848, and still makes superior MILLEFIORI examples with one cane showing its insigne and the date. Among the modern artists whose glassware has been made by the firm is LAURENCE WHISTLER. It formerly made important STAINED-GLASS WINDOWS (e.g. for St Paul's Cathedral), and recently has executed glassware with engraved figures copied from the windows of Coventry Cathedral by JOHN HUTTON who had engraved the window panels. The chief designer (since 1952) is Geoffrey P. Baxter (b. 1922) whose predecessor was James Hogan (1883–1948).

Wiederkomm(humpen) (German). Literally, come again beaker. A term used by A. Nesbit in his *Catalogue of the Collection Formed by Felix Slade, Esq.*, London, 1871, in referring to WILLKOMM(HUMPEN) and many other types of enamelled HUMPEN at the British Museum. It is a term not used by German museums (e.g. Veste Coburg) nor mentioned by Axel von Saldern, *German Enameled Glass*, New York, 1965, as to any of the many *Humpen* at the Corning Museum.

Wiedmann, Karl (1905–   ). A glass-maker born near Göppingen, Germany, who in 1921 became *maître-verrier* at the glassworks of the WÜRTTEMBERGISCHE METALLWARENFABRIK (W.M.F.), established in 1883. He developed there in 1925 the ART GLASS known as MYRA-KRISTALL. During 1947–48 he worked at DAUM FRÈRES and then returned to W.M.F. where he still conducts research in ancient glass.

Wiegel, Albert (1869–1943). A German glass-technician, engraver, and sculptor at Kassel who developed new glass techniques, but mainly did work in carving and engraving. On goblets and plaques, using glass of several hues, he executed engraved and cut portraits of statesmen and famed musicians (copying from photographs, unlike the portraitists of SPA GLASS who engraved from life). After *c*.1905 he made three-dimensional animal figures by chiselling and carving solid glass blocks. For a short period *c*.1900, he embraced the ART NOUVEAU style. For a history and illustrations of his work, see *Journal of Glass Studies* (Corning), XV (1973).

Wiener Stil (German). Literally, Vienna style. A style of glassware developed in Vienna from the early 20th century and executed in Bohemia (after 1918, Czechoslovakia). Its leading exponent was JOSEF HOFFMANN, an architect and designer who was a founder of the Wiener Werkstätte, which provided facilities for young designers in several fields of decoration. The style in glassware featured female figures and geometric patterns. Among the disciples was MICHAEL POWOLNY. The style became popular in Bohemia and elsewhere in Europe.

Wier, Don (1903–   ). An American glass-designer on the staff of the design department of STEUBEN GLASS WORKS. He made the designs for some major pieces of ornamental glassware, such as his 'David and Goliath'.

wig. A head-covering usually made of human hair, but at least two examples are known made of fine GLASS FIBRE threads, attached to a network in the usual manner. One was made by Antonio Lera of Milan, *c*.1740, and is in Copenhagen; the other is in Uppsala, Sweden. A glass bust in the British Museum has a lock of hair on the forehead, made of glass fibre. *See Journal of Glass Studies* (Corning), XVI (1974), p. 72.

wig-stand. A globular stand for a wig, supported by a stemmed FOOT. Some glass examples have been made by FALCON GLASSWORKS, with MILLEFIORI decoration. Some wig-stands made of metal with enamelled or *cloisonné* decoration have at the top a glass globe (sometimes silvered). *See* WIG-STAND VASE.

wig-stand vase. A type of VASE, sometimes bulbous with a stemmed FOOT, high neck, and everted rim, the characteristic feature of which is a large globular cover that fits into the mouth of the vase and serves as a wig-stand. The globular cover sometimes has an extended tube that descends midway into the vase to provide stability. Examples have been made of enamelled OPAQUE WHITE GLASS in France, *c.* 1715–20, and in the United States in the early 19th century.

Wild Rose Glass. A type of ART GLASS developed at the NEW ENGLAND GLASS CO. by Edward Drummond Libbey, probably assisted by JOSEPH LOCKE, and patented by Libbey in 1886. It shaded from creamy white to a rose colour, and had in some pieces a glossy finish, in others a mat finish. The name was adopted after MOUNT WASHINGTON GLASS CO. sued for infringement of its patent for PEACH BLOW GLASS.

William III glassware. English wine-glasses and other objects decorated with a portrait, usually equestrian, of William III (1650–1702) which were made to commemorate his victory over James II at the Battle of the Boyne, 1 July 1690. The pieces are usually decorated with DIAMOND-POINT ENGRAVING or WHEEL ENGRAVING. *See* WILLIAMITE GLASSWARE. For a discussion of such ware and many illustrations, see *Journal of Glass Studies* (Corning), XV (1973), p. 98.

Williamite glassware. English glassware with emblems (e.g., the Orange Tree) and inscriptions indicating sympathies with William III (1650–1702) in his struggle with James II (1633–1701) and celebrating his victory at the Battle of the Boyne, 1690. They were also used for loyalist toasts of a more general Protestant character long after William's death, while also being used for toasting his memory. The engraved decoration includes portraits, equestrian and otherwise, of William III. Some such glassware was almost certainly made or engraved in Ireland where sentiment was strong for William III. *See* JACOBITE GLASSWARE; WILLIAM III GLASSWARE.

Williams, Richard, & Co. A glasshouse of Dublin, 1764–1817, that started 16 years before Free Trade and so was a well-established factory. *See* DUBLIN GLASSWARE.

Willkomm(humpen) (German). Literally, welcome (beaker). A salutation drinking glass used to greet guests, decorated with an enamelled salutatory inscription and sometimes also having a portrait of the host. They were of gigantic size, some said to have a capacity of up to five litres. They were made in Germany in the 16th to 18th centuries. It is the converse of the so-called *Wiederkomm* glass said to be used to say *Au revoir* to departing guests; these were also made, with enamelled decoration, in the 16th/17th centuries in Germany and Lorraine. *See* WIEDERKOMM; ENAMELLING, *German*.

Winchester House Glasshouse. A glasshouse in Southwark, London, operated successively by Edward Salter, Sir Robert Zouch, and SIR ROBERT MANSELL. It made crystal glass but closed probably before 1624.

window. (1) An opening in the wall of a building to admit air and light, usually furnished with casements or sashes surrounding transparent panes of glass, and generally capable of being opened and closed. Apart

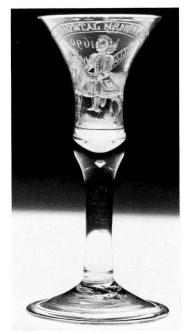

William III glass. Diamond-point engraved trumpet-bowl wine glass, with portrait of William III and inscription on reverse 'Boyne July 1st 1690', *c.* 1720. Ht 17·2 cm. Cinzano Glass Collection, London.

Willkommhumpen. Beaker enamelled as greeting glass of the Coburg Bakers' Guild, Lauscha, Thuringia, dated 1672. Ht 23·5 cm. Kunstsammlungen der Veste Coburg.

from fixed STAINED GLASS WINDOWS in churches, the earliest glass windows probably were those made apparently by CASTING, such as have been found at Pompeii. Later, windows with panes of glass were made in France and elsewhere in the Middle Ages; these were leaded windows of two types: (a) some with rows of small glass roundels with a 'bull's-eye' PONTIL MARK (*boudin*) in the centre, made from CROWN GLASS, with the spaces filled with small glass triangles (*coins*), this type being common in Italy and the Netherlands; and (b) some with flat panes, cut from BROAD GLASS or crown glass, and mounted in lead within a wooden frame. Window panes for homes and carriages were often decorated by GLAZIERS, frequently with a coat-of-arms, but some panes were decorated with WHEEL ENGRAVING, e.g., in Switzerland from the late 17th into the 19th century. Window panes were sometimes coloured, e.g., some were stained yellow in the Middle Ages, some were FLASHED with ruby glass, and some were coloured throughout, as in the case of the Colonial lavender glass panes still to be seen today in old houses on Beacon St., Boston, Mass. After the invention of CAST GLASS, *c*.1687, windows became widely used, and the production of large glass windows in France and elsewhere was greatly increased. Today plate glass windows and panels, sometimes of tinted glass, curtain the entire exterior of some large buildings and skyscrapers. *See* RULLO. (2) *See* PAPER-WEIGHT, *window*.

wine bottle. A BOTTLE made of brown, green or black glass, in various shapes, sizes, and styles; they were made in many factories, for storing or serving wine.

TYPES OF WINE BOTTLE

——, *English*. A glass bottle for wine, first made in England about the middle of the 17th century to supersede the earthenware bottles that were absorbent or subject to leaking, and even the later more secure German 16th-century stoneware bottles. Early examples were of thin green glass, but SIR ROBERT MANSELL introduced, *c*.1630, examples in dark-green or brown BOTTLE GLASS that popularized the use of glass wine bottles in England. They were used to transfer wine from the imported cask, and later for binning. The shape of the bottles evolved from the 'Shaft-and-Globe Bottle' (bulbous body with a tall tapering neck) to a bottle with straighter sides, more pronounced shoulder, and shorter neck, then the 'Onion-shaped Bottle' (squat and short-necked; binned upside-down), later, from *c*.1715, the 'Slope-and-Shoulder Bottle' (straight sides), and, *c*.1750, the 'Cylinder-Bottle' which continues today. The last types became desirable after 'binning' became customary following the use of a cork in the 16th and 17th centuries and straight-sided bottles could be stored on their sides (to keep the cork moist), and especially after the invention of the corkscrew (first recorded *c*.1681), had led to the use of the 'flush cork' to replace the tied-on cork secured by means of a STRING RIM. *See* wine bottle, *sealed*, below; *see* also HOGARTH BOTTLE.

——, *modern French*. Bottles that vary in shape and colour depending on the place of origin of the contents, thus facilitating identification; all have a KICK-BASE. Bordeaux red wines (clarets) are in dark-green bottles, white wines in clear bottles, both with vertical sides and a sharp shoulder; Burgundy red wines are in dark-green bottles, white wines in light-green bottles, both having a gently tapering shoulder and neck; Alsatian wines are in narrow green bottles with a markedly tapering shoulder and neck; and Champagne bottles are green but wider and heavier and with a tapering shoulder and neck, and a string rim. Other regional wines, e.g. from the Rhône Valley and Provence, are often in bottles of distinctive shapes. Still wines come normally in bottles of 70–75 cl capacity, but also in Magnums (2-bottle size). Bordeaux clarets come in large ceremonial bottles: Marie Jeanne (3 bottles), Double Magnum (4 bottles), Jeroboam (5 bottles), and Imperial (8 bottles). Champagne is bottled in 'splits' (half-pint or $6\frac{1}{2}$ oz.), 'pints' ($\frac{1}{2}$ bottle or 13 oz.) and 'quarts' ($\frac{1}{5}$ gallon or 26 oz.), and then the ceremonial sizes: Magnum (2 bottles), Jeroboam (4 bottles),

Rehoboam (6 bottles), Methuselah (8 bottles), Salmanasar (12 bottles), Balthazar (16 bottles), and Nebuchadnezzar (20 bottles).

——, *modern German*. Bottles that vary in shape and colour according to the place of origin of the contents, thus facilitating identification. Rhine red wines come in tall, thin, brown bottles with markedly tapering necks; Rhine white wines (hocks) and Moselle wines come in bottles similarly shaped but green. Franconian wines (*Steinwein*) come in a BOCKSBEUTEL.

——, *modern Italian*. The traditional Italian wine bottle (*fiasco*) is globular (originally often flattened) with a long thin neck; the rounded bottom required it to be wrapped in straw to provide a flat surface for standing on the table, and made binning impossible. Superior Italian wines are bottled today in cylindrical bottles, with a KICK-BASE, that permit binning while maturing.

——, *sealed*. Wine bottles made in England from the 17th century onwards that bore a seal in the form of a glass medallion affixed to the bottle on the body or the shoulder. The lettering impressed on the seal indicated the name of the owner or the wine merchant, or sometimes the brand of the contents. Many bore a year date, the earliest recorded date being 1652 on a detached seal and 1657 on a sealed bottle. *See* also wine bottles, *English* (above).

wine cooler. A bucket-shaped receptacle for holding ice to chill a single bottle of white wine. It may be with or without HANDLES; it sometimes has a COVER. It is a large ICE PAIL; examples in porcelain and in glass are known. The French term is *seau à bouteille*.

wine cup. A small cup (smaller than a TEA BOWL) without a handle, used in the Far East for drinking wine. They were usually made of porcelain, but some were made of glass in China in the Ch'ien Lung period (1736–95).

wine glass. A drinking glass for wine, made in various shapes (usually stemmed), sizes, and styles; those made in modern times are each said to be most suitable for a particular type of wine. *See* BRANDY GLASS; BURGUNDY GLASS; CHAMPAGNE GLASS; CLARET GLASS; HOCK GLASS; CORDIAL GLASS; PORT GLASS; SHERRY GLASS; TULIP GLASS; FLUTE; DOUBLE WINE GLASS; TOASTING GLASS; TOAST-MASTER'S GLASS.

wine-glass ale. An ALE GLASS in the form of a stemmed WINE GLASS that is decorated with engraved motifs of hops and barley.

wine-glass cooler. A BOWL to contain iced water for cooling the bowl of a wine-glass immersed in the water, usually by being suspended by its foot from notches or lips on the side of the rim. (*See* MONTEITH.) They were introduced in the late 17th century and became popular in the 18th century. It has been said of such bowls with two lips on opposite sides (instead of the single-lipped or cylindrical type), that the purpose was to place an emptied glass in the bowl for rinsing and cooling while a second glass was being used, preparatory to serving a different wine to be drunk from the first glass; this usage has been depicted in an 18th-century engraving. Such receptacles have sometimes been referred to as a FINGER-BOWL, and may have served both purposes. *See* LIPPED BOWL.

wine-glass lamp. A 19th-century type of SPARKING LAMP made in the form of a contemporary WINE-GLASS, on a stemmed FOOT and with a FONT, of various shapes, enclosed by glass curving inward from what would be the rim of the wine-glass to a central orifice where a metal collar is inserted or sometimes screwed on for holding the wick. They were made at several factories in the United States.

wine-label. *See* BOTTLE TICKET.

wine-siphon. A long (*c*. 40 cm.) thin glass tube having on one end a small bulb or a curved trumpet-shaped mouth, with a hole in each end.

wine cup. Semi-opaque white glass with enamel decoration in *famille rose* palette, Ch'ien Lung mark. Ht 4·3 cm. Percival David Foundation of Chinese Art, London.

It was used to transfer wine to a glass, by stopping the top end with the thumb. *See* TODDY-LIFTER.

wine-urn. A receptacle for serving wine by means of a spigot. Some were in the shape of a DECANTER with a stemmed FOOT, and had the spigot near the bottom of the decanter. They were made in the United States and elsewhere. *See* WEINHEBER.

wing-handled cup. *See* PTEROTOS CUP; SKYPHOS.

winged glass. A GOBLET, having a STEM elaborately decorated with applied THREADING in wing-like designs extending out on both sides, making ornamentation in the form of loops, twisted rope or snake-like patterns, occasionally in the form of sea-horses, dragons, or a double-eagle. Sometimes the threading was so elaborate that there was no independent stem. The wings were often of coloured glass, with additional pincered TRAILING or CRESTING of clear glass. Such goblets, of clear SODA GLASS, were made in Venice in the 16th and 17th centuries (called *calice ad alatte* or *calice a serpenti*), and were copied in Germany and the Netherlands in the 16th and 17th centuries. The German term is *Flügelglas*. *See* SNAKE GLASS; DOUBLE-EAGLE GLASS; DRAGON GLASS; EHRENFELD.

Winter, Friedrich (d. *c.* 1712). A Silesian engraver of glass who established in 1687 an engraving workshop under the patronage of Count Christoph Leopold von Schaffgotsch to engrave glass in the manner of ROCK CRYSTAL. In 1690–91 he set up, at Petersdorf in Silesia, a water-power mill (the first used for glass-engraving) and engraved glass in HOCHSCHNITT and sometimes TIEFSCHNITT in baroque style, with richly carved stems and massive KNOPS, and often including the Schaffgotsch coat-of-arms. Later he set up a workshop at WARMBRUNN, in Silesia, decorating similar baroque glass, and also a retail shop. He was the brother of MARTIN WINTER.

Winter, Martin (d. 1702). An engraver of glass from the Riesengebirge who in 1680 probably engraved glass from the glasshouse of JOHANN KUNCKEL at Potsdam, near Berlin, and was under the patronage of Friedrich Wilhelm, Elector of Brandenburg, from whom he received the privilege to engrave in HOCHSCHNITT in 1684. He set up in 1687 a cutting and engraving workshop, with a water-power mill, at Potsdam, to do *Hochschnitt* and TIEFSCHNITT glass, in conjunction with GOTTFRIED SPILLER, his pupil and nephew and (from 1683) partner. No glass has been definitely ascribed to Winter, but much work is recognized as in his style and has been attributed to him. He was the brother of FRIEDRICH WINTER.

Wirkkala, Tapio (1915–). A Finnish designer in several media, including glassware. His pieces are often asymmetrical and freely formed. From 1947 he has been with the IITTALA GLASSWORKS, and has also designed for VENINI at Murano and others. Among his well-known glassware are STACKING GLASSES. He has developed a glass surface with a frosted finish suggestive of Arctic ice.

wishbone decoration. TRAILED decoration in the form of a wishbone or merrythought. Such decoration is sometimes found on specimens of FRANKISH GLASS.

Wistar, Caspar (1696–1752). A glass-maker who emigrated in 1718 and founded in 1739 the first established glassworks in the United States, at Wistarberg, Salem County, in southern New Jersey. It made WINDOW GLASS and also JUGS, BOTTLES, and FREE-BLOWN vessels in clear and coloured glass, decorated with spiral TRAILING, LILY-PAD DECORATION, and opaque white stripes and loops. Other factories in southern New Jersey made similar ware, all of which came to be known

winged glass. Double-eagle goblet, with cover, Venezianerhütte, Kassel, 1538–84. Ht 46 cm. Löwenburg Collection, Verwaltung der staatlichen Schlösser und Gärten, Bad Homburg, Germany.

as 'South Jersey Glass'. After his son Richard took control upon Caspar's death, the quality of the ware deteriorated, and the workmen left to join other glassworks and make similar ware. The factory closed in 1780. No piece has been definitely attributed to Wistar. *See* SOUTH JERSEY TYPE GLASSWARE.

witch ball. A glass globe, about 8 to 18 cm. in diameter, made in England (some of NAILSEA GLASS) from the early 18th century, to be hung in a cottage to 'ward off the evil eye'. They were made of coloured glass (blue, green or red), and often had the interior silvered or daubed with a variety of colours that scintillated when it was hung or decorated by the DECALCOMANIA or POTICHOMANIA process; some were decorated with MARVERED loops of multi-coloured glass. Some had a small hole at the top for a peg and string by which they were suspended, and sometimes several hung vertically on a string. Early examples had a small neck at the top, and were called a 'witch bottle'. They were also made at Cadalso, in Catalonia, Spain. Modern imitations have been made with silvery lining.

Wolff, David (1732–98). A talented follower of FRANS GREENWOOD in the use of STIPPLING, using the same technique but also in his early period, 1770–75, using some linear work, e.g. in depicting hair. He was born in The Hague and lived there 1782–98. Originally a painter, he made his own designs, often with *putti*, and also did portraits. His later period, 1784–95, shows close stippling. He signed some of his work, but much glass was stippled in his style by others toward the end of the century, and has been called in general 'Wolff glass'.

Wolff, Peter (fl. 1660–77). A glass-engraver who did work in DIAMOND-POINT ENGRAVING in Cologne on RÖMERS and at least one WINGED GLASS, in a style influenced by Dutch engravers, his motifs being coats-of-arms and landscapes. Signed pieces that survive are dated 1660, 1669, and 1677.

wolf's tooth. A style of decoration found on FRANKISH GLASSWARE, in the form of zig-zag trailing arranged in a continuous series of long vertical loops. It occurs on some CONE-BEAKERS and DRINKING HORNS.

Woodall, George (1850–1925). and **Thomas** (1849–1926). English glass-engravers and cutters; the two brothers were apprentices of JOHN NORTHWOOD until 1874. With their associates, they worked in collaboration with THOMAS WEBB & SONS, later making modern glassware influenced by the ART NOUVEAU style, especially CAMEO GLASS. They retired in 1911. *See* IVORY-IMITATION GLASSWARE.

woods. A GLASS-MAKER'S TOOL having two wooden arms joined at one end by a metal spring handle. It was formerly used in shaping the BOWL of a glass. It was similar to a metal PUCELLAS which it superseded, *c.* 1830; it had the advantage of minimizing striation marks.

Wordsley Flint Glassworks. A glasshouse at Wordsley, near Stourbridge, England, founded in 1720 by a Mr Bradley who was succeeded by Bradley & Ensell and was operated by Wainwright Bros. until it was closed in 1825. It was taken over in 1829 by Thomas Webb I (1804–69) in partnership with William Haden Richardson and Benjamin Richardson I, and operated as Webb & Richardson. This partnership, after Webb inherited the WHITE HOUSE GLASSWORKS in 1835, was dissolved in 1836, and succeeded by W. H., B. & J. RICHARDSON, and some years later by HENRY RICHARDSON & SONS, LTD. In the 1930s the glasshouse was acquired by WEBB'S CRYSTAL GLASS CO., LTD. Glass manufacturing had already ceased but some glassware with the Richardson label was produced and sold until the late 1960s by THOMAS WEBB & SONS from its DENNIS GLASSWORKS.

Wirkkala glassware. Bowl designed by Tapio Wirkkala, 1968. Iittala Glassworks, Finland.

Woodall glassware. Vase, prune-coloured glass, white cameo decoration, *c.* 1880–85, signed 'T & G Woodall'; mark 'Thomas Webb & Sons – Gem Cameo'. Ht 33 cm. Courtesy, Delomosne & Son, London.

wrought button. A type of KNOP, found on short stems of some English wine glasses, that is spherical and ribbed. Such stems were made after the hollow BALUSTER stems of the mid-17th century and before the subsequent plainer stems.

wrythen. Decorated with spiral or swirling vertical REEDING.

Württembergische Metallwarenfabrik (W.M.F.). A German steelworks founded in 1853 at Gieslingen which in 1883 established a glassworks at Göppingen. Research in crystal glass was undertaken by Hugo Debach who joined the company in 1903 and became director c.1904. Among its products were IKORA-KRISTALL and MYRA-KRISTALL.

Würzburg Flask. A Syrian Islamic PILGRIM FLASK decorated with gilded scrolls and enamelled in eight colours with scenes depicting hunting and feasting. It is of ALEPPO GLASSWARE, c.1250–60. It is so-called as it was for long in the possession of a family in Würzburg, Germany. *See* ENAMELLING, *Islamic*; SYRIAN GLASS.

wrythen. Short ale glass, English, c.1760. Courtesy, Derek C. Davis, London.

Y

yacht jug. A type of CREAM JUG with an inverted rim to prevent slopping from the motion of a ship. They were made in ANGLO-IRISH GLASSWARE.

yard-of-ale. A special type of ALE GLASS, varying in length from 75 cm. to over 1 m., and having a capacity of about one pint. It is a long thin tubular vessel made of clear undecorated glass, with a TRUMPET MOUTH. Early examples from *c.*1685 had a FOOT like an ordinary DRINKING GLASS, but later examples have a globular receptacle at the end opposite the mouth. This bulb served to make these a TRICK GLASS, for when the glass was nearly empty and tilted high, air rushed into the bulb and caused the remaining ale to squirt out on the drinker. They were frequently made in England in the 19th century, perhaps originally for ceremonial occasions. Some examples *c.*1800 have a flared hollow base below the globe. There were also smaller sizes, called 'half-a-yard' or 'quarter-of-a-yard'. Modern versions have been made since *c.*1955.

yellow glass. Glass with yellow colouring; LEAD GLASS (usually opaque) was so coloured by the inclusion of oxide of ANTIMONY in the batch. Early Egyptian glass (e.g. the TUTHMOSIS JUG) included yellow TRAILING, and later Roman glass was made yellow throughout by the use of iron in a ferric state. Chromate of lead was used in the early 19th century to produce yellow, replacing an earlier mixture of silver, lead, and antimony. A silver compound was used to stain the surface of glass a deep yellow, e.g., by ANTON KOTHGASSER on STAINED-GLASS WINDOWS. In Bohemia JOSEF RIEDEL used URANIUM to produce greenish-yellow and yellowish-green shades (*see* ANNAGRÜN). The Chinese made monochrome opaque glassware coloured Imperial Yellow (*see* CHINESE SNUFF BOTTLES).

Ysart, Paul (1904–   ). A glass-maker born in Barcelona. His father Salvador (d. 1956), grandfather, and brothers Augustus (d. 1957) and Vincent (d. 1971) were glass-blowers. He worked as an apprentice at ST-LOUIS, then went to Scotland with his family in 1915 and worked with the Leith Flint Glass Co., Edinburgh, and later moved to Glasgow. In the 1920s the family moved to Perth, working there for the MONCRIEFF GLASSWORKS. When his father and brothers left in 1948 to form the Ysart Bros. Glassworks (later VASART GLASS, LTD.), near Perth, he remained, making MONART GLASS and PAPER-WEIGHTS. From 1961 to 1971 he worked for the CAITHNESS GLASSWORKS at Wick, making paper-weights, and thereafter he continued at his own small factory (Paul Ysart Ltd.) at Wick to make fine-quality paper-weights, some good examples sent to the United States having a cane with his initials 'PY'.

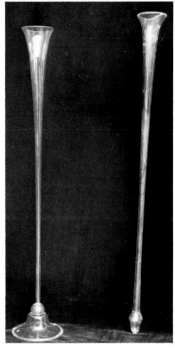

yard-of-ale. Ale glasses, English, Ht 0·93 m. and 1·01 m. British Museum, London.

Ysart, Paul. Paper-weight, with parrot on jasper ground, and incorporating a cane marked 'PY'. Courtesy, Spink & Son Ltd, London.

Z

zaffer (sometimes zaffre). An impure oxide of COBALT obtained by fusing cobaltite with sand. It is used in making SMALT and BLUE GLASS.

Zanesville glassware. Glassware made at Zanesville, Ohio, at several glasshouses from *c*.1815 to *c*.1840/45, often referred to collectively as 'Ohio glass' as none was signed or had been certainly attributed to any particular factory. The ware includes bowls, pitchers, tumblers, FLIPS, and SALTS, some being decorated in the style known as TEN-DIAMOND-MOULD. Among the principal glasshouses were: (1) Zanesville Glass Manufacturing Co. (sometimes called the White Glass Co.), founded in 1815 and having a long succession of owners until it failed in 1852; it is best known for its bottles and HISTORICAL FLASKS; (2) Peter Mills & Co., later Murdock & Casell, founded in 1816 and having various owners until it closed *c*.1852. *See Journal of Glass Studies* (Corning), II (1960), p. 113.

zanfirico (Italian). The present-day term used in Venice and MURANO for VETRO A RETORTI; the spelling used elsewhere in Italy is *sanfirico*.

Zangi flask. A flask of ISLAMIC GLASS having a spherical body with decoration of an encircling gilded frieze, attributed to the period between 1127 and 1146. It is so-called from its Arabic inscription referring to the reign of the Seljuq Atabeg, 'Imad al-Din Zangi, the ruler of Mosul from 1127 and of northern Syria from 1130 until his death in 1146. The gilded decoration is, rather than gold leaf, gold applied in suspension in a medium and fired. Fragments of the piece are in the British Museum. It and a few other rare examples of such Islamic gilded glass are thought to have been made at some centre in south-eastern Anatolia or northern Syria where the technique is said to have been probably introduced from Egypt.

Zecchin, Vittorio (1878–1947). A glass-maker at MURANO who *c*.1921 became artistic designer for Arte Vetraria Muranese and Fratelli Toso, both of Murano; he designed tableware in the *Wiener Stil*, being influenced by the work of JOSEF HOFFMANN. He was related to Francesco Zecchin who, with Napoleone Martinuzzi, operates a glasshouse at Murano.

Zechlin. The site, north of Berlin, of a German glasshouse where the successors of JOHANN KUNCKEL moved the POTSDAM GLASS FACTORY in 1736. From 1804 onwards Zechlin made black glass in imitation of Wedgwood's black basaltes. It remained as a State operation until 1890. *See* HYALITH.

Zeuner (fl. 1773–1810). An Amsterdam artist who decorated PANELS and objects of glassware in the style known as ÉGLOMISÉ. A visit he made to England is evidenced by a panel with a London scene. His work is in gold and silver leaf, sometimes supplemented with coloured backgrounds, and is signed. *See* GOLD ENGRAVING.

Zierglas (German). Literally, ornament glass. Glassware made primarily for ornamental rather than practical use, and intended for display, e.g. (1) the flower-shaped fragile liqueur glasses made by KARL KÖPPING and like glasses by SALVIATI, and (2) the VASE WITH FLOWER STALK.

Zitzmann, Friedrich (1840–1906). A German glass-maker from Thuringia who worked at Wiesbaden AT-THE-LAMP and became a pupil

Zanesville glassware. Stiegel-type flask with spiral ribbing, probably Zanesville, Ohio, early 19th century. American Museum in Britain, Bath.

Zierglas. Flower ornament by Salviati & Co., Murano.

of KARL KÖPPING, making types of glassware similar to the designs of Köpping. He also designed glassware for the Ehrenfelder Glashütte.

zuccarin (Italian). A type of glass vessel derived from and generally similar to a KUTTROLF, but made in Venice in the late 17th and 18th centuries.

Zwischengoldglas (German). Literally, gold between glass. A type of Bohemian drinking glass decorated with gold leaf by a process whereby the outer surface of one glass was decorated with ENGRAVED gold leaf, and then a bottomless glass was sealed over it (using a colourless resin) to protect the design. The inner glass had been ground down with great exactness for almost its whole height so as to permit the outer glass to be fitted precisely over it. On early rare Bohemian examples the joint showed at the top of the rim, but on later ones the rim of the inner glass was of double thickness for a distance of about 1 cm., so that a projecting flange fitted over the outer glass. The outer glass projected slightly at the bottom, and the space below, in the case of a BEAKER, was filled with a glass disk with similar gold engraving and sealed by transparent colourless resin. The outer surface of the double-walled glass was sometimes further decorated by cutting 12 to 18 narrow vertical FACETS or FLUTES. Silver leaf was sometimes used in place of gold leaf. A frequent form was a small straight-sided beaker, and popular decorations were hunting scenes, views of monasteries, Bohemian saints, and armorial bearings, all very delicately engraved.

Glass so decorated is very ancient, Hellenistic examples being known from c. 300 BC. (*See* CANOSA BOWL.) Examples of gold leaf protected between two layers of glass are known from the Parthian or Sassanian periods in Persia, perhaps 2nd–4th centuries AD, and from Egypt and Syria from the 12th century, when a piece of glass was decorated on its reverse and then fused to a layer of transparent glass. (*See* GOLD ENGRAVING.) The Bohemian technique is said to have been developed at an unidentified monastery in Bohemia, and the best examples date from the 1730s, but others were made until c. 1755. Such decoration was used not only on beakers but on GOBLETS and other double-walled vessels, sometimes with COLD PAINTING on coloured glass or with enamelling. Such ware is sometimes termed 'double glass'. JOHANN JOSEF MILDNER revived and elaborated the technique in 1787. *See* MEDAILLONBECHER.   [*Plate XVI*]

Zwischengoldglas. Beaker with faceted exterior, Bohemian, second quarter of 18th century. Ht 8 cm. Kunstgewerbemuseum, Cologne.